BASIC CONCEPTS OF

Environmental Chemistry

Des W. Connell

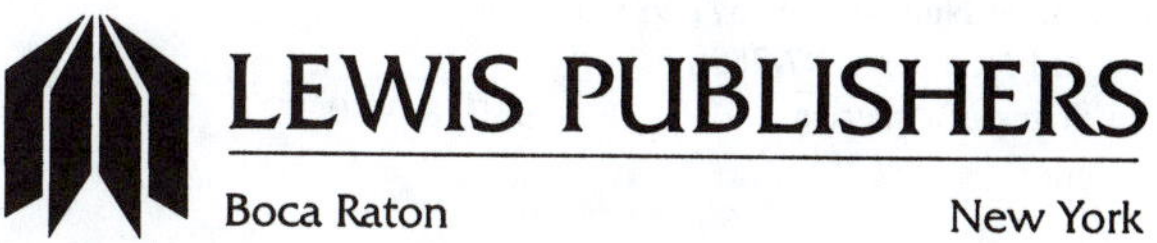

Acquiring Editor: Ken McCombs
Project Editor: Albert W. Starkweather, Jr.
Cover Designer: Dawn Boyd

Library of Congress Cataloging-in-Publication Data

Connell, Desley W.
Basic concepts of environmental chemistry/ Desley W. Connell, Darryl W. Hawker, Michael St. J. Warne, Peter P. Vowles
p. cm.
Includes bibliographical references and index.
ISBN 0-87371-998-0 (alk. paper)
1. Environmental chemsitry. I. Connell, D.W.
TD193.B37 1997
628.5—dc21 97-7698
CIP

International Standard Book Number 0-87371-998-0
Library of Congress Card Number 97-7698
Printed in the United States of America 2 3 4 5 6 7 8 9 0
Printed on acid-free paper

THE AUTHOR

Des W. Connell is Professor of Environmental Chemistry in the Faculty of Environmental Sciences at Griffith University in Brisbane, Australia. His broad interests are in environmental chemistry, ecotoxicology and water pollution and he has published several books on these topics. His research interests are in bioconcentration, quantitative structure-activity relationships (QSARs) and the mechanisms of nonspecific toxicity. He was the inaugural Chairperson of the Environmental Chemistry Division of the Royal Australian Chemical Institute and was awarded the Institute's Environment Medal in 1993. Professor Connell received his B.Sc., M.Sc., and Ph.D. in organic chemistry from the University of Queensland and a D.Sc. in environmental chemistry from Griffith University.

THE CO-AUTHORS

Darryl W. Hawker is a Senior Lecturer in environmental chemistry in the Faculty of Environmental Sciences at Griffith University in Brisbane, Australia. His research interests are in the physicochemical properties of chemicals, particularly organometallic compounds and how these properties influence environmental behavior. Dr. Hawker received his B.Sc. (Honors) and Ph.D. degrees in organic chemistry from the University of Queensland.

Michael St. J. Warne is the Senior Research Ecotoxicologist in the New South Wales Environment Protection Authority and Vice President of the Australian Society for Ecotoxicology. Research interests include the toxicity of mixtures and phosphate-free detergents, quantitative structure-activity relationships (QSARs) and earthworm ecotoxicology. He is involved in developing the national ecological risk assessment framework for contaminated sites and is developing an Australian ecotoxicology database. Before joining the NSW EPA, he was a Lecturer in environmental chemistry/toxicology in the Faculty of Environmental Sciences, Griffith University. Dr. Warne has a B.Sc. with Honors in Biology from Newcastle University and a Ph.D. in ecotoxicology from Griffith University.

Peter P. Vowles is a Lecturer in environmental chemistry in the Faculty of Environmental Sciences at Griffith University in Brisbane, Australia. His research interests are in physicochemical factors influencing the fate of persistent organic chemicals and the identification and source apportionment of urban aerosols. He is also interested in approaches and techniques for teaching chemistry, particularly environmental chemistry. Dr. Vowles has a B.Sc. (Honors) and Ph.D. in chemistry from the University of Melbourne and a Diploma in Education from the Melbourne State College.

PREFACE

In recent years there has been an extraordinary development in our knowledge of the behavior and effects of chemicals in the environment. In the 1960s our knowledge of environmental chemistry was confined principally to sets of data on the concentrations of chemicals in the environment. Now a body of specific knowledge has been developed that provides a theoretical basis for understanding distribution, transformation, toxicity and other environmental properties of chemicals. Out of this has emerged a new branch of the discipline of chemistry that can be defined as follows:

> Environmental chemistry is the study of sources, reactions, transport and fate of chemical entities in the air, water and soil environments as well as their effects on human health and the natural environment.

Environmental chemistry is probably the most interdisciplinary of all the branches of chemistry since it includes aspects of all the traditional chemistry branches as well as biochemistry, toxicology, limnology, ecology and so on. It has a practical focus on the environmental management of chemicals since that is a major area of concern. It also includes behavior of natural chemical entities in natural systems as well.

Environmental chemistry now has sufficient intellectual depth to provide the scope for undergraduate and graduate courses in chemistry. This book utilizes the fundamental properties of bonds and molecules as a framework for understanding the behavior and effects of environmental chemicals. The properties of common contaminants are illustrated using environmental behavior, rather than laboratory behavior, as the expression of the underlying bonding and molecular characteristics. It goes on to link these characteristics to biological effects in the environment such as toxicity. Also, these fundamental aspects are utilized in considering the great global environmental chemistry processes including respiration, photosynthesis, chemical evolution and so on. Recent developments in the management of hazardous chemicals, such as ecotoxicology and risk assessment, are treated as an aspect of environmental chemistry.

This book assumes only a basic prior knowledge of chemistry and is designed as a textbook for courses in environmental chemistry. Its objective is to provide a knowledge of environmental chemistry based on a series of theoretical principles rather than as a set of disjointed facts. It aims to give students not just a knowledge of environmental chemistry but an understanding of how and why processes in the environment occur. It can also be used as a basis for understanding chemistry, in general, since the principles apply generally as well as to the environment.

We have developed this textbook based on our experience of teaching graduate and undergraduate courses on environmental chemistry over the last 20 years. To some extent this represents a culmination of this experience and a distillation of the knowledge we have acquired in that time.

Many people have assisted us in preparing this book. Greg Miller was involved in discussions of the content and scope, Rahesh Garib and Lee Wallace typed the difficult manuscript while Aubrey Chandica and Magda Suto prepared the diagrams. To all of these people we are extremely grateful.

Des W. Connell

CONTENTS

CONTAMINANTS IN THE ENVIRONMENT

MANAGEMENT OF HAZARDOUS SUBSTANCES

Principles of Environmental Chemistry

Chapter 1

THE ROLE AND IMPORTANCE OF ENVIRONMENTAL CHEMISTRY

I. ALCHEMY

Modern chemistry has its early beginnings in ancient Greece and Egypt. The Greeks were great philosophers and thinkers and debated the ultimate nature of the material world that surrounded them. On the other hand, the Egyptians were skilled in the practical arts of applied chemistry. They had an intimate knowledge of such matters as embalming of the dead, dyeing of clothing and the isolation of some metals. In fact, the word *alchemy* is derived from the Arab word *al-kimiya.* This phrase was used to describe the art of transformation of materials and the practical use of chemicals in society. However, the lack of understanding of chemical processes led to alchemy being more of an art than a science. The art of alchemy was closely related to religion and involved the use of a set of hieroglyphics to represent the different metals. For example, gold was depicted as the sun, silver the moon, copper as venus and so on. The occurrence of chemical changes was interpreted in a mysterious fashion and took on a mythological significance. The major area of concern of alchemy was not the development of a science but the transmutation of metals, particularly the transmutation of base metals, such as lead, into gold.

The decline of alchemy was signaled by the works of such scientists as Agricola, in fact a German named Georg Bauer, who lived between 1494 and 1555. Agricola published a book titled *De Re Metallica* (Of Metallurgy) which took a practical approach and gathered together all of the knowledge available on metallurgy at the time. Also, the German scientist Andreas Libau (1540–1616), better known as Libavius, published *Alchemia* in 1597. Despite the name of this text, the book is written clearly and without resource to mysticism. The famous Swiss physician, Paracelsus (1493–1541), although an alchemist, believed that the processes that occur in the human body are basically of a chemical nature and that chemical medicines could provide remedies to illness. So this marks the start of the development of chemotherapy and doctors everywhere now use a vast array of chemicals to cure all kinds of illnesses.

Perhaps the final end to alchemy as a science occurred in 1661 when Robert Boyle, an Irishman, published his book called *The Sceptical Chymist.* Boyle attacked the ideas and approach of the alchemists and advocated a rational scientific approach. To mark the change from alchemy, Boyle also advocated the use of the term *chemistry* to describe the science of materials. The well-known Boyle's law was developed by Boyle and amply illustrates the scientific approach to chemistry that he promoted.

It could be said that *environmental chemistry* had its beginning with one of the great pioneers of chemistry, Antoine Lavoiser who was born in Paris in 1743. Lavoiser's experiments on the atmosphere mark a great advance in chemistry and also a great advance in understanding the chemistry of the atmospheric environment.

Not only did he discover fundamental information concerning the chemistry of air, but he also examined the use of air by animals and by so doing was investigating one of the major aspects of the chemistry of the environment. The use of oxygen by animals and their consequent release of carbon dioxide is a fundamental aspect of environmental chemistry. Lavoiser's book *Elementary Book of Chemistry* published in 1789 effectively marks the start of the systematic development of the then new science of chemistry.

II. THE CHEMICAL AGE

In the first half of the 19th century, early chemists were endeavoring to describe the nature of molecules. They were taking the early steps in assembling the structures of organic chemicals which we now take for granted as common scientific knowledge. As they were carrying out this early work, they were concerned with the reactions between organic chemicals and the nature of the products generated, thus embarking on the path to modern synthetic organic chemistry.

One of these early chemists, William Perkin (1838–1907), became aware that some of these synthetic organic chemical processes could possibly be turned to commercial advantage. Perkin left his academic studies at university and used money obtained from his family to start a factory to manufacture synthetic dyes. One dye he produced, Aniline Purple, was extremely popular in the textile industry and in heavy demand with textile manufacturers in Europe. In fact, Perkin had founded the first chemical industry to be based on studies of the nature of organic compounds conducted in a research environment. Not only had he founded the synthetic chemical industry, but he was able to retire a very wealthy man at an early age. This marks the start of the synthetic chemical industry which, from this small beginning, was to expand to a size and extent where its products now dominate modern society. This development had important implications for the environment since it resulted in the preparation and discharge of relatively large quantities of synthetic organic substances, previously unknown to occur in nature, and with unknown environmental effects.

Of course, chemical processes such as the smelting of ores and the manufacture of soap had been carried out on a large commercial scale prior to this time. But these processes were developed as a result of trial and error and were more in the nature of trade practices rather than the products of scientific chemical research.

Another chemist whose activities were to have important environmental implications was on the scene about this time. This was the German scientist, Fritz Haber (1868–1934). Haber demonstrated in the laboratory how nitrogen and hydrogen could be combined to yield ammonia. He further developed this onto a commercial scale so that commercial plants operating in Germany in the early part of this century were fixing nitrogen gas from the atmosphere and producing ammonia. This ammonia could be used as a fertilizer for the production of food crops, thereby eliminating the need to rely on the natural occurrence of nitrogen compounds in soil to stimulate plant growth. This process has continued to be used to the present day, thereby allowing the production of large quantities of plant fertilizer used to produce food crops and allowing for rapid expansion of the world's population. Without this process, the production of food would be limited by the availability of natural nitrogen compounds in soil for food plant growth. Thus, this chemical process has probably had the most impact on the environment through the rapid expansion of the human population with the subsequent environmental changes that have resulted.

It is traditional for human history to be divided into a series of Ages, each building on, and improving, the technology of the previous. Thus, we have a sequence of Ages going from the Stone Age to the Bronze Age to the Industrial Age and the Atomic Age. However, during the last approximately 100 years, the Age in which we are living could very well be described as the *Chemical Age*. Table 1.1 indicates the wide range of synthetic chemicals now used in human society. The use of these substances permeates almost every aspect of our society from life in the home, to transport, food production, availability of medicines and so on. The production and use of these many substances have improved human conditions and caused an enormous increase in the human population. These factors have many implications for the chemical processes that occur in the environment.

TABLE 1.1
Use of Synthetic Chemicals in Human Society

Substance	Examples of chemical classification	Uses
Plastics	Polystyrene, polyvinylchloride polypropylene, nylon	Textiles, car tyres, household fittings, furniture etc.
Pesticides	DDT, 2,4-D, malathion, glyphosate	Control of weeds, insects and other pests
Drugs	Aspirin, penicillin, valium, sulphanilamide, barbiturates	Medicinal uses
Petroleum products	Hydrocarbons	Motor fuel, lubricant
Crop fertilizer	Ammonia, ammonium sulphate, ammonium nitrate, ammonium phosphate	Stimulate food crop production

III. ENVIRONMENTAL CHEMISTRY

Environmental chemistry is not new but the use of this term to describe a body of chemical knowledge has become accepted by scientists in recent years. This principally reflects the importance attached to the behavior and possible adverse effects of chemicals discharged to the environment.

There are many scientific events that could now be identified as the starting points of environmental chemistry. Some of these are in the area of fundamental chemistry while others lie in applied chemistry. Also, the chemical aspects of other disciplines have played a major role. One of the starting points has been mentioned previously and that is the work of Lavoiser. His analysis of air and determination of the nature of combustion are fundamental to the development of modern chemistry and atmospheric chemistry. Turning to the applications of chemistry, perhaps pollution of the Thames River during the 1800s can be identified as another starting point of environmental chemistry. The bad odors and diseases associated with the Thames were investigated by various British Royal Commissions who weighed evidence on the chemistry of sewage treatment and water pollution. This culminated with the Royal Commission on Prevention of River Pollution, which reported in 1885 and recommended the use of the biochemical oxygen demand (BOD) test. This test is still extensively used today to evaluate the effects of sewage and other wastewaters on waterways.

The chemistry of the oceans, lakes and freshwater areas also has a comparatively long history. In 1872, HMS Challenger commenced its historic voyages with many of the most noted men of science at that time aboard. They conducted many investigations of seawater composition and chemical processes in the oceans. Similarly in limnology, the key to understanding the ecology of freshwater areas lies in chemical transformations of carbon, oxygen, nitrogen and phosphorus. This was recognized early in the development of this discipline, which has always had a strong emphasis on chemistry.

Since these early beginnings, environmental chemistry has expanded rapidly. Natural chemical processes in all sectors of the environment, particularly soil, water and the atmosphere, have been subject to investigation as well as the environmental behavior of contaminating chemicals. Not surprisingly, the management of chemicals discharged to the environment has become a major focus for environmental chemistry. Governmental agencies and industries employ large numbers of environmental chemists on a worldwide basis to monitor and manage the discharge of chemicals and their adverse effects.

Over recent years, a body of knowledge associated specifically with the behavior and effects of chemicals in the environment has been developing. A theoretical basis for understanding the distribution, transformation, toxicity and other biological properties of chemicals in the environment is now becoming established. This means that environmental chemistry can be appropriately seen as a subbranch or subdiscipline of chemistry. The recent development of environmental chemistry was initiated during the 1960s. Rachel Carson's book, *Silent Spring,* published in 1962, can be identified as a significant event stimulating worldwide interest and concern regarding chemical residues in the environment. This event was made possible by the development of analytical techniques capable of detecting chemicals at very low concentrations. In 1952, Richard Synge and Archer Martin were awarded the Nobel Prize for inventing the chromatography technique, which has principal application to the analysis of organic compounds. Later, the technique was extended and improved such that trace amounts of xenobiotic organic chemicals could be quantified in environmental samples. The flame ionization detector extended the sensitivity of the method so that levels of a few parts per million of organic compounds could be detected in samples. In addition, with the chlorinated hydrocarbons, the development of the electron capture detector further extended the sensitivity of the technique. In 1967, only 15 years after the Nobel Prize for chromatography was awarded, a considerable body of information was available on the concentrations of xenobiotic organic compounds in biota.

Environmental chemistry was dominated at this time by the collection of data on residues of synthetic compounds in biota but there was little understanding of the mechanisms how the residues accumulated or of their biological effects. Determination of residue levels remain an important aspect of chemical behavior in the environment, but now there is considerable interest in placing residues in a broader context of environmental effects. There have been major advances in understanding the distribution, transport and transformation of contaminants, as well as exposure and uptake by biota. But human health effects as well as the responses of natural ecosystems require further expansion of knowledge.

Over the last 15 years, some theoretical concepts have emerged that provide a sound conceptual basis for important aspects of environmental chemistry. Drawing on ideas already established in other fields, the introduction of partition and fugacity theory to explain environmental distribution of chemicals has occurred. In addition, the use of properties measured in the laboratory to assess behavior and effects in the environment has allowed the development of a clear understanding of many environmental processes. Theoretical methods to predict environmental properties of chemicals have been placed on a sound footing and provide a basis for expansion of knowledge in the future.

The global problems of the *greenhouse effect* and the effect of CFCs (chlorofluorocarbons) on the ozone layer now occupy the center stage of environmental management. The environmental chemistry of the processes leading to these problems provides the starting point in their resolution. A range of different scientists have pointed out the importance of developing a mechanistic chemical picture of the cycles of carbon and other elements and their interaction with land and sea surfaces in devising strategies for management of these global problems.

IV. THE SCOPE OF ENVIRONMENTAL CHEMISTRY

It is difficult to precisely define environmental chemistry since the topic has not yet reached a stage where there is universal accord in the chemical community on its scope. However, this is developing and the following definition provides a reasonably acceptable statement at this stage:

> **Environmental chemistry** is the study of sources, reactions, transport and fate of chemical entities in the air, water and soil environments, as well as their effects on human health and the natural environment.

Environmental chemistry is probably the most interdisciplinary of the many branches of chemistry. It contains aspects of related branches of chemistry such as organic chemistry, analytical chemistry, physical chemistry and inorganic chemistry, as well as more diverse areas such as biology, toxicology, biochemistry, public health and epidemiology. Figure 1.1 represents a diagrammatic illustration of some of the major aspects of these relationships. To place the area in some perspective, a set of topics that fall wholly or partially within environmental chemistry are listed in Table 1.2. Many of these topics are concerned with chemical pollutants in the environment, but environmental chemistry is not only concerned with pollution but also with the behavior of natural chemicals in natural systems. This is exemplified by topics such as oceanic and limnological chemistry, which are primarily concerned with natural systems.

Investigations within the scope of environmental chemistry provide a possible explanation for our very existence. The primitive earth's atmosphere contained simple gases which, on equilibration, followed by subsequent complex reaction sequences have led to the formation of proteins, carbohydrates, fats and other

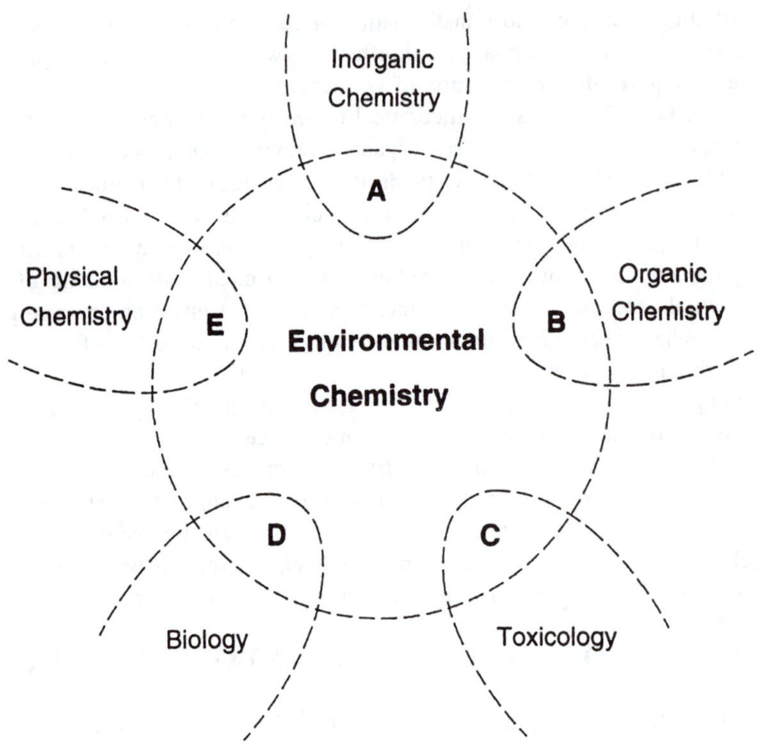

Examples of areas common to environmental chemistry and other areas of science:

A Adsorption of chemicals by soil & sediments

B Mechanisms of degradation of pesticides

C Effects of environmental chemicals on human health

D Effects of discharges on aquatic ecosystem structure

E Kinetics of chemical movement

FIGURE 1.1 Diagrammatic illustration of some of the relationships between environmental chemistry and other areas of science.

substances that are the basic molecules needed for life to develop. Later processes in this sequence may have led to the formation of cells and then onto more complex life forms. The mechanisms of these processes draw heavily on our present understanding of the fundamental properties of molecules and reaction mechanisms.

Environmental chemistry is basically concerned with developing an understanding of the chemistry of the world in which we live. Such investigations and knowledge have an intrinsic value of their own. Our world, with all of its many complex chemical processes, is a worthy topic for a well-rounded education and research.

TABLE 1.2
Some Topics be Considered to Be Wholly or Partially Within Environmental Chemistry

Aspect of the environment	Area within or relevant to environmental chemistry
Evolution	Chemical evolution
Chemicals processes in sectors of the abiotic natural environment	Oceanic, atmospheric, soil and limnological chemistry, global chemical systems
Chemical influences in natural ecosystems	Chemical ecology, pheromones, allelochemistry
Behavior of hazardous chemicals in the environment	Mathematical modeling of environmental distribution, degradation processes, waste disposal
Effects of toxic chemicals on individuals, populations and ecosystems	Environmental toxicology, ecotoxicology, QSARs, environmental analysis
Effects of chemicals on human populations	Environmental health, safety, occupational health, epidemiology

The chemical processes in soil, water and the atmosphere are central to our existence and through them we can better understand ourselves and our role in the environment. Thus, it could be argued that environmental chemistry should be a part of the chemistry curriculum wherever chemistry is taught. At a tertiary level, graduates in chemistry are in fact now expected to be familiar with many aspects of environmental chemistry. They are being asked by management in industry and within government to advise on the environmental and health effects of chemicals, management and safety of chemicals and so on.

V. CHEMISTRY IN ENVIRONMENTAL MANAGEMENT

There is little doubt that the principal applications of environmental chemistry are in the development of an understanding of the behavior and effects of discharged chemicals on human health and the natural environment. When chemical use results in environmental contamination, it is necessary to set standards for acceptable concentrations in water, air, soil and biota. Monitoring of these concentrations, and the resultant effects, must then be undertaken to ensure that the discharge standards are realistic and provide protection from adverse effects. Also, considerable attention is now being focused on regulation of the use of new chemicals and prevention of chemicals that may have adverse effects, from entering the marketplace and environment. Most of these actions are undertaken by government and industry; for example, the disposal of chemical wastes generated in highly concentrated form as well as occurring as trace contaminants in discharges. These chemical management issues involve political, social and economic problems as well as technical problems, and many new approaches to management will be needed. On the other hand, new industrial processes that generate less chemical waste will be required. Some of the activities undertaken in government agencies and industry, within the scope of environmental chemistry, are outlined in Table 1.3. In addition, there is a considerable volume of research in environmental chemistry being undertaken. A number of well-established research institutions specializing in environmental chemistry are operating in countries throughout the world.

TABLE 1.3
Some Activities of Government and Industry Related to Environmental Chemistry

Objective	Some actions taken
Management of industrial emissions	Setting emission standards, monitoring ambient concentrations, disposal of waste, modeling distribution of chemicals
Protection of worker health	Biochemical and physiological testing, epidemiology, monitoring ambient concentrations, evaluation of adverse effects
Protection of the natural environment	Monitoring contaminants in water, air, soil and biota, evaluation of adverse effects
Testing and evaluation of new chemicals	Modeling of potential distributions, testing toxicity and other effects

Environmental chemistry activities in government, industry and education are growing, to some extent in concert with the concerns expressed regarding chemicals in the environment by the community. At present, despite the enormous benefits that accrue from the many uses of chemicals, the community often sees chemicals as having a negative impact. A negative impact has occurred in many situations, but many of these have been eliminated and increased knowledge has provided a basis for enlightened future management and control. The further development of environmental chemistry will provide access to chemical products that will enhance the lives of people without resulting in detrimental effects. In this way, environmental chemistry will have a significant effect on the future of chemistry, chemists and chemical industry as well as the community in general.

VI. BASIC CONCEPT OF THIS BOOK

In the early 1970s, a new approach to explaining the environmental effects of chemicals was initiated. Hamelink and co-workers (1971) proposed that the bioaccumulation of organic compounds by biota was governed by the properties of the chemical rather than ecological factors as was previously thought. Since that time, this concept has been extended considerably and now it is clear that the distribution of chemicals in the environment relates to the properties of the chemical as well as characteristics of the environment (see Mackay and Patterson, 1981).

The relationship of biological effects, particularly toxicity, to properties of a chemical, such as the olive oil/water partition coefficient, was well established by Ernest Overton in the early part of this century and extended by Covin Hansch during a later period. These concepts were then applied, with considerable success, to explain toxicity, and other effects of chemicals, in the environment. In addition, the influence of characteristics of the molecule itself, such as molecular surface area, were found to be of considerable significance.

A concept of an interdependent set of properties of a chemical has now developed, as shown in Figure 1.2. This has been used as the basic underlying concept of this book. The basic characteristics of the molecule itself are the starting point for understanding the environmental chemistry of chemicals. These characteristics govern the physical chemical properties of the compound, such as its water solubility and vapor pressure, which in turn control transformation and distribution in the environment. Biological effects relate to the chemical in its transformed and dispersed state. This set of interrelationships also provides a basis for predicting the environmental properties of a chemical. The distribution, transformation and some biological properties can be predicted in many situations utilizing physical chemical properties.

CHARACTERISTICS OF THE MOLECULE
(e.g., surface area, molecular weight,
functional groups, chemical bonds, etc.)

↓

**PHYSICAL CHEMICAL PROPERTIES
OF THE COMPOUND**
(e.g., aqueous solubility, vapour pressure,
melting point, octanol/water partition coefficient, etc.)

**TRANSFORMATION & DISTRIBUTION
IN THE ENVIRONMENT**
(e.g., persistence, bioaccumulation, etc.)

BIOLOGICAL EFFECTS
(e.g., lethal toxicity, reduction in growth,
reduction in reproduction, etc.)

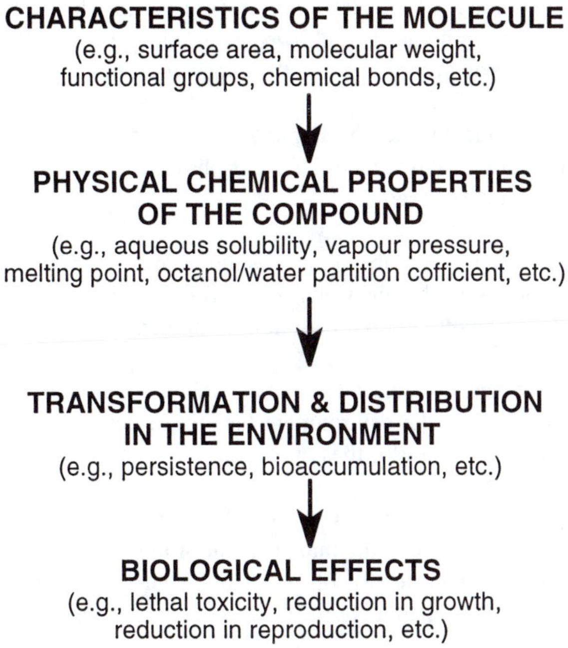

FIGURE 1.2 The environmental chemistry of a chemical can be seen as an interrelated set of characteristics as shown.

This concept provides a framework into which aspects of the environmental characteristics of a chemical can be logically placed. It allows these many diverse environmental characteristics to be rationalized into a set relationships rather than be seen as a disorganized collection of facts.

VII. KEY POINTS

1. The alchemists were the earliest practitioners of applied chemistry, but alchemy was more of an art than a science. Many alchemists were concerned with mysticism and the transmutation of base metals, such as lead, into gold.

2. Environmental chemistry had an early beginning with Antoine Lavoiser, a French scientist born in Paris in 1743. Lavoiser combined classical experiments on the composition of air and its use by animals to investigate the chemical nature of the atmosphere.

3. The commercial application of chemical knowledge derived from laboratory experiments was commenced by William Perkin in the late 1800s. Perkin founded the synthetic dye industry, which was the forerunner of the large and diverse synthetic chemical industry of today.

4. Perhaps one of the most important environmental applications of chemical knowledge has been the development, by Fritz Haber in the early part of this century, of a commercial process to combine atmospheric nitrogen with hydrogen to produce ammonia. This allowed the production of artificial fertilizers to enhance food crop production and the expansion of the human population.

5. The term "environmental chemistry" has become common only in recent times although environmental aspects of chemistry have been investigated since the beginning of chemistry itself.

6. Environmental chemistry is the study of sources, reactions, transport and fate of chemical entities in the air, water and soil environments, as well as their effects on human health and the natural environment.

7. Over the last 15 years, theoretical concepts of the distribution, behavior and effects of chemicals in the environment have been developed that provide a fundamental understanding of these processes.

8. Environmental chemistry plays a major role in government and industry in relation to many areas, including management of industrial emissions, protection of workers health, protection of the natural environment, as well as testing and evaluation of new chemicals.

9. The environmental properties of a chemical can be seen as an interrelated set of properties based on the characteristics of the molecule. These characteristics govern the physical chemical properties of the compound, which in turn control transformation, distribution, and biological effects in the environment.

REFERENCES

McGowen, T., *Chemistry — the Birth of a Science,* Franklin Watts, New York, 1989.

Lyman, W.J., Reehl, W.F., and Rosenblatt, D.H., *Handbook of Chemical Property Estimation Methods Environmental Behavior of Organic Compounds,* American Chemical Society, Washington, D.C., 1990.

Mackay, D. and Patterson, S., Calculating Fugacity, *Environmental Science and Technology,* 15, 1981, 1006.

CHAPTER 1 PROBLEMS

1. The terms Stone Age, Bronze Age, Industrial Age and Atomic Age are used to describe phases of knowledge and technology that human civilizations have passed through. Why is the term "Chemical Age" appropriate for today's society?
2. The applications of chemistry in industry have had a major effect on human society. Which chemical applications have had the greatest impact on the human and natural environment?
3. Environmental chemistry has developed from a diverse field of investigations conducted in relation to the environment. Which of these areas of investigation have been the most important in influencing the scope of environmental chemistry as it is perceived today?

CHAPTER 1 SOLUTIONS

1.
    - The chemical industry is one of the largest industries in our society.
    - Chemical products dominate the home, workplace, agriculture, transport, medicine and other areas of human activity.
    - Waste chemical control is now a major aspect of environmental management for industry and government.

2.
    - The artificial fixation of nitrogen to produce ammonia that is used as an agricultural fertilizer has allowed the expansion of the human population.
    - Commercial synthetic chemical operations produce products that are used in almost all human activities.

3.
    - Development of analytical techniques for trace organic compounds and other substances has revealed widespread trace contamination.
    - Attribution of adverse biological effects to contamination by trace chemicals.

Basic Properties of Chemicals in the Environment

Chapter 2

BONDS AND MOLECULES —THEIR INFLUENCE ON PHYSICAL CHEMICAL PROPERTIES IN THE ENVIRONMENT

I. INTRODUCTION

The behavior of chemicals in the environment is governed by their physical chemical properties as well as transformation and degradation processes, which are discussed in Chapter 3. The physical chemical properties of compounds include such characteristics as boiling point (bp), melting point (mp), solubility in water and similar properties. These properties are, in fact, measurements made in the laboratory of environmentally relevant characteristics. For example, compounds with low boiling points evaporate rapidly into the atmosphere, whereas compounds that are highly soluble in water disperse readily in streams and rivers. Thus, an understanding of these properties would be expected to give a clearer perception of how compounds will behave in the environment. This means that the measurement of the physical chemical properties of a compound could be used to provide an evaluation of its environmental distribution in air, water, sediments, soil and animals.

II. STATES OF MATTER IN THE ENVIRONMENT

A cursory examination of the nature of our environment reveals that matter exists in basically three states: **solid, liquid and gas**. Solids are present as soil, rocks and so on, whereas liquids are represented by the great water bodies of the oceans, lakes and rivers. The gaseous component of the environment is the atmosphere. These states exercise a basic influence on the nature and distribution of all substances, both natural and man made, in the environment

In basic chemical terms, **solids** have a definite volume and shape and consist of large numbers of particles that could be atoms, ions or molecules (see Figure 2.1). In fundamental terms, solids are matter where attractive forces between the particles present are more powerful than disruptive forces and these hold the particles in a fixed relationship to one another. This also means that the particles are closely packed together, with the consequent effect of making solids relatively dense. Some solids can be converted into liquids by heating. For example, ice melts to form liquid water. Heating actually increases the vibrational motions of the particles within the solids until the particles can no longer be held in a fixed relationship to one another. At this point, the substance is converted into a liquid and the temperature at which this occurs is referred to as the melting point (m.p.). For example, the melting point of ice at atmospheric pressure is 0°C.

Liquids have a definite volume but no definite shape. With liquids, the particles are close together, almost as close as solids but are free to move relative to one another (see Figure 2.1). This means that the cohesive forces holding the particles together are balanced by the dispersive forces. So the waves of the oceans and the

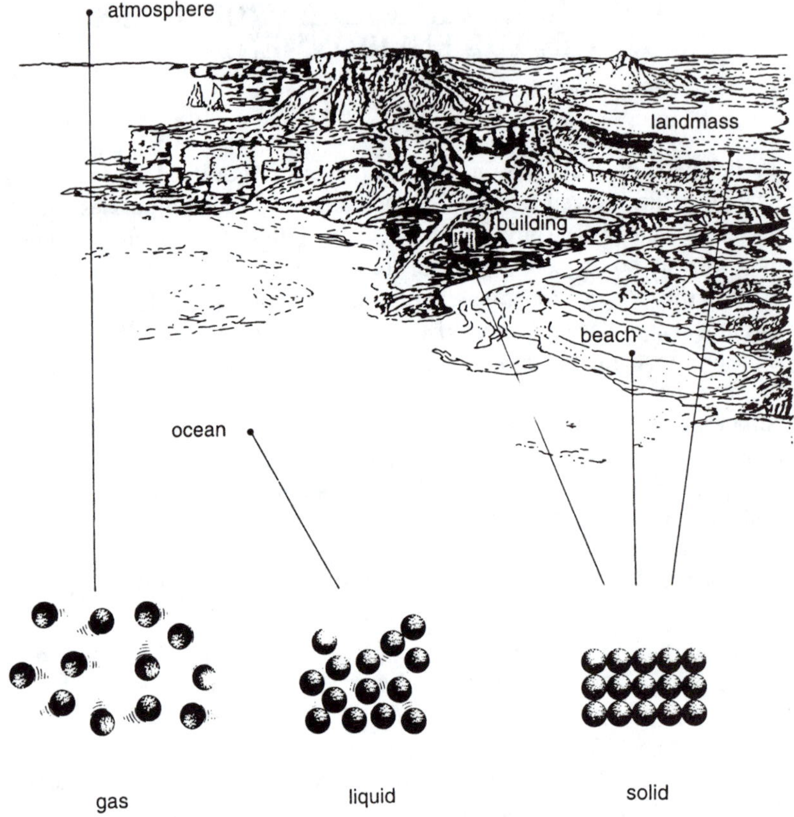

FIGURE 2.1 Illustration of the different states of matter in the natural environment.

flow of water in rivers represent the basic properties of liquids. An important characteristic of many liquids is their boiling point (b.p.), which is the temperature at which the liquid vaporizes to give a gas. For example, water has a boiling point of 100°C at atmospheric pressure. However, it should be kept in mind that water does not have to boil in order to vaporize. When the temperature is increased and the rate of vibration of the particles in a liquid increase, some particles bounce completely out of the liquid before the boiling point is reached. In this manner, liquids (and to a lesser extent solids) exhibit a pressure above the surface referred to as the vapor pressure (vp). Water evaporates from the surface of the oceans, lakes and rivers without reaching its boiling point. As the temperature increases, the rate of evaporation increases. Somewhat similarly, gases from the atmosphere may dissolve to some extent in water bodies although the water bodies are at much higher temperature than the boiling point of the gas. Thus, oxygen and nitrogen gases are present in low concentrations in seawater.

Gases and vapors are produced when the thermal energy input into the liquid reaches a level whereby the movements of the particles (kinetic energy) becomes greater than the cohesive energy. The particles move apart at this temperature, which is the boiling point, and move at relatively high velocity, colliding with one another and other substances present in the atmosphere. In this way, the molecules in the atmosphere are in constant motion and collide with one another and with the walls of any container, leading to the phenomenon known as pressure. At sea level, the pressure is about 1 atmosphere (atm), and this pressure falls with distance away from the earth's surface. As both the density of the atmosphere and the pressure fall, the distance between the molecules in the atmosphere increases and the mean free path between collisions increases. So, at sea level, a molecule has a mean free path of about 10^{-6} cm. At a height from the earth's surface of 100 km the pressure has fallen to 3×10^{-7} atm, with a molecule having a mean free path of about 10.0 cm. At a height of 500 km the mean free path has increased markedly, now exceeding 2×10^{6} cm (20 km) while the pressure is $<10^{-6}$ atm.

The properties of some substances in the environment are shown in Table 2.1. Oxygen and nitrogen have relatively low b.p. and m.p. values and are the major gaseous components in the atmosphere. Water has a higher b.p. and m.p. and exists as a liquid, with the molecules in close association but free to move in relation to one another. Quartz and common salt have very high b.p. and m.p. values reflecting the close and fixed relationship the atoms present have to one another.

TABLE 2.1
Properties of Some Substances in the Environment

Substances	Occurrence in the environment	Normal physical state	Boiling point (°C) (atmospheric pressure)	Melting point (°C)
Oxygen	21% of atmosphere	Gas	–183	–218
Nitrogen	78% of atmosphere	Gas	–196	–210
Water	Oceans, lakes, rivers	Liquid	100	0
Common salt	3.5% of seawater	Solid	1413	801
Quartz	Rocks, sand, geological strata	Solid	2230	1610

III. NATURE OF BONDS

To understand why matter exists as solids, liquids or gases and how they physically evaporate, dissolve and generally distribute in the environment, we must start by considering the molecular nature of chemical compounds and the way atoms are bound to one another. First, we will look at the nature of chemical bonds, and this will give an insight into physical chemical properties that will provide an understanding of environmental properties.

The most important type of bonding is the **covalent bond**. Usually with this type of bonding, two atoms react together, with each contributing one electron to form a bond. Thus, two atoms of hydrogen can react to form a hydrogen molecule with one covalent bond. Thus,

$$H\bullet + \bullet H \rightarrow H{-}H$$

Each covalent bond consists of two electrons moving rapidly between the hydrogen atoms in a defined space as illustrated in Figure 2.2. Each electron holds a full negative charge and, if the electrons spend equivalent times near the two atoms in the bond, this results in no difference in charge between the two ends of the bond. This can be interpreted as the electron density around the two atoms being symmetrical, leading to a nonpolar bond. However, a different situation applies with the hydrogen chloride bond. In this bond, the chlorine tends to attract electrons as shown in Figure 2.3. This attraction is not sufficient to cause the electron to remain permanently with the chlorine atom, but causes the electron to spend more of its time when it is moving between the two atoms toward the chlorine end of the bond. This results in a small partial negative charge (denoted as δ–) occurring on the chlorine atom and leading to a small partial positive charge (denoted as δ+) occurring at the hydrogen atom. These factors are illustrated with the hydrogen chloride bond in Figure 2.3. The outcome of this effect is that the bond becomes **polar** and has a **dipole moment** or **polarity**. Quantitatively, the polarity of bonds can be characterized by two factors: (1) the charge on each atom and (2) the distance between the two atoms.

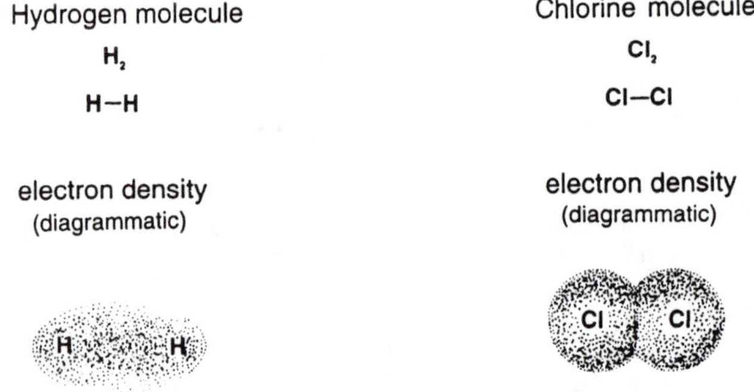

Points:

* the electron density is symmetrical

* there is no polarity

* the bond is nonpolar

FIGURE 2.2 Illustration of the characteristics of nonpolar covalent bonds in the hydrogen and chlorine molecules.

The combination of these two factors allows the development of a quantitative characteristic described as the **dipole moment,** which is defined as follows:

Dipole Moment (D.M.) = Electrostatic Charge × Distance Between Atoms in Debye units, D

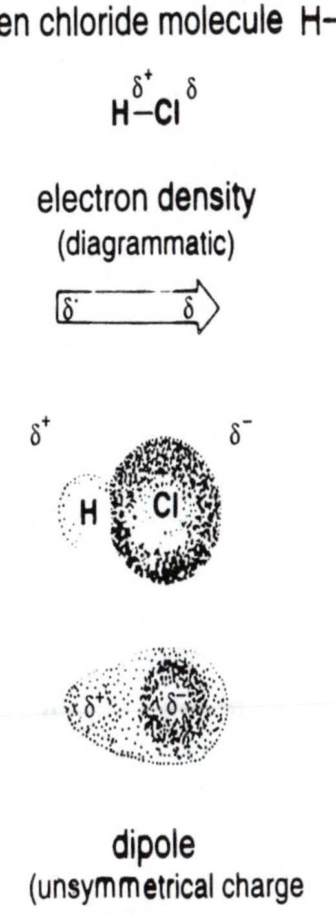

FIGURE 2.3 Illustustration of the characteristics of polar covalent bonds using hydrogen chloride as an example.

If both atoms, or ends of the bond, are the same, as with the hydrogen (H_2) and chlorine (Cl_2) molecules, then there can be no differences in the overall electron density in the bond and the dipole moment for such symmetrical bonds must be zero. However, most unsymmetrical bonds have a dipole moment. For example, the hydrogen chloride bond has a dipole moment of 1.03 Debyes (D) and there are examples of the dipole moments of bonds contained in Table 2.2.

It can be expected that a bond will be polar if the two atoms involved differ in their ability to attract electrons. This property is described as the **electronegativity** of an atom and can be estimated quantitatively. The order of magnitude of the electronegativity of some common elements is shown below:

$$F > O > Cl, N > Br > C, H$$

A table of quantitative measures of the electronegativities of various elements is shown in Table 2.3. There is an approximate correlation between the differences in the electronegativity of two atoms in a bond and its dipole moment. For example, the difference in electronegativity of H and F is 1.9, and the dipole moment is 1.91 Debyes; the H and O difference is 1.4 and DM is 1.5; and the C and O difference is 1.0 and DM is 0.7. The partial negative charge resides at the F and O atoms in these bonds.

TABLE 2.2
Dipole Moments of Some Chemical Bonds

Bond	Dipole moment (Debye, D)	Bond description
H-F	1.91	Highly polar
H-Cl	1.08	Polar
H-Br	0.80	Polar
H-O	1.5	Polar
H-N	1.3	Polar
C-H	0.4	Weakly polar
C-Cl	0.5	Weakly polar
C-Br	0.4	Weakly polar
C-O	0.7	Weakly polar
C-N	0.2	Nonpolar

TABLE 2.3
Electronegativities of Some Elements

Element	Electronegativity	Element	Electronegativity
Li	1.0	C	2.5
Na	0.9	Pb	1.9
K	0.8	N	3.0
Cr	1.6	P	2.1
Fe	1.8	O	3.5
Cu	1.9	S	2.5
Ag	1.9	F	4.0
Au	2.4	Cl	3.0
Zn	1.6	Br	2.8
Cd	1.7	I	2.5
		H	2.1

IV. POLARITY OF MOLECULES

The C-H bond is one of the most common in organic compounds. By looking at Table 2.3 it can be seen that the electronegativity difference between these two atoms is 0.4. On the other hand, if we look at the O and H atoms, we find that the electronegativity difference is 1.4. This leads to a description of the C-H bond as being **weakly polar** and O-H bond as being **polar**. By this means we can provide descriptive terms for the polarity of the various bonds, as shown in Table 2.2.

If we join sets of bonds and atoms together to form covalent compounds, this results in the overall polarity of the compound being due to a combination of the dipole moments of the bonds present. The overall dipole moment for the molecule is due to the spatial resolution or vector sum of the various dipole moments of the bonds present in terms of their size and direction.

Figure 2.4 contains a diagrammatic representation of the methane molecule. Since the hydrogen atoms in this molecule are directed symmetrically toward the corners of a tetrahedron, the spatial resolution of the dipole moments of the C-H bonds would result in a dipole moment for methane of zero. So, even though the C-H bond is a weakly polar bond, the overall dipole moment is zero. However, if we replace one of the hydrogen atoms in methane with a chlorine atom to give monochloromethane (see Figure 2.4), this gives an unsymmetrical molecule that has a dipole moment of 1.87 D. Similarly, the symmetrical compound tetrachloromethane, which is also known as carbon tetrachloride, has a dipole moment of zero since it is a symmetrical molecule; and trichloromethane, also known as chloroform, has a dipole moment of 1.02 D.

In this discussion we have looked at **atoms** in terms of their electronegativity, but common groupings of atoms and functional groups have consistent properties in terms of their effects on polarity of molecules. The polarity classifications of a variety of common groupings are shown in Table 2.4. It is interesting to note that the alkyl groups containing many C-H bonds are listed as nonpolar, whereas the C-H bond is listed as weakly polar. This is due to the spatial orientation of the C-H bonds present, which tends to reduce polarity. Thus, the n-alkanes such as hexane are nonpolar.

A variety of important chemicals widely distributed in the environment are weakly polar to nonpolar. For example, DDT with the structure shown in Figure 2.4 is weakly polar to nonpolar since the weakly polar C-Cl bonds are distributed in such a manner as to reduce the overall polarity of the molecule to a very low level. The polychlorinated biphenyls (PCBs) and most of the chlorohydrocarbon pesticides are also weakly polar to nonpolar. On the other hand, the herbicide 2,4-dichlorophenoxy acetic acid (2,4-D) (see Figure 2.4) is a polar molecule because the carboxylic group (–COOH) is polar (see Table 2.4).

One of the most important compounds in the environment is water since it is the major component of oceans, lakes and rivers and thus has a major impact on the distribution of chemicals. Water has a molecular structure that consists of two hydrogen atoms attached to an oxygen atom with both O-H bonds residing in a plane at an angle of 104° to each other. Spatial analysis of the direction and strength of these bonds gives a resultant dipole moment of 1.80 D, which is in agreement with the experimental value of 1.87 D (see Figure 2.5). As a result, water is a polar molecule with physical properties, and its ability to dissolve other substances, strongly influenced by its polarity.

V. IONIC COMPOUNDS

With some elements, the difference in electronegativities can be so great that one atom almost completely loses an electron to the other atom in the bond. For example, in the hydrogen chloride molecule, there can be a transfer of an electron from the hydrogen to the chlorine resulting in the formation of a chloride ion with a full negative charge and consequently a hydrogen ion with a full positive charge. This is illustrated below.

$$\overset{\delta+}{H} - \overset{\delta-}{Cl} \rightarrow H - Cl \rightarrow H^+ + Cl^-$$

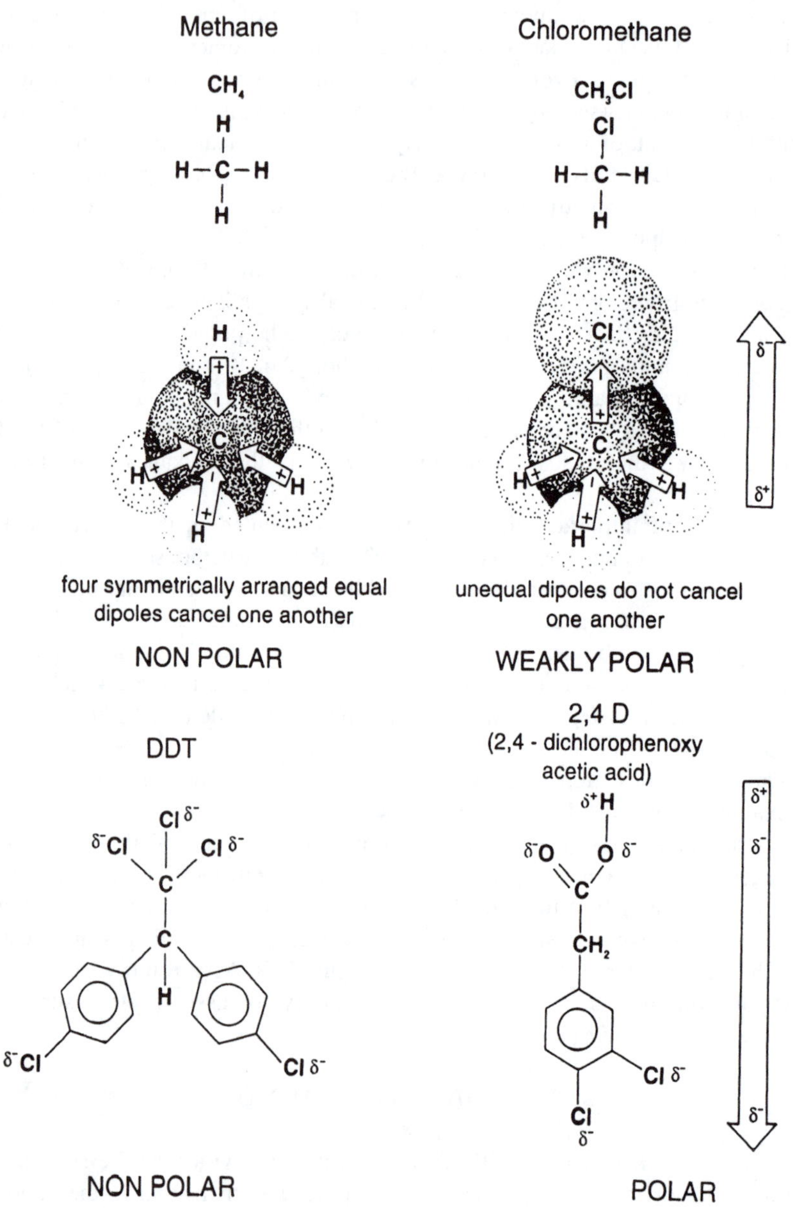

FIGURE 2.4 Chemical structures of some compounds with different polarities.

In this transfer, the arrow indicates that both electrons that form the covalent bond have moved to the chlorine atom. In this situation, the chlorine atom in the HCl molecule which held a small negative charge (δ-) now becomes a fully negatively charged free atom described as an **ion**. The covalent bond has been broken and no longer exists (see Figure 2.6). With the formation of ions, the hydrogen ion and chloride ion can move apart depending on conditions. In fact, in hydrogen

TABLE 2.4
Polarity of Common Groupings in Organic Molecules

Highly polar	Polar	Weakly polar	Nonpolar
$-COO^-$	$-O-H$	$\rangle C-Cl$	$-CH_3$
$-O^-$	$-COOH$	$\rangle C-Br$	$-CH_2CH_3$
$-NH_3^+$	$-NH_2$	$\rangle C-H$	
	$\rangle C=O$		

chloride there is a mixture of polar hydrogen chloride molecules together with hydrogen and chloride ions. Measurements indicate that in pure liquid hydrogen chloride, 17% of the compound is in the ionic form. With metal salts of some substances, there can be 100% of the substance in the ionic form. Examples of these substances are sodium chloride, potassium bromide and so on. Similar processes occur with these substances as with hydrogen chloride, as illustrated in Figure 2.6, but to a greater extent.

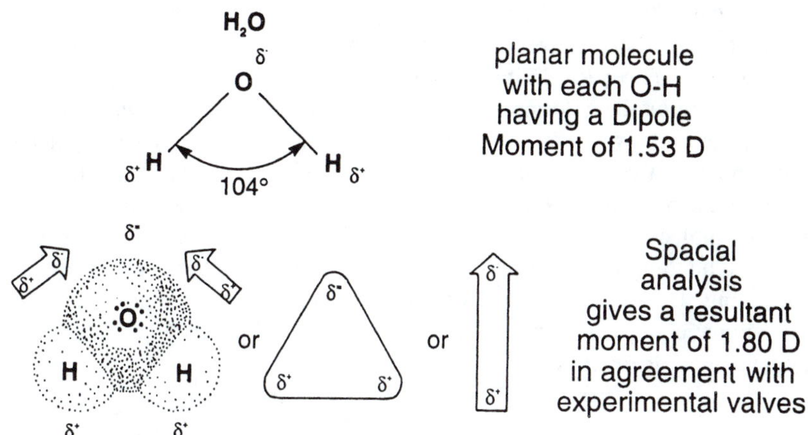

FIGURE 2.5 Representations of the water molecule and its polar characteristics.

VI. INTERMOLECULAR FORCES

In the previous section we considered charges that are generated within molecules. It can be appreciated that these charges within molecules will interact with adjacent molecules with the usual rules for the interaction of charges applying. This means that **like** charges result in repulsion and **unlike** charges result in an attraction. Thus, positive-to-positive and negative-to-negative forces result in the generation of a disruptive or repulsive force, and positive-to-negative charges result in an attraction

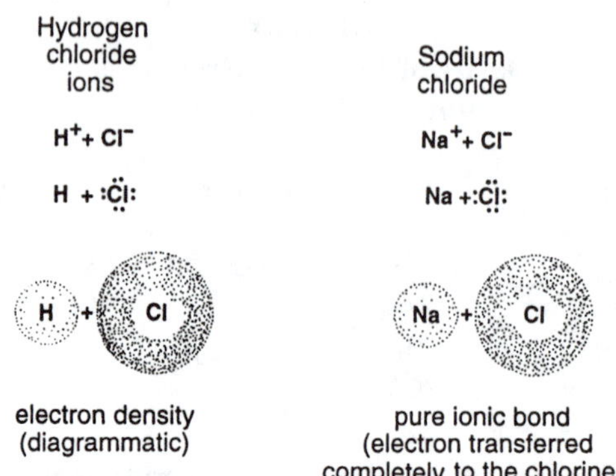

FIGURE 2.6 Formation of ions with hydrogen chloride and sodium chloride.

or a cohesive force. Intermolecular forces between covalent compounds are usually described as **van der Waals forces**. Van der Waal was a Dutch scientist who developed an understanding of intermolecular forces. These intermolecular forces have a strong influence on the physical chemical properties of compounds and how substances behave in the environment.

Molecules that are polar can be attracted to one another so that particular orientations of the positive and negative charges are formed. Examples of these interactions, referred to as **dipole bonding,** are shown in Figures 2.7 and 2.8. Here the δ– charge on the chlorine in hydrogen chloride is oriented toward the δ+ on the hydrogen atom in water (see Figure 2.7). In water, alcohols and acids, a network of orientations of the δ+ hydrogen toward the δ– of the oxygen in another molecule of the same compound is set up. Of course, in mixtures, these interactions occur between molecules of different compounds. The **hydrogen bond** is a special form of dipole bonding that is of particular significance because it is relatively common and strong. Hydrogen bonding with water, alcohol and acid molecules is shown diagrammatically in Figure 2.8. Somewhat similarly, polar molecules with dipole moments may interact in various ways with ions.

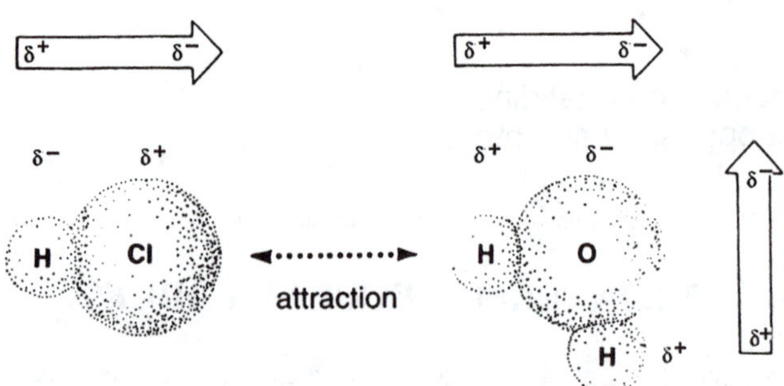

FIGURE 2.7 Attraction between the negative end of a hydrogen chloride molecule and the positive end of a water molecule.

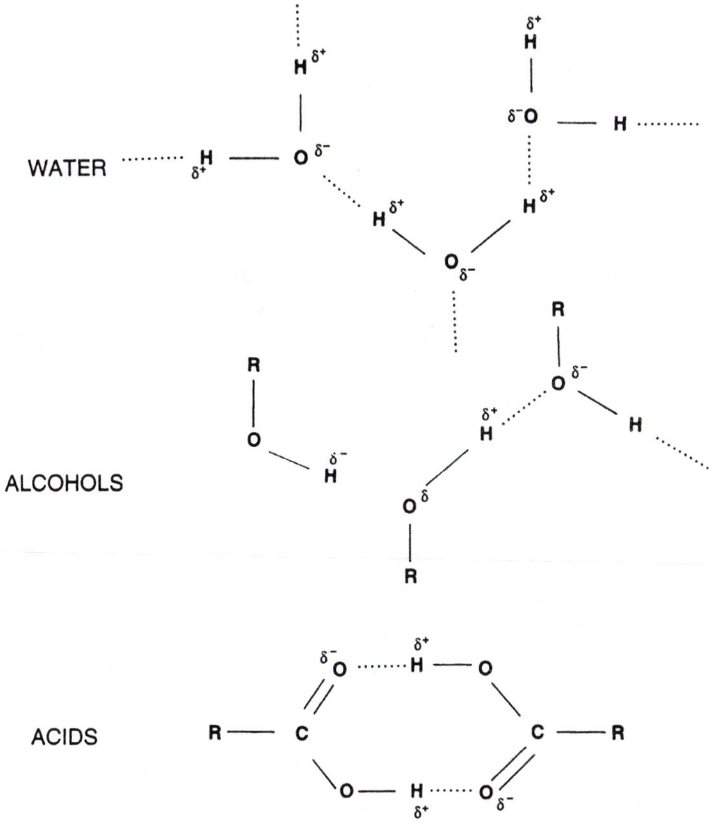

FIGURE 2.8 Hydrogen bonding in water, alcohols and acids.

Another important group of intermolecular interactions are described as **London forces** or **dispersion forces.** These operate over relatively short distances and are responsible for the aggregation of molecules that possess neither free charges nor dipole moments. London forces increase rapidly with molecular weight and are particularly strong with aromatic compounds; but with relatively low molecular weight compounds, these are often relatively weak when compared with dipole interactions.

VII. PHYSICAL CHEMICAL PROPERTIES OF COMPOUNDS

A. MELTING POINT (mp)

When a substance melts, it changes from a solid to a liquid. In molecular terms, a solid has the component particles in a fixed relationship to one another. There are strong forces between chemical entities as occur, for example, with ions, high molecular weight substances and highly polar substances. With ionic solids, each of the ions is held in a fixed relationship to other ions by the strong forces between the ions. This is illustrated diagrammatically with sodium chloride in

Figure 2.9. This compound is held in the solid form by the fixed and close relationship between the strongly attractive forces that exist between the sodium and the chloride ions with their full positive and full negative charges. It is also noteworthy that each ion is surrounded by ions of an opposite charge, leading to an overall neutral electrostatic situation. Because of these forces, it requires a high energy input to break this solid structure down to give a liquid. This means that this substance will exhibit a very high melting point and, in fact, sodium chloride exhibits a melting point of 801°C.

With nonpolar compounds of low molecular weight, including oxygen, nitrogen, and methane, there are no ionic or dipole forces holding separate molecules together, but there are the relatively weak London forces causing attachment between the molecules. Thus, it would be expected that these substances would exhibit a very low melting point. This is indicated by the melting point of methane, which is –183°C.

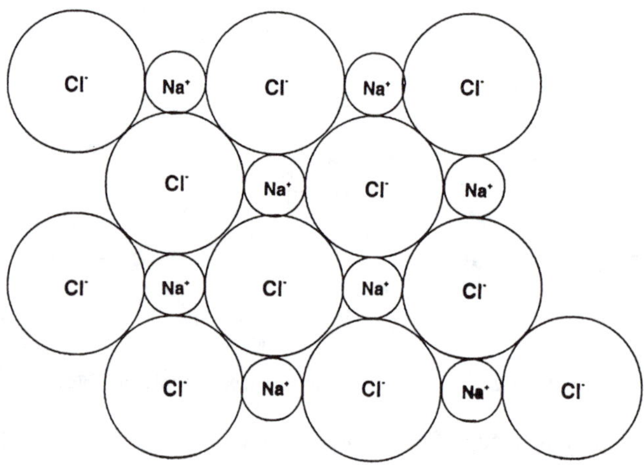

Points:

* this is a single layer of ions out of the solid

* other layers rest above & below this layer

FIGURE 2.9 Diagrammatic illustration of the ionic structure of sodium chloride.

B. BOILING POINT (bp)

The boiling point (bp) and related properties, such as vapor pressure (vp), have a major influence on such important environmental characteristics as evaporation and the distribution of compounds in the atmosphere. Boiling involves the conversion of a liquid with molecules in close, but random proximity, into a gas with molecules some distance apart. Gases at room temperature are substances with weak intermolecular forces. This means that generally they will be nonionic, low molecular weight, nonpolar substances. In a general sense, liquids can be seen as an intermediate state between gases and solids.

With ionic compounds, the ion charges hold the ions together with considerable force. Thus, we would expect that, in general, ionic compounds would have very high boiling points. For example, sodium chloride has a boiling point of 1413°C. With nonionic and nonpolar compounds, the major intermolecular forces are London forces that are weak at low molecular weight. This means that this group of compounds will have a much lower boiling point than ionic compounds. For example, methane has a boiling point of –161.5°C.

With covalent compounds, the boiling point is influenced by two factors: (1) the dipole moment and (2) the size of the molecule. This latter factor is related to the strength of the London forces, which are related to molecular weight. Thus, compounds of similar molecular weight have boiling points which are related to their dipole moment. There is a comparison of the boiling points of compounds of similar molecular weight but different dipole moments in Table 2.5. It can be seen that, since molecular weight is roughly constant, the boiling point in the series of compounds decreases as the dipole moment decreases.

TABLE 2.5
Change in Boiling Point with Compounds of Similar Molecular Weight

Compound	Molecular weight	Dipole moment (D)	Boiling point (°C)
H_2O	18	1.84	100
NH_3	17	1.46	–33
CH_4	16	0	–161.5

An illustration of the action of London forces is shown in Table 2.6. In this data set, the dipole moment of the normal alkanes is constant at zero, and the molecular weight increases from methane to butane. As the molecular weight increases, the boiling point also increases in accord with the molecular weight change. This is due to the increase in the strength of the London forces with increasing molecular weight. This relationship within a series of compounds having a regular change in size, such as in a homologous series, is useful for predicting properties of members of the series that are not available. For example, there is a plot of vapor pressure against number of chlorine atoms in Figure 2.10 for the chlorobenzenes. The vapor pressure of some members of the series (tetra- and pentachlorobenzenes) may not be known but can be obtained by interpolation as shown in Figure 2.10.

C. SOLUBILITY

An important environmental property of a substance is its solubility in water, and other solvents, since this property influences its dispersal in the open environment, including the oceans and aquatic systems generally. In addition, a range of important biological properties can be related to solubility in water and other solubility-related properties. An understanding of this property can give an understanding of many of the most important environmental properties of compounds.

When a substance dissolves to form a solution, the individual molecules of the solute are separated by the molecules of the solvent as shown in Figure 2.11. In

TABLE 2.6
Change in Boiling Point with Compounds of Different Molecular Weight

Compound	Dipole moment (D)	Molecular weight	Boiling point (°C)
Methane	0	16	–161
Ethane	0	30	–88
Propane	0	44	–42
Butane	0	58	0

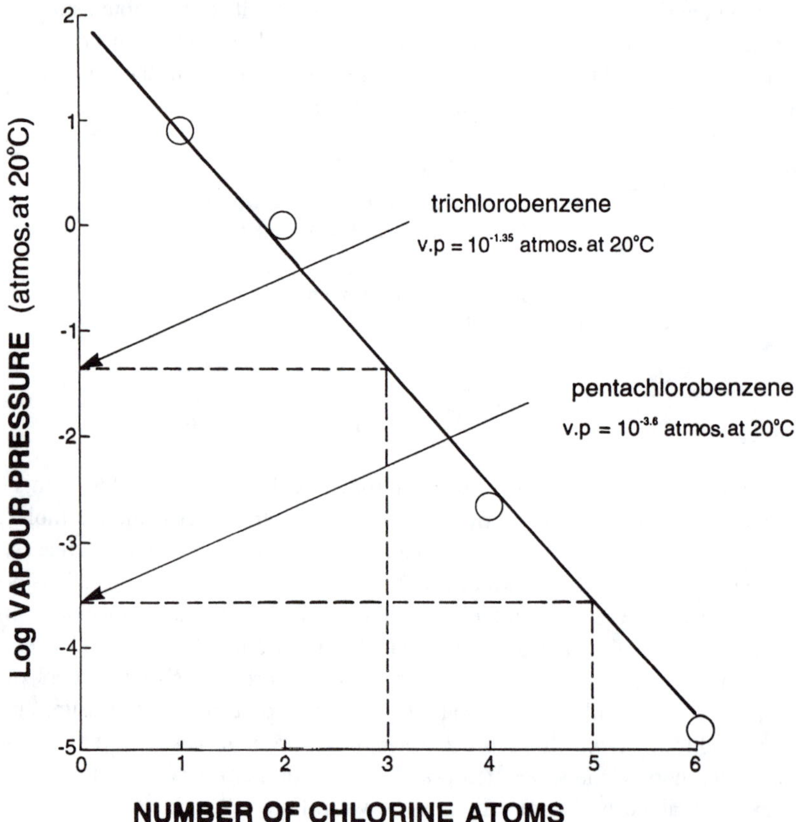

FIGURE 2.10 Plot of the relationship between log (vapor pressure) and number of chlorine atoms for the chlorobenzenes.

simple terms, for a substance to dissolve, the solvent molecules (s) must form a set of bondings with the solute molecules (M) which are, in total, stronger than the solute-to-solute bonds. If this doesn't occur, the solvent molecules (M) will move together due to mutual attraction and the substance will come out of solution and thus be insoluble.

The solution of an ionic compound in a polar solvent is shown diagrammatically in Figure 2.12. This is typified by the solution of sodium chloride in water. The ions

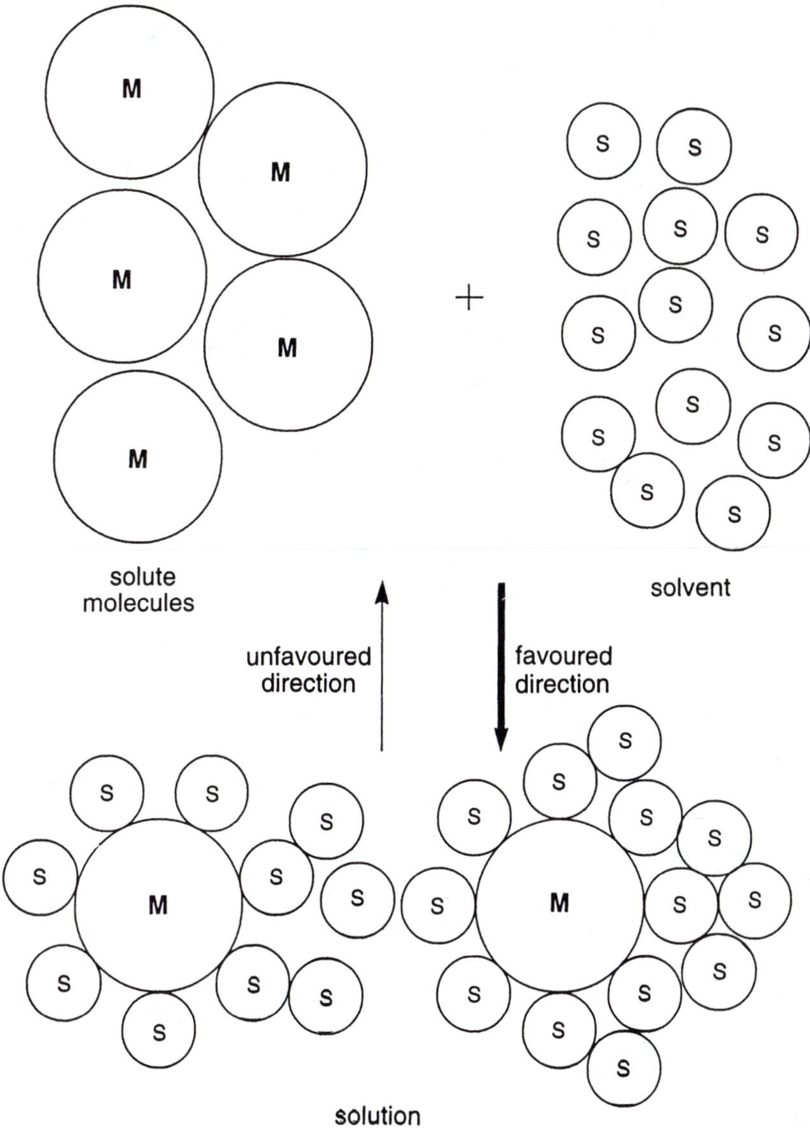

FIGURE 2.11 Interaction of solute (M) and solvent (S) molecules to produce a solution.

in solution are each surrounded by water molecules with the oppositely charged ends attracted to the ion. The forces of attraction between the water molecules and the ions are greater than the attraction between the sodium and chloride ions. In this way, a stable solution is formed and so generally ionic compounds are dissolved by polar and ionic solvents. A similar situation would be expected to apply to the solubility of polar covalent compounds by polar solvents.

If a nonpolar solvent was used with an ionic or polar solute, the bonds between the nonpolar solvent molecules and the ions or polar substances would be very weak and generally insufficient to form a stable solution. Thus, nonpolar solvents, such

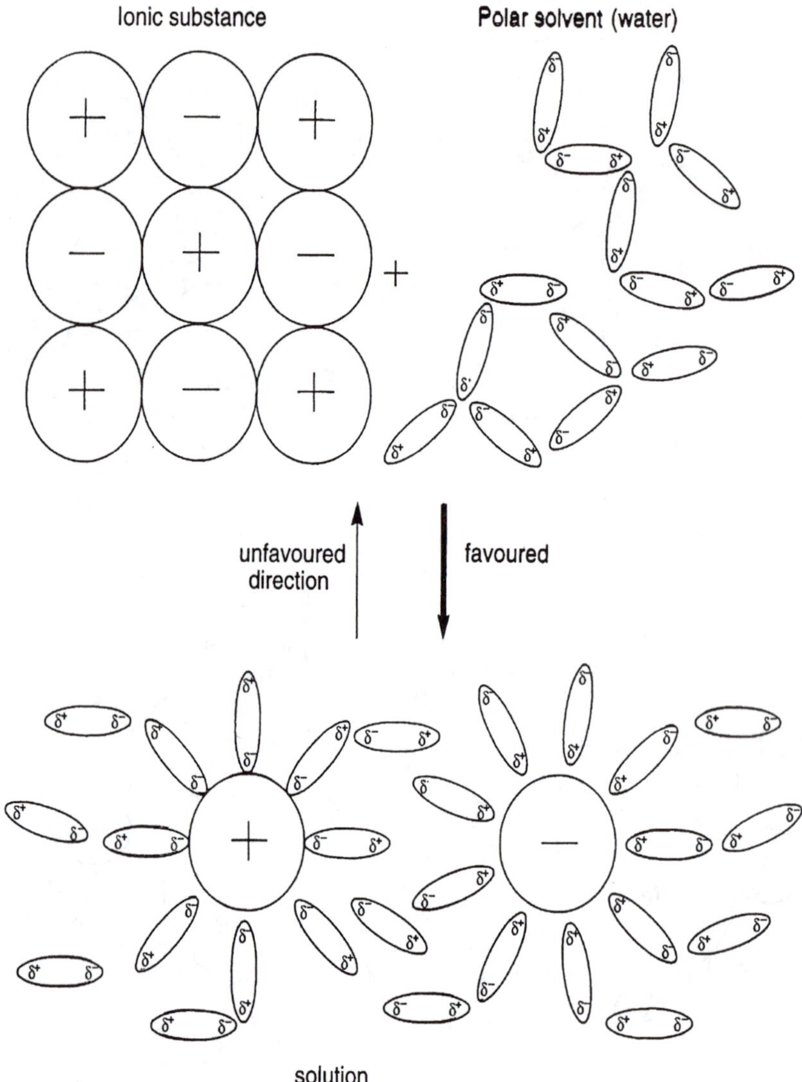

FIGURE 2.12 Solution of an ionic substance, such as sodium chloride, in a polar solvent, such as water.

as hexane, would not be expected to dissolve polar and ionic compounds to any significant extent. For example, water, which is a polar substance, dissolves ionic and polar substances such as sodium chloride, hydrogen chloride, ammonia and so on. On the other hand, liquid methane is a nonpolar substance and will not significantly dissolve these substances. This leads to the generalization that, in terms of polarity, *like* compounds dissolve and *unlike* compounds are insoluble in one another.

Many chemicals of environmental importance are weakly polar to nonpolar, and include compounds such as the dioxins, PCBs and the chlorohydrocarbon pesticides. The situation using a polar solvent (such as water) to dissolve these nonpolar compounds is illustrated in Figure 2.13. Since the solute is nonpolar, little bonding between water and the solute occurs. If a nonpolar solute molecule dissolves in water, its surrounding environment is quite different from that of an ion or a polar compound. The ion is surrounded by the ends of water molecules all of the same charge so as to stabilize and neutralize the opposite charge on the ion (see Figure 2.12). With the nonpolar solute, the water molecule orients to stabilize and neutralize the system by alternating the positive and negative ends facing the solute molecule, as in Figure 2.13. This creates a force on the inner surface of the cavity formed by the solute molecule that tends to contract and thus expel the solute molecule. As a result, the solubility of nonpolar substances in water decreases with increasing size of the solute molecule and the cavity it consequently forms in the solvent. This is illustrated by the plot in Figure 2.14. Here, the aqueous solubilities of the PCBs decline as the surface area of the molecules increases.

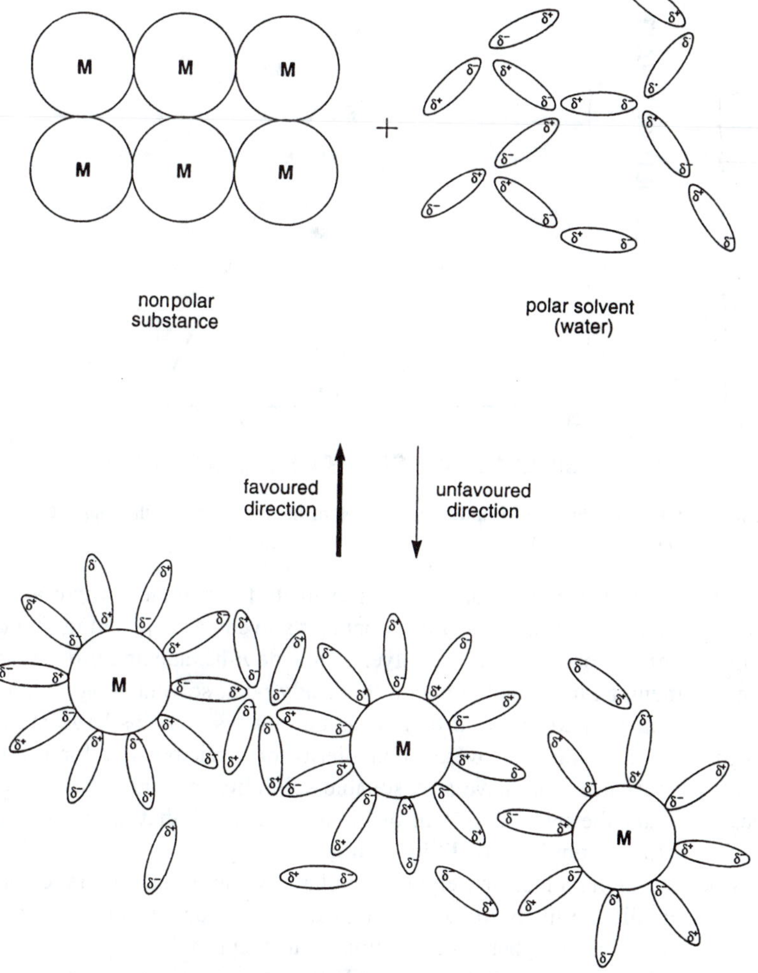

FIGURE 2.13 Solution of a nonpolar substance (M) in a polar solvent, such as water.

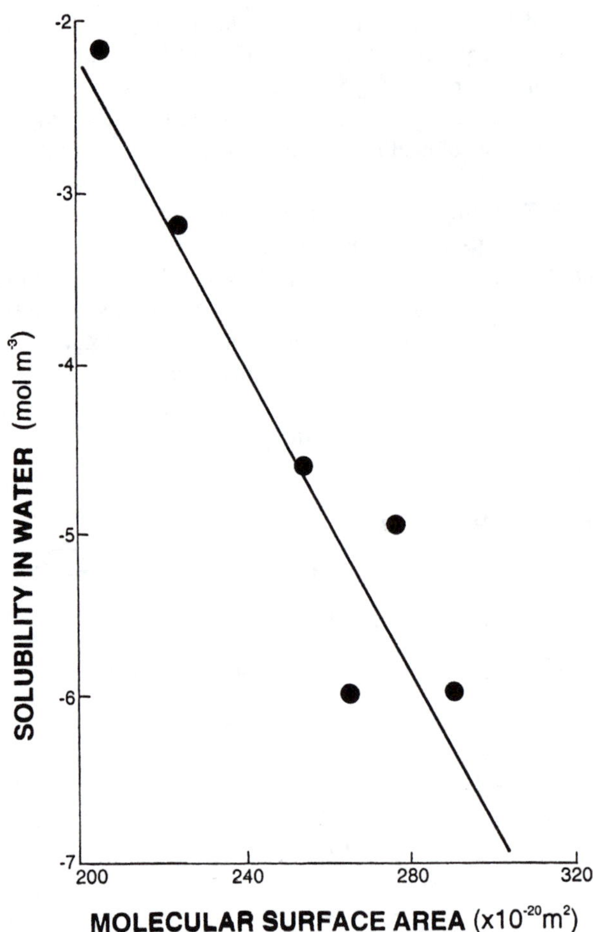

FIGURE 2.14 Relationships between the surface area of the molecule and the solubility in water for the polychlorobiphenyls (PCBs).

In contrast to these situations, we can evaluate the situation where a nonpolar or weakly polar solute is dissolved in a nonpolar solvent, as depicted in Figure 2.15. Common nonpolar or weakly polar solvents include *n*-hexane and *n*-octanol; but in the environment, an important nonpolar to weakly polar solvent is biota lipid. Biota lipid is a complex mixture in all organisms but always contains large numbers of nonpolar groups, leading to overall nonpolarity or weak polarity for the lipid in general. A substance must have a reasonable solubility in lipid to penetrate membranes and enter the internal system of biota. Substances that are soluble in lipid are referred to as **lipophilic** or **lipid loving**. The major components of lipids are fats, which are high molecular weight, nonpolar to weakly polar esters. Compounds that are lipophilic are likely to have biological effects, such as a tendency to accumulate in organisms to higher concentrations than occur in the external environment and a general toxicity to all biota. In addition, these substances are often also persistent in the environment and exhibit resistance to transformation and degradation. The general characteristics of lipophilic compounds are outlined in Table 2.7.

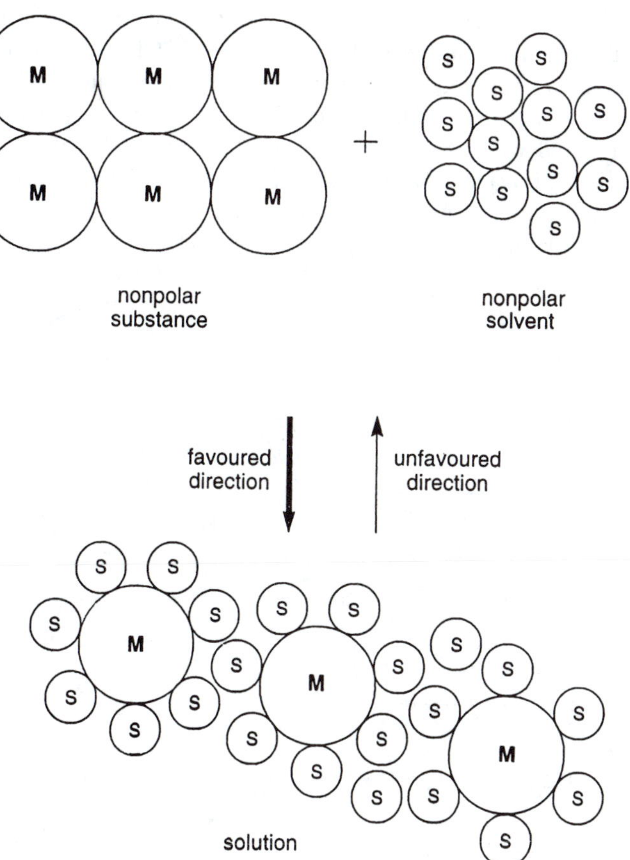

FIGURE 2.15 Solution of a nonpolar (or weakly polar) substance (M) in a nonpolar solvent such as biota lipid.

TABLE 2.7
Characteristics of Lipophilic Compounds

Characteristic	Features
Chemical structure	Nonpolar compounds such as chlorohydrocarbons, hydrocarbons, polychlorodibenzodioxins, polychlorodibenzofurans and related compounds
Molecular weight	100–600
Water solubility (mol/m^{-3})	0.002–18

In addition to having the property of lipid solubility, lipophilic compounds usually exhibit relatively low solubility in water. For this reason they are often also called **hydrophobic** compounds, which means that they are **water hating.**

VIII. PARTITION BEHAVIOR

Previously we looked at polarity and the physical chemical properties of organic compounds. The principle of solubility discussed was that in terms of polarity: **like** dissolves **like,** and **unlike** substances do not dissolve. However, we have many situations that are not clear-cut, and the compounds have both polar and nonpolar properties. For example, butyric acid (CH_3-$(CH_2)_2$- COOH) has polarity and in fact partially forms ions (depending on the pH of the solvent) but also has a nonpolar part, as illustrated in Figure 2.16. If placed in water, which is a polar solvent, some of the butyric acid will dissolve in the water because it is moderately soluble in water. However, if it is placed in a nonpolar solvent such as hexane, or diethyl ether, some of it will also dissolve in these nonpolar solvents. This means that it is moderately soluble in both polar and nonpolar solvents.

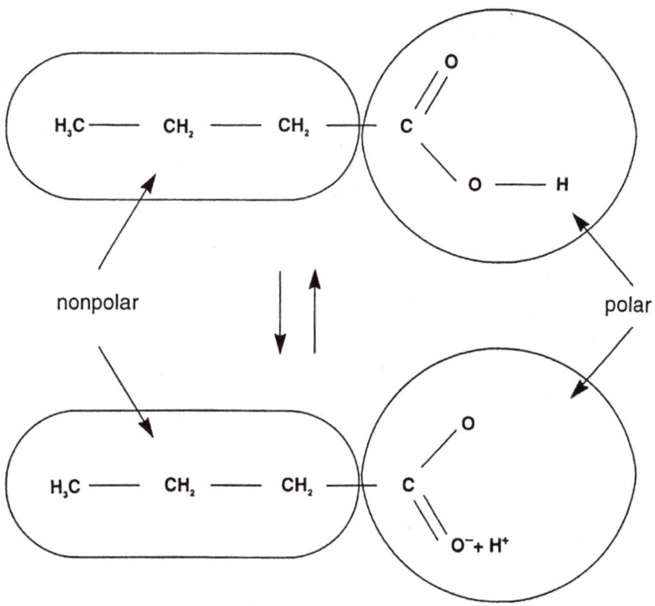

FIGURE 2.16 The butyric acid molecule, indicating polar and nonpolar sections.

If one places sufficient diethyl ether and water together, they will form a two-phase system with diethyl ether forming a separate layer floating above the water, as shown in Figure 2.17. The diethyl ether (CH_3-CH_2-OH-CH_2-CH_3) is a weakly polar to nonpolar liquid that has limited solubility in the polar water. If one places butyric acid in this system and shakes it thoroughly, one finds that the butyric acid has partitioned (or distributed) between the two liquids. In fact, at equilibrium, the diethyl ether acid contains a concentration of three units per volume and the water contains a concentration of one unit per volume. This partitioning can be quantitatively measured as the **partition coefficient**. The partition coefficient for butyric acid in a diethyl ether-water system is 3. The **partition coefficient** is constant for a given substance and two specific liquids under constant environmental conditions such as temperature and pressure. The partition coefficient is dependent on:

1. Polarity of the substance
2. Molecular weight
3. Relationship of the polarity of the solvents used

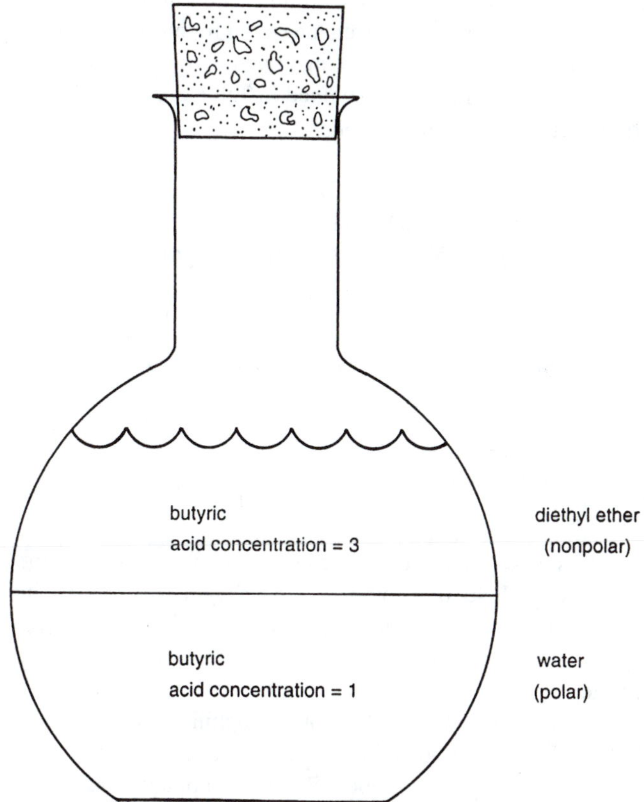

FIGURE 2.17 Partitioning of butyric acid between diethyl ether and water.

If one uses hexane instead of butyric acid in this system, and since hexane is less polar than butyric acid, hexane will have a different partition coefficient. Hexane has a higher concentration in the diethyl ether phase and a lower concentration in the water phase as compared to that of the diethyl ether/water system. In fact, the partition coefficient of hexane in this system is approximately 100.

One of the most widely used physical chemical characteristics of compounds applied in environmental chemistry is the *n*-octanol/water partition coefficient. This partition coefficient is important because it imitates the biota lipid/water partition process. Octanol (CH_3-$(CH_2)_7$-OH) is in many ways similar to biota lipid since it is fat-like in many of its physical chemical properties. The *n*-octanol/water partition coefficient (K_{OW}) is defined as

$$K_{OW} = C_O/C_W$$

where C_O is the concentration of the substance in octanol and C_W is the concentration of the substance in water at equilibrium and at a constant temperature. K_{OW} values can be measured in a **shake-flask** system, similar to the diethyl ether/water system in Figure 2.17, by adding the compound to the system and measuring the concentrations after shaking to reach equilibrium and then calculating K_{OW}. The K_{OW} values are dimensionless since both C_W and C_O are measured in the same units, i.e., units of mass per unit volume. Values have been measured for a wide variety of compounds of environmental importance and these range from about 0.001 to over 10,000,000. Some typical, illustrative values are contained in Table 2.8 (the values are expressed on a logarithmic scale for convenience).

TABLE 2.8
Some Octanol/Water Partition Coefficient (K_{OW}) Values for Various Compounds

Compound	K_{OW}	log K_{OW}
Ethanol	0.49	–0.31
2,4,5-T	3.98	0.60
2,4-D	37.20	1.57
Benzene	134.90	2.13
1,4-Dichlorobenzene	3310.00	3.52
DDT	2.29×10^6	6.36

The K_{OW} values are constant for a given compound and, apart from experimental errors, reflect the lipophilicity of a compound. As K_{OW} increases, the solubility in lipid (represented by octanol) increases relative to water. Thus, increasing K_{OW} values reflect increasing lipophilicity. An increasing K_{OW} value also usually reflects a decline in the solubility in water. A compound such as ethanol with $K_{OW} = 0.49$ actually has greater solubility in water than in octanol and is not a lipophilic compound. Similarly, 2,4,5-T and 2,4-D have K_{OW} values of 3.98 and 37.2, respectively, and these substances would be regarded as being of low lipophilicity. Compounds are usually considered lipophilic at values of $K_{OW} > 100$. As a general rule, lipophilic compounds have K_{OW} values in the range from 100 to 3,000,000, i.e., log k_{ow} 2 to 6.5. The lipophilic compounds in Table 2.8 are benzene, 1,4-dichlorobenzene and DDT.

The K_{OW} values can be predicted using relationships somewhat similar to those used to predict boiling point and vapor pressure (see Figure 2.10). As outlined previously, the surface area of the cavity formed by a solute in the solvent, or the surface area of the solute itself, is a useful parameter for evaluation of environmental properties (see Figure 2.14). The K_{OW} values for members of various homologous series have been found to be related to the molecular surface area of the compounds. The relationship between these characteristics for chlorobenzenes is shown in Figure 2.18, with graphical projections that can be used to provide data on some members.

Another useful partition coefficient used to describe the environmental properties of a chemical is the Henry's law constant (H), defined as:

$$H = P/C_W$$

where P is the partial pressure of the compound in air and C_W is the corresponding concentration in water. This characteristic quantifies the relationship between a contaminant in a water body (such as a lake or the oceans) and the atmosphere at equilibrium.

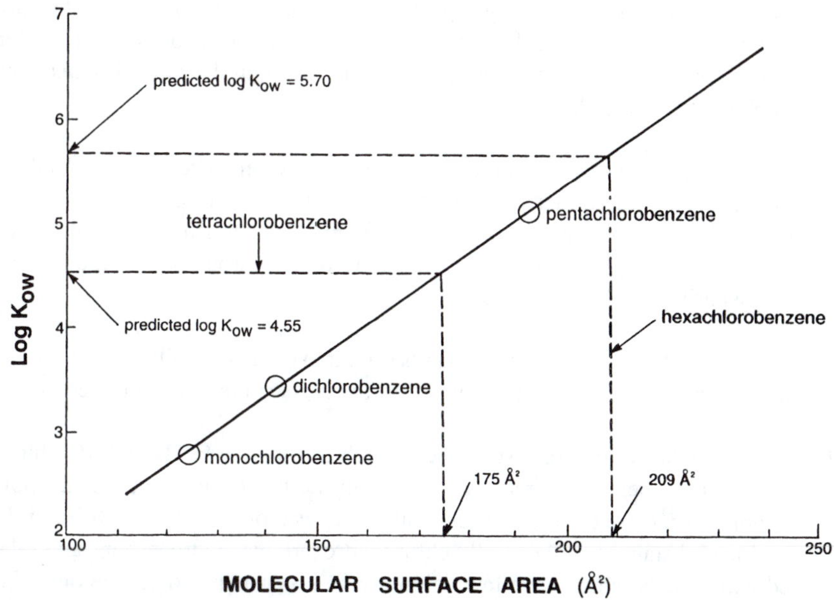

FIGURE 2.18 Relationship between log K_{OW} and molecular surface area for the chlorobenzenes with extrapolation and interpolation to estimate compounds with unknown properties.

The environment can be considered to be made up of a set of phases, such as air, water, soil, atmosphere and biota. A chemical enters the environment and partitions between these phases as the basic mechanism governing distribution. This is described in more detail in Chapter 18 "Distribution of Chemicals in the Environment."

IX. KEY POINTS

1. There are three states of matter in the environment — solids, liquids and gases. Solids such as common salt, sand, rocks and so on are relatively dense and consist of particles that are closely packed in a fixed relationship to one another. Liquids, such a water, have less closely packed particles but do not have a fixed relationship to one another. With gases, such as those in the atmosphere, the particles are in relatively rapid motion and are comparatively large distances apart.

2. Sharing of electrons by atoms to form a covalent bond can result in the polarization of the bond if the atoms are of a different element. The direction and size of the polarization of bonds in a molecule can result in compounds exhibiting overall dipole moments.

3. The intermolecular forces, generally described as van der Waal's forces, and measured by the dipole moment of molecules, have a major influence on properties such a melting point, boiling point and so on. London forces also play a major role in the properties exhibited by nonpolar compounds.

4. Ionic and polar solutes will generally dissolve in water since the water molecules form stronger attachments to the solute than occur between the solute molecules and ions themselves. This is in accord with the rule "**like** dissolves **like**" in terms of polarity.

5. Nonpolar substances will not readily dissolve in water since solute and solvent molecules are not attracted and stabilization of the solublized solute does not occur. This means that weakly polar to nonpolar chemicals important in the environment, such as dioxins, PCBs and chlorohydrocarbon pesticides, are only sparingly soluble in water.

6. The solubility of nonpolar compounds is related to the size of the cavity formed in the water mass and is also related to the surface area of the molecule.

7. Weakly polar to nonpolar substances, such as dioxins, PCBs and the chlorohydrocarbon pesticides, readily dissolve in the most important weakly polar to nonpolar solvent in the environment, which is biota lipid. **Lipophilic (lipid-loving)** substances are generally chlorohydrocarbons, hydrocarbons and related compounds which have low solubility in water and also can be described as **hydrophobic (water-hating)** compounds.

8. A substance must pass through nonpolar lipid-containing membranes to enter biota and so lipophilic compounds are often biologically active.

9. Covalent molecules usually have a mix of polar and nonpolar sections and will partition between a nonpolar phase (such as *n*-hexane) and a polar phase such as water. This partitioning can be quantitatively described by the partition coefficient, which is the concentration in one phase/concentration in the other phase. The octanol/water partition coefficient (K_{OW}) is a widely used environmental property and its magnitude can often be predicted using relationships with the molecular surface area and other characteristics.

REFERENCES

James, K.C., *Solubility and Related Properties*, Marcel Dekker, New York, 1986.

Reichardt, C., *Solvents and Solvent Effects in Organic Chemistry,* 2nd Ed., VCH, Weinheim, 1988.

Lyman, W.J., Reehl, W.F., and Rosenblatt, B.H., *Handbook of Chemical Property Estimation Methods, Environmental Behavior of Organic Compounds,* American Chemical Society, Washington, D.C., 1990.

CHAPTER 2 PROBLEMS

1. A range of different compounds and their physical chemical properties that influence environmental behavior are listed in Table 2.9. Based on these characteristics,

TABLE 2.9
Various Compounds and Some of Their Properties

Compound	bp (°C)	Water solubility (g 100 mL^{-1})	Dipole moment (D)
H_2	–253	2×10^{-4}	0
CH_4	–164	2×10^{-3}	0
CO_2	–78	0.15	0
HBr	–67	221	0.82
HCl	–84.9	82	1.08
NH_3	–33.5	90	1.3
H_2O	100	∞	1.85
HF	19.5	∞	1.82
LiF	1676	0.27	6.33

(a) Draw a horizontal line dividing the polar and nonpolar substances.
(b) Draw another horizontal line between the weakly polar and polar molecules.
(c) Can you see the reason why the ionic lithium fluoride has relatively low solubility in water?

2. There are listed in Table 2.10 the boiling points (bp) and water solubilities of an homologous series of *n*-alkanols. Based on these characteristics,

TABLE 2.10
Some Properties of the Alcohols

Alcohol		bp (°C)	Water solubility (g 100 mL^{-1})
Methanol	CH_3OH	64	∞
Ethanol	CH_3CH_2OH	78	∞
Propanol	$CH_3(CH_2)_2OH$	97	∞
Butanol	$CH_3(CH_2)_3OH$	118	8.1
Pentanol	$CH_3(CH_2)_4OH$	138	2.6
Hexanol	$CH_3(CH_2)_5OH$	156	0.6
Heptanol	$CH_3(CH_2)_6OH$	176	0.1

(a) Draw a horizontal line between the weakly polar and the polar compounds.
(b) How does the number of carbon atoms and molecular weight influence water solubility?
(c) Which intermolecular forces operate in this series of compounds and how does the relative importance change with the change in the number of carbon atoms?

3. The chemical structures of a number of different compounds are shown in Table 2.11. Use this (do not look up tables) to answer the following questions in general terms (e.g., gas, liquid, solid; very low, low, moderate, high).

TABLE 2.11
Question 3: Structures of Various Compounds

No.	Name	Structure	State	Water solubility	Lipid solubility
A	Ethane	$H_3C–CH_3$			
B	Benzene				
C	Citric acid	$HOOC–C(OH)(CH_2–COOH)–CH_2–COOH$			
D	Anthracene				
E	Glyphosate	$HOOC–CH_2–NH–CH_2–P(=O)(OH)–OH$			

(a) What is the physical state of these substances under normal conditions of temperature and pressure?
(b) Which of the compounds would you expect to be soluble in water?
(c) Which of the compounds would you expect to be soluble in biota lipid (lipophilic)?

Answers on Page 44.

CHAPTER 2 SOLUTIONS

1. (a) A line between CO_2 and HBr.
   (b) A line between NH_3 and H_2O.
   (c) The attraction between the Li^+ and the F^- is greater than the attraction between the water molecules and the ions; thus, Li and F ions cannot be separated by solution in water. This means that lithium flouride will not dissolve in water.

2. (a) A line between propanol and butanol.
   (b) Water solubility declines with increasing numbers of carbon atoms and molecular weight.
   (c) With methanol, the polar hydroxy group (OH) interacts with water and, because it is a large group in the molecule, its property of increasing water solubility dominates and methanol is soluble in water. With increasing molecular weight, the role of the hydroxy group diminishes as its dominance in the molecule declines. At the same time, London forces between molecules increase, reducing the capacity to dissolve in water and increasing the boiling point.

3.

No.	Name	State[a]	Water solubility	Lipid solubility
A	Ethane	Gas	Very low	High
B	Benzene	Liquid	Low	High
C	Citric acid	Solid	High	Low
D	Anthracene	Solid	Very low	High
E	Glyphosate	Solid	High	Low

[a] At normal conditions of temperature and pressure.

Chapter 3

ENVIRONMENTAL TRANSFORMATION AND DEGRADATION PROCESSES

I. INTRODUCTION

The environment contains many chemicals. Fox example, animals and plants comprise vast arrays of organic chemicals in addition to other substances. Chemicals from animals and plants, as well as organic chemicals added to the environment by actions of human society, occur in water, soil and the atmosphere. These chemicals may undergo **transformation,** which can be defined as any change in the molecular structure of the substance. This could be a rearrangement of the molecule into another form; alternatively, it could be the addition or loss of chemical groups by environmental processes. **Degradation** usually refers to the breakdown of the original molecule by the loss of the various component parts or by the fragmentation of the molecule into smaller substances.

Transformation and **degradation** processes may occur through interactions with other chemicals in the environment. This can be facilitated by the input of energy in the form of radiation or heat. Alternatively, biota may be involved, leading to the transformation and degradation of compounds through biological processes. With these processes, the organic compounds are chemically acted upon by other substances in the external environment or within the biota. Both oxygen and water are substances that are reactive and available in large quantities in the environment for transformation of organic compounds. Oxygen comprises about 20% of the atmosphere, and water occurs in high proportions in biota as well as existing in large quantities in the oceans, lakes and rivers.

Oxygen reacts with organic compounds by oxidation processes that often involve the addition of oxygen to the molecule. This can result in the formation of a molecule that is increased in size by the addition of oxygen or it may result in splitting the molecule into smaller oxygen-containing fragments. The oxidative degradation of an organic compound to the ultimate level results in the formation of carbon dioxide, water, ammonia, nitrate ion (NO_3^-), nitrite ion (NO_2^-), orthophosphate ion (PO_4^{3-}), hydrogen sulfide (H_2S), sulfate ion (SO_4^{2-}) and so on, depending on the conditions involved and the nature of the original compound. A simple example is the oxidation of methane by combustion:

$$CH_4 + 2O_2 \rightarrow CO_2 + 2H_2O$$

Hydrolysis is a process that results in the addition of water to a molecule. This often results in the fragmentation of the molecule into smaller fragments that may contain additional hydrogen and oxygen. An example of this is the hydrolysis of the ester functional groups in isooctylphthalate, a substance used to make plastics pliable, as represented by the following equation.

$$C_6H_4[C(=O)-O-(CH_2)_5-CH(CH_3)_2]_2 + 2H_2O \rightarrow C_6H_4[C(=O)-OH]_2 + 2\,HO-(CH_2)_5-CH(CH_3)_2$$

Thus, oxidation and hydrolysis may result in the production of smaller fragments that contain additional oxygen alone or both oxygen and hydrogen. Both oxidation and hydrolysis can occur when substances come in contact with oxygen and/or water under the appropriate conditions in air, water, soil and biota. These processes may occur without the intervention of biota as a result of abiotic processes. On the other hand, oxidation and hydrolysis can also occur through the facilitation of biota. Reactions facilitated by animals or plants are described as "biotic reactions." Both types of reactions are influenced by the prevailing temperature, the presence of oxygen and water, light and a variety of other factors.

II. ABIOTIC TRANSFORMATION AND DEGRADATION

A. OXIDATION THROUGH COMBUSTION

Often spectacular oxidation of organic matter by atmospheric oxygen occurs by combustion. The burning of trees, grass and petroleum are examples of oxidation through the combustion process. Many organic compounds can exist in the environment in the presence of the 20% of oxygen in the atmosphere without combustion occurring. However, ignition by a spark or flame initiates the occurrence of combustion. The burning or combustion of organic matter such as wood and petroleum, is a major source of energy in human society. This has a major impact on the occurrence of oxygen and carbon dioxide in the Earth's atmosphere. The energy produced by combustion of many organic compounds can be estimated. The production of this energy is critical to the use of organic compounds as fuels. First, we will look at combustion of hydrogen as a simple example of the oxidative combustion process. An equation can be written for the combustion of one molecule of hydrogen as follows:

$$H_2 + {}^1/_2\,O_2 \rightarrow H_2O$$

This occurs with the release of energy, which is the *heat of combustion.* Looking at the reaction in greater detail, we can develop a concept of the reaction as occurring in a number of steps. First, the molecules can be seen as dissociating into atoms if there is sufficient energy supplied. This is illustrated by the following equation:

$$H{-}H \rightarrow H\cdot + \cdot H$$

The point to note is that the bond energies are the additional energy needed to break the bond or the excess energy left after the bond is formed. This is illustrated diagrammatically in Figure 3.1.

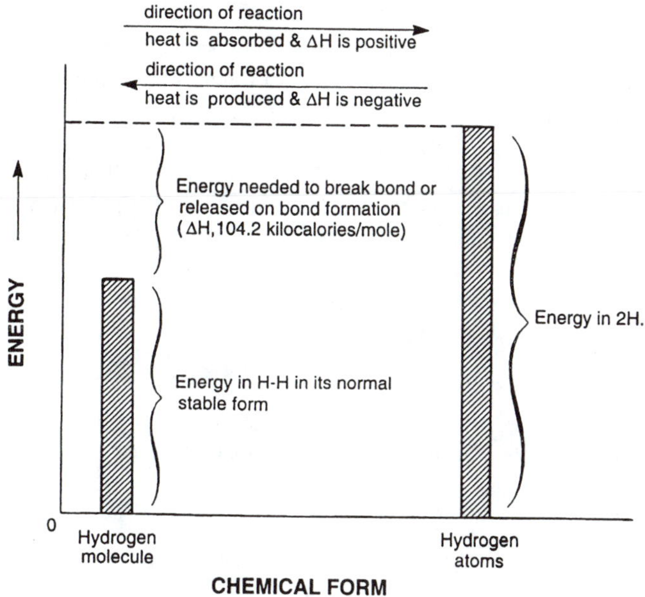

FIGURE 3.1 Diagrammatic representation of the energy present in forms of hydrogen.

The heat of reaction (ΔH) is defined as

$$\Delta H = (\text{heat content of products}) - (\text{heat content of reactants})$$

If a reaction, such as the dissociation of the H_2 molecule to two H atoms, requires an input of energy to the reactants, this means that the products are of higher energy than the reactants and ΔH will be positive (see Figure 3.1). This is known as an *endothermic* process. Alternatively, if a reaction, such as the reaction of two H atoms to form a H_2 molecule as the product, results in an output of energy, then the products are of lower energy than the reactants and ΔH will be negative (see Figure 3.1). This is an *exothermic* process. For the H_2 dissociation, ΔH is +436 kJ $mole^{-1}$.

Considering the other reaction, the O_2 molecule: in the initial dissociation step, the O_2 molecule dissociates into atoms. Thus,

$$\text{input 495 kJ/mol}$$
$$O{=}O \rightleftarrows O{:} + {:}O$$
$$\text{output 495 kJ/mol}$$

This means that 0.5 mole O_2 needs +247.5 kJ input to cause dissociation to the oxygen atoms. Thus, there is a total energy input required to produce atoms of oxygen and hydrogen from water of +683.5 kJ mole^{-1} (436 + 247.5). In the final step, the atoms of hydrogen and oxygen can be considered to recombine and form water vapor. Thus,

$$2H\cdot\ (g) + {:}O\ (g) \rightarrow H_2O\ (g)$$

The amount of energy released in forming H_2O is equivalent to two O-H bonds, i.e., 2 × 462.8 kJ mole^{-1} which equals –925.6 kJ mole^{-1} (see Table 3.1). This energy is released, which on balance means that ΔH for the overall process is equal to 683.5 – 925.6, which is –242.1 kJ mole^{-1}. Thus, the heat of combustion is 242.1 kJ mole^{-1}, which is equivalent to 121.05 kJ g^{-1} (242.1 ÷ 2 g mole^{-1}). This calculated value compares favorably with an actual experimental value for the heat of combustion of hydrogen of 241.8 kJ mole^{-1}.

TABLE 3.1
Bond Dissociation Energies for a Range of Bonds

Bond	Energy (kJ mole^{-1})
H–H	436
O=O	498
C–O	351
C–H	413
O–H	463
C=O (as in CO_2)	803
C–C	347

The hydrocarbons are important fuels in the form of natural gas, LPG, petrol, diesel, etc. The heat of combustion is an important characteristic of these fuels and can be estimated using the characteristic bond dissociation energies shown in Table 3.1. It is important to note that these energies are average values encountered and the energy in a particular type of bond may vary with its position in a molecule. This can lead to differences between the calculated and observed heats of combustion with specific compounds.

Methane is an important component in natural gas. The oxidative combustion of methane can be represented as:

$$CH_4\ (g) + 2O_2\ (g) \rightarrow CO_2\ (g) + 2H_2O\ (g)$$

The conceptual dissociation of the methane and oxygen into atoms, the recombination of these atoms to form carbon dioxide and water, and the associated energy changes are shown in Table 3.2. This indicates that the heat of combustion is calculated as –809.2 kJ mole^{-1}, which is close to the experimental measured heat of combustion of 802.5 kJ mole^{-1}. In a similar way, the heat of combustion can be estimated for many other fuel compounds.

TABLE 3.2
Energy Changes and Heat of Combustion Resulting from Oxidation of Methane by Combustion

Step 1: Dissociation of Methane and Oxygen Molecules to Atoms

Original molecule	Bonds involved	Energy involved (kJ mole^{-1})	Energy change (ΔH) (kJ mole^{-1})
H | H—C—H | H methane	4 × C-H	4 × 413	1652
O=O oxygen	2 × O=O	2 × 498	996
		Total energy input	2648

Step 2: Recombination to Form CO_2 and H_2O

CO_2	2 × C=O	2 × 803	–1606
H_2O	4 × O–H	4 × 463	–1852
		Total energy output	–3458
	Total energy produced	= –3458 + 2648	= –810 kJ mole^{-1}
	Heat of combustion	= +810/16	=+50.6 KJ g^{-1}

In general terms, the basis of the estimation procedure can be illustrated as is shown in Figure 3.2. Organic compounds in the form of fuels that are high in energy can be considered to be converted into their constituent atoms by the input of energy. The constituent atoms can then be seen as recombining by reforming bonds to form compounds of a lower energy content with the release of energy. The overall change in energy represents the heat of combustion, and a range of experimentally measured heats of combustion is shown in Table 3.3.

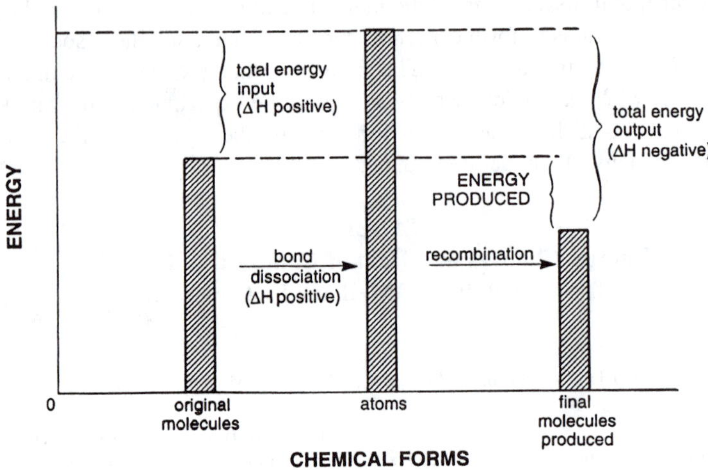

FIGURE 3.2 Diagrammatic illustration of the changes in chemical form and energy during oxidation by combustion.

TABLE 3.3
Experimentally Measured Heats of Combustion of Some Fuels

Substance	**Heat of combustion ($kJ\ g^{-1}$)**
Petroleum fuels (gasoline, kerosene, diesel etc.)	40–48
Lipids	38–40
Carbon	33
Proteins	18–23
Carbohydrates	15–18

The oxidation process with organic matter in the form of food occurs within biota, resulting in the production of carbon dioxide and water and smaller amounts of other products. Overall, this is chemically somewhat similar to oxidation through combustion and results in the release of energy that can be used by biota.

B. PHOTOTRANSFORMATION

Many organic chemicals are introduced into the environment and can absorb radiation and, as a result, undergo chemical transformation. All chemical processes require the reacting substances to have attained a certain energy. This energy can be obtained either from thermal energy (heat) as with combustion, which was considered in the previous section, or by absorption of radiation. The internal energy of a molecule is in several different forms. Thermal energy in a molecule results from being "jostled" by its neighbors, and it is manifested as translation, rotation and vibration of the molecules. Absorption of electromagnetic radiation with frequencies in the infrared (IR) range (see Figures 3.3 and 3.4) causes increased molecular rotation and vibration. A molecule exists in specific electronic energy level states, generally the lowest energy one or the ground state. Most molecules absorb IR radiation and thermal energy, but to absorb radiation to change the electronic energy state requires ultraviolet (UV) or visible (VIS) light and specific chemical groupings in an organic molecule (chromophores).

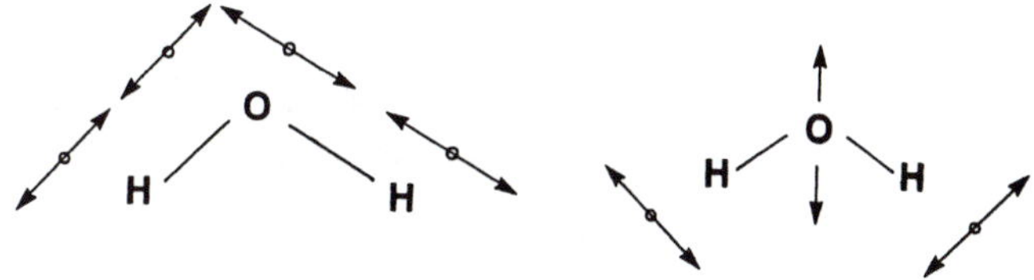

FIGURE 3.3 Some vibrational modes of a water molecule.

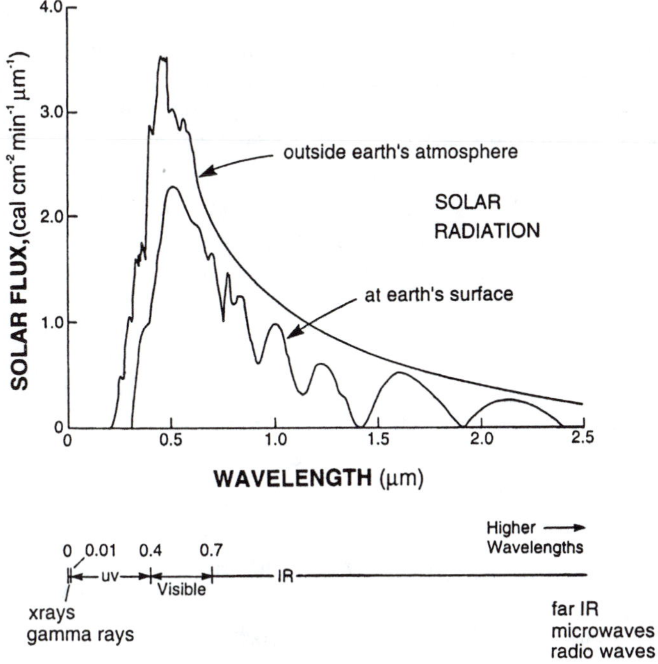

FIGURE 3.4 The spectrum of solar radiation and the electromagnetic spectrum.

Molecules activated by radiation absorption are described as photochemically activated molecules and they differ in a number of important aspects from those that are thermally activated. Radiation absorption that will cause a change in the electronic state of a molecule is only possible if the energy of an incident photon corresponds to the energy difference between two internal electronic energy states of the molecule. To initiate a photochemical transformation, the photon energy must be relatively large, which corresponds to radiation in the UV/VIS region of the electromagnetic spectrum (see Figure 3.4). If the light absorbed is of sufficient energy, the excited molecule produced can undergo various transformations that would essentially not occur under normal conditions. The basic characteristics of radiation can be described by relatively simple equations. Two important equations relate energy, frequency and wavelength of radiation. First,

$$E = h\nu$$

where E is energy in Joules, h is Plank's constant (6.63×10^{-34} J), and ν is frequency (Hertz, or cycles per second, s^{-1}). The second equation is:

$$S = \nu\lambda$$

where S is the velocity of light (assumed constant at 3×10^{8} m s^{-1}) and λ is the wavelength in meters. The velocity of light is dependent on the medium that is it passing through and reaches a maximum in a vacuum. The equations above indicate that in general terms, as the wavelength (λ) increases, the frequency (ν) will correspondingly decline and the energy of a photon of radiation will also decline since energy (E) is directly dependent on frequency. Accordingly, a photon of UV radiation has greater energy than one of VIS radiation, which has greater energy than an IR photon.

Reactions induced by UV/VIS include fragmentation, oxidation and polymerization. It is also possible that an excited molecule can return to its original state and in the process emit radiation at a different wavelength. The following are examples of photochemical reactions.

$$\mathrm{H{-}C({=}O){-}H} + h\nu \longrightarrow \mathrm{H_2} + \mathrm{CO}$$

acetone (λ 370 nm)

$$\mathrm{O_3} + h\nu \longrightarrow \mathrm{O} + \mathrm{O_2}$$

ozone (λ 320 nm)

Photoreactions can be divided into three stages:

1. Absorption, the stage in which the absorption of photons of light gives excited molecules
2. Primary photochemical process
3. Secondary photochemical processes

Phototransformation in the environment can only occur if the UV/VIS absorption spectrum of the compound and the solar admission spectrum overlap. Solar radiation contains a substantial amount of UV/VIS radiation when it reaches the Earth's surface, as shown in Figure 3.4.

Examples of chemical groups (chromophores) which absorb UV/VIS radiation in solar radiation are given in Table 3.4. These groups all contain double bonds, in some form, aromatic rings or series of double bonds separated by a single bond.

Some structures or functional groups are poor absorbers of solar radiation and these include alcohols (-OH), ethers (R-O-R), and amines ($R\text{-}NH_2$). For such groups, or chemicals containing only these functional groups, phototransformation is likely to be unimportant. In addition, phototransformation can only occur when the chemical is likely to be exposed to solar radiation. For example, the atmosphere, upper layers of water bodies and the surface of soil are likely locations in which chemicals would be exposed to solar radiation.

TABLE 3.4
Groups in Molecules (Chromophores) That Absorb Ultraviolet and Visible Radiation in Solar Radiation

Group	Wavelength of maximum absorption (l_{MAX} in nm)	Molar extinction coefficient (e) (l mol cm^{-1})
>C = O	295	10
>C = S	460	Weak
—N=—N	347	15
(naphthalene)	{311, 270}	{250, 5000}
C = C—C = O (with single bonds on the central carbons)	330	20

C. ESTIMATION OF PHOTOLYSIS RATE CONSTANTS OR PHOTOTRANSFORMATION RATE CONSTANTS

The Beer-Lambert law relates absorbance to concentration at a given wavelength for a chemical in solution (see Figure 3.5). This law can be expressed by the following equations.

$$\log (I_o/I) = A = \varepsilon Cl$$

$$\text{Also, } I/I_0 = T \text{ and } -\log T = A$$

where I_0 and I are incident and transmitted light intensity (quanta or energy s^{-1} or photons $cm^2\ s^{-1}$), respectively, ε is the molar extinction coefficient (L $mole^{-1}$ cm^{-1}), C is the concentration (mole L^{-1}) and l is the path length (cm). This means that if 10% transmission occurs, then $I/I_0 = 10/100$, $I_0/I = 100/10 = 10$ and $\log (I_0/I) = A = 1$. Similarly, if 1% transmission occurs, then A is equal to 2.

The Beer-Lambert law can be adapted for gas-phase situations with concentration usually in units of molecules/cm^3 (N), and the use of natural logarithms. Thus,

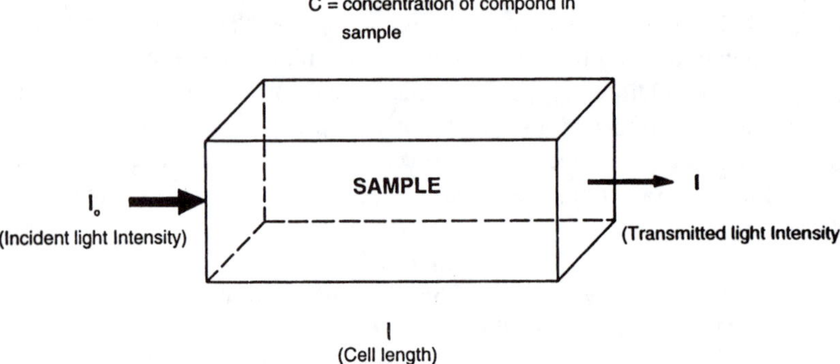

FIGURE 3.5 Passage of radiation through a sample.

$$ln\left(\frac{I_0}{I}\right) = \sigma N l$$

where σ is the absorption cross section (cm^2 molecule^{-1}) and l is the path length; so,

$$\frac{I}{I_0} = e^{-(\sigma N l)}$$

To determine the number of photons absorbed per cm^2 per second, consider a cube (1 cm^3) of atmosphere containing a compound at a concentration of N molecules cm^{-3} and having an absorption molecule cross-section of σ cm^2 molecule^{-1} (see Fig. 3.5).

The Beer-Lambert law can be used as follows. Considering a single wavelength, then:

$$I_{absorbed} = I_{incident} - I_{transmitted} = I_0 - I = I_0\left(1 - \frac{I}{I_0}\right)$$

but

$$\frac{I}{I_0} = e^{-(\sigma N l)}$$

So

$$I_{absorbed} = I_0\left(1 - e^{-(\sigma N l)}\right)$$

for weak absorption, $\sigma N l$ is small, which means that $1 - e^{-(\sigma N l)}$ is approximately $\sigma N l$.

This may seem unusual but can be verified using an actual example. Take the situation where $\sigma N l$ is small, for example 0.01. Then,

$$1 - e^{-0.01} = 1 - 0.99 = 0.01$$

This now means that:

$$I_{absorbed} = I_0 \sigma N l$$

but l is equal to 1 cm, so $I_{absorbed} = I_0 \sigma N l$, where $I_{absorbed}$ has units of photons cm^{-3} s^{-1}.

This gives a measure of the number of photons absorbed per unit volume per second on passing through the atmosphere. This makes sense since, on logical grounds, the amount absorbed should be proportional to the number of molecules present and the strength of the solar radiation. Assuming each photon that strikes a molecule initiates phototransformation of that molecule, then

$$\text{Phototransformation rate} = I_0 \sigma N l$$

However, in actual phototransformations induced by solar radiation, each photon does not necessarily initiate a phototransformation. So another factor that can be taken into account is the efficiency with which a compound converts radiation into chemical reactive energy. This is not related to the strength of the radiation absorption but is a characteristic of the compound and its chemical constitution. This is called the **quantum yield** and is designated by ϕ and has a value of unity if 100% translocation of the radiation into chemical conversion energy occurs as is assumed above. Values thus range from unity to zero and must be measured experimentally for each compound. Including this factor into the equation above yields:

$$\text{Rate} = I_0 \sigma N \phi$$

If values for ϕ are not available, a value of unity is often assumed.

In atmospheric chemistry, I_0 is often given the symbol J and depends on factors such as latitude and season. The phototransformation rate expressed above is for one wavelength only; so to estimate the total transformation rate, the individual rates for each wavelength must be summed. Thus,

$$\text{Rate} = \Sigma J \sigma N \phi$$

The values for J, ϕ and σ change with wavelength, but N does not. This means that the phototransformation rate constant, k, for solar radiation is expressed by:

$$k_p = \Sigma J \sigma \phi$$

This equation applies for all wavelengths where the solar radiation spectrum and the absorption spectrum of the compound overlap.

1. Example Calculation of k

Formaldehyde is a important atmospheric pollutant that can be produced in photochemical smog. The rate of photolysis in the atmosphere at sea level at a latitude of 40°N can be calculated using the equation previously developed. There are a variety of major cities throughout the world at a latitude of about 40°N including New York, Lisbon and Rome.

Formaldehyde undergoes two different photolysis reactions:

$$HCHO + h\nu \rightarrow H\cdot + \cdot HCO$$

(formyl radical)

$$HCHO + h\nu \rightarrow H_2 + CO$$

The first reaction is of particular importance since the hydrogen and formyl radicals formed can play a major role in photochemical smog production. The second reaction can be assumed not to play a significant role in atmospheric pollution.

Formaldehyde has a number of bands of absorption of UV radiation in the range from 260 to 360 nm. The spectrum of solar radiation shown in Figure 3.4 indicates that there is strong radiation in this range that will be absorbed by formaldehyde. Solar radiation varies in intensity with factors such as season and time of day. For noon on July 1 on a cloudless day at 40°N latitude, the J values shown in Table 3.5 are applicable. This table also includes values for σ and ϕ, and the values are calculated for k over each radiation wavelength range. The overall rate constant, k_p, is:

TABLE 3.5
Data for the Estimation of the Rate Constant for Photolysis of Formaldehyde for Noon July 1 on a Cloudless Day at 40°N Latitude

Wavelength range (nm)	J ($\times 10^{14}$ photons cm^{-2}/s^{-1})	σ ($\times 10^{-20}$ cm^2 $molecule^{-1}$)	ϕ	k (s^{-1})
290–295	0.0	2.51	0.71	0
295–300	0.031	2.62	0.78	6.3×10^{-8}
300–305	0.335	2.62	0.78	6.8×10^{-7}
305–310	1.25	2.45	0.77	2.3×10^{-6}
310–315	2.87	2.45	0.77	5.4×10^{-6}
315–320	4.02	1.85	0.62	4.6×10^{-6}
320–325	5.08	1.85	0.62	5.8×10^{-6}
325–330	7.34	1.76	0.31	4.0×10^{-6}
330–335	7.79	1.76	0.30	4.3×10^{-6}
335–340	7.72	1.18	0	0
340–345	8.33	1.18	0	0
345–350	8.33	0.42	0	0
350–355	9.45	0.42	0	0
355–360	8.71	0.06	0	0
360–365	9.65	0.06	0	0

$$k_p = \Sigma\sigma\phi J = 2.7 \times 10^{-5}\ s^{-1}$$

This is a valuable characteristic for the evaluation of the rate of photochemical transformation of formaldehyde in the atmosphere.

D. HYDROLYSIS

Water is present in large quantities in the environment in oceans, rivers and streams. It is also available in substantial quantities within all biota and in the vapor form in the atmosphere. In biota, it is the basic fluid used for the transfer of substances in biological processes. The chemical reaction of a compound with water is described as **hydrolysis** and was briefly outlined previously. It is one of the most important chemical processes that can act upon the many types of organic compounds occurring in the environment arising from both natural and man-made sources. The general reaction below defines the hydrolysis process.

$$\underset{\text{organic compound}}{R - X} + H_2O \rightarrow R - OH + H - X \text{ (or } H^+ + X^-)$$

In this reaction, an organic molecule R-X reacts with water, forming a new C-O bond and cleaving a C-X bond in the original molecule. The overall effect is a displacement of the X group by an hydroxyl group.

A wide variety of functional groups and compounds are potentially susceptible to hydrolysis, including peptides, the glycosidic linkage in polysaccharides and the ester group in fats. These reactions are shown in a generalized form in Table 3.6; it can be seen that carbohydrates, proteins and fats, which are the major components of living tissue, are all susceptible to hydrolysis. Hydrolysis also occurs with a wide range of synthetic compounds, including many pesticides and other substances. These reactions can be mediated by biota or can occur without the need for biological assistance. However, when they occur abiotically, the rates of reaction can be very slow. Some chemical structures and groups tend to be resistant to hydrolysis, including alkanes, polycyclic aromatic hydrocarbons, alcohols, aldehydes and ketones.

The importance of hydrolysis from an environmental fate view is that the reaction introduces an hydroxyl group into the parent molecule and may fragment the molecule into smaller groups. The hydroxyl group is a polar group (see Chapter 2) and usually tends to increase the polarity of the molecule, but this depends on the nature of the group that is removed. The products of hydrolysis are usually more susceptible to biotransformation and the hydroxyl group makes the chemical more water soluble, as does the smaller size of molecular fragments that may be produced. With compounds having high K_{OW} values (see Chapter 2), the K_{OW} values of the products will be less and the biological activity altered. Furthermore, the product is usually less toxic than the initial starting material but there are some exceptions to this. For example the butoxy ethanol ester of 2,4-dichlorophenoxy acetic acid (2,4-D) exhibits an EC_{50} of >100 mg L^{-1} to daphnia; whereas, the hydrolysis product (which is 2,4-D itself) exhibits greater toxicity at 47 mg L^{-1}. This is illustrated in Figure 3.6. Hydrolysis reactions are commonly catalyzed by H^+ or OH^- ions. This results in a butoxyethanol ester of 2,4D — HYDROLYSIS $(+H_2O)$ → 2,4 D + ROH (alcohol)

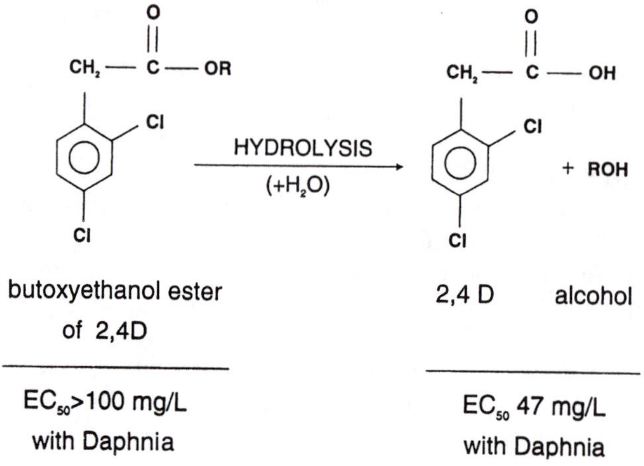

butoxyethanol ester of 2,4D	2,4 D
EC_{50}>100 mg/L with Daphnia	EC_{50} 47 mg/L with Daphnia

FIGURE 3.6 Toxicity of butoxyethanol ester of 2,4D compared to its hydrolysis product 2,4D.

TABLE 3.6
Groups which are Susceptible to Hydrolysis

Group		Products
$R_1-C(=O)-NH-R_2$	→	$R_1-C(=O)-OH$ + $H-NH-R_2$ (acid, amine)
R_3–O–R_4 (disaccharides, oligosaccharides, polysaccharides)	→	R_3–OH+HO–R_4
$R_4-C(=O)-O-R_5$ (fats)	→	$R_4-C(=O)-OH$ + $HO-R_5$ (fatty acids, glycerol)
R–X (pesticides which are alkyl halides, amides, carbamates and esters)	→	R–OH+H–X

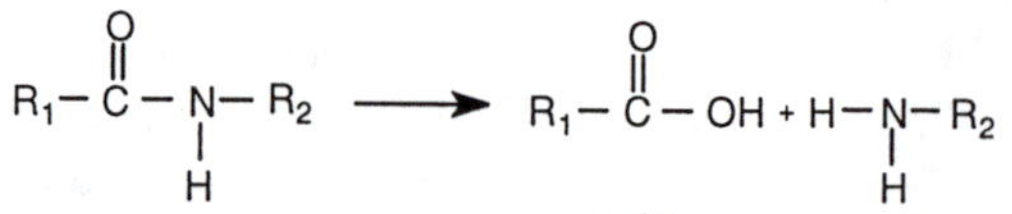

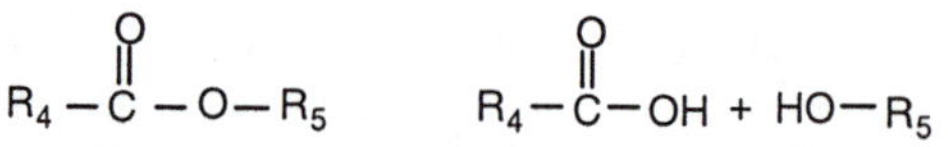

Example:

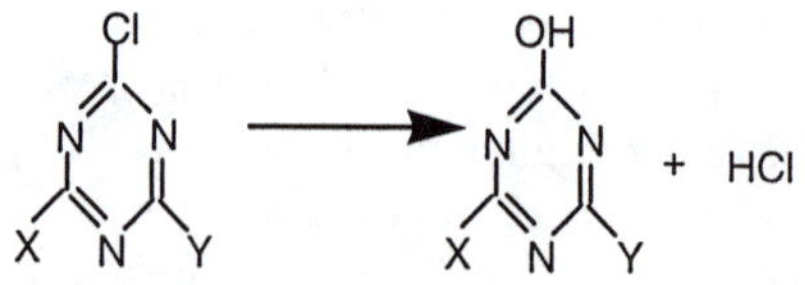

atrazine → hydroxy atrazine + HCl

strong dependence of the rate of hydrolysis on the pH of the water in which the reaction occurs.

III. BIOTRANSFORMATION AND BIODEGRADATION

Biotransformation and biodegradation of chemical compounds by the action of living organisms is one of the major processes that determines the fate of organic chemicals in aquatic and terrestrial environments. These processes can be divided into two broad categories: (1) microbial transformations and (2) transformation by higher organisms. These two different groups are described below.

A. MICROBIAL TRANSFORMATION

Microorganisms are ubiquitous in the environment, as is shown by the populations of bacteria in waterbodies illustrated in Table 3.7. Generally, as the amount of organic matter increases in a water body, the microbial population also increases. Microorganisms play a major role in the biogeochemical cycles of various elements that occur in the environment. Frequently, microbial transformation is the most important and possibly the only significant process that can decompose an organic xenobiotic chemical in the environment. Microorganisms include bacteria which are small, single-celled organisms, fungi which are nonphotosynthetic organisms, algae which are photosynthetic organisms, protozoans which are unicellular, eucaryotic microorganisms and viruses which are parasitic microorganisms unable to multiply outside the host tissues.

TABLE 3.7
Size of Typical Bacterial Populations in Natural Waters

Environment	Bacteria numbers (ml^{-1})
Clear mountain lake	50–300
Turbid, nutrient-rich lake	2000–12,000
Lake sediments	8×10^9–5×10^{10}
Stream sediments	10^7–10^8

Most bacteria fall into the size range 0.5 to 3.0 μm. In general, it is assumed that a filter with a 0.45-μm pore size will remove all bacteria in water passing through it. The majority of bacteria in aquatic environments are nutritionally heterotrophic. Fungi are aerobic organisms (require atmospheric oxygen) which can be uni- or multicellular and generally can thrive in more acidic media than bacteria. Perhaps the most important function of fungi in the environment is the breakdown of cellulose in wood and other plant materials. To accomplish this, fungal cells secrete an enzyme (cellulase) that breaks insoluble cellulose down to soluble carbohydrates that can be absorbed by the fungal cell. Although fungi do not grow well in water, they play an important role in determining the composition of natural water because of the large amounts of their decomposition products that enter natural water bodies. A particularly important example of these decomposition products are the humic substances (discussed in Chapter 17, Soil Contamination) that occur in soil and runoff water entering natural water bodies. Algae are photosynthetic and are abundant in both fresh and saline waters, soil and other sectors of the environment.

B. TYPES OF MICROBIAL DEGRADATION

The rate at which a compound is biotransformed and biodegraded by microorganisms depends upon its role in microbial metabolism and a variety of other factors. Heterotrophic bacteria, which are capable of using complex carbon compounds as their principle source of energy, can degrade organic compounds to provide the energy and carbon required for growth. This is known as metabolism of growth substances; these substances are identifiable by their ability to serve as the sole carbon force for a bacterial culture. Many toxic and synthetic substances function as growth substrates for bacteria in a manner similar to naturally occurring organic compounds. Metabolism of growth substances usually results in relatively complete degradation or mineralization to carbon dioxide, water and inorganic salts.

C. PATTERNS OF GROWTH

Before the degradation of a compound can begin, the microbial community must adapt itself to the chemical, in many cases, which results in a lag phase initially when little growth occurs (see Figure 3.7). With growth substances, both laboratory and field investigations have shown that this adaptation results in a lag time of 2 to 50 days before the microbial community adjusts. Frequently, specific organisms within a community with specific enzymes systems are required for degradation to occur. The adjustment period can involve species selection and numbers increase as well as the production of specific enzymes systems to match the substrate. This production of enzymes is referred to as **enzyme induction**. These factors are outlined below.

1. **Prior exposure to the organic compound:** Prior exposure to the organic compound substrate reduces the adaption or lag time. Thus lagtimes in pristine environments should generally be much longer than in locations which have been previously exposed.
2. **Initial numbers of suitable species:** Areas with larger microbial communities should require relatively short lag times to develop a viable population of degrading microorganisms (see Table 3.7).
3. **The presence of more easily degraded carbon sources:** The presence of more easily degraded carbon sources may delay the adaptation of the microbial community to more persistent contaminants. For example, it has been found that microorganisms degraded added glucose completely before degrading hydrocarbons in lake water.
4. **Concentration of the organic compound:** There may be concentration thresholds below which adaptation does not occur. On the other hand too high a concentration of the organic compound may be toxic to the microorganisms.

When the lag phase is completed, population growth occurs rapidly and usually increases at an expotential rate. At the completion of this phase, the microorganism population effectively establishes an equilibrium with growth substances available,

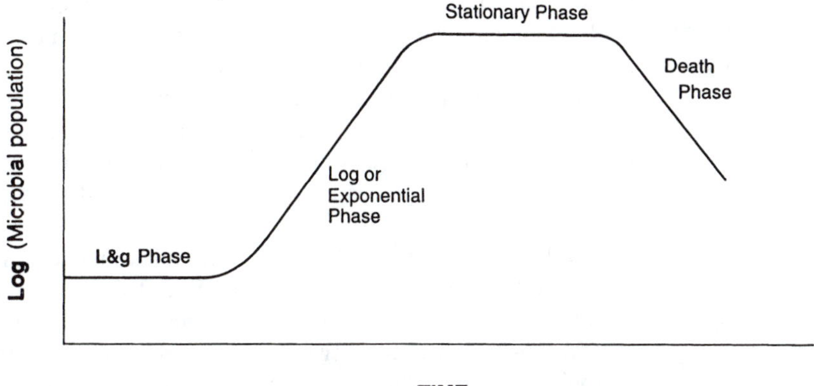

FIGURE 3.7 Typical pattern of growth of a microbial population.

and a stationary phase in terms of population occurs. Finally, the growth substances are exhausted and wastes accumulate, leading to a decline in the population number as is illustrated in Figure 3.7.

D. CO-METABOLISM

Compounds that co-metabolize usually degrade only in the presence of another carbon source. This is in contrast to growth substances that are able to serve as the sole carbon source for a microbial community. Microorganisms can degrade compounds that they apparently cannot use for growth or energy via co-metabolism. Co-metabolism is believed to occur when enzymes of low specificity alter or degrade a compound to form products that other enzymes in the organism cannot degrade. Substances that undergo co-metabolism are usually similar in structure to natural substrates, but are altered or degraded without necessarily providing significant amounts of energy to the microorganism. Often, this is an important mechanism for the degradation of pesticides in soil.

The kinetics of co-metabolism differ significantly from that of growth substances. Often, no lag period occurs before co-metabolism begins, and accumulation of intermediate products resulting from partial degradation is likely. Generally, slower rates of degradation are observed compared to metabolism of growth substances.

E. CHEMICAL PROCESSES

In aerobic aquatic environments, microorganisms utilize molecular oxygen (O_2) as a major reactant in the degradation of organic compounds. The concentration of oxygen dissolved in water depends on factors such as temperature, salinity, biological activity and reaeration rate. Usually, the dissolved oxygen concentration in natural waters ranges from 0 to about 10 mg L^{-1} due to the low solubility of oxygen in water. With sufficient oxygen, the complete aerobic degradation of carbohydrates can be simply expressed as follows:

$$\underset{\text{Carbohydrate}}{6\ CH_2O} + 6\ O_2 \rightarrow 6\ CO_2 + 6\ H_2O + \text{Energy utilized by organism}$$

If the organic compound contains nitrogen, the normal products of carbon dioxide and water are formed but the nitrogen present is converted into ammonia (NH_3) and nitrate ion (NO_3^-). It is noteworthy that nitrate is not produced directly but rather, with aerobic, heterotrophic microorganisms, ammonia is the first product formed and nitrate is formed by oxidation of this ammonia by different microorganisms. From this general type of reaction, bacteria and other microorganisms extract the energy needed to carry out their metabolic processes, to synthesize new cell material, for reproduction and for movement.

As the concentration of dissolved oxygen (DO) is depleted, microbial degradation pathways change and degradation reactions are modified. When DO is removed, anaerobic degradation without atmospheric oxygen occurs, having generally lower energy yields and microbial growth rates. Most organic substances are biodegraded more slowly under anaerobic conditions although there are a number of exceptions to this general rule. Rate constants derived for degradation in oxygenated systems do not apply to anaerobic systems. The dissolved oxygen content of waters in various parts of the environment can vary considerably. In zones such as sediments and bottom waters that lack a mechanism for aeration, DO can be in short supply and organic compounds are typically microbially degraded utilizing a sequence of available oxidizing agents. Initially, aerobic degradation occurs; but when the DO falls to 0.5 to 1.0 mg/L or less, nitrate begins to substitute for molecular oxygen in the breakdown process as an oxidizing agent. This process can be simply expressed by the following equation.

$$\underset{\text{carbohydrate}}{2\ CH_2O} + \underset{\text{nitrate}}{2\ NO_3^-} \rightarrow 2\ CO_3^{2-} + 2\ H_2O + N_2$$

If nitrogen was present in the original organic compound undergoing oxidation, then ammonia could also be produced. As the nitrate is consumed, other oxidizing agents such as SO_4^{2-} are utilized, with the production of carbon dioxide, water and hydrogen sulfide (H_2S). With nitrogen-containing compounds, ammonia (NH_3) is produced as well. Generally, different microorganisms are responsible for, or associated with, the use of the various oxidizing agents.

In the complete absence of dissolved oxygen, the degradation of organic compounds can still proceed. This can be represented for carbohydrates by the following equations.

$$\underset{\text{Carbohydrate}}{6\ CH_2O} \rightarrow \underset{\text{Methane}}{3\ CH_4} + CO_2$$

$$\underset{\text{Carbohydrate}}{6\ CH_2O} \rightarrow \underset{\text{Ethanol}}{2\ C_2H_5OH} + 2\ CO_2$$

These processes release a relatively low amount of energy, as can be seen by the production of methane and ethanol, which are capable of further oxidation with the release of relatively large amounts of energy.

Microorganisms require nutrients, such as nitrogen and phosphorus, in order to metabolize organic substances. The availability of nutrients is one of the most important factors controlling the activity of heterotrophic microorganisms in aquatic environments. Nutrient limitation has been reported to retard the biodegradation of organic compounds, particularly in marine environments, which are generally relatively deficient in both nitrogen and phosphorus.

As well as the major nutrients (C, N and P), microbial growth and activity can be affected by essential micronutrients (growth factors, trace metals). For example, the degradation of petroleum oil in seawater can be enhanced by the addition of iron, which is consistent with the role of iron as a cofactor in some of the enzymes responsible for hydrocarbon oxidation.

F. TERRESTRIAL SYSTEMS

Soils are heterogeneous systems composed of varying proportions of organic and mineral matter, and are the result of many complex processes such as the disintegration and weathering of rocks and the decomposition of plant and animal material. In an average soil, mineral matter accounts for about 50% of the soil volume, with organic material typically representing less than 10%. Air and water occupy the remaining volume of the soil in a complex system of pores and channels. Well-drained soils are generally aerobic in nature, with a fairly ready supply of oxygen from the atmosphere.

Microorganisms are generally considered to be aquatic in nature but soil microorganisms tend to exist in a sorbed state attached to the soil solid matter. Decreasing soil moisture content usually decreases the rate of degradation of organic compounds such as pesticides. This is particularly true at lower soil moisture levels approaching air-dried soil conditions. At excessive soil moisture levels, the soil may change from an aerobic to an anaerobic condition with generally reduced microbial degradation rates.

G. DEGRADATION BY HIGHER ORGANISMS

Higher organisms often tend to degrade endogenous (i.e., arising from within the body) chemicals such as bile acids, fatty acids, steroids and other hormones as well as xenobiotic compounds such as petroleum hydrocarbons. One of the major enzyme systems for facilitating degradation of xenobiotic lipophilic compounds is the **mixed function oxidase** (MFO) system. The name **mixed function oxidase** is used because the system acts as an oxygenase and an oxidase. The key enzyme in the system is Cytochrome P_{450}, which is based on a heme structure. **Cytochrome P_{450}** is not one or two enzymes, but a family of closely related enzymes called *isozymes*. This enzyme system facilitates the insertion of oxygen into C-H bonds of substrates containing aliphatic groups. This is illustrated by the reaction below.

$$\underset{\text{Substrate}}{\text{R—H}} + O_2 \xrightarrow{\text{Cytochrome } P_{450}} \text{R—OH} + H_2O$$

With aromatic compounds, epoxidation occurs, but overall an oxygen atom is inserted into substrates.

Most enzymatic reactions are highly specific for only one substrate or a small series of closely related substrates. However, the MFO enzyme series is distinguished by the fact that there is a lack of specificity. These enzymes are able to act on xenobiotic compounds to which an organism has never been exposed before. The enzyme system can be generated as a result of exposure and this is described as **enzyme induction**. Cytochrome P_{450} is found most abundantly in the liver of vertebrates. This reflects the liver's role as the body's primary site for degradation of xenobiotic lipophilic compounds. The occurrence of this enzyme system is not restricted to mammalian tissues, but also occurs in fish, birds, yeast, some plants and some bacteria as well. Induction is triggered by exposure to certain xenobiotic lipophilic compounds described as *inducers.* Not all xenobiotic compounds are equally effective as inducers, but the best inducers are chemicals that are lipid soluble and have a relatively long half-life in the organism.

The most potent inducing agent yet discovered is dioxin, more accurately described as 2,3,7,8-tetrachlorodibenzodioxin (see Chapter 6). This substance is often described by the abbreviation TCDD. Other inducers include DDT-type insecticides, barbiturates, PAHs such as benzo-(a)-pyrene and ethanol. It is interesting to note that cigarette smokers generally have higher Cytochrome P_{450} levels that non-smokers. In fact, elevated levels of Cytochrome P_{450} have been proposed as an indicator of previous exposure to xenobiotic substrates such as petroleum hydrocarbons.

The degradation or biotransformation process often occurs in two steps. The activity of the Cytochrome P_{450} isozymes is referred to as Phase I reactions. Phase I reactions typically result in the insertion or introduction of an OH group in place of a simple hydrogen. This results in the insertion of a relatively reactive, polar functional group into a compound. This makes the product more soluble in water, with a lower K_{OW} value and also places a reactive site on the molecule. Phase II reactions are known as conjugation reactions since they conjugate or join together the product mentioned above with another substance. These reactions usually involve the coupling of a Phase I product with a naturally occurring derivative of a carbohydrate, peptide or a sulfate.

The product of this reaction also is of higher polarity, greater water solubility and thus is more easily eliminated than the original xenobiotic compound. Neither humans nor animals are born with a full compliment of P_{450} isozymes. In addition, it may take several years for human infants to acquire adult levels of biotransformation activity.

IV. KINETICS OF TRANSFORMATION AND DEGRADATION

It is frequently important to determine how rapidly transformation and degradation occur in the environment so that the extent and period of contamination can be evaluated. As a result, numerous methods have been developed to simulate transformation of organic compounds in the various compartments of the environment. One of most widely used evaluations in environmental management is a test of water and wastewater that measures the capacity for biodegradation of organic matter present and consequent demand for oxygen. This is the **biochemical oxygen**

demand (BOD); it is a measure of the amount of dissolved oxygen consumed by microorganisms in degrading the organic matter in a water sample over a 5-day period. Not only does it measure the degradation of organic matter present but, at the same time, it evaluates the possible dissolved oxygen effects of organic wastes present when discharged to a water body.

A. KINETIC EXPRESSIONS

Environmental transformation and degradation involve reaction of an organic compound with another substance. As discussed previously, the other substance usually involved is either water or oxygen or, more often, both of these substances. A kinetic expression that can be used is as follows:

$$-\frac{dc}{dt} \propto CX$$

where C is the concentration of organic compound present, t is the elapsed time period and X is the concentration of oxygen or water. This is a second-order kinetic expression where the rate of degradation is proportional to the concentration of two reactants. In most environmental situations, the amount of oxygen or water available is very large and effectively constant. This means that the concentration of organic matter is relatively low and the following expression applies.

$$-\frac{dc}{dt} \propto C$$

This expression describes first-order kinetics, and the rate of reaction is proportional to the concentration of the organic reactant present. Assuming this to be the normal situation in the environment, kinetic expressions for degradation of growth substances by microorganisms can be relatively complex, but the following expression for degradation is often used.

$$-\frac{dC}{dt} \propto BC$$

where B is the size of the bacterial population. This expression is a second-order rate expression, similar to the expression above, in which the population of bacteria was not considered. In this case, the second-order rate of reaction is proportional to the population of microorganisms and the concentration of the organic reactant present. Both B, the viable microbial population, and C, the concentration of the organic compound, will change with time. Most environmental chemical transformations and degradation facilitated by microorganisms occur in an open water, soil, air or other phase. This situation often mitigates against the increase in population of microorganisms in a specific area and tends to keep the population relatively constant. This means that the first-order expression indicated previously can generally be applied to most environmental degradation processes; that is,

$$-\frac{dC}{dt} \propto C$$

and

$$-\frac{dC}{dt} = kC$$

where k is the first-order transformation or degradation rate constant with units of time^{-1}. On integration and rearrangement, the following equation can be derived.

$$\ln (C_0/C_t) = kt \text{ or } C_t = C_0\, e^{-kt}$$

where C_0 is the concentration of the organic substance at time zero, and C_t is the concentration of the substance after a time period t. This can be plotted as shown in Figure 3.8 with ln (C_0/C_t) against the elapsed time period t. The slope of this plot represents the rate constant k. From the equation above, it can be shown that:

$$\ln C_t = \ln C_0 - kt$$

Thus,

$$\ln C_t = \text{constant} - kt$$

This relationship can be plotted as shown in Figure 3.9 and a measure of persistence of the compound is the rate of loss that can be determined as the rate loss constant, k, which is the reverse slope of the plot in Figure 3.9; thus,

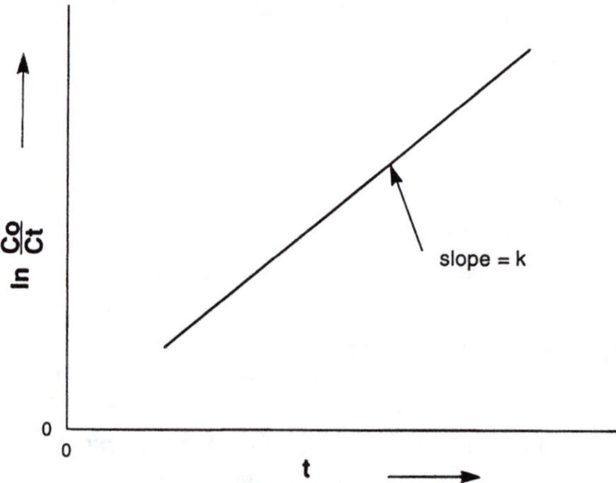

FIGURE 3.8 Theoretical plot of ln (C_o/C_t).

$$\text{slope} = -k = -\left(\frac{\ln C_t}{t}\right)$$

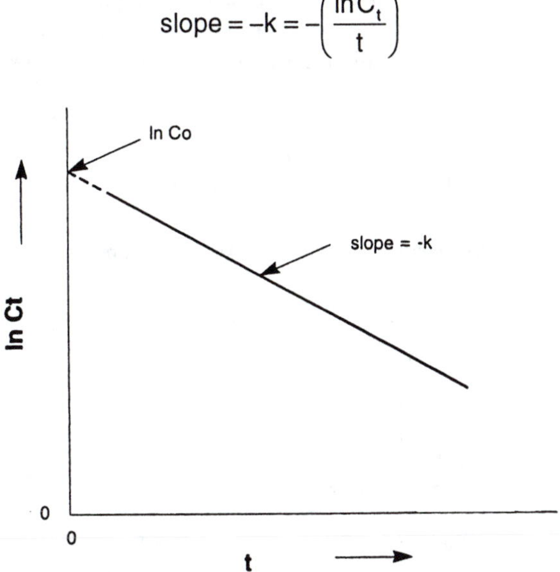

FIGURE 3.9 Theoretical plot of ln C_t against the elapsed time period.

A more convenient measure of the persistence of a chemical in the environment is the half-life, $t_{1/2}$. An expression for this can be derived as follows. It can be shown that:

$$t = \frac{\ln(C_0 / C_t)}{k}$$

Starting at any point in time when half of the original substance has been degraded, then

$$\frac{C_0}{C_t} = \frac{1}{0.5} = 2$$

Thus,

$$t_{1/2} = \frac{\ln 2}{k} = \frac{0.693}{k} = \text{constant}$$

This means that the half life ($t_{1/2}$) is constant for a given compound and a given environmental degradation process that occurs under specific conditions. This has found to be generally true for most environmental processes since these obey approximate pseudo-first-order kinetics. The actual degradation rate constants measured in the environment show a high level of variability. The pH, temperature, availability of water, availability of oxygen and other factors all influence persistence in a particular situation. These factors vary within different phases of the environment, causing this high degree of variation. Examples of some half-lives that have been measured are shown in Table 3.8.

TABLE 3.8
Half Lives of Some Pesticides in Soil

Compound	Half life (years)
Chlorohydrocarbons	3–10
DDT	1–7
Dieldrin	10
Toxaphene	
Organophosphorus	
Dyfonate	0.2
Chlorfenfos	0.2
Carbophenothion	0.5
Carbamates	
Carbofuran	0.05–1

B. VOLATILIZATION

In some situations, disappearance from an environmental compartment involves loss to another compartment rather than degradation. These processes also often follow first-order kinetics. For example, votilization of pollutants from water to the atmosphere is a very important physical loss process. Even substances with very low water solubility and having very high boiling points and low volatility, such as the PCBs and DDT, volatilize at a significant rate over relatively short periods of time. The processes involved in volatilization are:

1. Diffusion to the air/water interface from within the water mass
2. Movement of the substance away from the air/water interface into the atmosphere

These processes are illustrated diagrammatically in Figure 3.10. Volatilization generally follows first-order kinetics and it can be shown that:

$$C_t = C_0\, e^{-(Kt/D)}$$

where C_t is the concentration after an elapsed time period t, C_0 is the concentration at time zero, K is the liquid exchange constant and D is the depth of an individual compound molecule taken from the surface. The half-life in water is a useful environmental characteristic and, from the expression above, can be shown to be represented by the following equation.

$$t_{1/2} = \frac{\ln 2D}{K}$$

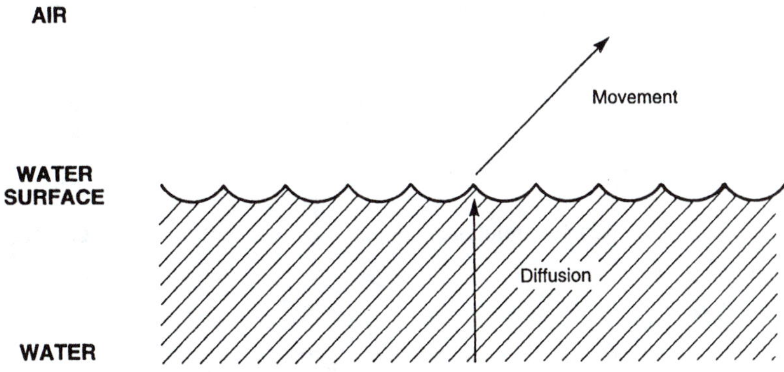

FIGURE 3.10 Processes involved in the volatilization of a compound from water.

Examples of half-lives for volatilization of various compounds in water are shown in Table 3.9.

TABLE 3.9
Examples of the Half-Lives of Some Compounds Due to Volatilization from Water

Compound	Half-life (in hours, depth of 1m)
n-Octane	5.6
DDT	73.9
Mercury	7.5

V. KEY POINTS

1. Substances in the environment, both synthetic and natural, are transformed and degraded by abiotic and biotic processes.

2. There are two major substances in the environment that participate in degradation and transformation processes. These are oxygen in the atmosphere, or originating from this source, leading to oxidation processes; and water, in water bodies, the atmosphere and biota, leading to hydrolysis.

3. The energy produced by oxidative combustion of hydrocarbons can be estimated from the bond dissociation energies.

4. Environmental phototransformation of molecules occurs as a result of the absorption of UV/VIS wavelengths present in solar radiation. Absorption occurs when certain chromophores are present in the molecular structure of the organic compound.

5. Two important equations relating energy to frequency and wavelength of radiation are

$$E = h\nu$$

$$S = \nu\lambda$$

6. The Beer-Lambert law relates absorbance to concentration of a chemical and takes the form:

$$\log\left(\frac{I_0}{I}\right) = A = \varepsilon C l$$

This can be used to derive the following equation for the rate (k_p) of transformation of a chemical in the atmosphere due to solar radiation:

$$k_p = \Sigma J\sigma\phi$$

7. The following general equation describes hydrolysis reactions with organic compounds in the environment.

$$\underset{\text{Organic compound}}{R - X} + H_2O \rightarrow R - OH + H - X \text{ (or } H^+ + X^-\text{)}$$

8. Many biotransformation and biodegradation reactions are facilitated by microorganisms that are common in most sectors of the natural environment.

9. Oxidation reactions of carbohydrates take different forms depending on the availability of oxygen and can be expressed by the following equations.

Aerobic conditions:

$$\underset{\text{Carbohydrate}}{6\ CH_2O} + 6\ O_2 \rightarrow 6\ CO_2 + 6\ H_2O$$

Oxygen in water at about 0.5–1.0 mg L^{-1} or less:

$$\underset{\text{Carbohydrate}}{2\ CH_2O} + \underset{\text{Nitrate}}{2\ NO_3^-} \rightarrow 2\ CO_3^{=} + 2\ H_2O + N_2$$

Absence of Oxygen

$$\underset{\text{Carbohydrate}}{6CH_2O} \rightarrow \underset{\text{Methane}}{3CH_4} + 3CO_2$$

10. The **mixed function oxidase** enzyme system (Cytochrome P_{450}) is induced in organisms due to exposure to xenobiotic lipophilic compounds. This system facilitates oxidation by insertion of oxygen into C-H bonds as follows:

$$\underset{\text{Lipophilic compound}}{R{-}H} + O_2 \xrightarrow[\text{P 450}]{\text{Cytochrome}} R{-}OH + H_2O$$

11. Most biotransformations and biodegradation processes in the environment follow pseudo-first-order kinetics according to the following equation:

$$\ln \frac{C_0}{C_t} = kt$$

12. For first-order processes, the persistence of a chemical in the environment is best expressed by the half-life:

$$t_{1/2} = \frac{0.693}{K}$$

REFERENCES

Finlayson-Pitts, B.J. and Pitts, J.N., *Atmospheric Chemistry: Fundamentals and Experimental Techniques,* John Wiley & Sons, New York, 1986.

Somasundaram, L. and Coats, J.R., *Pesticide Transformation Products: Fate and Significance in the Environment,* American Chemical Society, Washington, D.C., 1991.

Connell, D.W. and Miller, G.J., *Chemistry and Ecotoxicology of Pollution,* John Wiley & Sons, New York, 1984.

Maki, A.W., Dickson, K.L., and Cairns, J., *Biotransformation and Fate of Chemicals in the Aquatic Environment,* American Society for Microbiology, Washington, D.C., 1980.

CHAPTER 3 PROBLEMS

1. When hydrocarbons undergo combustion in air, small quantities of nitrogen dioxide (NO_2) and nitric oxide (NO) are produced and can play an important role in atmospheric pollution. Hydrocarbons do not contain any nitrogen at all, so why are nitrogen dioxide and nitric oxide produced? Write equations for reactions that may be involved.

2. Ethane and ethanol are both commonly used fuels, but ethane is usually in petroleum-based fuels whereas ethanol is obtained by fermentation of plant matter. In addition, these two substances are closely related chemically since ethanol has the same structure as ethane but has an additional oxygen atom in the molecule to form an alcohol group.

```
    H  H                 H  H
    |  |                 |  |
 H—C—C—H              H—C—C—OH
    |  |                 |  |
    H  H  ethane         H  H  ethanol
```

   Compare these two compounds as fuels on the basis of calculated heats of combustion and production of carbon dioxide from complete combustion.

3. Explain the fundamental reasons for the difference in the heats of combustion of ethane and ethanol in terms of their bond dissociation energies.

4. In July, a fire occurred in a factory located in a major city at about a latitude of 40°N. Large volumes of formaldehyde were used and a proportion was vaporized into the atmosphere. Measurements in the surrounding atmosphere indicated that formaldehyde attained a concentration of 2500 $\mu g\ m^{-3}$. It would be expected that winds would disperse the chemical; but if photodegradation were the only process involved, how long would it be before a concentration of 50 $\mu g/m^3$ is attained in the atmosphere?

Answers on Page 74.

CHAPTER 3 SOLUTIONS

1. Air involved in the combustion of hydrocarbons contains about 78% nitrogen and 21% oxygen. Under the conditions produced by combustion, N_2 and O_2 react to produce NO and NO_2. The reactions involved can be expressed as follows:

$$N_2 + O_2 \rightarrow 2\ NO$$

$$N_2 + 2\ 0_2 \rightarrow 2\ NO_2$$

2. In evaluating the complete combustion of ethane, the following equation can be obtained:

$$\underset{(MW = 30)}{C_2H_6} + 3^1/_2\ O_2 \rightarrow \underset{(MW = 44)}{2\ CO_2} + 3\ H_2O$$

The heat of combustion can be calculated as follows

(a) Ethane has 6 × (C–H) and oxygen $3^1/_2$ × (O=O)
(b) (6 × 413) = 2478 kJ mole^{-1} + ($3^1/_2$ × 498) = 4221 kJ mole^{-1}
(c) Total energy input = 4221 kJ mole^{-1} ethane
(d) The products comprise 4 × (C=O) to form CO_2 + 6 × (O–H) to form 3 H_2O
(e) (4 × 1803 = –3212 kJ mole^{-1} ethane + (6 × 463) = –2778 kJ mole^{-1} ethane
(f) Total energy release = –5990 kJ mole^{-1} ethane

Overall energy change = –1769 kJ mole^{-1} ethane
Thus, heat of combustion/unit mass =1769/30 kJ mole^{-1}

Heat of combustion = 59.0 kJ mole^{-1}

The amount of CO_2 produced is 88 g mole^{-1} or 88/30 g/g = 2.9 g/g ethane

CO_2 produced is 2.9 g/g ethane

Considering the complete combustion of ethanol, the following equation is obtained:

$$\underset{(MW = 46)}{C_2\ H_5OH} + 3O_2 \rightarrow \underset{(MW = 44)}{2\ CO_2} + 3\ H_2O$$

The heat of combustion can be calculated as follows:

(a) Ethanol has 5 × (C–H), 1 × (C–O), 1 × (O–H), 1 × (C-C) and oxygen 3 × (O=O)
(b) (5 × 413) + (1 × 351) + (1 × 463) + (1 × 347) + (3 × 498) = +4720 kJ mole^{-1} ethanol input

(c) The products comprise 4 × (C=O) to form 2 CO_2 and 6 × (O–H) to form 3 H_2O
(d) Early release is –(4 × 803) – (6 × 463) = –5990 kJ $mole^{-1}$ ethanol

Overall energy change = –1270 kJ $mole^{-1}$ ethanol

Heat of combustion = 1270/46 = 27.6 kJ g^{-1}

The amount of CO_2 produced is 88g $mole^{-1}$ or 88/46 g/g ethanol = 1.9 g/g ethanol.

CO_2 produced is 1.9 g/g ethanol

Thus, in summary:

1. Ethane gives 59.0 kJ/g of heat on combustion and 2.9 g CO_2/g ethane.
2. Ethanol gives 27.6 kJ/g of heat on combustion and 1.9 g CO_2/g ethanol.

These calculations indicate ethane could be a much better fuel than ethanol but CO_2 production is much higher.

3. Diagrammatically, the energy changes occurring in ethane and ethanol on combustion can be represented as shown in Figure 3.11.

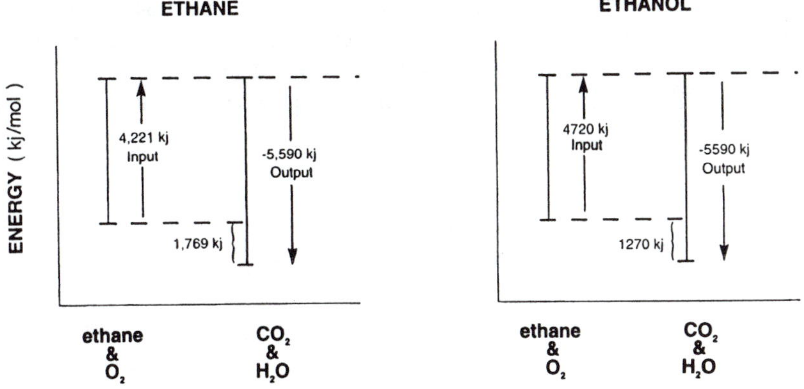

FIGURE 3.11.

This indicates that for each mole of ethane and ethanol, there is the production of the same amount of energy output on the formation of the products but there is a difference in the energy input to each initial set of reactants. Ethane and O_2 require 4221 kJ $mole^{-1}$ to form atoms from thier components, whereas ethanol and O_2 require 4720 kJ $mole^{-1}$ due to the presence of extra bonds resulting from the additional O atom.

4. The major photodegradation process involved is

$$H-CHO \rightarrow H\cdot + H\cdot CO$$

At noon on a cloudless day in July at 40°N latitude, this process would have a rate constant (k_p) of 2.7×10^{-5} s^{-1}. The expression for the kinetics of the first-order process is:

$$t = \frac{\ln(C_0/C_t)}{k_p}$$

Thus, t for this photodegradation of formaldehyde is:

$$t = \frac{\ln(2500/50)}{2.7 \times 10^{-5} s^{-1}}$$

$$= \frac{3.9}{2.7 \times 10^{-5} s^{-1}}$$

$$= 1.45 \times 10^{+5} s$$

This means it would take 16.7 days to reach a concentration of 50 μg m^{-3}. This rate occurs at midday and photodegradation would not be expected to occur during the night period without solar radiation. Thus, the total time would be approximately 33.4 days (assuming 12 hours of darkness per day).

Chapter 4

ENVIRONMENTAL TOXICOLOGY

I. INTRODUCTION

The history of toxicology dates from ancient times and is a major theme in the development of chemistry. The ancient papyrus document titled "Ebers" dates from 1500 BC and contains the earliest known reference to the beneficial and toxic effects of a range of medicinal products. Other early references to toxic substances were by Hippocrates and Aristotle. It's interesting to note that the famous Roman historian Pliny, the Greek physician Nikander, and the Roman architect Vitravius all commented on the harmful effects of lead ingestion. Thus, this can be considered as one of the earliest observations of harmful effects of chemicals in the environment.

The next major breakthrough in toxicology was around 1500 by the Swiss chemist Phillipus von Hohenheim Paracelsus, commonly called Paracelsus. He is credited with moving alchemy away from a prime concern with the transmutation of elements into gold, and toward chemicals in medicine. Paracelsus is often quoted in modern times with his statement that there is a relationship between dose of toxicant and the response. He expressed it as follows:

> All substances are poisons; there is none that is not a poison. The right dose differentiates a poison and a remedy.

Other key breakthroughs occurred in 1755 when Percival Pott discovered the link between soot and scrotal cancer in chimney sweeps and in 1815 with the publication of *General System of Toxicology* by Mattieu Orfila. In 1962, Rachel Carson's book *Silent Spring* first alerted society to the potential danger of many organic chemicals, particularly DDT and related pesticides, when discharged to the environment. Since then the field of environmental toxicology has undergone quite dramatic growth.

Environmental toxicology is the study of the toxic or poisonous effects that chemicals in the environment exert when individual organisms or relatively small numbers of individuals are exposed. The effects of chemicals on whole ecosystems are a related, but different, topic described as **ecotoxicology** and outlined in Chapter 20. Strictly speaking, **environmental toxicology** is only concerned with the effects of chemicals, although in a broader sense other forms of pollution such as thermal pollution are often included. Toxic effects are by definition not beneficial but deleterious, and the type and severity of toxic effects that can occur is very broad, ranging from temporary sublethal effects to lethality.

Toxicology, including environmental toxicology, is principally concerned with effects on animals although toxic effects on plants can be significant in some situations. In broad terms, toxicology can be considered to consist of three areas:

Clinical toxicology: the investigation of the effects of pharmaceutical chemicals on human health.

Forensic toxicology: concerned with the illegal use of chemicals such as elicit drugs.

Environmental toxicology: covers the effects of chemicals in the environment on organisms.

Environmental toxicology presents many new challenges to the field to toxicology. Toxicology, up to recent times, has been principally concerned with relatively large doses of chemicals with short exposure periods, usually of the order of days, leading to lethal effects; for example, possible lethal effects of medicinal compounds on human beings. Short-term lethal effects are also of interest in environmental toxicology. A typical investigation might include an area being studied for the lethal effects of agricultural pesticides on fish. However, there are additional aspects that relate to low exposures over long periods, perhaps decades, with more subtle, but still very important sublethal effects. Some areas of interest here are the influence of low levels of chlorine in drinking water on human health and effects of trace petroleum contamination in aquatic systems. To cope with these aspects, environmental toxicology has been extended by a range of new approaches to evaluation of the deleterious effects of chemicals. Of particular importance are **ecotoxicology** covered in Chapter 20, **risk assessment** covered in Chapter 21 and **genotoxicology** covered in Chapter 19.

A toxicant in the environment could occur in water, air, soil or food as a result of direct discharges, transport from elsewhere or biotransformation of another substance. To exert a toxic effect on an organism, it must be transferred to the site of action within the organism by the processes generalized in Figure 4.1. This means chemical is taken up by the organism through the stomach, gills, lungs, etc. and then distributed throughout the body, where it may be biotransformed and excreted. Some of the original compound, or its biotransformed product, may be transferred to the site of action. With carbon monoxide poisoning, the site of action is the hemoglobin to which the carbon monoxide attaches, leading to a loss of the oxygen-carrying capacity and subsequent toxic effects. When the toxicant is attached to the site of action, the observed toxic effects occur.

II. ROUTES AND MECHANISMS OF TOXICANT ENTRY TO ORGANISMS

A. MECHANISMS OF ENTRY TO ORGANISMS

In order for a chemical to exert a toxic effect, it must move from the ambient environment into the organism. For any compound to enter an organism, regardless of the route of entry, it must cross cell membranes. There are three principal mechanisms by which toxicants cross cell membranes: passive diffusion, facilitated diffusion and active transport. Of these, passive diffusion appears to be the predominant mechanism for toxicants.

Passive diffusion is largely governed by the difference in toxicant concentration on either side of the membrane, i.e., the concentration gradient. The basic structure

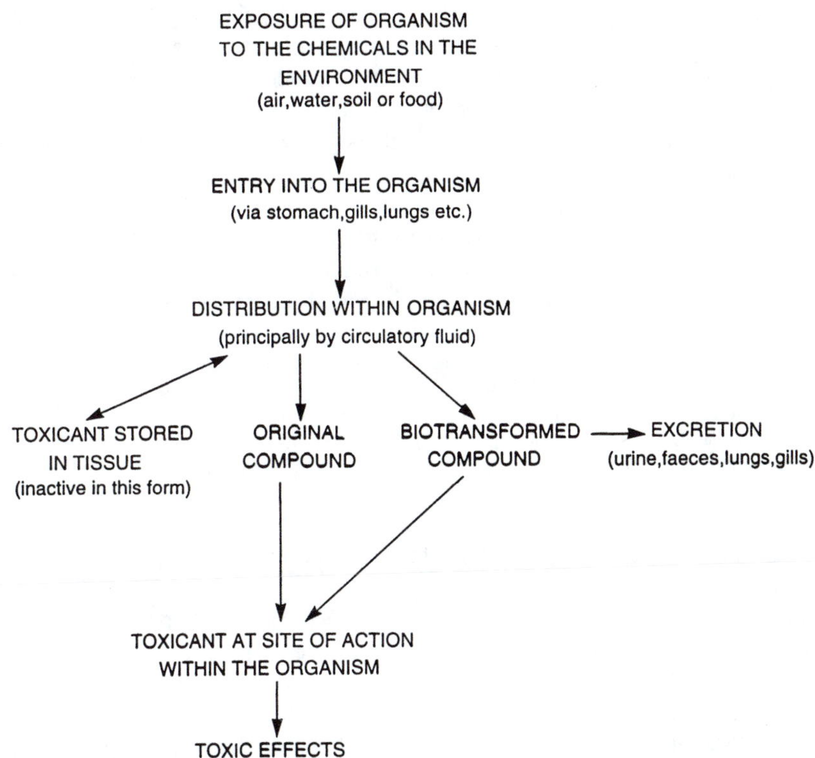

FIGURE 4.1. Generalized transfer processes for a chemical from the environment to the site of toxic action.

of cell membranes consists of a lipid bilayer; thus, where passive diffusion is the mechanism of entry, a toxicant will cross the membrane only when the compound is lipid soluble and the concentration of the toxicant is lower on the inside of the membrane than on the outside.

Facilitated diffusion is also a diffusion process, so a favorable concentration gradient is required for the compound to cross the membrane. The diffusion is facilitated by chemical processes that aid in the transport and thus the rate of transfer across the membrane is higher than for simple diffusion. In active transport, the compound is transported via a carrier; however, this mechanism requires an energy input. Because of the expenditure of energy, toxicants can transfer across a membrane regardless of the direction of the concentration gradient or any other gradient (e.g., electrical).

B. CHEMICAL PROPERTIES OF COMPOUNDS THAT CROSS CELL MEMBRANES

There are a number of physical and chemical properties that affect the penetration of chemicals through lipid membranes. Some of the most important properties are ionic state, molecular size, lipophilicity — most commonly measured by the octanol-water partition coefficient (K_{OW}, described in Chapter 2), viscosity and concentration. The influence of these properties on cross-membrane transport is shown in Table 4.1. Lipophilic (fat-loving) compounds have K_{OW} values in the range 10^2 to $10^{6.5}$, with lipophilicity increasing with increases in the K_{OW} value. For example, DDT is a highly lipophilic compound with a K_{OW} value of $10^{6.2}$ and can readily cross through cell membranes.

TABLE 4.1
Physical and Chemical Properties that Affect Toxicant Transfer Across Membranes

Parameter	Effect on cross-membrane transfer
Ionic state	Dissociated molecules in the ionic state are not transferred readily
Molecular size	Smaller compounds are transferred more readily
Lipophilicity	Lipophilic compounds are transferred more readily ($K_{OW} = 10^2$ to $10^{6.5}$)
Concentration	Compounds at higher concentrations are transferred more readily

The ionic state can have a major effect on the ability of toxicants to cross cell membranes. Ions attract a surrounding shell of water molecules (see Chapter 2, Bonds and Molecules, Their Influence on Environmental Properties) effectively preventing their passage through the lipophilic layer in cell membranes, which is hydrophobic (water-hating) in nature. Less ionized compounds pass through slowly, whereas non-ionized compounds with their greater lipid solubility, cross readily.

The pH of the solution is a variable that influences the extent of ionization of chemicals and thus their ability to cross membranes. The relationship between pH and the extent of ionization has been quantified by the Henderson-Hasselbalch equation:

$$pH = pK_a + \log [\text{ionized form/non-ionized form}]$$

where pH is the negative logarithm of the hydrogen ion concentration ($-\log[H^+]$) and pK_a is the negative logarithm of the acid ionization constant ($-\log K_a$).

When the pK_a of a chemical is numerically equal to the pH, then, according to the Henderson-Hasselbalch equation, there is an equal number of ionized and non-ionized forms of the compound (i.e., log [ionized form/non-ionized form] = 0 and ionized form/non-ionized form = 1). Any change in pH will change the extent of ionization. A variation of one pH unit will cause a 10-fold change in the ratio of ionized to non-ionized forms. This will, in turn, have significant effects on the ability of chemicals to be transferred across lipid membranes (see Table 4.1).

C. ROUTES OF ENTRY TO ORGANISMS

There are three principal routes of entry for toxicants into an organism. These are:

1. Inhalation through the respiratory system (lungs or gills)
2. Absorption through the skin
3. Ingestion through the gastrointestinal tract (GIT)

If plants are under consideration, a different set of entry routes would exist.

The gills of aquatic organisms are essentially the aquatic equivalent of the lungs of terrestrial animals. They have similar structures and use similar features to maximize the uptake of oxygen, which makes them excellent sites for toxicant uptake. The mechanism of toxicant transfer from gills or lungs to blood is diffusion. In order to maximize the rate of diffusion, gills have a very large surface area, typically areas 2 to 10 times the body surface area of the fish. There is a very small distance through which the toxicant must diffuse (i.e., 2 to 4 μm) and blood passes through the gills in the opposite direction to the water flow over the gills. The counter-current flow of blood and water maximizes the difference in toxicant concentration between the two phases, and thus increases the rate of diffusion. In aquatic organisms, this is the predominant form of toxicant entry, except perhaps in some cases, for food.

The lungs have a similar purpose to the gills and are structured to maximize-contact between atmospheric oxygen and blood to facilitate exchange. This provides an excellent medium to also exchange toxicants in the vapor or gaseous form in the atmosphere. Toxicants can also be attached to particles in high concentrations. The respiratory system of terrestrial mammals has several mechanisms to remove or prevent the entry of particulate pollutants into the body. The defense mechanisms are: impedence to movement through the nasal passage, the inertia of large particulates, and the lungs' self-cleaning system. Many of the cells that line the respiratory system secrete mucus and have cilia. This mucus is thick and sticky and entraps particles that strike it. The cilia all beat in one direction, pushing the particle-contaminated mucus up to the mouth where it is then generally swallowed and enters the gastrointestinal system prior to excretion. There are no mechanisms to prevent the entry of vapors or gases. However, many gases and vapors are respiratory irritants or have an odor that can offer a degree of protection if avoidance behavior results.

Skin acts as a selective barrier to chemicals; some can penetrate it while others cannot. The barrier properties of skin are primarily due to only one of the seven layers of cells that toxicants must cross to enter the body, which is described as the *stratum corneum.* This is composed of densely packed, biologically inactive cells and is one of the cell layers that comprise the epidermis, the outer most part of the skin. The barrier qualities of skin vary with the location of the skin due to changes in the thickness and quality of the *stratum corneum* and the abundance of either sweat glands and/or hair follicles.

The gastrointestinal tract (GIT) consists of the mouth, esophagus, stomach, intestines and the rectum (Figure 4.2). A toxicant inside the GIT has not yet entered the human body. To enter the body, it must cross the lining of the GIT and enter the bloodstream. The mouth, esophagus and rectum play a minor role in the absorption of food or toxicants. In contrast, the stomach and intestine walls are designed to maximize the absorption of food and thus are ideal sites for absorption of toxicants. However, the characteristics of the absorption in the GIT may well be different from that for the same compound if it were absorbed elsewhere. The stomach is acidic with a pH of approximately 1.5. If the pH of the stomach is different from the pH of the toxicant carrier, often food, the chemical form of the toxicant (i.e., the ratio of ionized to non-ionized form) will change in accordance with the Henderson-Hasselbalch equation (see Section IIB) which, in turn, will affect the extent of absorption across the GIT. The length of time that a toxicant stays in the GIT is also very important, as increased exposure can lead to increased concentrations of the toxicant in the body.

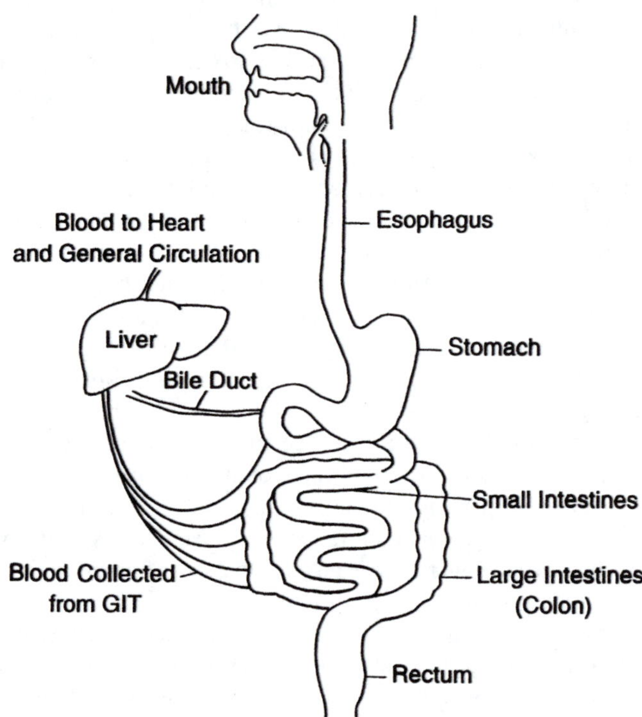

FIGURE 4.2. The anatomy of the gastrointestinal tract.

III. DISTRIBUTION OF TOXICANTS WITHIN THE ORGANISM

Once a toxicant crosses a cell membrane, it has entered the body but it does not stay in those cells at the entry point and is distributed throughout the body (see Figure 4.1). The principal distribution mechanism is through the blood circulatory system. Blood is not homogeneous; its components belong to three different fractions: water, lipids and proteins. A toxicant will partition into each of these three phases according to the physical and chemical properties of the toxicant and the different blood fractions. If the toxicant is water soluble, the bulk of it will be found dissolved in the aqueous phase of blood. However, if the toxicant has a high affinity for either lipids or proteins, it will bind reversibly to that fraction. The toxicant in the aqueous phase is not bound to water but is simply dissolved and is hence described as "free." It is this "free" toxicant that can exert toxic effects, whereas bound forms are biologically inactive (i.e., they cannot exert a toxic effect). Examples of the distribution of chemicals between free and bound forms are presented in Table 4.2. DDT has very low aqueous solubility and a high K_{OW} value, and thus is highly lipophilic and expected to be located principally in the lipid fraction. Parathion is somewhat similar, but less lipophilic, and so has a higher proportion in the aqueous fraction. Nicotine is very water soluble and correspondingly occurs mainly in the aqueous fraction.

TABLE 4.2
The Relative Distribution of Several Chemicals Between the "Free" and "Bound" Forms in Blood

Chemical	Log K_{ow}	Aqueous solubility (mg L^{-1})	% Free	% Bound
DDT (an organochlorine pesticide)	6.2	0.003	0.1	99.9
Parathion (an organophosphate pesticide)	3.8	6.5	1.3	98.7
Nicotine	1.3	Very soluble	75.0	25.0

The reversible binding of toxicants to either lipid or protein biomolecules can be represented by the following equation:

$$T_f + \text{Biomolecule} \leftrightarrow T_b$$

where T_f is the "free" (unbound) toxicant and T_b is the bound toxicant. The term "reversible" means that toxicant molecules are continuously moving from being free to bound and can become unbound (or "free") at some point in time. As the formation of bound toxicants is reversible, an equilibrium can be established between the bound and "free" forms of the toxicant. The position of the equilibrium is governed by the relative affinity of the toxicant to the various blood components or tissue. In most cases, the toxicant forms an attachment with the tissue in which it is stored and thus becomes biologically inactive. Such attachments are also reversible, so equilibria are established between compartments of the body that store toxicants and the blood (see Figure 4.3) and hence all the compartments are in equilibrium.

Due to the equilibria, a change in the concentration in any compartment will lead to concentration changes in the other compartments. For example, if a person has been routinely exposed to a toxicant so that a steady concentration was always in the blood and then the exposure source was removed, the toxicant level in the blood will drop. However, the toxicant will move out of the other compartments where it was stored and into the blood until equilibrium is once again established.

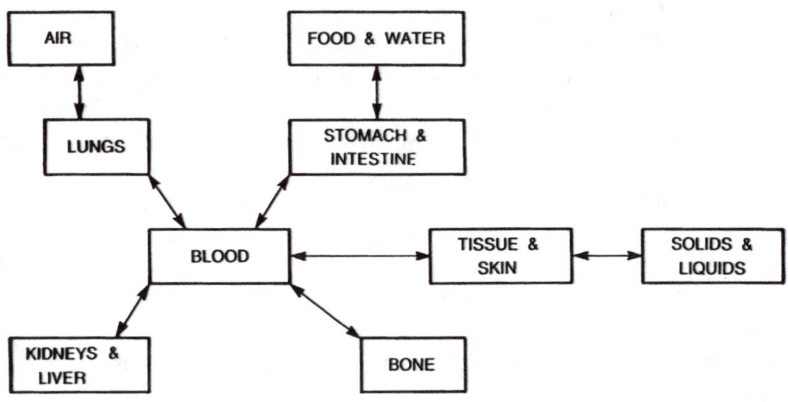

FIGURE 4.3. The routes of entry and distribution of toxicants between the various compartments in the body.

IV. BIOTRANSFORMATION OF TOXICANTS

(also see Chapter 3 — Transformation and Degradation Processes — Reactions and Kinetics)

A. PRINCIPLES OF BIOTRANSFORMATION

Many environmental toxicants are substances that are resistant to chemical transformation and degradation. Often, these substances are lipophilic with K_{OW} values in the range of 10^2 to $10^{6.5}$, or log K_{OW} from 2 to 6.5. Some examples are DDT, dieldrin and polychlorinated biphenyls (PCBs).

The very property that allows these lipophilic toxicants to pass through lipid membranes and thus enter an organism is also a characteristic that limits their excretion. The normal means by which organisms rid themselves of dangerous chemicals is by solubilization in urine and feces, and here the water solubility of the chemicals has a strong influence on their capacity to be removed. As many environmental organic toxicants are reasonably lipophilic, and therefore have low aqueous solubilities (see Chapter 2), they cannot be readily excreted via the urinary or fecal routes. Thus, organisms have developed means to increase the aqueous solubility of toxicants by various biotransformation processes. Biotransformation normally occurs in two stages: Phase I and Phase II. However, compounds with sufficiently high aqueous solubilities and reactivities may skip or eliminate Phase I altogether. All biotransformation reactions are facilitated by enzyme systems.

B. PHASE I TRANSFORMATIONS

Phase I reactions generally tend to involve such enzymes as mixed function oxidases (MFO), reductases and hydrolases, which oxidize, reduce or add either the hydroxyl (OH^-) or hydronium (H_3O^+) ions to reactants, respectively. This increases the polarity, water solubility and reactivity of the toxicant and lowers the K_{OW} value, reflecting decreased lipophilicity. Of the MFO enzymes, the cytochrome P_{450}-dependent monooxygenase system is the most important.

C. PHASE II TRANSFORMATIONS

The reactions that occur in Phase II are called *synthetic* or *conjugation reactions,* as metabolites from Phase I are combined (conjugated) with naturally occurring metabolic intermediates. Examples of such metabolic intermediates include UDP-glucuronic acid, amino acids such as glycine and glutamine, and other substances. This process further increases the aqueous solubility, lowers the K_{OW} value and greatly facilitates the excretion of toxicants via the urinary system.

D. SITES OF BIOTRANSFORMATION

The most important organ associated with biotransformation is the liver (hepatic tissue), followed in order of decreasing importance by the kidneys, lungs and intestine. In non-liver tissue (extrahepatic tissue), only certain types of cells contain the necessary enzymes for biotransformation. Such cells have a limited range of enzymes and those enzymes present generally have lower activities compared to liver tissue.

V. EXCRETION OF TOXICANTS

There are three principal routes for the excretion of toxicants: urinary, fecal and elimination through the lungs or gills. The relative importance of each of these routes depends on both the type of organism and the chemical involved. However, it is important to realize that all bodily secretions (e.g., saliva, breast milk, semen, sweat and tears) and all cells lost from the body (e.g., dead skin cells, hair, finger, toenails and the molted skin or exoskeleton) lead to the removal of toxicants. This occurs, as mentioned previously, because toxicants are distributed widely in an organism.

If a toxicant is biotransformed to a water-soluble compound, the bulk of it will move into the aqueous component of the blood. Eventually, the blood passes through the kidneys where the aqueous component and the associated biotransformation product are removed and excreted in the urine. The fecal route of excretion receives toxicants from two main sources. First, toxicants that are not absorbed from food and liquids that enter the gastrointestinal tract are automatically present in the feces. The other source is bile. When toxicants reach the liver, they are biotransformed and removed by either entering urine or bile. The principal properties that determine whether a toxicant or transformation product enter the bile or urine are the aqueous solubility and molecular size. The larger and less water-soluble compounds enter the bile, which is then released into the intestine to mix with the feces.

Elimination by the lungs is only important for chemicals that are gases or are highly volatile, i.e., have high vapor pressures. The means by which such compounds are transported from the blood to air in the respiratory system appears to be passive diffusion. An example of a compound eliminated by this mechanism is ethanol. With humans, approximately 2 to 4% of ethanol consumed is excreted in exhaled breath. Police take advantage of this and use instruments (breathalyzers) that measure the concentration of alcohol in exhaled air in order to approximate the blood alcohol concentration, assuming an equilibrium is reached in the partition process between air and blood.

Elimination by the gills can be a very important route of elimination of nonpolar, lipophilic compounds for aquatic organisms, particularly fish. For example, half of the pentachlorophenol in fish is eliminated by the gills. The mechanism of exchange from blood to water is passive diffusion. Thus, passive diffusion is the predominant mechanism for the uptake and elimination of lipophilic toxicants across the gills.

VI. CLASSES OF POISONS BASED ON EFFECT

It is important to recognize that the type of toxic effect exerted, varies with the type of chemical. Chemicals can be classified into different groups based on the type of deleterious effect they exert:

Toxicants: Any substance that causes a deleterious biological effect when living organisms are exposed to it. This is the largest class of compounds, as all compounds can exhibit toxic effects, depending on the dose as postulated by Paracelus (see Section I, Introduction). In fact, the other categories of chemicals discussed below, are really subsets of this group.

Teratogens: Any substance that causes defects in the reproduction process by either reducing productivity or leading to the birth of offspring with defects.

Mutagens: Any substance that leads to inheritable changes in the DNA of sperm or ovum cells.

Carcinogens: Any substance that causes a cell to lose its sensitivity to factors that normally regulate cell growth and replication. Such cells replicate without restriction to form a growing mass called a tumor.

In this chapter, only toxic properties without teratogenic, mutagenic or carcinogenic aspects are discussed. Further information on the other classes of toxic properties can be found in Chapter 20.

VII. QUANTITATIVE PRINCIPLES OF TOXICOLOGY

A. BACKGROUND

For most chemicals, there are relationships between the biological effect observed and either the toxicant concentration in the ambient environment (e.g., water with fish) or dose (e.g., occurrence of lead in food with humans). These relationships are collectively termed "*dose-response relationships.*" An aspect of environmental toxicology that is of particular importance is that some environmental toxicants are in fact essential to growth and development at low concentrations, but become toxic at higher concentrations. These substances are mainly metals such as iron, magnesium, zinc, copper and a variety of other substances. The general nature of the relationship between dose (or concentration) and biological response is different for chemicals essential to the organism's metabolism and those that are nonessential (as illustrated in Figure 4.4). For essential substances, the relationship is parabolic. When present at concentrations well below those required by the organism for normal growth and development, deleterious effects, and possibly death, occur due to their absence. At increased concentrations, the effects become less harmful, i.e., first irreversible and then reversible effects occur, until the toxicant is present at a concentration range suitable for normal metabolism, at which no deleterious effect are exerted. As the concentration of an essential substance increases beyond this range, toxic effects become apparent and become increasingly deleterious, leading eventually to death.

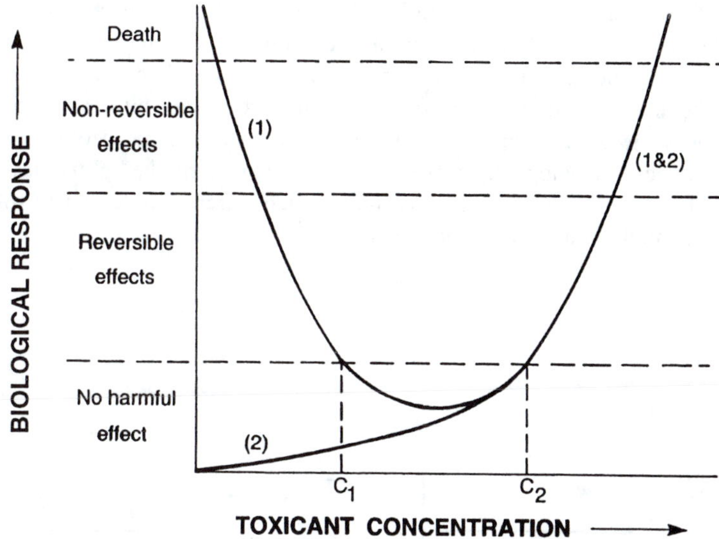

FIGURE 4.4. Biological responses to differing concentrations of essential chemicals (1) and nonessential chemicals (2). Normal metabolic activity with an essential chemical occurs between concentrations C_1 and C_2, which below C_1 the chemical is in deficiency and above C_2 is in excess.

The relationship for nonessential compounds is somewhat similar to an exponential growth curve (see Figure 4.4). At zero and low concentrations, no measurable toxic effects are apparent. However, as the concentration increases, the severity of the deleterious effects increase, with initially reversible effects occurring, then irreversible effects and finally death results. An example of a nonessential chemical is toluene (C_6H_5–CH_3). This substance is a neurotoxin (i.e., it affects nerves) and has been implicated in central nervous system depression, dizziness, nausea and vomiting, respiratory depression and death. It is from general observations of doses and the responses which occurred that the well known principle "the right dose differentiates a poison and a remedy" was formulated by Paracelsus. Even such innocuous compounds as water, oxygen and sodium chloride can exert toxic effects and even be fatal at high enough doses.

The chemical form and route of entry of toxicants affects the toxicity. For example, pure metallic mercury (Hg) is much less toxic than methylated mercury [$(Hg9CH_3)_2$]. Curare (the poison used by some South American Indians to kill prey) is a very rapid-acting poison if introduced directly into the blood, whereas consumption in food has no effect. Another example is amygdalin, which is found in the pits of peaches and apricots. If consumed orally, it is approximately 40 times more toxic than by intravenous injection; this is because the intestines contain bacteria with enzymes that convert the amygdalin to more harmful products.

B. MEASURES OF TOXICITY

Individuals from a population of the same species have different susceptibilities (or tolerances) to toxicants. The distribution of tolerance is generally accepted to be a normal or gaussian distribution when plotted against logarithm of the toxicant concentration or dose (Figure 4.5). Those individuals affected at low levels are termed "intolerant" or "susceptible," while those that are only affected at much higher levels are termed "tolerant" or "nonsusceptible." The majority of individuals have a tolerance between these two extremes. Examples of the variation in tolerance include the different amounts of alcohol that can be consumed by various individuals before the same effects appear, and the observation that some people can smoke heavily and never develop lung cancer and yet others who smoke lightly may develop cancer. This variation in tolerance is due to such differences as innate tolerance, rates of metabolism and body composition.

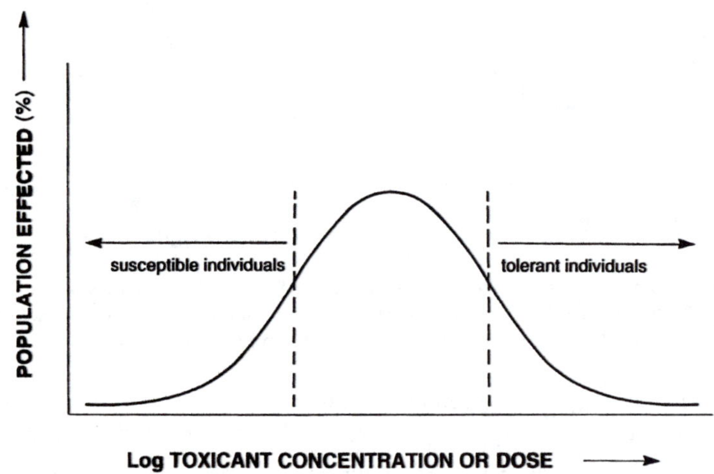

FIGURE 4.5. The susceptibility, as tolerance, of a population of individuals of a single species to different concentrations of a toxicant, plotted as percentage responding against log [toxicant concentration].

All measures of toxicity refer to either a concentration or a dose (amount) that causes a toxic effect and it is important to clearly differentiate between these. *Concentration-based* measures of toxicity state the concentration in the surrounding environment (e.g., water for aquatic toxicity tests, air in inhalation tests, and soil for terrestrial toxicity tests) needed to exert the particular toxic effect and are expressed in units such as mg/L, mol/L or mmol/L. On the other hand, *dose-based* measures of toxicity are expressed on a mass per mass of organism tested basis, e.g., grams of toxicant per kilogram of tissue. Dose-based measures of toxicity are used when a known mass of a toxicant is administered directly to the test organism. This is done by exposing the animals to the chemical by injection, direct introduction to the stomach (gavage) or by application to the skin. However, concentration-based measures of toxicity are used when the toxicant is dissolved in water. Thus, concentration is nearly always used in aquatic and inhalation toxicity tests, where the toxicant is dissolved in water or air, respectively.

The data in Figure 4.5 can be replotted instead of the proportion (as %) affected at each concentration range, but as the total proportion that has responded up to that concentration. Thus, this new curve (Figure 4.6) should start at zero cumulative percentage of the population affected at zero concentration and increase to a maximum of 100% cumulative percentage for some higher toxicant concentration.

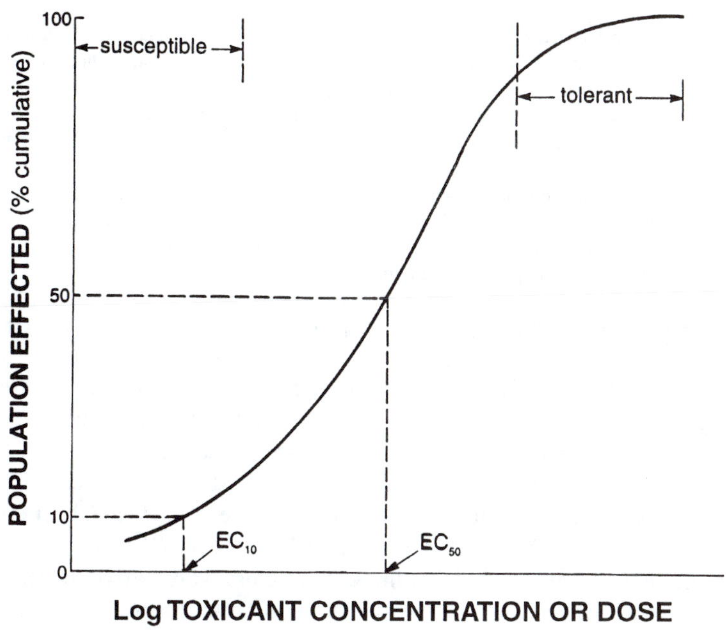

FIGURE 4.6. The susceptibility, or tolerance, of a population of individuals plotted as the cumulative percentage responding against log [toxicant concentration].

The principal measure of the toxic effects used in toxicity studies is the 50% effect level, where 50% of the individuals are more tolerant and 50% are less tolerant. This represents the average organism in the population and exhibits the greatest consistency in experimental measurements. Measures of toxicity with different magnitudes, for example, 90% and 10% of the test organisms affected are used, but less frequently. The concentration, or dose, corresponding to the percent affected is referred to as the **effective concentration** (EC) for that percent affected; for example EC_{90}, EC_{50} and EC_{10}.

The biological effects measured as percent of population affected in a toxicity test are called "**biological endpoints**." A vast number of different endpoints have been used. They can be subdivided into two principal categories: lethal and sublethal. Lethality is simply the quantification of the number of test animals that have died or are still alive. There are a much greater number of sublethal endpoints, which include: changes in the rate of respiration or the activity of an enzyme; inhibition of growth, mobility, metabolic rate, bioluminescence, reproductive capability; and a stimulation of deformities. Some common toxicity measures are shown in Table 4.3.

TABLE 4.3
The Abbreviations of Some Common Toxicity Measures

LC_{50}	The concentration of a chemical, in the ambient environment, that is lethal to 50% of the test organisms.
IGC_{50}	The concentration of a chemical, in the ambient environment, that inhibits growth of the test organisms by 50% when compared to the growth of the control under standard conditions and over a specified exposure period.
EC_{50}	The concentration of a chemical, in the ambient environment, that exerts a 50% change of the measured effect on the test organisms when compared to the control under standard conditions and over a specified exposure period.
NOEL	The maximum concentration of a toxicant used in a toxicity test, that has no statistically significant effect on the biological response when compared to the control.
LOEL	The lowest concentration of toxicant used in a toxicity test that has a statistically significant effect on the biological response when compared to the control.

Note: The above measures have equivalent measures expressed in terms of dose e.g., LD_{50}.

As mentioned previously, measures of lethality are useful in environmental toxicology only where lethal levels are likely. In most situations in the environment, sublethal levels are present, and so techniques are needed to address this. Some of these utilize other measures of toxicity based upon statistically significant differences between treatments and the controls. Such measures are termed **No Observed Effect Level** (NOEL) or **No Observable Adverse Effect Level** (NOAEL), and **Lowest Observable Effect Level** (LOEL) or the **Lowest Observable Adverse Effect Level** (LOAEL). These values can be expressed in terms of concentrations or dose. The relationship between the values is indicated in Figure 4.7. The NOAEL is a maximum value, as indicated, and is not represented by the level corresponding to point A. Similarly, the LOAEL is a minimum level as indicated in Figure 4.7 and not represented by the level at point B or those at higher concentrations.

In evaluating mechanisms of toxicity, it is often useful to compare the toxicity of different chemicals in terms of the number of molecules (e.g., mole L^{-1} or mmole L^{-1}). For example, if two compounds A and B both had toxicities of 45 mg/L, they may not necessarily have the same toxicity in molecular or molar terms. Assuming these compounds had molecular weights of 126 amu and 50 amu, respectively, the toxicity expressed in mole L^{-1} would be 0.37 and 0.9 mole L^{-1}, respectively, which means that compound A is more toxic.

C. FACTORS INFLUENCING TOXICITY

The toxicity of a compound varies with the period of exposure. An experiment could be carried out to determine the concentration needed to exert a particular toxic effect, e.g., LC_{50} at different exposure times. A typical plot of the toxicity values (e.g., LC_{50} against the exposure time) would resemble Figure 4.8. This clearly

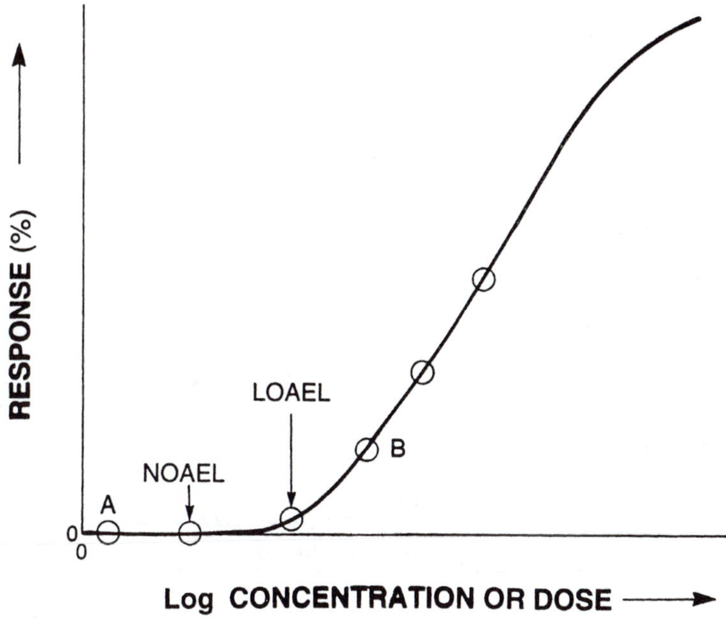

FIGURE 4.7. Diagrammatic illustration of the lowest observed adverse effect level (LOAEL) and no observed adverse effect level (NOAEL).

illustrates that as the period of exposure increases the amount of toxicant required to exert the required effect decreases. The minimum concentration that will exert a toxic effect, regardless of the period of exposure, is called the *incipient* or *threshold* toxicity of a chemical. To obtain the threshold toxicity, a vertical line is drawn from the point where the toxicity curve becomes effectively parallel with the y-axis (time) down to the x-axis (see Figure 4.8). If the toxicity curve does not become effectively parallel to the y-axis, it should be extrapolated until it does and then the incipient toxicity can be determined.

Different species and even different strains of the same species can have markedly different susceptibilities to the same compound. For example, the amounts of dioxin (i.e., 2,3,7,8-tetrachlorodibenzo-*p*-dioxin) required to kill 50% of exposed guinea pigs and hamsters are 1 and 5000 μg/kg, respectively, and there is only 65% agreement between the results of carcinogenic tests for mice and rats. This difference in susceptibility can cause problems, as most toxicity tests are not conducted on the organisms of concern since this may not be suitable or available; rather, they are conducted on surrogate species. For example, rainbow trout *(Salmo gairdnerii)* and fathead minnow (*Pimephales promelas*) are commonly used to represent all freshwater fish.

The sex of an organism can also affect its tolerance to chemicals. This is largely related to the rate of metabolism and body composition. The rates of metabolism of chemicals between the sexes is often different. These differences in metabolism appear at puberty and are usually maintained throughout adult life. It is not possible

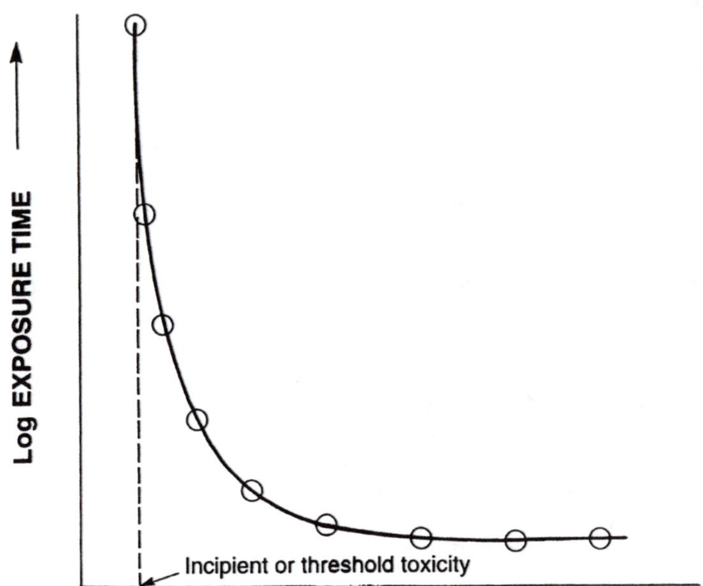

FIGURE 4.8. The variation in toxicant concentration to cause lethal toxicity with exposure time.

to generalize which sex metabolizes chemicals faster than the other. The effect of body composition on tolerance is often a result of the partitioning of compounds between the lipid and water phases of organisms. For example, females generally have more fat than males of the same mass, so they will have higher concentrations of fat-soluble compounds and lower concentrations of water-soluble compounds on a gram-per-kilogram (g kg^{-1}) basis.

There is a wide range of other factors that affect or modify the toxicity of chemicals. Some of these include:

1. Composition of diet
2. Age and general health of the test organisms
3. Hormone levels (e.g., pregnancy)
4. Experimental conditions under which the toxicity was determined

An important point concerning the stated toxicity of a chemical to an organism, is that it is not a constant. Under different conditions, the toxicity may change. The stated value is specific for that particular species, with that particular body composition based on sex, age, health, etc. under the stated experimental conditions for the given exposure period and route of exposure.

VIII. EXPERIMENTAL TESTING FOR TOXICITY — BIOASSAYS

A. TYPES OF BIOASSAYS

Bioassays are experiments, usually conducted in the laboratory, where toxic effects to specific toxicants are measured under controlled conditions. Apart from being able to vary the test organism, there are many other variables that can be modified to produce different bioassays. Two of the variables that can be modified are the duration of exposure to the toxicant and the constancy of toxicant concentration. Variation in the exposure period produces four main types of bioassays: **acute, subacute, multigenerational**. and **chronic**. *Acute* toxicity experiments are conducted for short periods of time, i.e., a maximum of 96 hours. In contrast, *chronic* tests are conducted so that the organism is exposed for a significant proportion of their normal expected lifetime. In between these two are *subacute* tests in which the exposure to the toxicant should not exceed 10% of the normal life expectancy of the species. Other bioassays that are designed to determine if toxicants affect reproduction or are teratogens must be conducted over several generations and are thus called *multigenerational* bioassays. The length of chronic, subacute and multigenerational bioassays vary extensively, depending on the test organism. For instance, a chronic bioassay for rats lasts for 6 months to 2 years, but only 28 days for shrimps; and a multigenerational (three generations) bioassay for cladocerans (e.g., *Daphnia magna*) takes 7 to 21 days, while for rodent species, it would take approximately 20 months.

Experiments with aquatic organisms are usually conducted in aquaria and the toxicant is introduced into water. The toxicant concentration is maintained using three types of experimental systems: **static, semistatic** and **flow-through**. In static experimental conditions, the test organisms are exposed initially to the toxicant at the required concentration. Thus, with time, the chemical is absorbed by the animal, adsorbed by the test vessel, biodegraded or volatilized, and the concentration in the ambient environment will decrease (see Figure 4.9). In semistatic bioassays, the toxicant is introduced at regular intervals throughout the bioassay by replacing the depleted aquarium water with water at the specified concentration. The toxicant concentration will therefore tend to oscillate due to the removal and replenishment of the toxicant (Figure 4.9). Flow-through tests are carried out where the toxicant is continuously introduced to the test animals with a corresponding outflow. Under such a scenario, the toxicant concentration should remain essentially constant (Figure 4.9).

The test system that most closely resembles the likely exposure regime for the toxicant and environment in question should be used. However, flow-through bioassays are generally viewed as the best form of bioassay since the test organisms will exhibit a toxic effect resulting from exposure to a constant known concentration. Such experiments are more reproducible and thus the results are more reliable.

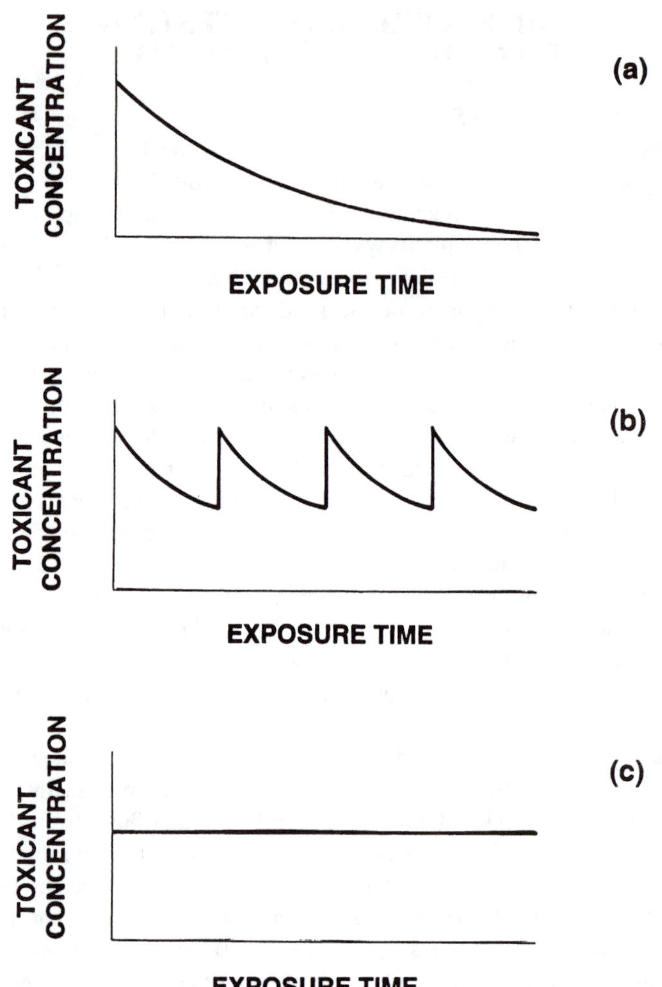

FIGURE 4.9. The change in toxicant concentration with time for different aquaria systems: (a) static, (b) semistatic, and (c) flow-through.

B. SELECTION OF TEST ORGANISMS AND THE ROUTE OF TOXICANT EXPOSURE

The choice of test organisms is a subject of constant debate amongs toxicologists. Some argue that it is always preferable to conduct toxicity tests using the species that are likely to be affected. Others argue that it is better to use standard laboratory animals, for which there is better knowledge of the variation in sensitivity and comparison with other research is easier.

In many cases, it is not possible to use species native to the area of interest, as they are not available in sufficient numbers or suitable for laboratory experiments. In such cases, surrogate animals must be used. The choice of surrogate test animals should almost solely be based on the degree of similarity between the test organism and the animal being modeled, in terms of the toxicity of chemicals, the rates and routes of absorption, biotransformation and excretion. However, in reality, the decision is usually based on other factors such as cost, ease of use, ease of maintenance in the laboratory and the availability of suitable facilities and expertise.

The route of exposure to a toxicant can affect the toxicity of a compound; thus, it is crucial to choose appropriate routes. The chosen route of exposure should be either the same as the normal or expected route of exposure for the species and chemical in question, or the route needed to test a specific hypothesis.

C. CONDUCTING A BIOASSAY

Prior to commencing a bioassay, all organisms to be used should be acclimatized to the proposed experimental conditions. The length of acclimation varies from a couple of hours to weeks, depending on the organism. Any animals that appear unhealthy should be removed and not used in subsequent bioassays. Generally, if greater than 10 to 20% of the organisms die during acclimation, then that batch is not suitable for experimental use. Control experiments are used in all bioassays. Controls are normally exactly the same as the test experiments without the toxicant or at toxicant concentration of zero. Normally, controls would be expected to exhibit zero toxic effects.

The next step is to conduct a range-finding bioassay to find the approximate toxicity. A second, more accurate bioassay, termed the *definitive bioassay,* is then conducted in which the test organisms are exposed to five or six concentrations determined from the range-finding bioassay. These toxicant concentrations are usually related geometrically, (i.e., ar, ar^2, ar^3, ar^4, etc., where ar is an appropriate constant, and is generally between 1.5 and 2). The number of test organisms exposed at each concentration and the extent of replication are decided according to the objectives.

Wherever possible, the concentration of the toxicant in the test chamber, should be measured at regular intervals throughout the bioassay or at least at the beginning and end. This is important as the measured concentration may be significantly different from the nominal value (i.e., the mass of toxicant added to the system divided by the ambient phase volume) or it may change over time. This is particularly relevant where highly lipophilic, volatile or degradable compounds are used in static or semistatic bioassays.

Depending on the type of test and the specific information being sought, a range of factors other than lethality can be examined. For instance, all animals on completion of the bioassay could have individual organs examined histologically and pathologically to ascertain the organs affected and the type and extent of cellular damage.

D. CALCULATING TOXICOLOGICAL DATA

The simplest way to estimate a measure of toxicity is to plot the logarithm of concentration or dose of the toxicant (x axis) against the biological response (y-axis). This can be easiest demonstrated by referring to Figure 4.10 where logarithm of toxicant concentration is plotted against the percentage of test organisms affected.

To estimate the *effective concentration* for a toxicological effect (EC_{50}) from this graph, a horizontal line is drawn from the point of 50% effect to where it intersects the toxicity curve (point A, Figure 4.10). From this intersection a vertical line is drawn down until it intersects the x-axis. This point, point B, is the concentration of toxicant that affects 50% of the test organisms exposed to the toxicant under the stated experimental conditions (i.e., EC_{50}). Lethality is a common endpoint and so the LC_{50} or LD_{50} can be estimated by this technique.

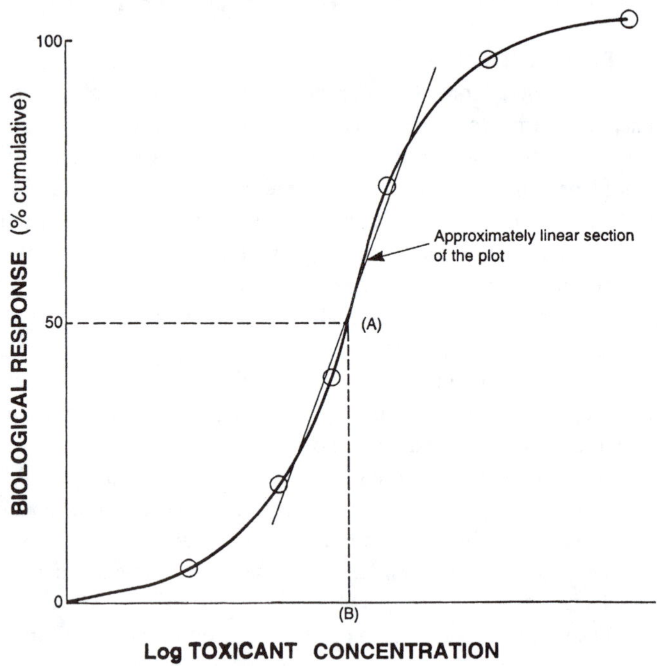

FIGURE 4.10. Plot of cumulative percent of organisms giving a response against the log [toxicant concentration].

Other information, besides an estimate of toxicity, can be derived from dose-response plots. The gradient of the linear portion, shown in Figure 4.10, indicates the potency of the toxicant or the sensitivity of the organism to the toxicant. For a toxicant with a steep gradient, small changes in the concentration of toxicant may cause quite large changes in the magnitude of the biological response. For toxicants with small gradients small changes in the concentration may cause very little change in the biological response. This information can be utilized in deriving safe exposure levels for toxicants.

IX. ALTERNATIVE, MORE HUMANE METHODS FOR TOXICITY ASSESSMENT

Over the last two decades there has been increasing pressure from the public and animal liberationists to cease all animal testing or at least to minimize the suffering involved and the number of test organisms used. As a result, many countries have passed legislation banning certain toxicological tests, such as the Draize test where chemicals are added to the eyes of animals. In other countries, government approval must be obtained for every lethal toxicity test using vertebrate animals, the rationale being that such animals can feel pain. Great progress has been made in the development of new more humane methodologies. For example, new range finding methods have been developed that use less organisms and yet provide sufficiently accurate information. These include the up and down, moving average and British Toxicological Society (BTS) methods (further information on these methods can be found in Brown, 1988).

Other advances have been the rapid increase in the development and use of:

1. *In vitro* bioassays where cell cultures are used to assess the toxicity of chemicals to particular organs
2. Bioassays using microorganisms or other simple organisms and relating these to the toxicity of higher animals
3. Relationships that attempt to predict the toxicity of compounds based on physicochemical properties and descriptors of the compounds. These are called quantitative structure-activity relationships (QSARs) or structure-activity relationships (SARs)

An example of a QSAR is shown in Figure 4.11, where the relationship between lethal toxicity, as log $1/LC_{50}$, and the log K_{OW} value is plotted for a range of chemicals. This relationship can be used to estimate the toxicity of other compounds of unknown toxicity.

The use and development of these more humane and less animal-intensive bioassay tests will increase in the future. Despite this, most toxicologists would say that these new methods cannot fully replace animal studies; rather, they are a complement.

X. KEY POINTS

1. All substances will exert toxic effects as long as the dose is high enough.

2. To exert a toxic effect, the toxicant must enter the body and be distributed to the site of action (target organ).

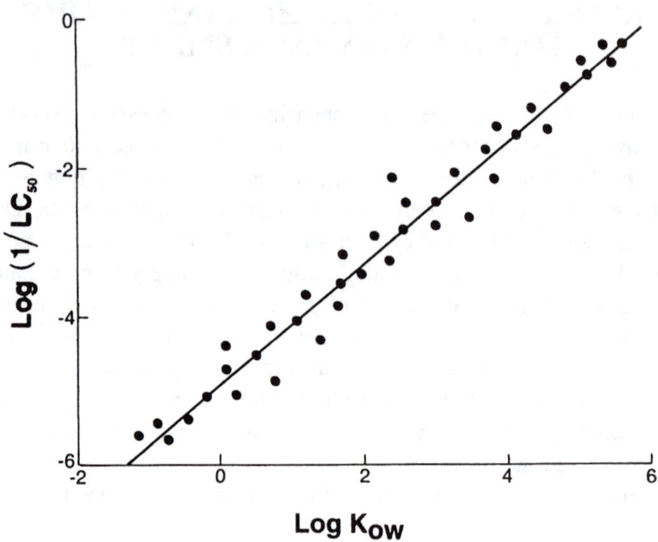

FIGURE 4.11. Plot of log K_{OW} against log $[1/LC_{50}]$ for fish for a range of chemicals.

3. The toxicity of a compound depends on several factors: (a) the rate of entry and excretion; (b) the physical chemical properties of the chemical; (c) duration of exposure; (d) the compounds susceptibility to biotransformation; and (e) the species, age, sex, and health of the test animal.

4. Biotransformation of lipophilic chemicals is generally due to induction of the mixed function oxidase (MFO) enzyme system in biota. Biotransformation generally increases the aqueous solubility and decreases the K_{OW} value and therefore decreases the time the chemical stays in the body.

5. Elimination of chemical is a critical measure in minimising toxic effects. Chemicals are eliminated from the body by urine, fecal and respiratory excretion.

6. Toxicity is measured by exposing animals or cells to chemicals. Numerous endpoints can be measured in toxicity tests, such as: death, fertility, movement, growth, photosynthesis, rate of activity and mutations.

7. The main measure of toxicity is the concentration that exerts a 50% effect on the test population, e.g., LC_{50}. Other measures include the No Observed Effect Level (NOEL) and Lowest Observed Effect Level (LOEL).

8. The methods used to determine the toxicity of compounds are undergoing radical change in an attempt to obtain the necessary information more humanely. QSARs can be used to estimate toxicity from physical chemical properties, such as the K_{OW} value.

REFERENCES

Amdur, M.O., Doull, J.D., and Klaassen, C.D., *Casarett and Doull's Toxicology.* 4th ed. Pergamon Press, New York, 1991, 1033.

Brown, V.K., *Acute and Sub-Acute Toxicology.* Edward Arnold, London, 1988, 125.

Salsburg, D.S., *Statistics for Toxicologists.* Marcel Dekker, New York, 1986, 196.

Zar, J.H., *Biostatistical Analysis.* 2nd ed. Prentice-Hall, Englewood Cliffs, NJ, 1984, 718.

CHAPTER 4 PROBLEMS

1. What is the scientific rationale for the 50% effect level being the most common measure of toxicity?

2. What is the difference between the data analysis methods used to determine the toxicity measures described as the LC_{50} and NOEL?

3. What is the key physicochemical factor property of toxicants that will affect their uptake across the skin and the stomach?

4. What properties of toxicants are modified by biotransformation reactions to facilitate their excretion, and how is this achieved?

5. A bioassay has been carried out on shrimp in aquaria and the following data attained after 24 hours:

No. of shrimp surviving (out of 200/aquarium)	Concentration in aquarium water (μg/L)
200	11.0
164	14.5
124	19.1
72	21.9
22	30.2
0	57.5

Plot this data out and graphically estimate the LC_{50} and LC_{80}.

6. Using the plot in Figure 4.11, estimate the 24-hour LC_{50} for fish for compound A with $K_{OW} = 3500$, and compound B with $K_{OW} = 135{,}000$.

Answers on Page 102.

CHAPTER 4 SOLUTIONS

1. This represents the level of toxic effect that occurs with the average organism, with 50% of the individuals being more tolerant and 50% being less tolerant. Also, the error in measurement is least at this level and therefore this estimate of toxicity is most accurate.

2. Measures such as the LC_{50} are determined using linear relationships or similar methods that are then used to calculate the concentration that causes the toxic effect. NOEL values are determined by evaluating if there is a significant difference between the toxicity of the control and different concentrations of the toxicant.

3. The key factor is lipid solubility, measured as the K_{OW} value. Toxicants in the stomach are subject to quite acidic conditions which, in accordance with the Henderson-Hasselbalch equation, can lead to quite large changes in the extent of ionization. As ionized compounds have lower K_{OW} values, the uptake will decrease.

4. The biotransformation reactions increase the polarity of the toxicants and hence the aqueous solubility. This is done by adding hydroxyl functional groups or combining with normal body metabolic products.

5. The data can be plotted out as in Figure 4.12, allowing the estimation of the log LC_{50} as 1.30 and the log LC_{80} as 1.43, thereby giving LC_{50} and LC_{80} values of 19.99 and 26.9 μg L^{-1}, respectively.

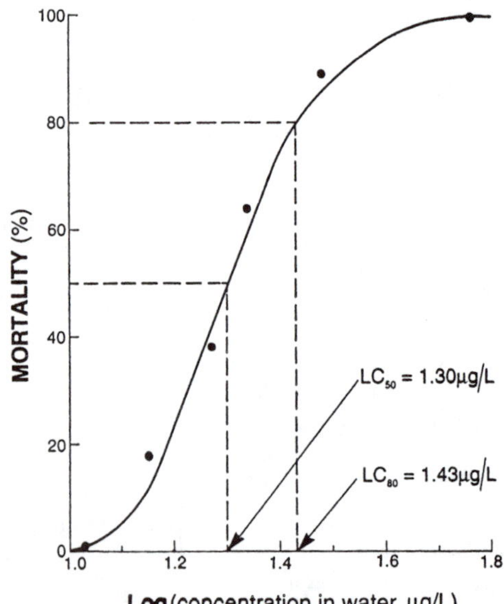

FIGURE 4.12. Plot of data on lethality to shrimps from Question 5.

6. Compound A: K_{OW}= 3500

    thus, log K_{OW}= 3.54

    Estimating from Figure 4.11,

    log $1/LC_{50}$= –2
    and $1/LC_{50}$= 0.01

    Compound A: LC_{50}= 100 mg L^{-1}

    Compound B: K_{OW}= 135,000

    thus, log K_{OW}= 5.13

    Estimating from Figure 4.11,

    log $1/LC_{50}$= –0.7
    and $1/LC_{50}$= 0.20

    Compound B: LC_{50}= 5.0 mg L^{-1}

Contaminants in the Environment

Chapter 5

PETROLEUM HYDROCARBONS

I. INTRODUCTION

Petroleum is a major fuel in contemporary society. For automobiles, cargo ships, planes, etc., petroleum fuels are the predominant energy source. Petroleum is also the source for a range of chemical products such as plastics, lubricating oils and bitumen. Crude and processed petroleum is transported on sea and land in large quantities. Discharges to the oceans are generally of the order of several million tons per year while discharges due to incomplete combustion are generally placed at tens of million tons per year. A qualitative indication of sources of petroleum to the environment is shown in Table 5.1.

The combustion of petroleum fuels is usually incomplete, and unchanged petroleum as well as partly combusted products are discharged to the atmospheric environment. This comprises the major discharges not due to the spectacular accidents resulting in spills, but rather the low concentrations present in municipal wastes or sewage and urban runoff. Sewage contains low concentrations due to the activities of the many urban industries that use petroleum. The petroleum in urban runoff is principally due to petroleum on road surfaces. Petroleum hydrocarbons are distributed throughout the world's oceans and the atmosphere. These substances are major global contaminants. On the other hand, petroleum from natural sources (see Table 5.1) has been entering the environment over long spans of geological time. This means that some of the contamination is natural and some is generated from human sources. There is a tendency for petroleum contamination to occur to the greatest extent adjacent to urban complexes.

II. CHEMICAL NATURE OF PETROLEUM

Liquid petroleum is a naturally occurring, oily and flammable liquid comprised principally of hydrocarbons. Unrefined liquid petroleum is generally termed "crude oil." The name **petroleum** literally means "rock oil" and gives some indication of its origin. Petroleum is commonly found in sedimentary rocks, where it is generated from the deposition of organic matter derived from microorganisms, plants and animals at the bottom of ancient water bodies. Burial and compaction occurs, destroying microorganisms and providing modest temperature (generally <200°C) and pressure. Over geological time spans of millions of years, these conditions convert the decayed organic matter into petroleum.

A. HYDROCARBONS

The chemical composition of crude oil or petroleum is quite complex and variable. This reflects the variety of organic matter from which it was originally formed, as well as the complex biochemical and chemical processes that occur during the transformation to petroleum. Typically, hydrocarbons (i.e., compounds composed

TABLE 5.1
Sources of Petroleum to the Environment

Source	Quantity
Marine environment	
Municipal waste (sewage)	+++
Urban run-off	++
Industrial waste	++
Accidents	++
Natural marine seeps	++
Terrestrial environment	
Incomplete combustion	+++
Evaporation	++
Industrial operations	++

of carbon and hydrogen only) constitute 50 to 90% of petroleum. The types of hydrocarbons found are *n*-alkanes, branched alkanes, cycloalkanes and aromatics. The *n*-alkanes are also known as normal paraffins, and are found in all crude oils. Depending on the sample, the *n*-alkanes range in size (and boiling point) from C_1 (methane) up to C_{40} or more.

The branched alkane content of petroleum is at least as great as the *n*-alkane content, and usually larger. The branches are typically methyl (CH_3) groups. This branching inhibits microbial transformation. The mass percentage of normal and branched alkanes in particular samples of petroleum from southern Louisiana and Kuwait are shown in Table 5.2. Structures of typical alkanes found in petroleum are shown in Figure 5.1.

Cycloalkanes (or naphthenes, as they are also known) usually comprise the largest group of hydrocarbons by mass in petroleum. The cycloalkanes most abundant in petroleum contain 5 or 6 carbon atoms in the ring, i.e., cyclopentanes and cyclohexanes. Compounds with 7- and 8-membered rings also occur in small amounts. This group of cycloalkanes also includes alkyl-substituted compounds, e.g., methyl cyclopentane (see Figure 5.1). Aromatic hydrocarbons are usually present to the extent of 20% by mass or less. As well as benzene and alkyl-substituted benzenes, fused ring polycyclic aromatic hydrocarbons (e.g., benzo[a]pyrene) are characteristic components.

Benzo [a] pyrene

Some of these are carcinogenic and, as a group, the aromatics are probably the group that is of greatest environmental significance of the different sorts of hydrocarbons present. Alkyl derivatives of aromatic hydrocarbons are also present in petroleum. It is important to note that crude oil or petroleum does not usually naturally contain alkenes or other similar unsaturated hydrocarbons. These occur in

TABLE 5.2
Mass Percentage of Chemical Components in Particular Samples of Petroleum from Southern Louisiana and Kuwait

Chemical Component	Southern Louisiana Crude	Kuwait Crude
Alkanes (normal and branched)	28.0	34.1
Cycloalkanes[a]	44.8	20.3
Aromatics[a]	18.6	24.2
Polar (O,N,S-containing) components	8.4	17.9
High boiling, high molecular weight material	0.2	3.5

[a] Includes compounds with alkyl side chains.

TABLE 5.3
Overall Fate of Nonrecovered Petroleum from the Exxon Valdez Approximately 3 Years after the Spill, Expressed As a Mass Percentage

Fate	Mass Percentage of nonrecovered spilled petroleum
Associated with subtidal sediments	15
Associated with intertidal strata	2
Evaporation and subsequent photolysis	24
Photolysis and microbial transformation in water	59

[a] Estimated mass of spill is 3.55×10^4 tonnes.

refined petroleum products, often as a result of catalytic cracking processes in refineries. These processes change the size and, in some instances, the hydrogen: carbon ratio of molecules, with the aim of producing a higher yield of profitable transportation fuels (gasoline, jet fuel and diesel).

B. NON-HYDROCARBONS

Sulfur occurs in petroleum in various forms, some chemical structures of which are illustrated in Figure 5.2. Some petroleum contains elemental sulfur, and sulfur can also occur as hydrogen sulfide (H_2S) and carbonyl sulfide (COS). Sulfur is also present in a wide range of hydrocarbons, largely in the form of mercaptans (or thiols), sulfides and thiophene derivatives.

Mercaptan R–SH

Sulfide R–S–R′

Thiophene

S

ALKANES

Methane	CH_4
n - Hexane (C_6H_{14})	$CH_3(CH_2)_4CH_3$
n - Triacontane ($C_{30}H_{62}$)	$CH_3(CH_2)_{28}CH_3$
2 - Methylpentane (C_6H_{14})	$CH_3CH(CH_3)(CH_2)_2CH_3$
Phytane ($C_{20}H_{42}$)	$CH_3CH(CH_3)(CH_2)_3CH(CH_3)(CH_2)_3CH(CH_3)(CH_2)_3CH(CH_3)CH_2CH_3$

CYCLOALKANES

Cyclopentane (C_5H_{10})

Cyclohexane (C_6H_{12})

Methylcyclopentane (C_6H_{12}) — CH_3

AROMATICS

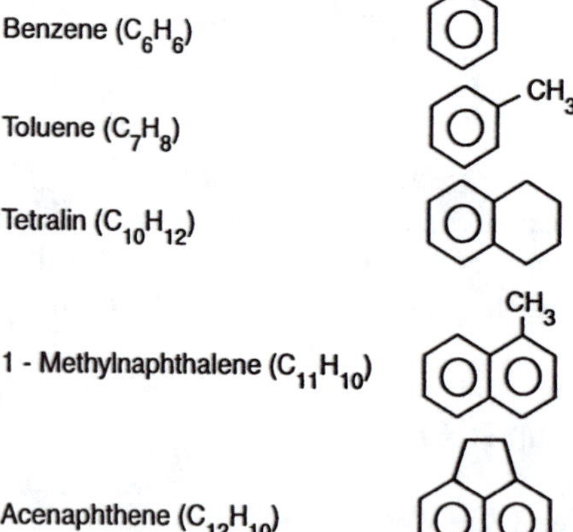

FIGURE 5.1 Structures of typical hydrocarbons found in petroleum samples.

SULPHUR CONTAINING COMPOUNDS

Ethanethiol (C_2H_6S) $CH_3CH_2{-}SH$

Dimethylsulphide (C_2H_6S) $CH_3{-}S{-}CH_3$

Thiacyclohexane ($C_5H_{10}S$)

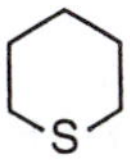

Dibenzothiophene ($C_{12}H_8S$)

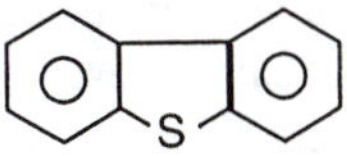

NITROGEN CONTAINING COMPOUNDS

2-Methylpyridine (C_6H_7N)

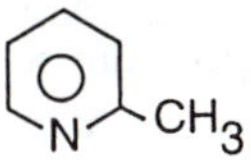

Quinoline (C_9H_7N)

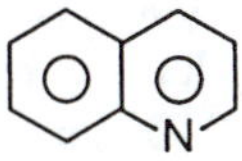

Indole (C_8H_7N)

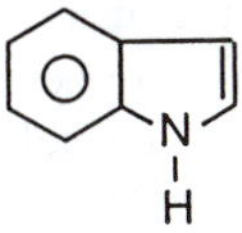

OXYGEN CONTAINING COMPOUNDS

Stearic acid ($C_{18}H_{36}O_2$) $CH_3(CH_2)_{16}{-}C(=O){-}OH$

Cyclopentanecarboxylic acid ($C_6H_{10}O_2$)

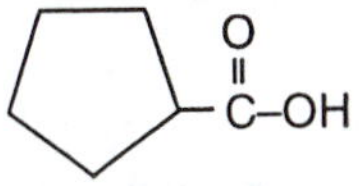

Phenol (C_6H_6O)

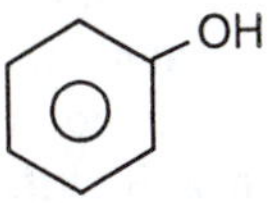

FIGURE 5.2 Some sulfur, nitrogen- and oxygen-containing compounds typically found in petroleum samples.

Sulfur compounds tend to be concentrated in the higher boiling fractions of petroleum. Sulfur-containing compounds represent unwanted impurities in petroleum since they are generally corrosive to metals, malodorous and may poison various catalysts used in refineries. In addition, on combustion, the sulfur present in many compounds is converted to sulfur dioxide (SO_2), an unwanted air pollutant. As the demand for petroleum increases, reserves of high-quality, low-sulfur material diminishes.

Nitrogen-containing compounds are present in most petroleum, and like the sulfur-containing compounds, tend to be concentrated in the high-boiling fractions. The presence of oxygen-containing compounds in most petroleum is relatively minor. The oxygen is usually present in the form of various functional groups, particularly carboxylic acids. Typical examples of nitrogen- and oxygen-containing compounds found in petroleum are shown in Figure 5.2.

Most samples of petroleum are dark in appearance, the color ranging from black to yellow-brown. The origin of this color is relatively large particles of a complex nature that are colloidally dispersed in the petroleum hydrocarbons. As well as having a high sulfur, nitrogen and oxygen contents, these particles contain metals, particularly vanadium, nickel, cobalt and iron. With few exceptions, the chemical form in which these metals occur is unknown. Some exist as water-soluble salts (e.g., NaCl) enclosed within the colloidal particles. Despite probable contact between the petroleum and water underground for extended periods of time, extraction of these salts into water has not occurred. Petroleum from Venezuela in particular, and also Canada, contains relatively high levels of vanadium (up to 1000 ppm) in the form of vanadyl cations (VO^{2+}) complexed to a porphyrin molecule that acts as a tetradentate group. On high-temperature combustion, the vanadium is likely to be discharged in the form of vanadium pentoxide (V_2O_5) and associated with particulate material in the atmosphere or any solid residue.

II. PETROLEUM HYDROCARBONS IN THE ENVIRONMENT

A. DISPERSION

Petroleum hydrocarbons are emitted or discharged into all phases of the environment: air, water and soil. Most petroleum has a density less than 1 g cm^{-3}, and so discharged material will initially float on the surface of marine waters as a slick. Typical seawater at 298°K (25°C) has a density of 1.03 g cm^{-3}. Dispersion or spreading is probably the most significant process affecting petroleum hydrocarbon slicks in the first 6 to 10 hours following discharge. Factors favoring an increase in the area of a slick include gravity, which promotes a decreasing thickness of oil. Frictional forces, among others, tend to oppose this spreading. Eventually, most slicks form a layer about 0.1-mm thick or less on the water. Drift is a large-scale movement phenomenon, and is a measure of the movement of the center of mass of an oil slick. Drift is governed by wind direction and speed, together with wave action and currents. If drift is likely to result in landfall, coral and mangrove communities, for example, can be impacted and intertidal benthic strata fouled. If, on the other hand, the slick is heading for the open sea, then environmental effects are much less.

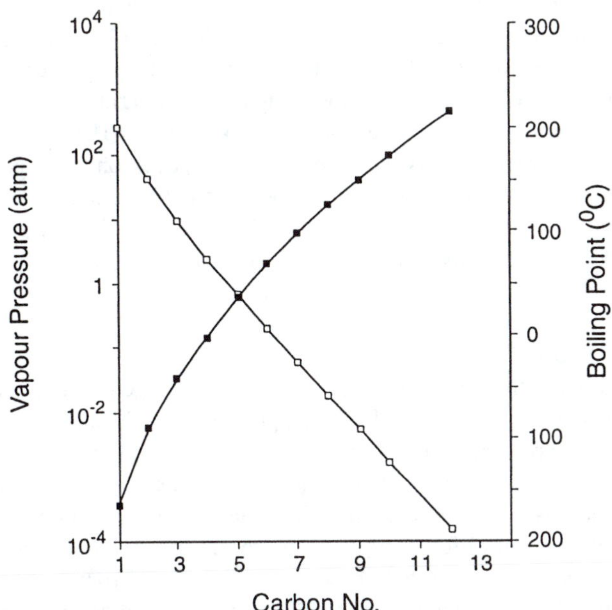

FIGURE 5.3 A plot of pure component vapor pressure (atm) and atmospheric pressure boiling points (°C) versus carbon number for a series of *n*-alkanes.

B. EVAPORATION

Evaporation is the process in which molecules escape from the surface of a liquid into the vapor phase or atmosphere. This can be a major fate of material on the surface of marine waters, particularly in the first few days following spillage.

Evaporation removes the lower molecular weight, more volatile components of the petroleum mixture. The rate of evaporation depends on the concentration of the compound in the slick and its vapor pressure. Other factors include the prevailing air and sea temperatures, slick surface area and wind velocity. The vapor pressures and boiling points of *n*-alkanes as related to carbon number are plotted in Figure 5.3. For those *n*-alkanes that are gaseous at room temperature, their vapor pressures at that temperature are greater than 1 atm (or 101.625 kPa). As the molecules become larger, their vapor pressures decrease and boiling points increase. Increasing boiling points reflect stronger intermolecular forces in the liquid phase that must be overcome in order to vaporize. Since the *n*-alkanes are all nonpolar, the intermolecular forces must be dispersion or London forces. The overall strength of these forces/attractions is related to molecular size. It is not surprising then that as the size of the *n*-alkanes increases in a regular manner (by increments of CH_2 groups), so does the boiling point.

Hydrocarbons with vapor pressures equal to that of *n*-octane (0.019 atm at 298 K) or greater will be lost quickly via evaporation. Pentadecane ($C_{15}H_{32}$) is the smallest *n*-alkane commonly found in weathered petroleum. The volatile fraction that may be lost makes up between 20 and 50% of most crude oils. For refined petroleum products such as gasoline and kerosene, this percentage can be even higher. Table 5.3 contains details of the fate of unrecovered petroleum from the Exxon Valdez, which grounded in Prince William Sound in March 1989. It can be seen that a significant amount of petroleum was subject to evaporative loss.

As evaporation proceeds, and the lower molecular weight, more volatile components are removed, the properties of the slick change. The volume becomes smaller, and viscosity and density increase. Eventually, over long time periods (>100 days), the petroleum may become sufficiently viscous that diffusion of potentially volatile molecules to the surface of the slick becomes rate limiting. Once in the atmosphere, the hydrocarbons can exist as aerosols that may be redeposited, or in the vapor phase, where degradation may occur.

C. DISSOLUTION

As well as being the most volatile, lower molecular weight hydrocarbons also tend to be the most water soluble. For example, lower molecular weight aromatic hydrocarbons are more water soluble than larger ones. Similarly, smaller *n*-alkanes have a greater aqueous solubility than larger ones. Generally, for lower molecular weight hydrocarbons, evaporation is more important than dissolution. In comparing the aqueous solubility of different types of hydrocarbons, we should compare compounds of a similar size. Table 5.4 shows that, among C_6 hydrocarbons, the order of aqueous solubility is aromatic > cycloalkane > branched alkane > n-alkane. In fact, benzene is some 200 times more soluble than n-hexane! There are several reasons for these differences in the closely related compounds and some of these are discussed in Chapter 3. The size of the surface of the molecule and the dipole moment are important factors involved.

TABLE 5.4
Aqueous Solubilities of Various C_6 Hydrocarbons

Compound	Structure	Aqueous Solubility ($g\ m^3$)	Aqueous Solubility ($mole\ m^{-3}$)
n-Hexane	$CH_3(CH_2)_4CH_3$	9.5	0.11
2-Methylpentane	$CH_3CH(CH_3)(CH_2)_2CH_3$ (drawn as $CH_3CH(CH_2)_3CH_3$ with CH_3 branch)	13.8	0.16
Cyclohexane	[hexagon]	55	0.65
Benzene	[hexagon with inner circle]	1780	22.8

Aqueous solubility values are usually for solution in pure water. Seawater contains dissolved ions and is a phase that hydrocarbons have even less affinity for than water. As a general rule of thumb, hydrocarbons are about 75% as soluble in seawater as pure water. Molecules containing N, S or O atoms are generally more polar than hydrocarbons. Such molecules are likely to be among the more soluble components of slicks, particularly if the molecules are small.

D. EMULSION FORMATION

Emulsions are a colloidal state where fine droplets of one liquid are dispersed in another. With petroleum and water, two kinds of emulsion are possible: an oil-in-water (O/W) emulsion and a water-in-oil (W/O) emulsion. With an O/W emulsion, fine droplets or globules of petroleum (typically 0.1 to 1 μm in diameter) are dispersed in water. These fine droplets can have, in total, a relatively large surface area. For example, 1 cm^3 oil in the form of a spherical droplet has a surface area of about $4.8 \times 10^{-4} m^2$. However, when this volume is in the form of fine spherical droplets of diameter, say 1 μm, there are more than 1×10^{12} droplets, with a total surface area of 6 m^2. Because of the large interfacial area, rates of processes such as microbial transformation and photodegradation are often maximal.

These O/W emulsions are often inherently unstable and tend to coalesce back to a slick. However, they can be stabilized by the presence of emulsifiers, particulates (e.g., clay) and continuous agitation. Emulsifiers are compounds with both hydrophobic and hydrophilic components. They coat the surface of the nonpolar or weakly polar oil droplets, with their polar hydrophilic sections orientated outwards toward the water. This helps keep the droplets dispersed. Natural emulsifiers present in petroleum include small N-, S-, and O-containing compounds with polar, hydrophilic side chains. In some circumstances, surfactants such as detergents or dispersants are deliberately added to slicks to act as emulsifiers and enhance O/W emulsion formation. The action of surfactants is described in more detail in Chapter 10, Soaps and Detergents.

In the petroleum/water system, the other sort of emulsion that can be formed is the water-in-oil (W/O) emulsion. Here, water is the dispersed phase, forming fine droplets within the oil, which is the dispersing medium or continuous phase. The mechanical agitation provided by wind and wave action promotes formation of W/O emulsions (as well as O/W emulsions). The ease of formation and stability of W/O emulsions depend primarily on the composition of the petroleum. In particular, the presence of asphaltene (a high molecular weight, high boiling material) and natural surfactants have been shown to favor the formation of W/O emulsions.

The water content of W/O emulsions is variable. With 30 to 50% water, they are fluid and resemble the parent petroleum. As the water content increases, the viscosity increases, until for material with >80% water, the viscosity and color are similar to chocolate mousse and it is referred to as a mousse. These emulsions have a volume greater than the petroleum component alone, and apart from viscosity have other modified physical properties, e.g., adhesive characteristics. A mousse is more likely to adhere to solid material, e.g., sand on a beach, than the original oil, but less likely to penetrate it.

Within W/O emulsions and particularly mousses, microbial transformation of petroleum components is usually reduced significantly. This is because diffusion of oxygen and nutrients to the aqueous droplets within the emulsion is minimal.

E. PHOTOOXIDATION

Of the processes affecting petroleum hydrocarbons we have considered thus far, none have actually chemically transformed them, or removed them from the environment as a whole. In the presence of oxygen, sunlight however can transform or degrade components of petroleum mixtures. **Photooxidation** is oxidation initiated by light, particularly solar UV light. On exposure to this light, a number of physical and chemical changes occur in petroleum. Compounds with oxygen-containing functional groups, e.g., alcohols, ketones, aldehydes, carboxylic acids and peroxides, are usually formed. Petroleum invariably contains compounds with benzylic methylene groups. These are CH_2 groups that are adjacent to aromatic rings, as shown below.

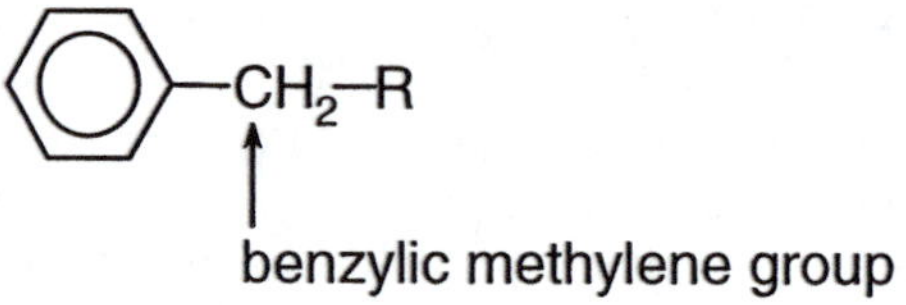

CH_2—R

benzylic methylene group

Benzylic methylene groups that are part of a ring are particularly prone to hydroperoxide formation. This can occur with common petroleum components such as tetrahydronaphthalene (or tetralin) and indan as shown below.

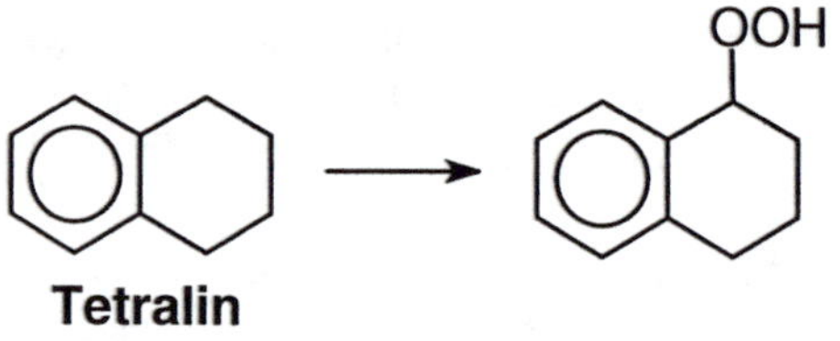

OOH

Tetralin

Hydroperoxides are more water soluble than the original parent hydrocarbon and in addition are relatively toxic to a variety of marine organisms. Whatever the mechanism of their formation, the O–O bond of hydroperoxides is quite weak. It can be split homolytically, thermally or by UV light as shown below.

These radicals can attack other hydrocarbons, ultimately forming a variety of oxygenated derivatives with increased polarity due to the presence of oxygen leading to increased aqueous solubility. The turbidity and apparent color of petroleum often increases on exposure to sunlight. This is due to the accompanying formation of polar, oxidized higher molecular weight material formed via radical recombination or reactions between various oxygen-containing compounds.

F. MICROBIAL TRANSFORMATION

As shown in Table 5.2, a significant fate for spilled Exxon Valdez oil was aqueous microbial transformation. In principle, the majority of the components of petroleum are able to be transformed by a variety of microorganisms. Whether or not this occurs in practice in the marine environment depends on factors such as the number and type of microorganisms present as well as the availability of oxygen and nutrients. The term "**bioremediation**" is now widely used and, in the context of petroleum spills, refers to the use of microorganisms to degrade the hydrocarbons to CO_2 and H_2O. Petroleum is a natural and long-standing part of our environment and it should come as no surprise that hydrocarbon-oxidizing bacteria, fungi, algae and yeasts exist.

The microbial transformation of hydrocarbons occurs either within the aqueous phase on dissolved or dispersed hydrocarbon molecules or at the hydrocarbon/water interface. Microorganisms do not exist within the hydrocarbon phase. Generally, *n*-alkanes, particularly smaller ones, are most rapidly transformed by microorganisms under aerobic conditions. In anoxic environments (e.g., sediments) saturated hydrocarbons are effectively resistant to microbial transformation.

The first step in the bacterial transformation of an *n*-alkane, such as hexane, is the insertion of an oxygen atom into the molecule, usually on a terminal carbon atom. Because only one of the two oxygen atoms of molecular oxygen is inserted into the hydrocarbon, the enzyme that catalyzes this process is known as a monooxygenase. Also necessary is NADH, which acts as a source of electrons and reducing power. NADH is a hydrogen carrier in the overall respiration process in which organisms transform and degrade organic compounds. In general, for n-alkanes,

$$R{-}CH_3 + O_2 + NADH + H^+ \rightarrow R{-}CH_2OH + NAD^+ + H_2O$$

Figure 5.4 shows a likely pathway for the bacterial oxidation of n-hexane. Eventually, n-hexanoic acid is formed, which can then undergo β-oxidation, which is a normal degradation process for fatty acids derived from triglycerides (or fats). Essentially, β-oxidation involves the successive cleavage of two carbon fragments from the main chain. The two carbon fragments are in the form of acetyl-CoA, which can be fed into the Krebs (tricarboxylic acid) cycle of respiration and used for energy. The carbon ultimately appears as CO_2.

Branch alkanes and cyclic alkanes tend to be less readily oxidized by microorganisms. For some branched hydrocarbons, the reason can be seen in Figure 5.4. If an alkyl group is attached to the β-carbon of hexanoic acid, a ketone cannot be formed at this position later in the sequence. If branched hydrocarbons such as these are to be transformed by microorganisms, other pathways that are probably relatively slow must be followed. The extent of microbial transformation of discharged or spilt petroleum, and hence its possible age, is often inferred from the ratio of a relatively degradable component such as an alkane and a more resistant component such as a branched alkane, e.g., pristane and phytane.

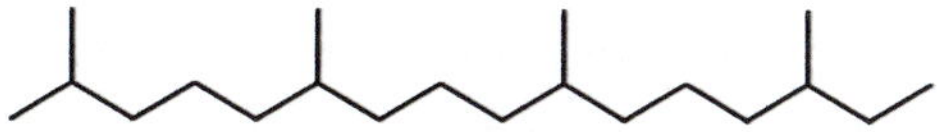

Phytane ($C_{20}H_{42}$)

Eventually, however, even compounds such as pristane and phytane are biotransformed, and even more resistant petroleum components are used to quantify the extent of microbial transformation.

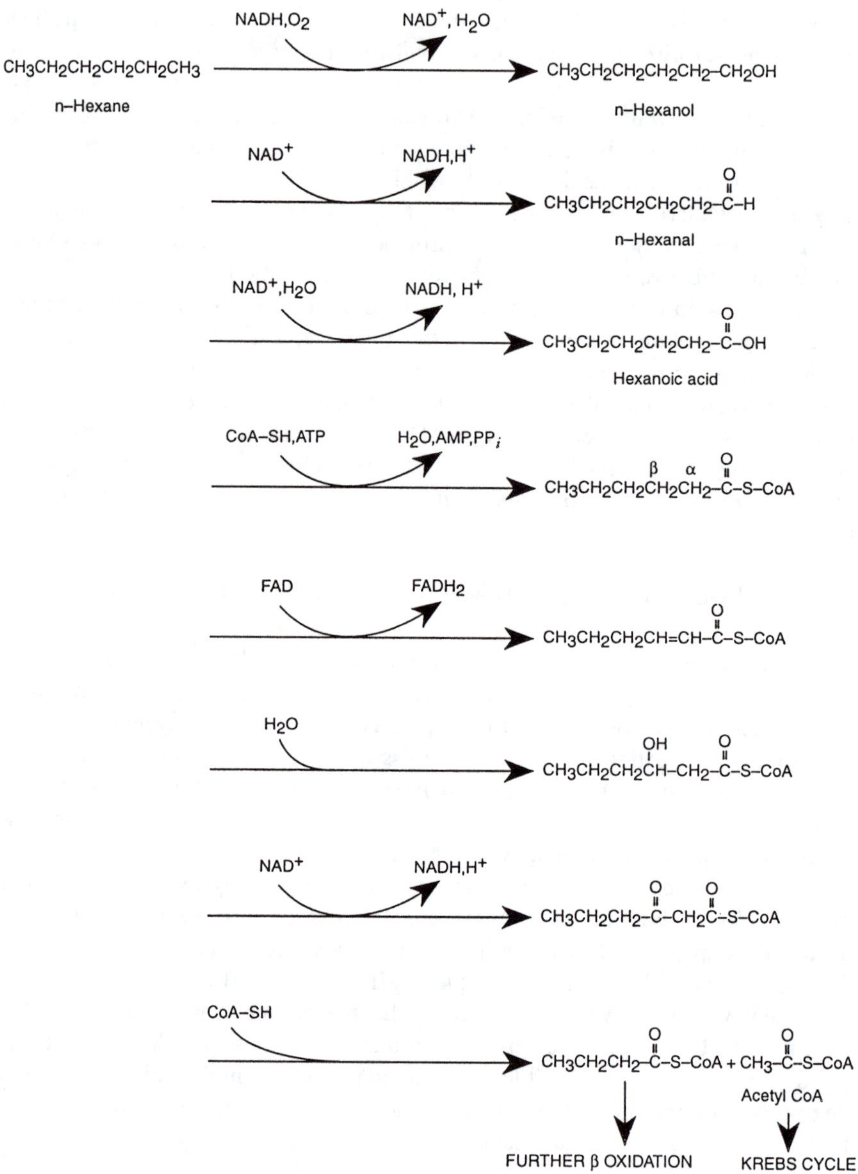

FIGURE 5.4 A pathway for the microbial transformation of an *n*-alkane.

Aromatic compounds can also be transformed by microorganisms (under both aerobic and anoxic conditions). As an example, a pathway for the aerobic transformation of benzene is shown in Figure 5.5. Each step is catalyzed by a particular enzyme or group of enzymes. The first step involves a dioxygenase-type enzyme inserting both oxygen atoms from molecular oxygen into the benzene molecule, subsequently forming in the next step, a *cis*-diol. Although there are exceptions,

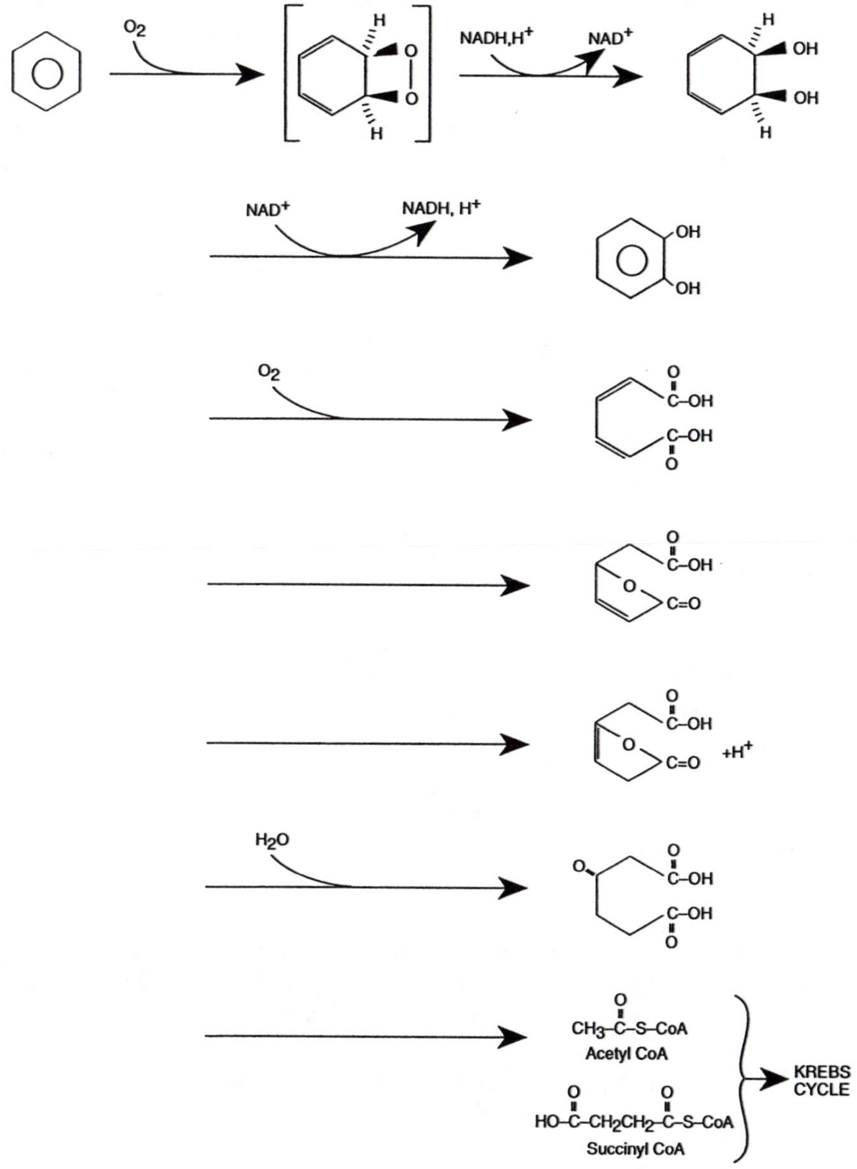

FIGURE 5.5 A pathway for the microbial transformation of benzene.

bacteria and higher organisms transform aromatic hydrocarbons by different mechanisms. Bacteria tend to produce *cis*-diols and use the hydrocarbon as a carbon source and for energy. From Figure 5.5, the products of bacterial transformation of benzene, i.e., acety- coenzyme A (acetyl-CoA) and succinyl-CoA are both components of the Krebs cycle.

Higher organisms (including humans) transform aromatic hydrocarbons initially using a monooxygenase enzyme, subsequently forming *trans*-diols. Higher organisms excrete the *trans*-diols or their derivatives as the original aromatic hydrocarbon is not used as a carbon source. These processes are illustrated below.

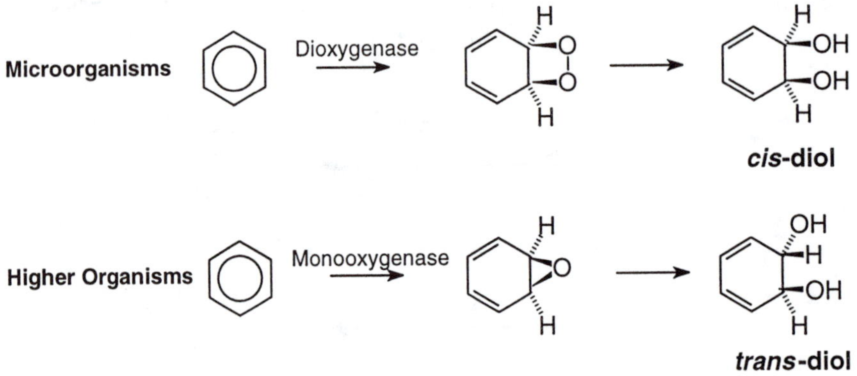

Microorganisms → Dioxygenase → *cis*-diol

Higher Organisms → Monooxygenase → *trans*-diol

Higher molecular weight aromatic compounds may also be transformed. For example,

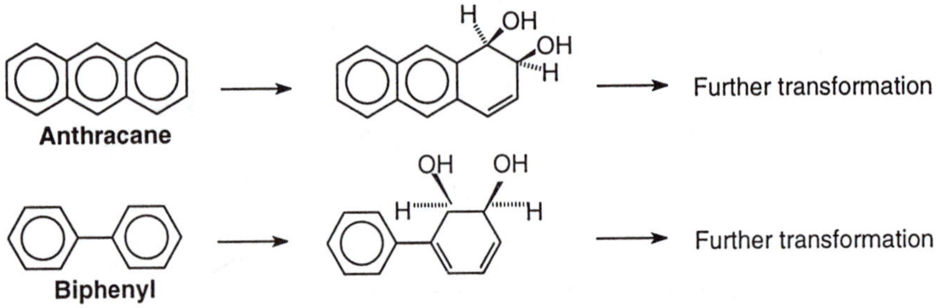

Anthracane → → Further transformation

Biphenyl → → Further transformation

It must be reemphasized though that for the processes mentioned to actually occur at a reasonable rate, microorganisms, in particular bacteria, need to be present in sufficient numbers. This is not always the case in marine systems.

III. PETROLEUM AND AQUATIC ORGANISMS

The toxicity of petroleum varies according to its composition. As a chemical group, aromatic hydrocarbons are generally more toxic than other types of hydrocarbons. A summation of toxicity of the water-soluble aromatic fraction from various petroleum products is shown in Table 5.5. As a general rule, larvae and juveniles of all species are more sensitive than adult life forms. Levels of 0.1 to 10 mg L^{-1} have been found to exert lethal effects on these groups. The adults are less sensitive, but lethal toxicity occurs generally in the range 0.1 to 100 mg L^{-1}.

If sublethal levels of petroleum components are present in water, the most prevalent effects with marine fish species are behavioral changes such as swimming performance and ability. Physiological effects include changes in heart rate, respiration rate and growth. Past chronic exposure to low levels of hydrocarbons can be

TABLE 5.5
Toxicity of Water-Soluble Aromatic Compounds from Petroleum to Marine Organisms

Organism class	Range of LC_{50} values ($mg\ L^{-1}$)
Fin fish	1–50
Crustaceans	0.1–10
Bivalves	1–50
Gastropods	1–100
Larvae (all species)	0.1–5
Juveniles (all species)	1–10

detected by elevated levels of enzymes responsible for metabolism, such as the MFO enzyme system.

Benthic intertidal communities are particularly vulnerable to physical coating with oil and subsequent adverse effects. Floating oil contacts the shoreline in the intertidal zone, i.e., the area between the high and low water marks. Resident organisms such as plants and small invertebrates are relatively immobile and cannot escape. Much public attention regarding oil spills is focused on the plight of seabirds. If hydrocarbons contaminate the plumage of seabirds, the trapped air providing thermal insulation and buoyancy is lost. Death can result from a number of causes. Loss of buoyancy can lead to drowning, or pneumonia can be caused by loss of thermal insulation. Starvation may occur due to an increased metabolic rate compensating for loss of body heat and also decreased feeding activity. Alternatively, toxic levels of petroleum components may be ingested as a result of excessive preening. Seabirds reproduce relatively slowly and so losses are difficult to replace.

IV. KEY POINTS

1. Petroleum is transported and used in large quantities and is a major contaminant in the environment. The major sources are the incomplete combustion of fuels in the atmospheric environment and sewage and urban runoff in the marine environment.

2. The chemical components of petroleum and derived products are hydrocarbons, including *n*-alkanes, cycloalkanes and aromatic compounds, ranging in physical state from gases to various liquids. In addition sulfur-, nitrogen- and oxygen-containing compounds are present.

3. Petroleum spilled into water bodies is acted on by the following physical processes: dispersion, evaporation, dissolution and emulsion formation. These processes are governed by the physical chemical properties of the compounds present, including aqueous solubility and vapor pressure.

4. Chemical transformation and degradation processes act on petroleum compounds in the environment. Photooxidation and microbial transformation are of particular importance.

5. Straight-chain compounds, such as n-alkanes, are more susceptible to microbial transformation by oxidation than branched-chain compounds. This is due to the oxidation process requiring the formation of a ketone on the β-carbon atom in the alkanoic acid initially formed by oxidation. A ketone cannot be formed if an alkyl group is attached at this position.

6. Aromatic hydrocarbons are oxidized by biota to form *cis*- and *trans*-diols as the initial product of reaction with oxygen.

7. The aromatic components of petroleum are the most toxic compounds present. These substances are usually lethal to larvae and juveniles in the range 0.1 to 10 mg L^{-1} and adults in the range 0.1 to 100 mg L^{-}.

REFERENCES

Anonymous, Sources, Effects and Sinks of Hydrocarbons in the Aquatic Environment. *Proceedings of the Symposium, American University,* Washington, D.C. 9–11 August, 1976. American Institute of Biological Sciences.

Jordan, R.E. and Payne, J.R., *Fate and Weathering of Petroleum Spills in the Marine Environment.* Ann Arbor Science, Ann Arbor, MI, 1980.

Payne, J.R. and Phillips, C.R., Photochemistry of petroleum in water. *Environ. Sci. Technol.*, 19, 569–579, 1985.

CHAPTER 5 PROBLEMS

1. A spill of a specialized petroleum product consisting of equal amounts of n-butane, n- hexane, n-octadecane and benzene has occurred in a marine area subject to moderate wave action. Using information in the text, evaluate the following:

   (a) Possible toxic effects in the marine system.
   (b) The possible composition of the product after 1 month in the marine area.

2. A common constituent of many petroleum products is indan with the structure:

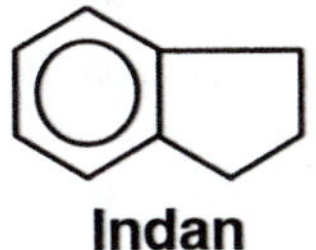

Indan

Would you expect this compound to be susceptible to photooxidation and, if so, what would the initial chemical product be?

CHAPTER 5 SOLUTIONS

1. (a) The only aromatic compound present is benzene. Table 5.4 indicates its water solubility as 22.8 mole m^{-3} or 22.8 × 78 (MW) = 1780 g m^{-3} or 1.78 g L^{-1} or 1780 mg L^{-1}. The LC_{50} values for adults and juveniles range from 0.1 to 100 mg L^{-1}. With moderate wave action the benzene in water concentration may initially approach its aqueous solubility. This greatly exceeds the toxicity range, so mortalities of marine organisms would be expected.

   (b) The vapor pressures of n-butane and n-hexane are >0.019 atm (see Figure 5.3) and so would be expected to evaporate rapidly and disappear over 1 month. On the other hand, n-octadecane has a vapor pressure considerably lower than 0.019 atm, as indicated by Figure 5.3 and would persist over 1 month. The benzene is water soluble and would be expected to have a vapor pressure in the range of n-hexane (i.e., >0.019 atm); with these two properties, benzene would be expected to disappear within 1 month. Thus, after 1 month, only n-octadecane would remain.

2. Indan contains a benzene ring and therefore will absorb UV radiation from sunlight, which will activate the molecule. In the activated state, the methylene groups in the α-position from the ring are susceptible to the addition of oxygen. Thus,

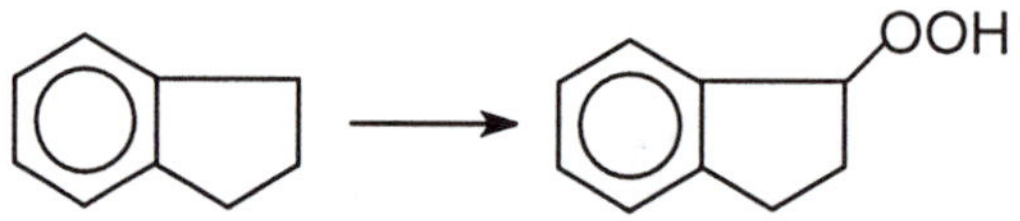

OOH

So, an *indan hydroperoxide* would be the initial product of oxidation.

Chapter 6

POLYCHLORINATED BIPHENYLS (PCBs) AND DIOXINS

I. INTRODUCTION

The polychlorinated biphenyls (PCBs) and dioxins are important environmental pollutants that have a different history of environmental contamination. These substances have similar chemical structures and share a wide range of chemical, biological and environmental properties. It is helpful to consider them together for these reasons.

The manufacture of PCBs has been carried out in many countries and the compounds marketed under trade names such as Aroclor, Chlophen, Kanechlor and Fenclor. The applications of these substances have been quite diverse, ranging from use as plasticizers to components of printer's ink. In the 1960s and 1970s, many investigations found that PCBs occurred widely in the environment and adverse biological effects were occurring. As a result, voluntary restrictions were introduced by many industries, and many governments have acted to curb the use of these substances. The PCBs were first produced commercially in the U.S. in 1929, and approximately 600,000 tons were produced up until 1977 when manufacture in North America was halted.

Industrially, PCBs are produced by chlorination of biphenyl with molecular chlorine, using iron turnings or ferric chloride as a catalyst and giving a reaction that can be described by

Biphenyl + Cl_2 —(Fe or $FeCl_3$)→ Cl–biphenyl–Cl + HCl

As reaction time increases, more of the 10 hydrogens on biphenyl are replaced by chlorine. A mixture of different individual PCBs is invariably produced however, and it is as mixtures that PCBs have been used commercially, and are encountered in the environment.

The first dioxin was synthesized in the laboratory in 1872, but unlike the PCBs, the dioxins or PCDDs, which stands for **polychlorinated dibenzo [1,4] dioxins**, have no known uses, and have never been deliberately manufactured, except on a laboratory scale. Dioxins are generally formed as unwanted by-products in chemical processes involving chlorine (e.g., 2,4-D and 2,4,5-T manufacture) and combustion processes. This family of compounds has come to public attention because of the relatively high toxicity of certain members, such as **2,3,7,8-tetrachlorodibenzo [1,4] dioxin (2,3,7,8-TCCD or TCDD)**, to some test organisms. The structure of TCDD is:

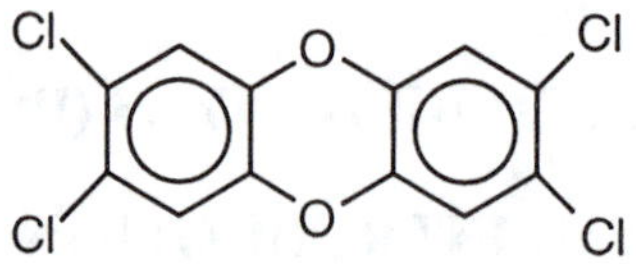

2,3,7,8-Tetrachlorodibenzo [1,4] dioxin

Like the PCBs, dioxins are usually found in the environment as mixtures. They are widely distributed environmental contaminants found, albeit in relatively low concentrations, in the atmosphere, soils, sediments, plants and animals, including humans.

II. THE NAMING SYSTEM FOR PCBs AND PCDDs — NOMENCLATURE

Altogether, there are some 209 different individual PCB compounds, although only about 130 of these are found in commercial mixtures. PCBs with the same number of chlorine substituents, but different substitution patterns are referred to as isomers, while PCBs with different numbers of chlorines are described as congeners. It is important to have a nomenclature system for the 209 PCB congeners so that they can be unambiguously identified. First, the system is somewhat complex in order to accommodate all the potential variations and permutations. A few of the more important rules are outlined below. The position of attachment of the chlorine atoms is denoted by numbering the carbon atoms of the phenyl rings as follows.

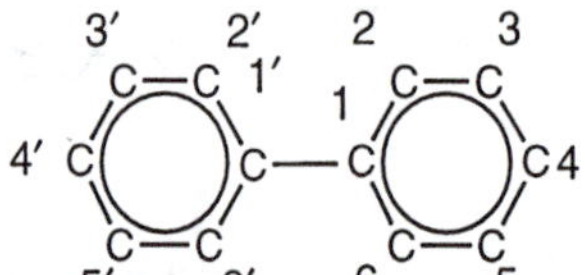

The ring with the smaller number of chlorine substituents is the one given the prime or dashed number scheme. The carbon atoms of each ring are numbered in such a way that the chlorine substituents are attached to carbon atoms with the lowest possible number. When writing the name of a particular congener, the numbers are arranged in ascending order, with dashed numbers following the equivalent non-dashed numbers, e.g., **2,2′,4,4′,5,6′-hexachlorobiphenyl**. Examples of named PCB congeners are shown in Figure 6.1.

On a worldwide scale, the Monsanto Corporation was the principal manufacturer of PCBs. Their **Aroclors** were a series of mixtures, largely comprising PCB congeners, each with a four digit code. For example, **Aroclor 1242** is a mixture of di-, tri-, tetra- and pentachlorinated biphenyls, with the 12 indicating PCBs as the

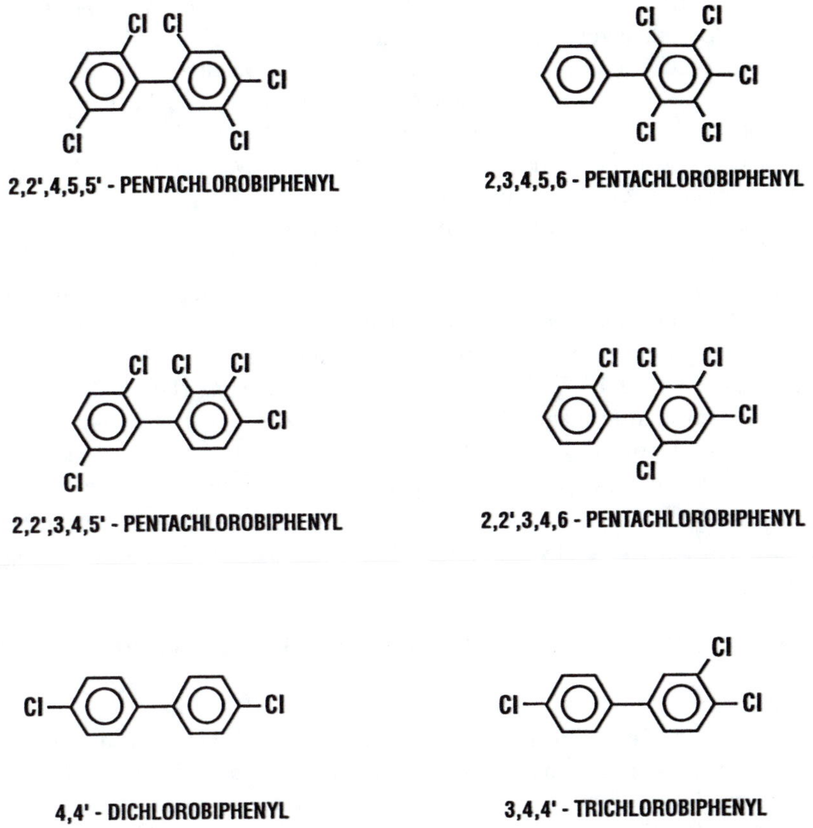

Cl Cl Cl Cl Cl

2,2',4,5,5' - PENTACHLOROBIPHENYL

Cl Cl Cl Cl Cl

2,3,4,5,6 - PENTACHLOROBIPHENYL

Cl Cl Cl Cl Cl

2,2',3,4,5' - PENTACHLOROBIPHENYL

Cl Cl Cl Cl Cl

2,2',3,4,6 - PENTACHLOROBIPHENYL

Cl Cl

4,4' - DICHLOROBIPHENYL

Cl Cl Cl

3,4,4' - TRICHLOROBIPHENYL

FIGURE 6.1. Molecular structures and names of some pentachlorobiphenyl isomers, and other PCB congeners.

principal component and the last two digits of the code indicating the approximate chlorine content as a mass percentage, i.e., 42% in the case of Aroclor 1242.

Most correctly, dioxins are polychlorinated derivatives of dibenzo[b,e][1,4] dioxin **(PCDDs)**. The [1,4] dioxin portion of the name refers to the middle ring of the tricyclic structure. The molecular structure of [1,4] dioxin itself as well as dibenzo[b,e][1,4] dioxin is shown below.

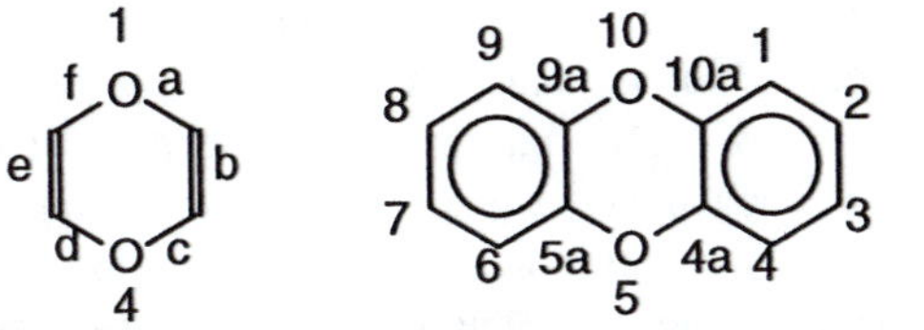

1 f O a e b d O c 4

1,4 dioxin

9 10 1 9a O 10a 8 2 7 3 6 5a O 4a 4 5

dibenzo[b,e][1,4] dioxin

The [b,e] denotes the sides of the [1,4] dioxin ring to which the phenyl rings are fused. As with the PCBs, the carbon atoms of the parent structure are numbered. However, **dibenzo [b,e][1,4] dioxin** also contains oxygen atoms, and these are also numbered. Note that these oxygen atoms are numbered 5 and 10, but the name itself still contains the [1,4] pattern from the original dioxin molecule. Throughout this chapter, dioxins will be referred to as chlorinated derivatives of dibenzo [1,4] dioxin, a slight simplification of their correct name.

In total, there are 75 different individual chlorinated dibenzo[1,4] dioxin compounds. These range from the monochloro isomers up to octachlorodibenzo[1,4] dioxin where all eight hydrogens on the parent molecule have been replaced by chlorine. In naming these compounds, the pattern of chlorine substitution is denoted by listing, in ascending order, the number of each carbon to which a chlorine atom is attached, e.g., 2,3,4,6,7,8-hexachlorodibenzo[1,4] dioxin. The number assigned to each carbon atom is obtained from the scheme shown previously. Further examples of the chemical structures of polychlorinated dibenzo[1,4] dioxins and their names are found in Figure 6.2. Also found in Figure 6.2 are examples of polychlorinated dibenzofurans (PCDFs), a another family of chlorinated compounds closely related to the dioxins but based on furan. These substances often co-occur with the dioxins, but in lower concentrations.

III. SOURCES OF ENVIRONMENTAL CONTAMINATION

A. PCBS

The sources of the PCBs found in the environment today are related to their uses both now and in the past, since a substantial proportion of PCBs from past uses is still present in the environment. PCBs have found a wide variety of commercial uses, particularly in electrical equipment such as capacitors and transformers. In these applications, PCBs have been employed as a heat transfer fluid, transferring heat to the outer casing of the equipment, where it can be dissipated to the atmosphere. This reflects the relatively high thermal stability of PCBs. They have also been used because they are an insulating medium or dielectric, and because their nonflammability reduces the risk of explosions and fires in the event of a spark. Frequently, the fluids contained not only PCBs, but also chlorinated benzenes.

Such uses, particularly with large capacitors and transformers, are termed "controllable closed-system" uses. The quantities of PCBs involved are generally such that there is an incentive for recovery and ultimately disposal of the material. In addition, unless there is an accidental leakage, the PCBs are not in direct contact with the environment. Smaller capacitors, for example those associated with lighting systems, air conditioners, pumps and fans, may also contain PCBs. Recovery of relatively small volumes of PCB mixtures from widely dispersed units is unlikely. Small PCB-containing capacitors are disposed of, often to landfills, along with the equipment in which they are incorporated, when the equipment reaches the end of its working life. This use is termed a "noncontrollable closed system" — closed in theory, but in practice there is little recovery and proper disposal. Other uses for

1,2 - DICHLORODIBENZO[1,4]DIOXIN

1,2,6 -TRICHLORODIBENZO[1,4]DIOXIN

1,3,6,8- TETRACHLORODIBENZO[1,4]DIOXIN

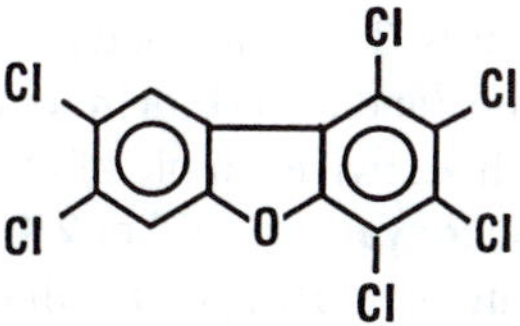

1,2,3,4,7,8 - HEXACHLORODIBENZO[1,4]DIOXIN

2,3,7,8 - TETRACHLORODIBENZOFURAN

1,2,3,4,7,8 - HEXACHLORODIBENZOFURAN

FIGURE 6.2. Molecular structures and names of some dioxin and PCDF congeners.

PCBs that can be classified as noncontrollable closed systems, i.e., not contained in a closed system but in direct contact with the environment, include hydraulic fluids and lubricants.

In 1971, closed system applications, whether controllable or noncontrollable, accounted for some 90% of PCB usage in the U.S. Open uses, with the direct environmental contamination in applications such as plasticizers and fireproofing agents in paints, plastics, adhesives, inks and copying paper, have largely ceased. Table 6.1 summarizes closed and open usage categories of PCBs. Given these uses, the principal sources of PCBs encountered in the environment include:

- Open burning or incomplete combustion of PCB-containing solid waste. Refuse incinerators can emit PCBs because of inadequate combustion conditions. Chemical incinerators have more rigorous combustion requirements, and a PCB destruction efficiency of 99.9999% for incinerators burning PCB-containing materials is mandatory in some countries.
- Vaporization of PCBs in open applications.

TABLE 6.1
Usage Categories of PCBs

Controllable closed systems	Noncontrollable closed systems	Open uses
Large capacitors Large transformers	Small capacitors Hydraulic fluids Lubricating fluids	Plasticizer and/or fireproofing agent in paints, plastics, adhesives, inks and copying paper

- Accidental spills or leakages of PCBs in closed system applications.
- Disposal into sewerage systems and subsequent dispersal of sewage sludge.

B. DIOXINS

Sources of dioxins in the environment can be subdivided into two types:

1. Chemical processes involving chlorine
2. Combustion processes

Chemical processes involving chlorine include manufacture of organochlorine compounds, bleaching of pulp in pulp and paper mills as well as other industrial processes. The manufacture of some organochlorines may result in the formation of **dioxins** (and **polychlorinated dibenzofurans** or **PCDFs**) as unwanted by-products. A well-known example of this occurs in the synthesis of the pesticide **2,4,5-trichlorophenoxyacetic acid** or **2,4,5-T**. One method of preparing this compound involves the high-temperature alkaline hydrolysis of 1,2,4,5-tetrachlorobenzene to 2,4,5-trichlorophenoxide (a derivative of 2,4,5-trichlorophenol formed under alkaline conditions), followed by reaction of the phenoxide with chloroacetic acid, as shown below.

1,2,4,5-Tetrachlorobenzene $\xrightarrow[\text{Ethylene Glycol}]{\text{NaOH}}$ 2,4,5-Trichlorophenol + NaCl

$\xrightarrow[\text{Ethylene Glycol}]{\text{NaOH}}$ 2,4,5-Trichlorophenoxide $\xrightarrow{ClCH_2CO_2H}$ 2,4,5-T (OCH_2CO_2H)

Unfortunately, a side reaction can also occur in which two phenoxide molecules react together to form 2,3,7,8-TCDD.

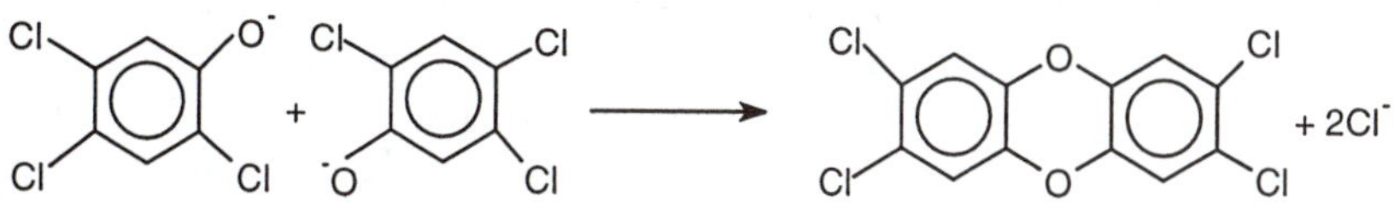

The higher the reaction temperature, the greater the amount of the dioxin formed as a by-product. Related phenoxy acids, used as pesticides, such as **MCPA (2-methyl-4-chlorophenoxyacetic acid)** and **2,4-D (2,4-dichlorophenoxyacetic acid)** may also potentially contain trace amounts of dioxin congeners.

Chlorophenols themselves are employed as fungicides, mold inhibitors, antiseptics, disinfectants and in leather processes, and whether prepared by the direct chlorination of phenol or the alkaline hydrolysis of chlorobenzenes, they may contain dioxins as contaminants. In general, *ortho*-halogenated phenols, compounds that contain this structure or compounds that involve *ortho*-halogenated phenols in their preparation are highly likely to be associated with the presence of dioxins. This is the chemical structure of an *ortho*-halogenated phenol:

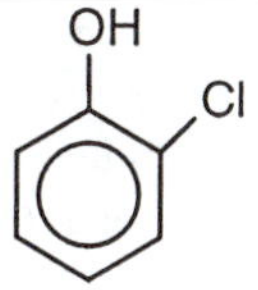

This is particularly true when reaction conditions such as high temperature, alkalinity or the presence of free halogens are involved. PCBs tend to contain PCDFs as contaminants introduced during manufacture.

In pulp and paper mills, bleaching of pulp with chlorine may result in the formation of dioxins from naturally occurring phenolic compounds such as lignin. As well as being found in the paper products, the effluent from mills employing a chlorine bleaching process may contain dioxins. Some reports indicate that a characteristic congener pattern is observed with the dioxins formed. The pattern tends to be dominated by 2,3,7,8-TCDD, and its chlorinated dibenzofuran analogue 2,3,7,8-TCDF. Use of alternative bleaching agents such as chlorine dioxide (ClO_2) and hydrogen peroxide (H_2O_2) or modification of the traditional process leads to substantial reductions in dioxin levels.

A number of industrial processes have also been identified as sources of dioxins. These generally involve high temperatures, and the presence (often accidentally) of chlorine containing-compounds. An example is the melting of scrap metals contaminated with PVC (polyvinylchloride). Another is dioxin contamination of the sludge formed when graphite is used as an anode in the electrolysis of brine to produce chlorine, a process known as the *chloralkali process*.

Apart from chemical processes involving chlorine, the second major source of dioxins is combustion processes. Combustion sources produce a variety of PCDD and PCDF congener profiles. Even for a single source, the profile will vary with the fuel used and the combustion temperature. Combustion sources include municipal solid waste incinerators, coal-burning power plants, fires involving organochlorine compounds (including PCBs) or even burning of vegetation treated with phenoxyacetic acid herbicides. Vehicle exhaust emissions are another source. This is particularly true when vehicles run on leaded fuel, presumably because of the organochlorine fuel additives such as 1,2-dichloroethane.

The presence of dioxins and related compounds in combustion emissions is thought to arise in three ways:

1. They are already present in the combusted material and are not destroyed by the combustion process.
2. They may be formed from organochlorine precursors (e.g., chlorophenols, PCBs) present in the combusted material, during the combustion process.
3. They are formed from high-temperature reactions between nonchlorinated organic molecules and chloride ions.

The last cause suggests that dioxins could have been present in the environment for a relatively long period, certainly since anthropogenic combustion of coal and wood commenced. It is conceivable that burning of wood contaminated with chloride ion could constitute a natural source of trace levels of dioxins. Analysis of archived soil samples from as long ago as the 1840s, as shown in Figure 6.3, tends to confirm this. Tissue analysis of 2800-year-old Chilean mummies or of Eskimos trapped in ice for more than 400 years might be expected to show an influence from this natural source, or from burning of wood or oil in cooking and heating fires. Analysis of the Eskimo tissue has in fact revealed trace (pg/g or parts per 10^{12}) levels of hepta- and octachlorodibenzo [1,4] dioxin. It is generally believed however that the vast majority of the PCDDs and PCDFs found in the environment today have an anthropogenic source. In the U.S., studies of sediment cores show relatively low concentrations up until the 1940s. After this time, increasing concentration, possibly due to combustion of material containing chlorinated aromatic compounds, is observed.

IV. PHYSICAL CHEMICAL PROPERTIES OF PCBS AND DIOXINS

Individually, most PCB congeners are colorless, crystalline solids at room temperature and atmospheric pressure, but some of the less chlorinated congeners are liquids under these conditions. PCBs have a number of physical and chemical characteristics that have contributed to their widespread use.

These include:

Low aqueous solubility
Nonflammability
Resistance to oxidation
Resistance to hydrolysis
Low electrical conductivity

As the degree of chlorination increases, aqueous solubility, flammability and reactivity tend to decrease. This is shown in Table 6.2 in which some properties of

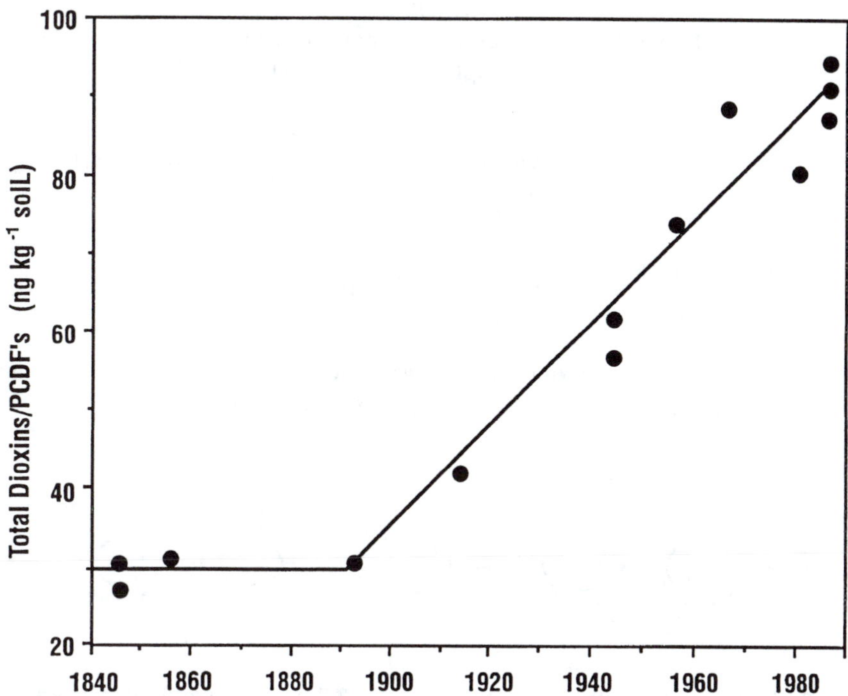

FIGURE 6.3. Trends in total dioxin and PCDF concentration in archived soil samples with time. (From Kjeller, L., Jones, K.C., Johnston, A.E. and Rappe, C. *Environ. Sci. Technol.*, 25, 1619, 1991. Copyright American Chemical Society. With permission.)

biphenyl, a pentachlorobiphenyl and decachlorobiphenyl are compared. The lack of flammability is important in terms of disposal by chemical incineration. PCBs need to be mixed with a large excess of combustible material (e.g., hydrocarbons) in order to generate the high temperatures necessary for decomposition. While PCBs are much more soluble in lipid material than water, lipoidal solubility also tends to decrease with increased chlorine substitution, though not to the same degree as that observed for aqueous solubility. Despite this, K_{OW} values increase with the number of chlorine substituents, as seen in Table 6.2. Among PCBs with equal numbers of chlorines, those with substituents in one or more of the 2,2′,6 and 6′ positions possess relatively low K_{OW} values and high aqueous solubility.

Unlike the individual PCB congeners, commercial PCB mixtures have different physical appearances. Aroclors 1221, 1232, 1242 and 1248 are colorless or almost colorless mobile oils, due to mutual depression of melting points of their components. The more highly chlorinated Aroclor 1254 is a light yellow viscous oil, while Aroclor 1260 and 1262 are sticky resins. The density of these mixtures also increases with the extent of chlorine substitution but is always greater than 1 g cm^{-3}. PCB mixtures can usually be distinguished from lower density mineral oils by the fact that PCBs settle to the bottom of water and do not float as pure hydrocarbon-based mineral oils do.

TABLE 6.2
Some Physical Chemical Properties of Biphenyl, 2,2′4,5,5′-Pentachlorobiphenyl and Decachlorobiphenyl

	Biphenyl	2,2′4,5,5′-Pentachlorobiphenyl	Decachlorobiphenyl
Molecular weight	154.2	326.4	498.7
log K_{OW}	3.76	6.38	8.20
Aqueous solubility (mole m^{-1})	4.4×10^{-2}	4.7×10^{-5}	1.3×10^{-9}
Vapor pressure at 25°C (Pa)	9.4×10^{-1}	1.1×10^{-3}	6.9×10^{-9}
Melting point (K)	344	349.5	578.9

Combustion of PCB mixtures at relatively low temperatures (<700°C), in the presence of oxygen, can lead to the formation of PCDFs. For example,

Cl Cl Cl Cl Cl Δ O_2 Cl Cl Cl O

2,2′,4,5,5′-pentachlorobiphenyl **2,3,8-trichlorodibenzofuran**

Thermal events such as fires and explosions with PCB-containing oils, for example, can result in the formation of PCDFs in any remaining oil, in combustion effluent and in soot that is formed.

The physical chemical properties of the 75 chlorinated dioxins are generally similar to those of the PCBs. Individually, most are colorless solids of decreasing volatility and aqueous solubility with increasing molecular size. Using log K_{OW} values as a measure of hydrophobicity, they vary from 4.20 for the parent nonchlorinated compound dibenzo[1,4] dioxin itself up to 8.20 for octachlorodibenzo[1,4] dioxin. Characteristics of the most toxic dioxin congener, 2,3,7,8-tetrachlorodibenzo[1,4] dioxin (2,3,7,8-TCDD), are shown in Table 6.3. It's interesting to note that the log K_{OW} value of 6.80 is similar to that of DDT at 6.20, which suggests maximum biological activity around this value for chlorhydrocarbons and closely related compounds.

V. ENVIRONMENTAL DISTRIBUTION AND BEHAVIOR

PCBs were first recognized as widespread or ubiquitous environmental contaminants in 1966. One estimate is that about 400,000 tons of PCBs are presently in the environment, which represents about 30% of total world production. Of this total burden, approximately 60% is to be found in the hydrosphere, principally the oceans, about 1% in the atmospheric and the remainder in the terrestrial environment. Despite

TABLE 6.3
Some Physical Chemical Properties of 2,3,7,8-TCDD

Molecular weight	322
log K_{OW}	6.80
Aqueous solubility (mole m^{-1})	6×10^{-8}
Solubility in methanol (mole m^{-1})	3×10^{-2}
Solubility in benzene (mole m^{-1})	1.8
Vapor pressure at 25°C (Pa)	2×10^{-7}
Henry's law constant (Pa m^3 $mole^{-1}$)	3.34
Melting point (K)	578
Decomposition temperature (K)	>973

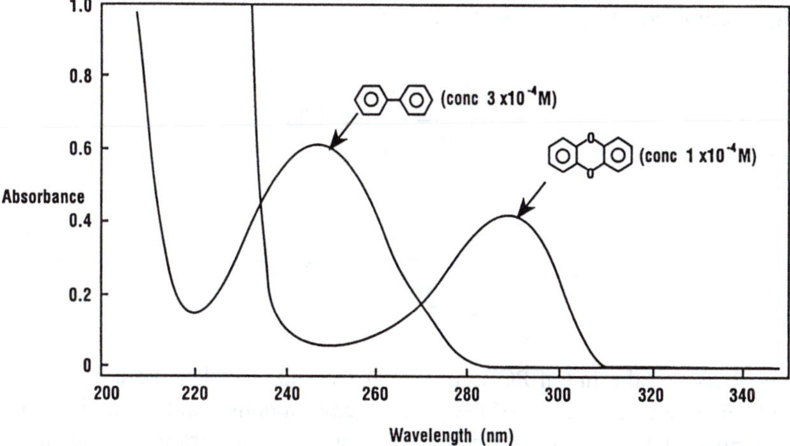

FIGURE 6.4. Ultraviolet absorption spectra of methanol solutions of biphenyl and dibenzo [1,4] dioxin.

relatively low vapor pressures, atmospheric transport and dispersal are important and account for the occurrence of PCBs in areas as remote as the polar regions. Individual PCB congeners are colorless materials so they do not absorb visible (VIS) radiation, though they do absorb slightly at the low wavelength end of the ultraviolet (UV) solar radiation received on the earth's surface. A UV-VIS absorption spectrum of biphenyl itself is shown in Figure 6.4. This shows a main absorption band near 200 nm and a weaker absorption around 240 nm. The wavelength for UV absorption varies according to the number and location of the chlorine substituents. Phototransformation is not a major fate for vapor-phase PCBs, except perhaps for highly chlorinated congeners.

A more important fate of vapor-phase PCBs is reaction with hydroxyl radicals ($OH^{\bullet}$). The principles of the kinetics of environmental transformation and degradation processes are described in Chapter 3. Since the rate of reaction is proportional to the product of the reactant concentrations, then with the reaction of PCBs with $OH^{\bullet}$, the rate of reaction may be described by:

$$\text{Rate} = -\frac{d[\text{PCB}]}{dt} = k[\text{PCB}][\text{OH}^\bullet] \tag{6.1}$$

where k is a second-order rate constant with units of concentration time (e.g., cm^3 molecule s^{-1}). The concentration of $OH^\bullet$ in the atmosphere can be considered to be effectively constant and so the kinetic situation can be treated as pseudo-first-order, where the rate is essentially dependent only on vapor-phase PCB concentrations. Equation (6.1) becomes:

$$\text{Rate} = -\frac{d[\text{PCB}]}{dt} = k'[\text{PCB}] \tag{6.2}$$

where k′ is a pseudo-first-order rate constant with units of $time^{-1}$.

Integrating Equation (6.2) and rearranging, we have:

$$\ln\left(\frac{[\text{PCB}]}{[\text{PCB}]_0}\right) = -k't$$

or (6.3)

$$\left(\frac{[\text{PCB}]}{[\text{PCB}]_0}\right) = e^{-k't}$$

where $[PCB]_O$ is the initial PCB concentration, and [PCB] is the concentration at some later time t. A measure of the persistence of a compound is the lifetime (T_L). A lifetime is defined as the time required for the concentration to fall to 1/e or approximately 37% of its initial value. After one lifetime when t is T_L, then [PCB] = $[PCB]_0/e$, and by substitution into Equation (6.3)

$$\left(\frac{[\text{PCB}]}{[\text{PCB}]_0}\right) = \frac{1}{e} = e^{-1} = e^{-k'T_L}$$

Therefore, the lifetime (T_L) = 1/k′

Although $[OH^\bullet]$ was considered to be constant and thus incorporated into k′, the magnitude of k′ is k′ = $k[OH^\bullet]$ and

$$\frac{1}{k'} = \frac{1}{k[\text{OH}^\bullet]}$$

Using this approach, and an $OH^\bullet$ concentration of 5×10^5 molecules cm^{-1}, it was found that the more highly chlorinated PCB congeners are less reactive toward the hydroxyl radical $OH^\bullet$. Calculated tropospheric lifetimes range from 5 to 11 days, depending on the particular isomer, for dichlorobiphenyls, and up to 60 to 120 days for pentachlorobiphenyls.

Dioxins are a ubiquitous, widespread family of compounds, generally present in the environment at trace levels. They appear to be more common in industrialized areas of the world. Dioxins associated with particules from incinerator emissions and other combustion processes may reside in the atmosphere for considerable periods, which may partly account for their widespread occurrence.

Compared with PCBs, a larger fraction of dioxins found in the troposphere is associated with particules, with the fraction increasing with the extent of chlorine substitution. Octachlorodibenzo [1,4] dioxin is found in the troposphere almost entirely sorbed to particules.

For dioxins (and also PCDFs), phototransformation of vapor-phase material is a much more important fate than for PCBs. The UV-VIS spectrum of the parent or unsubstituted dioxin, dibenzo [1,4] dioxin, is shown in Figure 6.4. In comparison with biphenyl, there is much more absorbance at longer wavelengths. As the number of chlorine substituents increases, the absorbances tend to shift to longer wavelengths. This means increasing overlap with the solar emission spectrum and increasing absorption of solar energy, leading to greater reactivity of the molecule.

Products of phototransformation of PCBs in the troposphere are often hydroxylated biphenyls. These probably arise by an initial homolytic cleavage of a C-Cl bond to form a radical intermediate, followed by reaction with oxygen. For example:

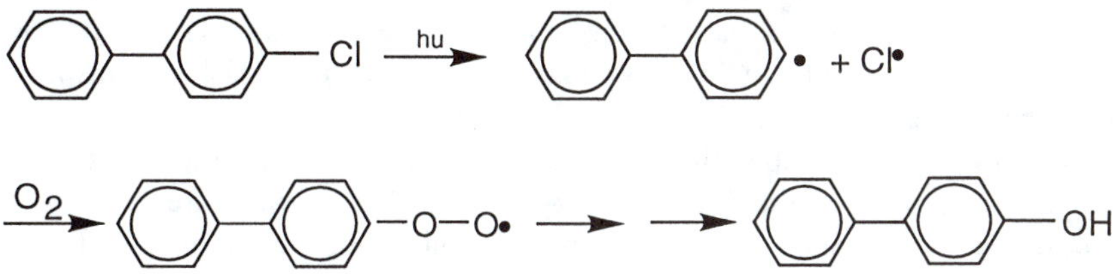

Cl hu • + Cl•

O_2 —O—O• OH

While an analagous process is possible with dioxins, a more important mechanism is probably C-O bond cleavage. As for reaction of OH• with dioxins, an estimate of the tropospheric lifetime for the parent compound is approximately 7 hours, while for 2,3,7,8-TCDD an estimated lifetime is about 3 days. The reaction of OH• with molecules such as dioxins is complex. The first step is usually addition of OH• to an aromatic ring, as shown.

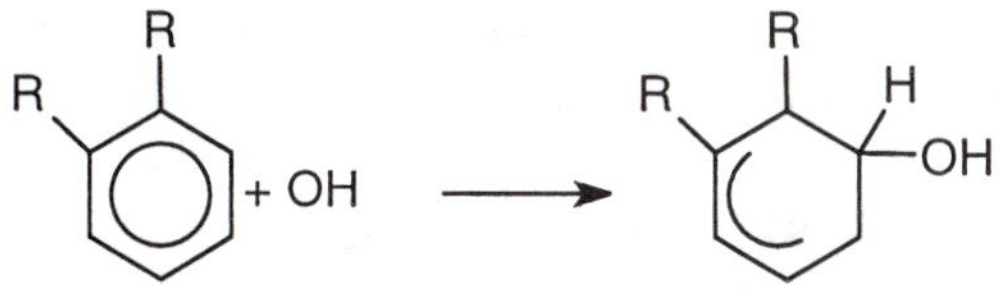

R R + OH R R H OH

Products may include hydroxyl derivatives and molecules in which an aromatic ring has been cleaved open.

Both PCBs and dioxins become less soluble in water and have increased K_{OW} values as the number of chlorine substituents increases, as shown in Table 6.2. In water bodies, these molecules may sorb to bottom sediments or suspended particules, volatilize from water and, in the upper regions where light penetrates, undergo phototransformation. The half-life for phototransformation of 2,3,7,8-TCDD in surface water in summer at 40°N latitude has been calculated to be about 6 days. The C-Cl bond found in PCBs and dioxins tends to be resistant to hydrolysis. On account of the relatively low aqueous solubility and high K_{OW} values, most of these substances will be found sorbed to sediment near discharges. Typically, sediment samples show octachlorodibenzo [1,4] dioxin to be the most prevalent dioxin congener.

The transport of PCBs and dioxins within rivers is by desorption into solution and subsequent resorption downstream, as well as by movement of sediment itself. If sufficient quantities of PCBs are discharged, pools or pockets of relatively pure material may be found at the sediment/water interface. As an example, in 1974 an electrical transformer accidently fell into the Duwamish River near Seattle, Washington during loading operations, resulting in a spill of some 950 L of Aroclor 1242 (density 1.38 to 1.39 g cm^{-3}). Divers observed pools of this material on the bottom being moved about by current and tidal activity. Dredges, including hand-held suction dredges, ultimately recovered about 90% of the spilled PCB mixture.

In the terrestrial environment, PCBs and dioxins tend to sorb strongly to the organic matter fraction of soils. Greatest sorption is observed with the larger, more highly chlorinated congeners. Mobility of these compounds within soils depends on soil characteristics such as bulk density and moisture content. It is also inversely related to the hydrophobicity of a compound, as measured by log K_{OW}. This means the larger congeners also tend to be the least mobile.

Within the soil, microbial biotransformation may occur. Bacteria degrade PCBs and probably dioxins by the action of dioxygenases. Dioxygenases are enzymes that catalyze the incorporation of two oxygen atoms into a substrate molecule. In order for this to occur, there should be at least two adjacent non-chlorine-substituted carbon atoms on an aromatic ring. The possession of two adjacent unsubstituted carbon atoms is less likely with the more highly chlorinated congeners, which are usually observed to be the most resistant toward biotransformation. A possible mechanism for the microbial biotransformation of 2,4-dichlorobiphenyl involving formation of a dihydroxy derivative and then ring cleavage is shown in Figure 6.5.

Compounds such as PCBs may also be transformed by bacteria under anaerobic conditions by a process of reductive dehalogenation. This essentially means the replacement of a chlorine substituent by a hydrogen from water. In this way, hexachlorobiphenyls are converted into pentachlorobiphenyls and so on. This process may occur in anaerobic river sediments and also during the anaerobic digestion of sewage sludge. The products are less chlorinated PCBs that are more susceptible to subsequent aerobic biotransformation than the original more highly chlorinated compounds.

Higher organisms such as mammals are also able to biotransform some of these compounds. Such organisms possess a monooxygenase system also known as the MFO, or mixed function oxidase, system. The incorporation of one atom of molecular oxygen into the substrate PCB or dioxin results in the formation of a three-membered oxygen-containing ring, which then undergoes further reaction. For example,

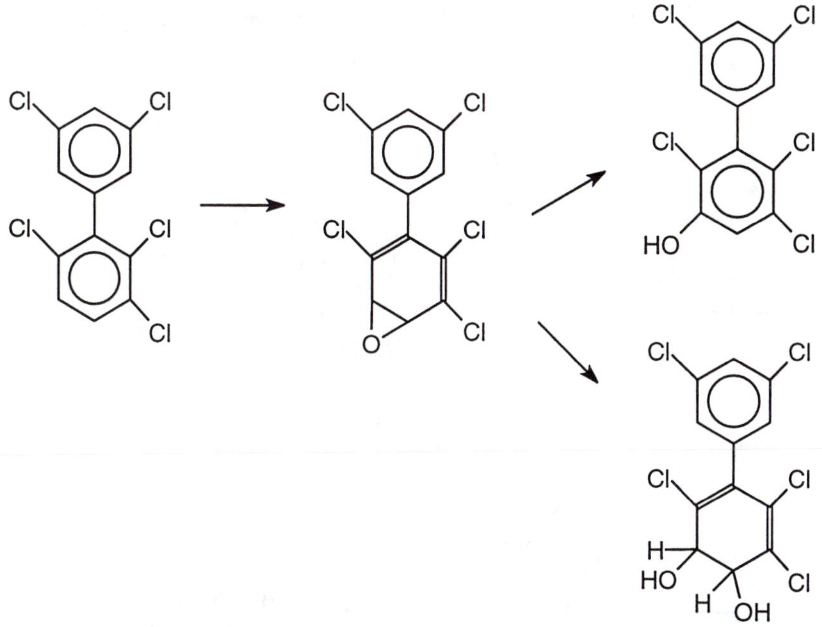

The unsubstituted parent dioxin, dibenzo [1,4] dioxin may undergo biotransformation by an analagous process, but 2,3,7,8-TCDD is relatively resistant. It is interesting to note that this compound does not possess two adjacent carbon atoms without chlorine substituents. This implies that the toxicity of this congener is due to the unmetabolized compound, rather than a product of biotransformation.

Overall, biotransformation tends to yield hydroxyl derivatives that are more polar than the original substrate. These derivatives are readily conjugated (joined) with compounds such as glucuronic acid to form water-soluble molecules that can be readily eliminated from the organism. These processes are described in more detail in Chapter 3. The structure of glucuronic acid is:

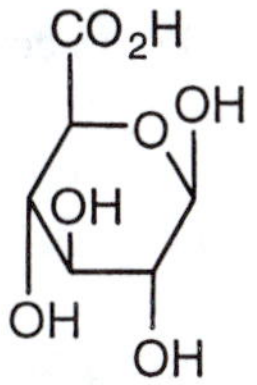

Glucuronic acid

The many polar groups on glucuronic acid confer high polarity on the conjugation product.

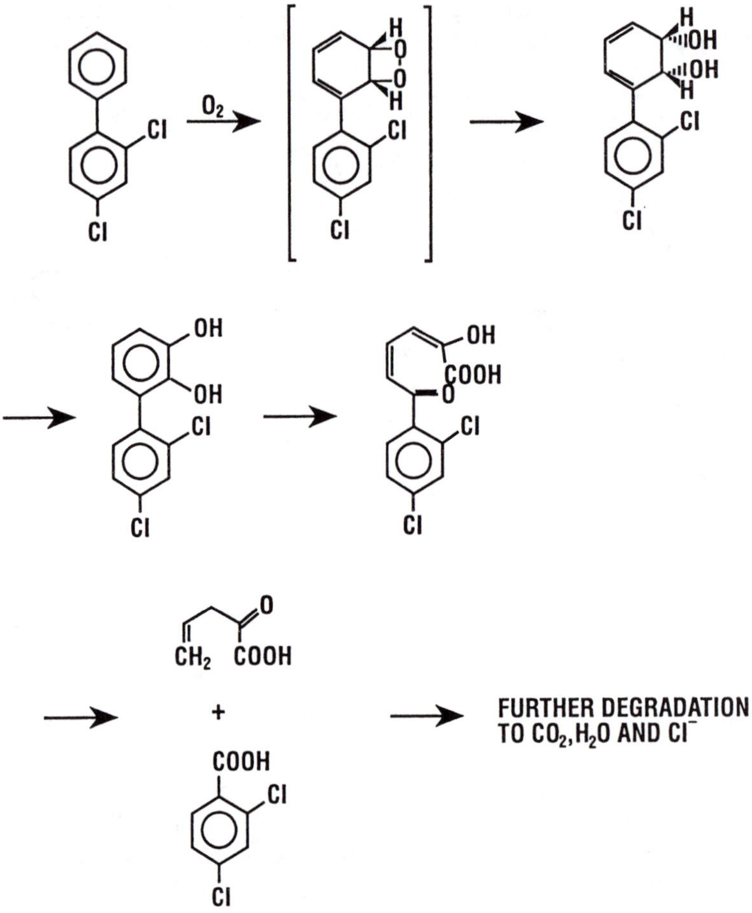

FIGURE 6.5. Possible microbial biotransformation pathway for 2,4-dichlorobiphenyl.

VI TOXICITY

A. PCBs

It is generally agreed that the acute toxicity of PCBs is relatively low. Acute (i.e., single-dose) oral LD_{50} values for a series of Aroclor mixtures with rats are reportedly of the order of 10 g kg^{-1} body weight. Put in context, this means that these mixtures are less toxic than DDT and other chlorohydrocarbon pesticides, and the LD_{50} values are similar to that of aspirin. One report suggests that the minimum dose necessary to produce clinical effects in humans is approximately 0.5 g. It is important to remember though that commercial PCB mixtures may vary in their toxicity according to the relative proportion of the various congeners present, and the amount of impurities such as PCDFs present. Events such as the Yusho incident of 1968, when rice oil contaminated with PCBs was accidentally ingested by people in Japan, have come to public attention. More than 1500 people suffered adverse health effects, but most were probably due to the presence of PCDFs, along with the PCBs in the rice oil. The PCDFs could have arisen during manufacture of the PCBs, or else were formed from the PCBs when contaminated rice oil was used for cooking.

PCBs have a tendency to produce chronic toxic effects. Chronic toxicity describes long-term, low-level exposure effects. Chronic exposure to PCBs produces damage to the liver as well as a condition known as chloracne. Chloracne is a skin condition that has long been known to be associated with chronic exposure to chlorinated aromatic compounds, including PCBs. It is characterized by the appearance of comedones (blackheads) and cysts, largely on the face and neck, but in severe cases, also on the chest, back and genitalia. As well as actual dermal (skin) contact, chloracne can also result from oral ingestion of these compounds. Its duration is dependant upon the degree of exposure, but is often prolonged, lasting several years. Other symptoms of chronic exposure to PCBs include nausea, vomiting, jaundice and fatigue. Biochemical changes involved include vitamin A depletion, alteration to lipid metabolism and hormonal changes. Available evidence suggests that if PCBs themselves are, in fact, carcinogenic, activity is weak.

Individual PCB congeners demonstrate a range of toxicities. The most toxic have chlorine substituents in some or all of the 3,3′,4,4′,5,5′ positions, as shown below.

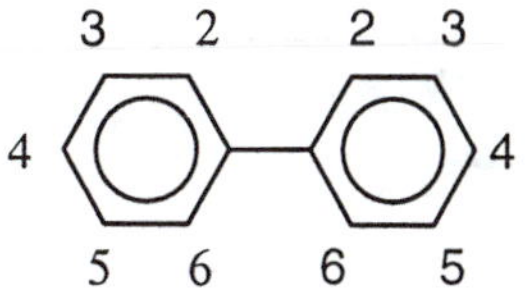

3 2 2 3

4 4

5 6 6 5

These congeners are also known as non-*ortho* or coplanar PCBs, as shown below. Non-ortho means that the molecules lack chlorine substituents on the carbon atoms adjacent to the carbon-carbon bond joining the two phenyl rings.

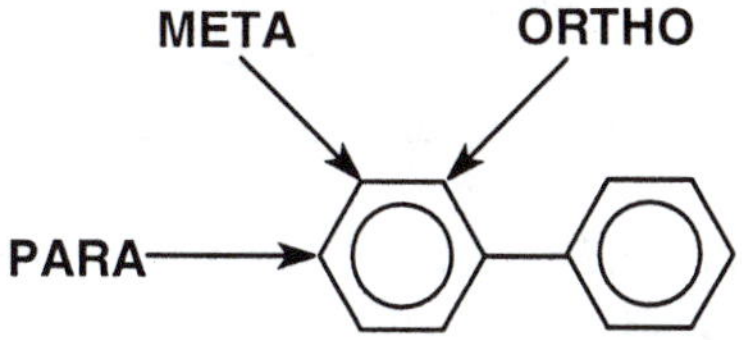

META ORTHO

PARA

Altogether, there are four *ortho*-positions (2,2′,6,6′) in a PCB molecule. A phenyl ring with two substituents separated by one "free" position to which a hydrogen atom attached is said to have a ***meta*** substitution pattern. When the two substituents are opposite each other on the ring, the substitution pattern is termed ***para***. Therefore, chlorines in the 3,3′,5,5′ and 4,4′ positions may be referred to as ***meta*** chlorines or substituents, and ***para*** chlorines or substituents, respectively.

Non-*ortho*-chlorine-substituted PCB congeners are referred to as coplanar because the phenyl rings, each of which is essentially flat, can rotate separately about the carbon-carbon bond joining the two rings. In doing so, the whole molecule may achieve coplanarity, where the phenyl rings are in the same plane, though this is not the most stable configuration. Chlorine atoms are much larger than hydrogen atoms. *Ortho*-substituted PCB congeners have one or more chlorine substituents in the 2,2′,6,6′ positions. Separate complete rotation of the phenyl rings is hindered because of a steric

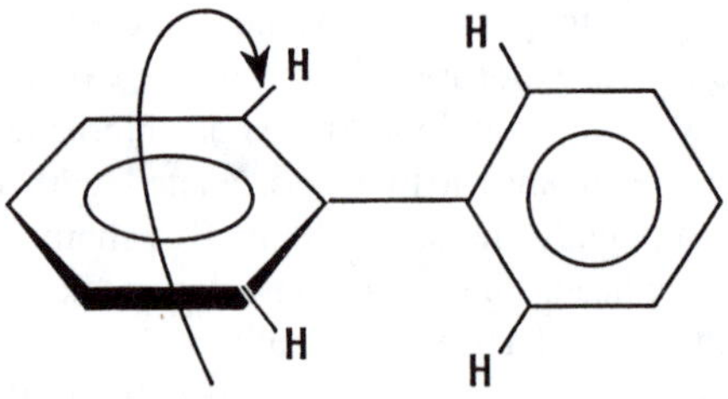

NON-ORTHO SUBSTITUTED PCB CONGENER

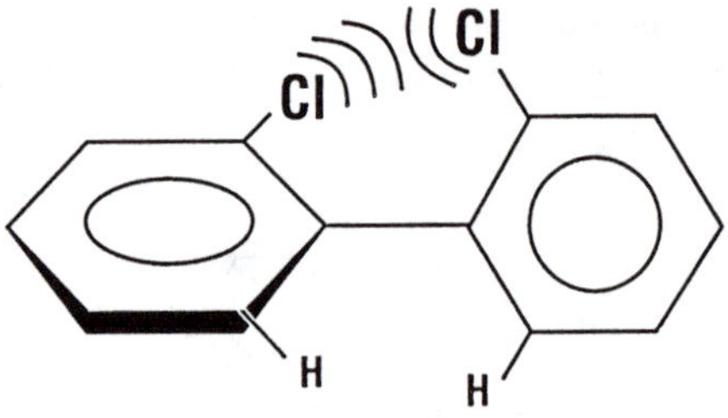

ORTHO SUBSTITUTED PCB CONGENER

FIGURE 6.6. Separate rotation of phenyl rings in a non-*ortho*-substituted PCB congener, but hindered rotation in an *ortho*-substituted PCB congener.

interaction (crowding) between an *ortho*-chlorine substituent on one ring and an *ortho*-substituent (hydrogen and particularly chlorine) on the other, as shown in Figure 6.6. Coplanarity is difficult, if not impossible, to achieve for these molecules.

The non-*ortho* or coplanar congeners may be the most toxic because PCBs can only interact with a receptor or active site in cells when in a coplanar or flat conformation. The dioxin 2,3,7,8-TCDD is some 200 times more toxic (single oral dose) to the guinea pig than the non-*ortho* 3,3′,4,4′,5,5′-hexachlorobiphenyl, which is one of the more toxic PCBs. It is interesting to note from Figure 6.7, however, the close correspondence in molecular structure between the planar conformation of this PCB and 2,3,7,8-TCDD. The same receptor may be involved with both molecules, but the dioxin binds more strongly, or fits the active site better. *Ortho*-substituted PCBs may be less toxic because they cannot achieve coplanarity and interact with the receptor, or because they exert their toxicity by a different mechanism. These compounds are the major components of commercial mixtures. Non-*ortho* congeners are generally present in only small amounts.

B. DIOXINS

Like the PCBs, the individual dioxin or PCDD congeners possess a range of toxicities, with some more toxic than others. The most toxic congener is 2,3,7,8-TCDD. (Among the PCDFs, 2,3,4,7,8-pentachlorodibenzofuran is the most toxic.) The compound 2,3,7,8-TCDD is among the most toxic compounds made by humans,

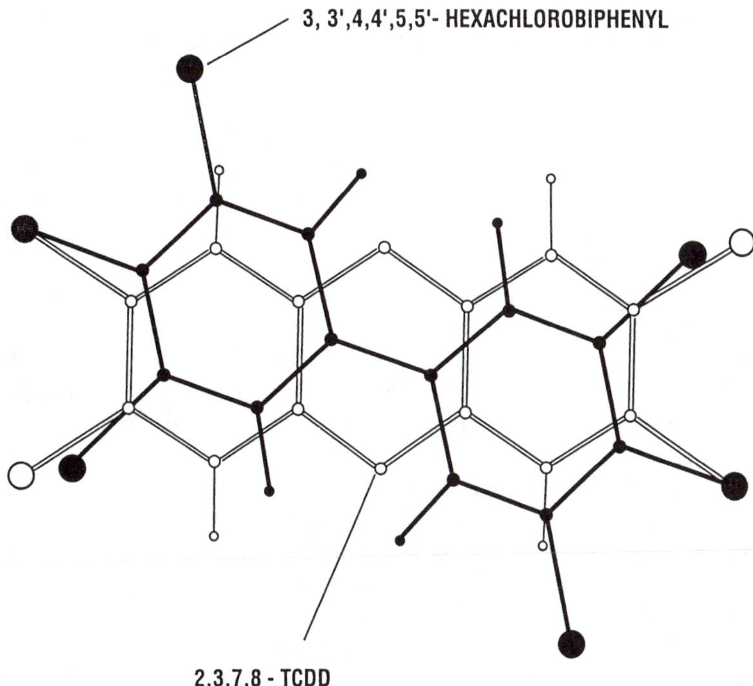

FIGURE 6.7. Correspondence in molecular dimensions as shown by an overlay of the planar conformation of 3,3′,4,4′,5,5′-hexachlorobiphenyl (l) with 2,3,7,8-TCDD (**O**). (From McKinney, J.D. and Singh, P. *Chem. Biol. Interactions,* 33, 277, 1981. Copyright Elsevier. With permission.)

although this is usually inadvertent. However, toxins produced by bacteria associated with tetanus and botulism (a kind of food poisoning) appear to be more toxic on the basis of minimum lethal dose (moles kg^{-1} test organism).

The acute toxicity of 2,3,7,8-TCDD shows considerable variation between species. Guinea pigs are among the most sensitive, while hamsters are comparatively resistant and mice, rats and rabbits show intermediate susceptibility. Available evidence indicates that humans are relatively insensitive. However, the toxicity of this dioxin was demonstrated by laboratory workers involved in the first reported syntheses of both 2,3,7,8-TCDD and its brominated analog 2,3,7,8-tetrabromodibenzo [1,4] dioxin being hospitalized as a result of accidental exposure.

Another feature of 2,3,7,8-TCDD is its delayed lethality. Following administration of a single lethal dose, test animals lose weight and death does not occur for a number of days or even weeks. In some test species, the main target organ is the liver.

It is thought that 2,3,7,8-TCDD exerts its toxic effects by binding to an intracellular protein molecule that is soluble and not bound to the cell membrane. This protein is known as the Ah receptor. This name arises because, as we shall see, this receptor causes the organism to become extremely responsive toward the presence of aryl (or aromatic) hydrocarbons. It seems that 2,3,7,8-TCDD binds extremely strongly to this receptor molecule. This particular dioxin congener optimally fits a binding site on the surface of the protein. Other dioxins, PCDFs and PCBs bind less strongly, fit less well and are observed to be less toxic. A dioxin receptor is found in many species, including humans, most mammals and even some fish. Although there is some structural variation of this receptor between different species, this fact alone does not account for the observed varying toxicity.

Following binding of 2,3,7,8-TCDD to the Ah receptor, a poorly understood transformation process takes place during which the structure of the receptor is modified. The dioxin–Ah receptor complex then moves into the nucleus of a cell, and binds to DNA at specific sites with high affinity. These sites are just upstream of a cytochrome P_{450} gene. Binding to the DNA induces the transcription of specific messenger RNA molecules, which direct the synthesis of new cytochrome P_{450}. As mentioned in Chapter 3, cytochrome P_{450} is a family of closely related enzymes (or isozymes) that catalyze the biotransformation of many lipophilic substances, including hormones, fatty acids, some drugs, and hydrocarbons such as polyaromatic hydrocarbons (PAHs). However, the action of cytochrome P_{450} on PAHs may produce reactive intermediates that are carcinogenic or exhibit toxicity by interacting with critical target molecules in various organs. This relatively complex mechanism of action of 2,3,7,8-TCDD, summarized in Figure 6.8, may in part explain the delayed lethality observed in test organisms. Despite the ability of 2,3,7,8-TCDD to trigger the biotransformation of many other compounds, it is itself relatively resistant to biotransformation.

Recent evidence suggests that binding of the dioxin-receptor complex to DNA may bend or distort the DNA helix, making it more accessible to other proteins associated with induction of other enzymes. It is known that 2,3,7,8-TCDD induces the formation of other enzymes apart from cytochrome P_{450}, although it can also suppress the formation of some enzymes.

Currently, there is considerable debate as to whether exposure of humans to dioxins such as 2,3,7,8-TCDD, results in cancer or birth defects. Epidemiological studies finding increased occurrences have been vigorously contested. It is often difficult, for example, to rule out the possibility that observed effects are due to coincidental exposure to other chemicals. There is little doubt however that chronic exposure of humans can result in effects such as some liver damage, nerve damage and chloracne. In addition, since dioxins are hydrophobic compounds, they may accumulate in adipose or fatty tissue of humans, and breast milk of nursing mothers.

Octachlorodibenzo[1,4] dioxin (structure below) is usually the congener found in greatest concentration in breast milk. Typically, more than 90% of the dioxin burden accumulated by an adult human is from food.

Cl Cl Cl Cl O O Cl Cl Cl Cl

Octachlorodibenzo[1,4] dioxin

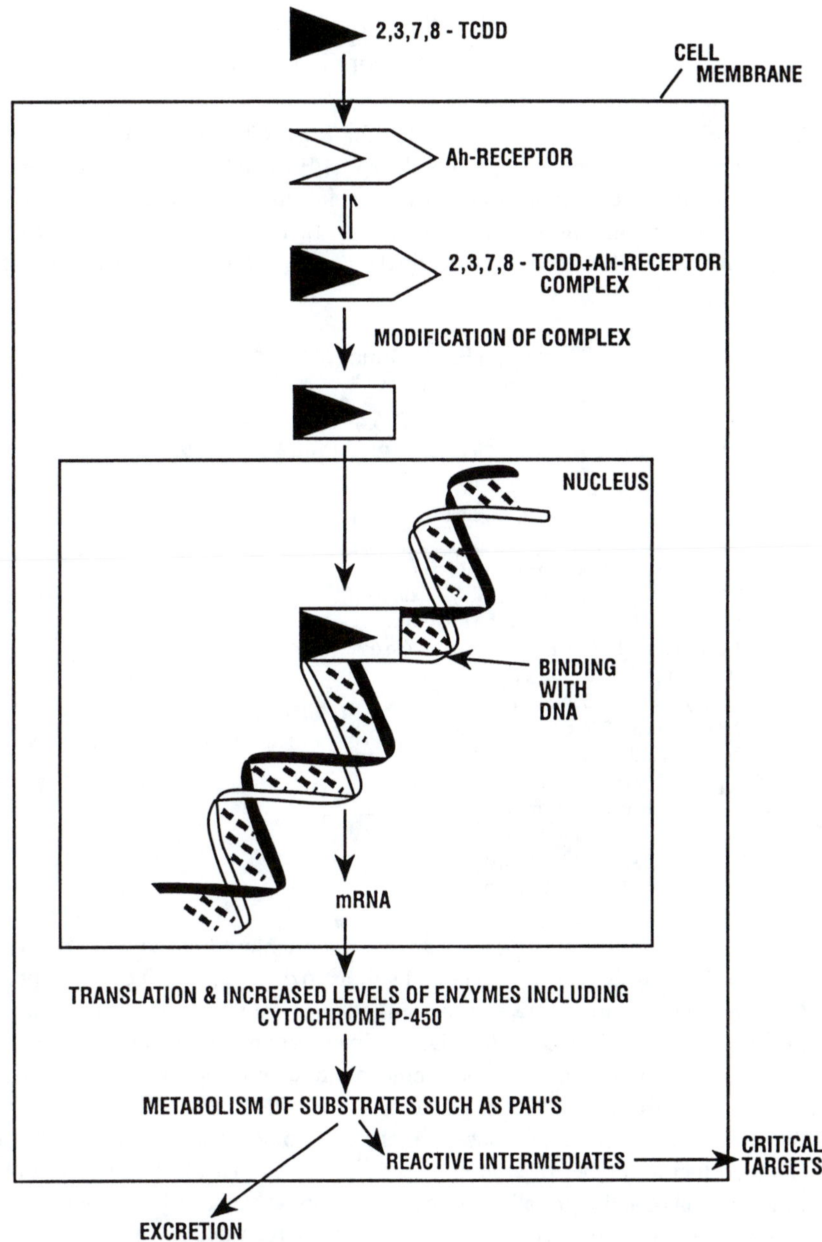

FIGURE 6.8. Possible mechanism of toxic action of 2,3,7,8-TCDD.

Dioxins are invariably encountered in the environment as mixtures of congeners, and the hazard potential of these mixtures is difficult to assess. It is also difficult to compare the hazard potentials of two different mixtures. One approach to this problem is the concept of International Toxic Equivalence Factors (or I-TEFs) in which a toxic measure of a given congener (e.g., EC_{50} with a test organism) is compared with the same toxic measure for 2,3,7,8-TCDD.

$$\text{I-TEF} = \frac{EC_{50}(2,3,7,8-\text{TCDD})}{EC_{50}(\text{congener of interest})}$$

By definition, therefore, the I-TEF value for 2,3,7,8-TCDD is unity. The other congeners are less toxic, and accordingly their I-TEF values are less than unity. The PCDF family and PCBs can also be considered in this way. Table 6.4 contains I-TEF values for some dioxin, PCDF and PCB congeners. The I-TEF values are then multiplied by the concentration of the particular compound to give a toxic equivalent (TEQ) of 2,3,7,8-TCDD.

$$\text{TEQ} = \text{I-TEF} \times \text{Concentration.}$$

TABLE 6.4
I-TEF Values for Some Dioxin, PCDF and PCB Congeners

Congener	I-TEF value
2,3,7,8-Tetrachlorodibenzo [1,4] dioxin	1
2,3,4,7,8-Pentachlorodibenzo [1,4] dioxin	0.5
1,2,3,4,7,8-Hexachlorodibenzo [1,4] dioxin	0.1
1,2,3,4,6,7,8-Heptachlorodibenzo [1,4] dioxin	0.01
Octachlorodibenzo [1,4] dioxin	0.001
2,3,7,8-Tetrachlorodibenzofuran	0.1
2,3,4,7,8-Pentachlorodibenzofuran	0.5
Octachlorodibenzofuran	0.001
3,3′,4,4′-Tetrachlorobiphenyl	0.01
2,3,3′,4,4′-Pentachlorobiphenyl	0.001
3 ,3′,4,4′,5-Pentachlorobiphenyl	0.1
3,3′,4,4′,5,5′-Hexachlorobiphenyl	0.05
2,3,3′,4,4′,5-Hexachlorobiphenyl	0.001

For example, the I-TEF value of 1,2,3,4,7,8-hexachlorodibenzo [1,4] dioxin is 0.1. This means it is estimated to be one tenth as toxic as 2,3,7,8-TCDD. Suppose it was present in adipose tissue at a concentration of 10 pg g^{-1}; this concentration is equivalent to 1 pg g^{-1} of 2,3,7,8-TCDD. This procedure can then be repeated for other mixture components, until the whole mixture has been converted into an equivalent concentration of 2,3,7,8-TCDD in terms of toxicity.

When converting observed concentrations into toxic equivalents, it should be remembered that a number of assumptions are involved. These include a common toxic mechanism of action for all mixture components as well as additive interactions between components. Toxic equivalent data should be regarded as approximate only because assumptions involved in calculations may not always be valid.

VII. KEY POINTS

1. The PCBs and dioxins have similar chemical structures, physical chemical properties and environmental behavior.

2. The PCBs are industrial chemicals used in a wide range of applications from electrical equipment to lubricants. Current uses are restricted, but past usage has caused current contamination due to the persistence of the compounds.

3. The dioxins are not deliberately produced as environmental contaminants, but are inadvertently produced by combustion and in some syntheses of commercial chemicals.

4. The physical chemical properties of the PCBs include low aqueous solubility [4.4×10^{-2} (biphenyl) to 1.3×10^{-9} (decachlorobiphenyl) mol m^{-3}] and declining solubility with increasing molecular weight and high log K_{OW} values [3.76 (biphenyl) to 8.20 (decachlorobiphenyl)].

5. The most important dioxin from an environmental perspective is 2,3,7,8-tetrachlorodibenzo [1,4] dioxin (2,3,7,8-TCDD or dioxin). It has low aqueous solubility (6×10^{-8} mol m^{-3}) and a high log K_{OW} value (6.80).

6. An important process for the transformation and degradation of PCBs is reaction with the hydroxyl radical in the atmosphere. With increasing chlorination, the PCBs become increasingly resistant to this process. The tropospheric lifetimes of these compounds range up to many days.

7. The dioxins have a UV-VIS absorption spectrum that results in significant absorption from solar radiation as compared to the PCBs. As a result, shorter lifetimes in the atmosphere occur. For example, 2,3,7,8-TCDD has a tropospheric lifetime of about 3 days.

8. The acute toxicity of the PCBs is relatively low, with LD_{50} values about 10g kg^{-1} body weight of rats. However, a range of long-term chronic effects occur.

9. The dioxin 2,3,7,8-TCDD is a highly toxic compound, with the other congeners exhibiting lower toxicities.

10. The dioxins always occur as mixtures and toxicity can be estimated in 2,3,7,8-TCDD equivalents described as International Toxic Equivalence Factors (I-TEFs).

REFERENCES

Connell, D.W. and Miller, G.J., *Chemistry and Ecotoxicology of Pollution,* John Wiley & Sons, New York, 1984.

Esposito, N.P., Teirnan, T.O., and Dryden, F.E., *Dioxins,* Industrial Environmental Research Laboratory, Office of Research and Development, U.S. Environmental Protection Agency, Cincinnati, OH, 1980.

Niimi, A.J., PCBs, PCDDs and PCDFs, in *Handbook of Ecotoxicology,* Vol. 2, P. Carlow, Ed., Blackwell Scientific, Oxford, 1994.

CHAPTER 6 PROBLEMS

1. Using the rules outlined in the text, devise names for the following compounds:

A.

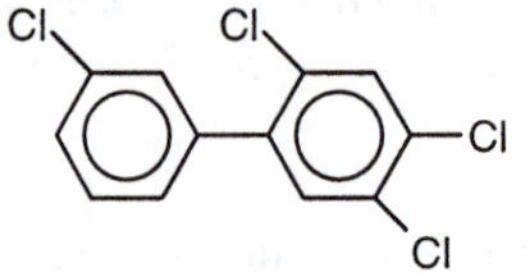

B.

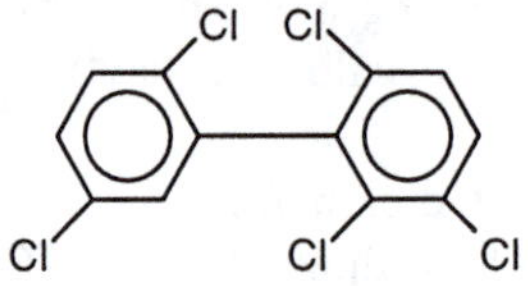

C.

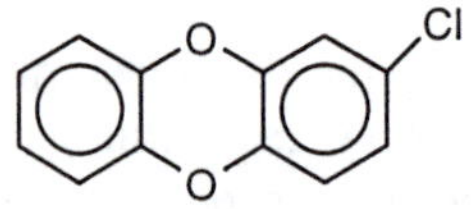

2. Using the data in Table 6.2, estimate the approximate log K_{OW} value of a hexachlorobiphenyl and a trichlorobiphenyl.

3. Which dioxin congener could potentially be formed if 2,4-D (2,4-dichlorophenoxy acetic acid) is prepared by alkaline hydrolysis of 1,2,4-trichlorobenzene, followed by reaction with chloroacetic acid?

4. How many half-lives does a lifetime correspond to?

5. You are evaluating the toxicity of two PCB isomers with the following structures:

A.

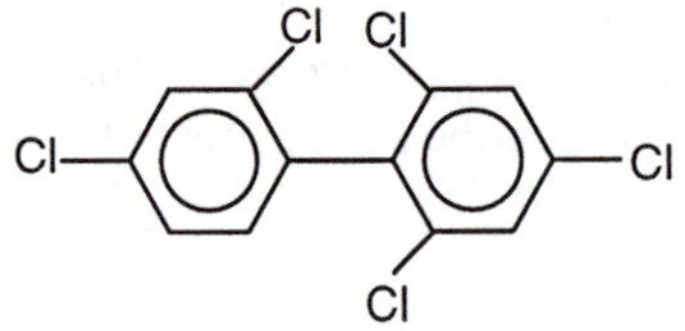

B.

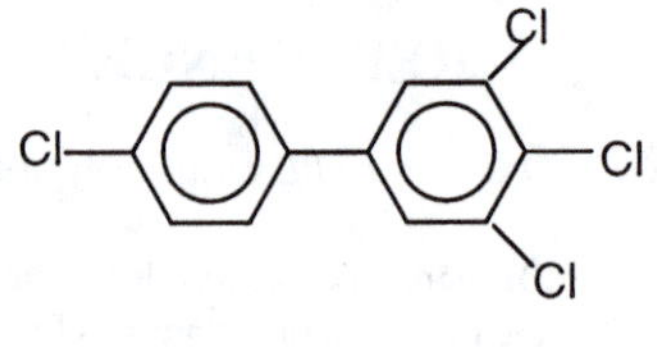

Which of these would you expect to be the most toxic and why would you expect this?

Answers on Page 151.

CHAPTER 6 SOLUTIONS

1. A. 2,3′,4,5-tetrachlorobiphenyl
   B. 2,2′,3,5′,6-pentaclorobiphenyl
   C. 2-chlorodibenzo [1,4] dioxin

2. The data can be plotted and estimations made as shown in Figure 6.9, giving:
   - Hexachlorobiphenyl: log K_{OW} = 6.55 (experimental values for isomers 6.2–7.2).
   - Trichlorobiphenyl: log K_{OW} = 5.25 (experimental values for isomers 5.16–5.58).

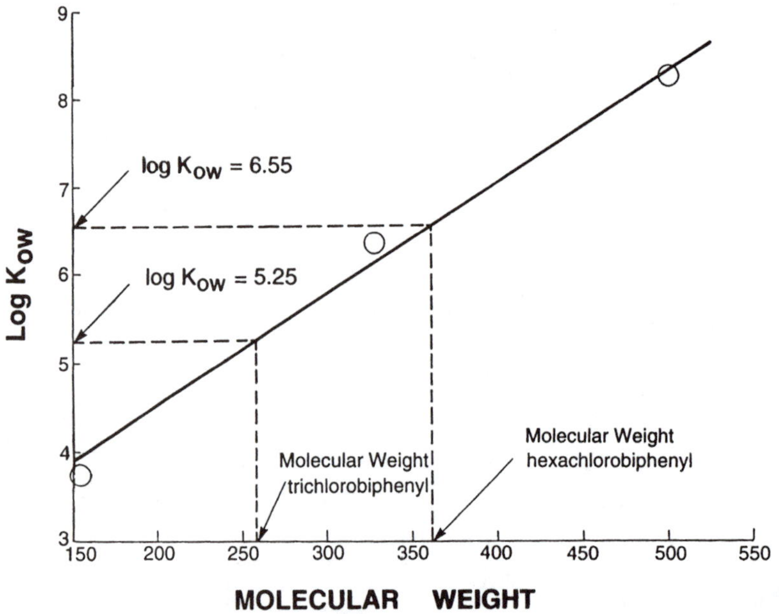

FIGURE 6.9. Plot of molecular weight against log K_{OW} for the PCBs in Table 6.2 with projections to estimate the log K_{OW} values of a hexa- and trichlorobiphenyl (answer to Question 2).

3. The following reaction sequence for the formation of a dioxin congener during synthesis of 2,4-D can be formulated as:

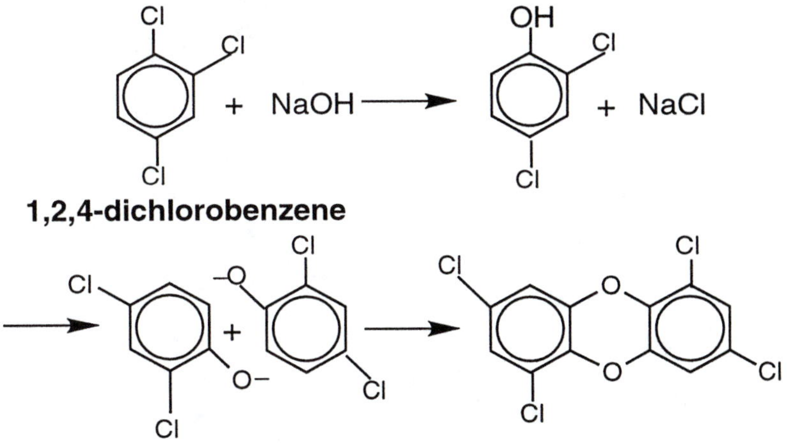

1,2,4-dichlorobenzene

So, 1,3,6,8-tetrachlorobibenzo [1,4] dioxin could be formed in this process.

4. Half-life($t_{1/2}$) = 0.693/k′ (see Chapter 3)
   Lifetime(T_L) = 1/k′
   Number of half-lives = $T_L/t_{1/2}$ = 1.44

5. Compound A would be expected to be the most toxic for the following reasons:
   - The molecular weights are the same and this factor would not be expected to influence the comparative toxicity.
   - Compounds A and B both have the same number of chlorine atoms and so this would not influence the comparative toxicity.
   - Compound A has three chlorines in the 2,2′,6,6′-positions; compound B has none in these positions; and compound B would be expected to be coplanar.
   - Coplanar PCBs are usually the most toxic, so compound B would be expected to be the more toxic.

Chapter 7

SYNTHETIC POLYMERS — PLASTICS, ELASTOMERS AND SYNTHETIC FIBERS

I. INTRODUCTION

In our society, synthetic polymers are among the major chemical products used. Most of these substances were unknown 50 years, but now are used in almost every aspect of our daily life. In the home, plumbing, textiles, paint, floor coverings and many other items are made from **synthetic polymers**, some of which are described in everyday language as **plastics**. A summary of some plastics and their uses is shown in Table 7.1.

TABLE 7.1
Some Common Plastics and their Uses

Plastic	Uses
Polyethylene	Packaging, electrical insulation
Polyvinyl chloride (PVC)	Credit cards, floor coverings, rain wear
Polyvinyl acetate (PVA)	Latex paints
BUNA rubbers	Car tires, hoses
Poly(methyl methacrylate) (Plexiglas, Lucite)	Transport equipment
Polyamides (Nylon)	Clothing

One of the properties of synthetic polymers that enhances their usefulness is their resistance to biotic and abiotic processes of transformation and degradation in usage situations. This means that plastic plumbing, paint, floor coverings and so on tend to last longer than other products. Most of these products then become solid wastes and are discarded in solid waste disposal facilities or some are disposed of directly into the environment and become litter. Their resistance to transformation and degradation processes can be an environmental management problem. So the chemistry and properties of plastics and other synthetic polymers is an aspect of environmental chemistry.

Synthetic polymers principally in the form of plastics, are being increasingly used in all areas of life. The volume of plastics consumed each year now exceeds the volume of steel. Most modern automobiles now contain over 100 kg plastics. The widespread use of plastics occurs because they are lightweight yet do not break like glass, can be made virtually inert to chemical and microbiological breakdown, and are versatile and cost effective. Packaging is the largest and fastest growing single market for plastics. Packaging material, however, typically has a lifetime of less than a year before it appears as discarded waste.

II. THE NATURE OF SYNTHETIC POLYMERS

The term **polymer** is derived from Greek words meaning "many parts." Polymers are, in fact, macromolecules built up from smaller molecules linked together covalently. The small molecules that are the basic building blocks for these macromolecules are known as **monomers**. The process of combining monomers together to form a polymer is known as **polymerization.** The number of monomers in the polymer, or the degree of polymerization may be in the hundreds or thousands. Monomers may be linked together such that they form a linear, branched or cross-linked polymer structure, as shown below:

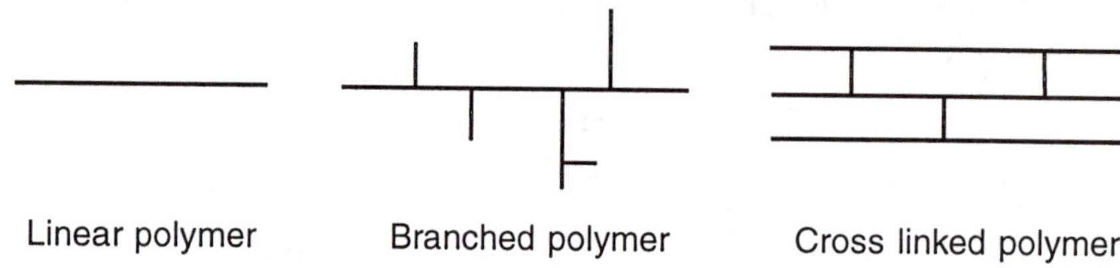

Although there is no clear threshold, polymers typically possess molecular weights or molar masses of 1000 or more. A striking difference between polymers and smaller, relatively simple molecules is the inability to assign an exact molecular mass to a given polymer. When synthesizing linear polymers, for example, molecules with a range of chain lengths, and hence molecular weights or molar masses, are generally formed. Such molecules are written as $[CRU]_n$, where CRU is the constant repeating unit in the chain.

It is important to recognize that there are many naturally occurring polymers. Cellulose, for example, is the main structural component of the cell walls of plants. It is a linear polymer of glucose. One of the most important and familiar commercial natural polymers is rubber, obtained as an aqueous suspension called latex, largely from the rubber tree, *Hevea brasiliensis*. Rubber is a polymer of isoprene, as shown below.

$$\left[-CH_2-C(CH_3)=CH-CH_2- \right]_n$$

Rubber or *cis*-1,4-poly(isoprene)

Other natural polymers include polynucleotides such as DNA and RNA and proteins that are essentially linear polyamides formed from α-amino acids.

The first synthetic polymers were produced over 100 years ago, but initially were often modifications of natural polymers such as cellulose. As understanding of the fundamental nature of polymers increased, so did the development of synthetic polymers. The vast majority of polymers in commercial use are organic compounds.

They may comprise, in addition to carbon and hydrogen, elements such as oxygen, nitrogen, sulfur and the halogens. A notable exception are the silicone polymers, linear macromolecules with a backbone consisting of alternating silicon and oxygen atoms.

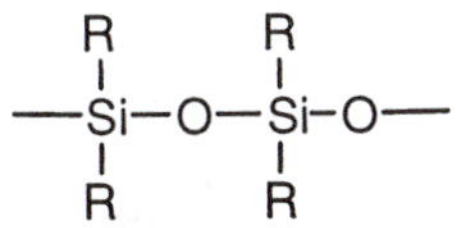

$$-\underset{R}{\overset{R}{Si}}-O-\underset{R}{\overset{R}{Si}}-O-$$

The physical and chemical properties of a polymer (e.g., melting point, chemical reactivity, flexibility, resistance to solvents) depend, among other factors, on its:

- Molecular structure
- Molecular weight
- Component monomers

Monomers can be combined in different ways so that polymers of different properties can be produced from a single monomer.

III. CLASSIFICATION AND DEFINITIONS

Polymers possess a diverse array of chemical structures, properties and behaviors. It is useful to classify them into groups based on some of these characteristics. The classification scheme for synthetic polymers shown in Figure 7.1 is based on physical properties and behavior. They may be divided into three major classes: elastomers, fibers and plastics. It should be remembered though that some polymers may belong to more than one class. For example, a polyester (a polymer where the monomer units are joined by ester functional groups) may form a useful plastic, but if melted, drawn and stretched, may also make a good fiber.

Elastomers may be defined as materials that exhibit rubbery or elastic behavior. Following application of stretching or bending forces, these materials can regain their original shape upon removal of the distorting forces (provided they have not been distorted beyond some elastic limit). An example of a synthetic polymer that is an elastomer is neoprene, which is a polymer of 2-chloro-1,3-butadiene, as shown below.

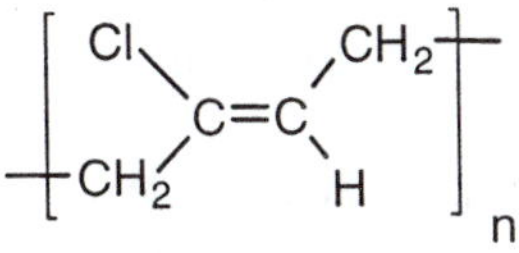

$$\left[-CH_2-C(Cl)=CH-CH_2-\right]_n$$

Neoprene

Neoprene tends to be more resistant toward attack by organic solvents than natural rubber. It has a similar chemical structure except that there is a ***trans***-

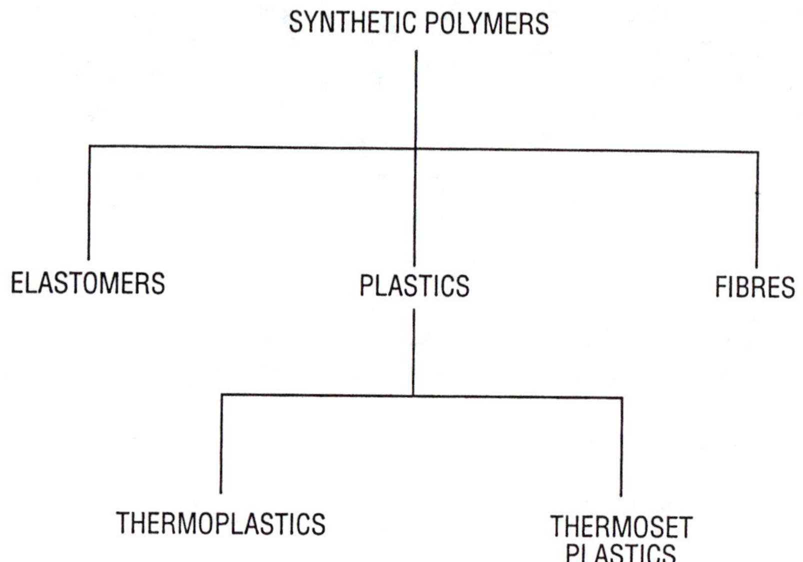

FIGURE 7.1 A classification scheme for the synthetic polymers.

arrangement around the carbon-carbon double bond compared to the *cis*-arrangement in natural rubber.

A **fiber** is a polymer with a high length-to-diameter ratio. Most polymers capable of being melted or dissolved can be drawn into filaments or fibers. However, to make a commercially useful fiber, the material should possess high tensile strength (i.e., resistance to breaking under tension), pliability and abrasion resistance. **Nylons**, which are **linear polyamides,** make excellent fibers, as do **linear polyesters** such as **terylene** [also known as dacron or poly(ethylene terephalate)].

$$\left[-NH-(CH_2)_6-NH-\overset{\displaystyle O}{\overset{\|}{C}}-(CH_2)_4-\overset{\displaystyle O}{\overset{\|}{C}}- \right]_n$$

Nylon -6,6

$$\left[-\overset{\displaystyle O}{\overset{\|}{C}}-C_6H_4-\overset{\displaystyle O}{\overset{\|}{C}}-O-(CH_2)_2-O- \right]_n$$

Terylene

The word **plastic** comes from a Greek word meaning to mold. In the context of polymers, plastics are defined as polymeric materials capable of changing their shape on application of a force, but retaining this shape on removal of the force. Plastics may be subdivided into those that are thermoplastic and those that thermoset.

Thermoplastics are polymers that will soften and eventually melt when heated. They can then be molded and shaped, and on cooling will harden in this shape.

Thermoplastics will repeatedly soften when heated, and harden when cooled. Important thermoplastics include polyethylene, polypropylene, perspex or poly(methylmethacrylate), and polystyrene. Although classed as a thermoplastic, teflon or polytetrafluoroethene demonstrates unusual melting behavior. It does not flow at its melting point (372°C), but forms a very viscous, transluscent material that fractures rather than flows when a force is applied.

The chemical structures of some specific thermoplastic polymers are shown below.

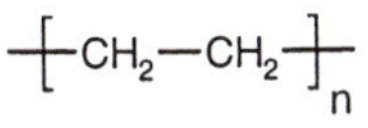

$$\left[CH_2-CH_2 \right]_n$$

Polyethylene

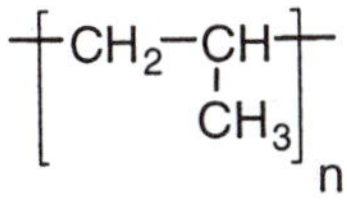

$$\left[CH_2-CH(CH_3) \right]_n$$

Polypropylene

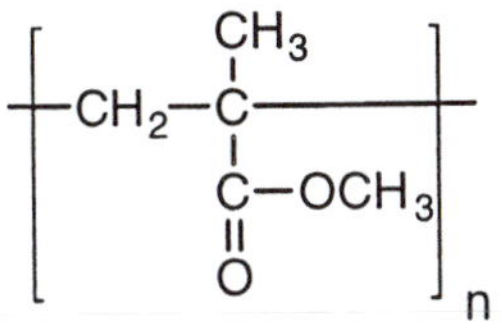

$$\left[CH_2-C(CH_3)(C(=O)-OCH_3) \right]_n$$

Poly(methyl methacrylate)

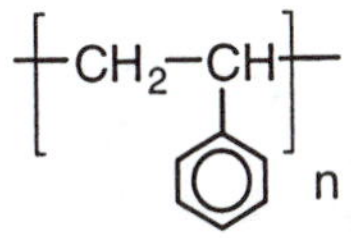

$$\left[CH_2-CH(C_6H_5) \right]_n$$

Polystyrene

$$\left[CF_2-CF_2 \right]_n$$

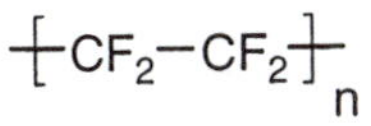

Teflon

Thermoplastic polymers are generally linear or only lightly branched in structure. Most of the plastics produced commercially in many countries are thermoplastics.

In contrast to thermoplastics, **thermosets** usually have a cross-linked structure. Because of this structure, thermosets do not melt and so must be shaped during the cross-linking process. The cross-linking may be initiated by heating, UV light or a chemical catalyst. Thermosetting polymers are used in paints and surface coatings where oxidation by molecular oxygen (O_2) during drying forms a tough, resistant, cross-linked film. Because they do not melt, thermosets are often employed in high-temperature applications, e.g., handles for cooking utensils, automotive transmission and brake components. With extremely high temperatures, however, thermosets decompose irreversibly, blistering as a result of gas evolution and eventually charring. Thermosetting plastics are generally rigid, and bending usually results in a sharp fracture of the material. In contrast, thermoplastics are usually less rigid, but if bent too far will crack and then separate into fragments.

IV. SYNTHESIS OF SYNTHETIC POLYMERS

A. CONDENSATION AND ADDITION POLYMERIZATION

Polymerization reactions were historically classified into two types.

1. Condensation polymerization, where the monomers are linked together with the simultaneous elimination of a small molecule such as water at each step. An example is the synthesis of terylene.

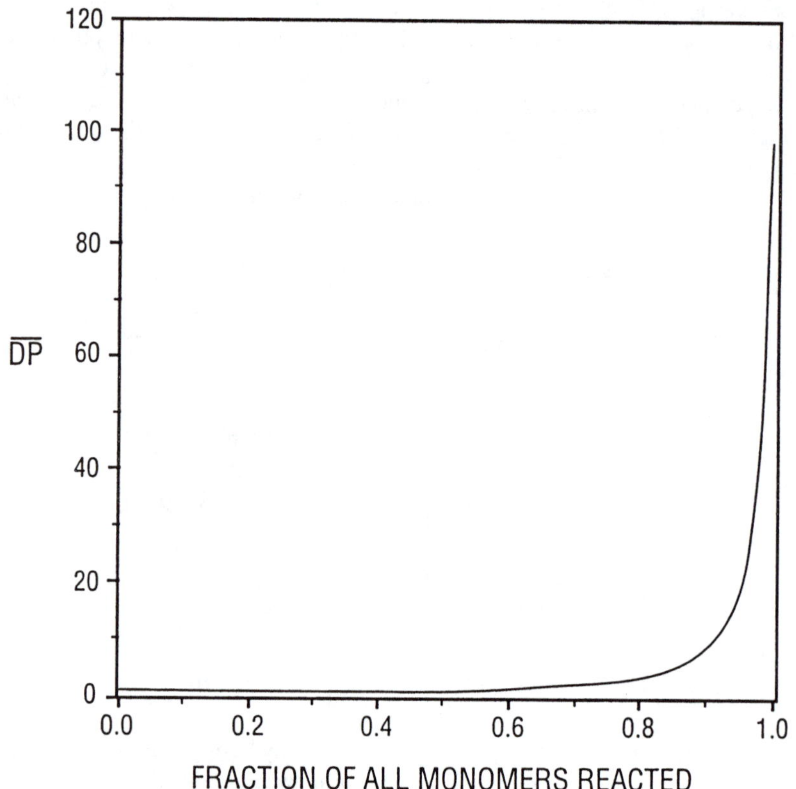

FIGURE 7.2 The relationship between the degree of polymerization $\overline{DP}$ and the fraction of all monomer reacted.

$$nHO-\overset{\overset{O}{\|}}{C}-C_6H_4-\overset{\overset{O}{\|}}{C}-OH + nHOCH_2CH_2OH$$

$$\longrightarrow -\overset{\overset{O}{\|}}{C}-C_6H_4-\left[\overset{\overset{O}{\|}}{C}-OCH_2-CH_2\cdot O\right]_n- + (2n-1)\,H_2O$$

Polymers prepared in this way were called **condensation polymers**.

2. Addition polymerization where the monomers are literally added together, so that the empirical formulae of monomer and polymer are the same. An example is the synthesis of poly(ethylene).

$$n\,CH_2{=}CH_2 \longrightarrow \left[CH_2-CH_2\right]_n$$

Polymers prepared in this way were called **addition polymers**.

Unfortunately, this classification scheme can lead to confusion. For instance, some polyesters can be formed by both addition and condensation reactions, and therefore, it is difficult to know how to classify the polymer.

A more modern and useful classification scheme is based on how the chain actually grows. Thus, polymers may be grouped according to whether the polymerization occurs in a stepwise manner (**step growth**) or from a growing chain (**chain growth**). This scheme eliminates the problems and anomalies identified above. Let us now examine in more detail the step growth and chain growth processes.

B. STEP GROWTH POLYMERIZATION

Step growth polymerization usually occurs with monomers containing functional groups such as hydroxyl (–OH), amino ($-NH_2$) and carboxylic acid (–C(=O)–OH). Polymers such as polyesters and polyamides are produced by a series of condensation reactions, so step growth polymers are essentially the same as the condensation polymers of the early classification scheme. However, molecules such as polyurethanes are also step growth polymers. Some typical step growth polymers are shown in Table 7.2. These polymers may be formed if the initial reaction mixture includes two different polyfunctional (usually bifunctional) monomers, or a monomer containing the appropriate two different functional groups. For example, polyesters can be formed from a diol and a dicarboxylic acid, or from a monomer containing a hydroxyl group and a carboxylic acid.

n HO—□—OH + n HO—C(=O)—■—C(=O)—OH ⟶ [—□—O—C(=O)—■—]$_n$ + (2n-1) H_2O

Diol **Dicarboxylic acid** **Polyester**

2n HO—C(=O)—□—OH ⟶ [—□—O—C(=O)—□—]$_n$ + (2n-1) H_2O

Monomer containing both alochol and carboxylic acid groups **Polyester**

In the first case, initially a diol molecule would react with a dicarboxylic acid molecule to form a dimer (a molecule formed from two monomer molecules). This dimer would contain one ester functional group.

HO—□—OH + HO—C(=O)—■—C(=O)—OH ⟶ HO—□—O—C(=O)—■—C(=O)—OH + H_2O

Can react with dicarboxylic acid here Dimer Can react with diol here

This dimer contains both hydroxyl and carboxylic acid functional groups and can thus react with either monomer (at the appropriate end) to form a trimer containing two ester functional groups.

TABLE 7.2
Examples of Step Growth Polymers

Monomers	Constant repeating unit	Polymer type
HOCH$_2$CH2OH + HO–C(=O)–C$_6$H$_4$–C(=O)–OH	–[O–CH$_2$CH$_2$–OC(=O)–C$_6$H$_4$–C(=O)]–	Polyester (e.g., Terylene, PET)
H2N(CH2)6NH2 + HO–C(=O)(CH$_2$)$_4$C(=O)–OH	–[NH–(CH$_2$)$_6$–C(=O)–(CH$_2$)$_4$–C(=O)]–	Polyamide (e.g., Nylon)
HO–Si(CH$_3$)$_2$–OH	–[Si(CH$_3$)$_2$]–	Polysiloxane or Silicone polymer
HO(CH2)4OH + OCN–C$_6$H$_4$–CH$_2$–C$_6$H$_4$=NCO	–[O–(CH$_2$)$_4$–O–C(=O)–NH–C$_6$H$_4$–CH$_2$–C$_6$H$_4$–NH–C(=O)]–	Polyurethane
Cl–C(=O)–Cl + HO–C$_6$H$_4$–C(CH$_3$)$_2$–C$_6$H$_4$–OH	–[O–C$_6$H$_4$–C(CH$_3$)$_2$–C$_6$H$_4$–O–C(=O)]–	Polycarbonate

$$HO-\square-O-\overset{O}{\overset{\|}{C}}-\blacksquare-\overset{O}{\overset{\|}{C}}-OH + HO-\square-OH \longrightarrow HO-\square-O-\overset{O}{\overset{\|}{C}}-\blacksquare-\overset{O}{\overset{\|}{C}}-O-\square-OH + H_2O$$

Dimer **Trimer**

This process can continue, forming longer chains. The term "step growth" is used because the polymer's molecular weight increases in a slow, step-like fashion as reaction time increases. Cross-linking is possible by including trifunctional monomers, e.g., a triol in the reaction mixture. Thus, both thermoplastic and thermoset polymers can be produced with step growth polymerization.

In step growth polymerizations, each step or reaction is essentially independent of the preceding one. Polymers are formed simply because molecules can undergo reaction at more than one site. Each step is practically identical in rate and mechanism to the first initial step or reaction where two monomers are linked together. Monomer molecules are consumed rapidly, but the average chain length or average degree of polymerization, $\overline{DP}$, remains quite low initially.

If there are N_0 monomer molecules initially and a total (monomer + polymers) of N molecules after a given reaction time period, then the number of monomers that have reacted is $N_0 - N$. The fraction of monomer molecules originally present that have undergone reaction (p) is given by

$$p = \frac{N_0 - N}{N_0}$$

$$\text{Because } \overline{DP} \text{ is equal to } \frac{N_0}{N}, \text{ then } \overline{DP} = \frac{1}{1-p}$$

This simple expression, which relates average chain length in a reaction mixture to the fraction of monomers that have undergone reaction, shows that after 25% of monomers have reacted ($p = 0.25$), then $\overline{DP}$ is only 1.33. After 75% reaction, the $\overline{DP} = 4$, and average chain length does not reach 20 until 95% of monomer molecules have reacted. A plot of this relationship highlighting the slow build-up of $\overline{DP}$ with fraction of monomers reacted is shown in Figure 7.2. From the diagram of step growth polymerization in Figure 7.3, it is clear that even after half of the monomers have reacted, the average chain length is small because the molecules in solution comprise 50% unreacted monomers, with the remaining 50% consisting mainly of dimers and trimers. Some forms of nylon require $\overline{DP}$ values of between 50 and 60 to form a satisfactory fiber, which in turn requires reaction of greater than 95% of monomers. It is important to remember that $\overline{DP}$ can still increase after most of the monomers have reacted. The oligomers (molecules made up from a few monomers) formed still have reactive functional groups that can link molecules together. In step growth polymerization, almost any two molecules in the mixture can generally react. Average chain length increases steadily as reaction proceeds, though long reaction times and almost complete reaction of monomers is necessary to produce polymers with large $\overline{DP}$ values. Eventually, however, the viscosity of the reaction mixture may increase to such an extent that it is difficult for reactive functional groups on the ends of chains to interact with each other, and reaction ceases.

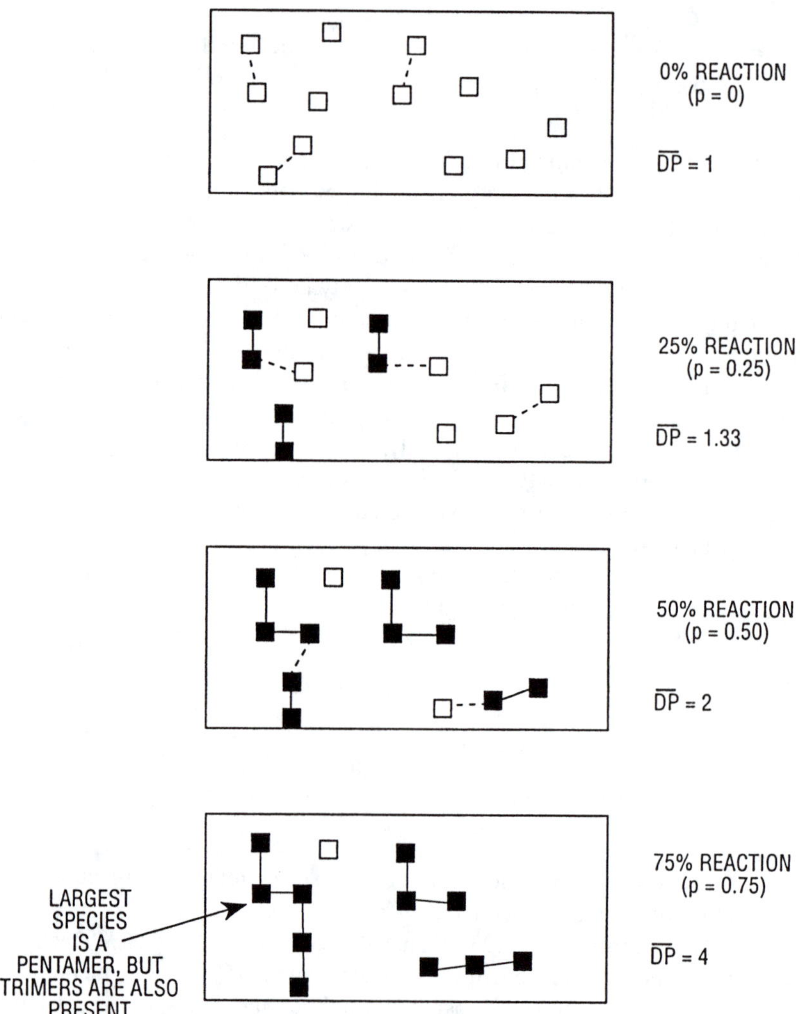

FIGURE 7.3 Diagrammatic representation of step growth polymerization.

C. CHAIN GROWTH POLYMERIZATION

Chain growth polymerization begins with the generation of reactive centers on a relatively small number of monomer molecules. The reactive center may be a radical, a carbocation (a cation where carbon carries the positive charge), or a carbanion (an anion where carbon carries the negative charge). This process constitutes initiation, the first of the three phases of chain growth polymerization:

1. Initiation
2. Propagation
3. Termination

Chain growth polymers are essentially equivalent to the old addition polymer classification. Chain growth polymerization tends to produce thermoplastics only.

The monomers involved are typically unsaturated, possessing a carbon-carbon multiple bond, e.g., $CH_2{=}CR_1R_2$.

The most important type of chain growth polymerization, both in terms of versatility and the amount of polymers produced commercially on an annual basis, is the radical-initiated variety. Here, a small trace of initiator substance that readily fragments into radicals is used. The most common initiators are benzoyl peroxide and azobisisobutyronitrile (or AIBN), which fragment when heated to around 100°C, or when irradiated with near-UV light, as shown below.

$$C_6H_5{-}\overset{O}{\overset{\|}{C}}{-}O{-}O{-}\overset{O}{\overset{\|}{C}}{-}C_6H_5 \xrightarrow[\text{Light}]{\text{Heat}} C_6H_5{-}\overset{O}{\overset{\|}{C}}{-}O^{\bullet}$$

Benzoyl Peroxide

$$\underset{\displaystyle CN}{(CH3)2C}{-}N{=}N{-}\underset{\displaystyle CN}{C(CH_3)_2} \xrightarrow[\text{Light}]{\text{Heat}} 2\underset{\displaystyle CN}{(CH_3)_2C^{\bullet}} + N_2$$

AIBN

Once produced, the radical reacts rapidly with a monomer molecule, forming a new larger species that is also a radical, and is called the chain carrier.

$$R^{\bullet} + \begin{matrix} H & & R_1 \\ & C{=}C & \\ H & & R_2 \end{matrix} \longrightarrow R{-}\underset{\displaystyle H}{\overset{\displaystyle H}{C}}{-}\underset{\displaystyle R_2}{\overset{\displaystyle R_1}{C}}\,{\bullet}$$

In the process of propagation, the radical center at the end of the growing chain reacts with another monomer molecule. There are two ways in which this could occur: either (1) head-to-tail, or (2) head-to-head.

$$R{-}CH_2{-}\underset{\displaystyle R_2}{\overset{\displaystyle R_1}{C}}{\bullet} + CH_2{=}\underset{\displaystyle R_2}{\overset{\displaystyle R_1}{C}} \longrightarrow R{-}CH_2{-}\underset{\displaystyle R_2}{\overset{\displaystyle R_1}{C}}{-}CH_2{-}\underset{\displaystyle R_2}{\overset{\displaystyle R_1}{C}}{\bullet} \quad \textbf{Head-to-tail}$$

$$R{-}CH_2{-}\underset{\displaystyle R_2}{\overset{\displaystyle R_1}{C}}{\bullet} + \underset{\displaystyle R_2}{\overset{\displaystyle R_1}{C}}{=}CH_2 \longrightarrow R{-}CH_2{-}\underset{\displaystyle R_2}{\overset{\displaystyle R_1}{C}}{-}\underset{\displaystyle R_2}{\overset{\displaystyle R_1}{C}}{-}\underset{\displaystyle H}{\overset{\displaystyle H}{C}}{\bullet} \quad \textbf{Head-to-head}$$

Normally, the head-to-tail mode of addition occurs. This is due to a number of more stable than that from head-to-head reaction. In addition, it is easier for the chain carrier to interact with the tail of the monomer, particularly if R_1 or R_2 are large bulky groups. Reaction of the chain carrier with a monomer molecule can occur repeatedly, forming a polymer chain of increasing length. Each individual reaction depends on the previous one. Chain formation occurs very quickly, since the average lifetime of a growing chain is relatively short. Several thousand monomers can be linked together in just a few seconds.

With chain growth polymerization, high molecular weight polymer chains are reached with relatively low percentages of monomer reaction (as seen in Figure 7.4). This is because the reaction mixture contains only the growing polymer chains and monomer molecules, and the concentration of growing chains is maintained at a low level. This is in contrast to step growth polymerization where in addition to monomer and large polymer molecules, many species of intermediate size are also present. A comparison between this and other features of step growth and chain growth polymerization is shown in Table 7.3.

Propagation or chain growth does not continue until all the monomers present have reacted. Termination of growth occurs when the reactive radical center on two different chains interact together, or when a radical center reacts with a solvent molecule, an initiator radical or impurities in the reaction mixture, e.g., oxygen. The first process is the most important, with the interaction being one of two possible types.

1. *Combination:* this is where two radical centers couple together, forming a covalent C–C bond, and one long chain that does not have a radical center and cannot grow.

$$\sim CH_2{-}\dot{C}R_1R_2 \;+\; \dot{C}R_1R_2{-}CH_2\sim \;\longrightarrow\; \sim CH{=}CR_1R_2 \;+\; H{-}CR_1R_2{-}CH_2\sim$$

2. *Disproportionation:* this is where a hydrogen atom is abstracted from one chain, and added to another. One saturated and one unsaturated chain is formed, with growth ceased in both.

$$\sim CH_2{-}\dot{C}R_1R_2 \;+\; \dot{C}R_1R_2{-}CH_2\sim \;\longrightarrow\; \sim CH{=}CR_1R_2 \;+\; H{-}CR_1R_2{-}CH_2\sim$$

The type that is predominant in any system depends on the monomer involved and the polymerization conditions. For example, growth of polystyrene chains is terminated largely by combination; whereas, with poly(methylmethacrylate), termination occurs only by disproportionation at temperatures above 60°C.

With radical-initiated chain growth polymerization, the number of unreacted monomers usually far outweighs the number of growing polymer chains. These

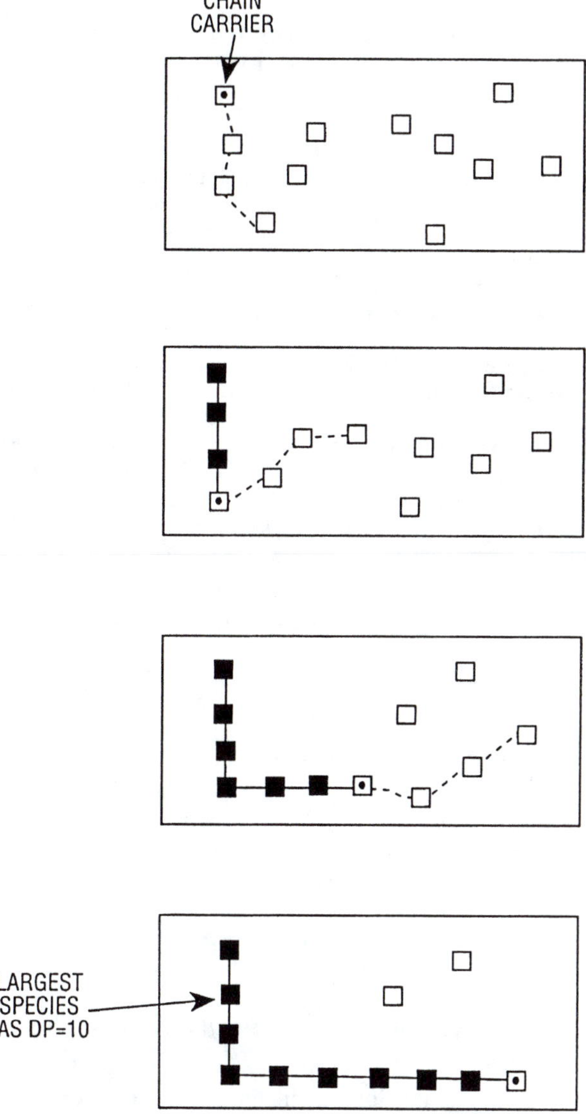

FIGURE 7.4 Diagrammatic representation of chain growth polymerization leading to high molecular weight chains with relatively low percentages of monomer reaction.

chains experience the same conditions, and undergo termination after similar periods of growth. Thus, both the average chain length and the molecular weight distribution of polymers remain approximately constant throughout the polymerization. Longer reaction times increase the yield of polymers but not their average chain length.

Polymers typically produced by a chain growth polymerization method include polyethylene, polystyrene and poly(vinyl chloride) or PVC.

TABLE 7.3
A Summary Comparison of Step Growth versus Chain Growth Polymerization

Step growth	Chain growth
Linear and cross-linked polymers can be formed	Linear (and branched) polymers can be formed
Forms thermosets and thermoplastics	Forms thermoplastics
No initiator necessary	Initiator necessary
Same reaction mechanism throughout	Initiation and propagation mechanisms different
Any two monomer and/or polymer species in mixture can react so long as they possess the appropriate functional groups	Only the active center can react with monomer molecules to add them to the growing chain
Monomer consumed rapidly, but molecular weight of polymerized species increases slowly	Monomer consumed relatively slowly, but molecular weight of polymerized species increases rapidly
High conversion of monomer and long reaction time necessary for the production of polymers with high $\overline{DP}$	Long reaction times increase the yield of polymers, but not their average chain length
No termination step, as end groups are still potentially active	Chain termination step usually involved

$$n\,CH_2{=}CH_2 \longrightarrow -\!\!\left[CH_2-CH_2\right]\!\!-_n$$

Ethylene **Polyethylene**

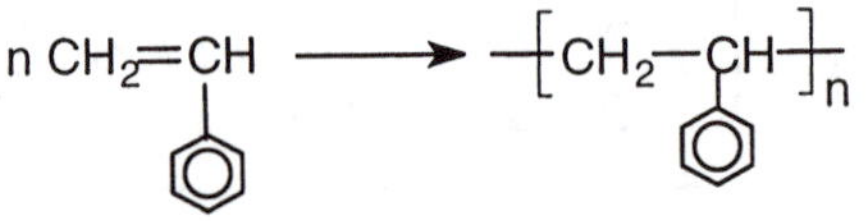

$$n\,CH_2{=}CH(C_6H_5) \longrightarrow -\!\!\left[CH_2-CH(C_6H_5)\right]\!\!-_n$$

Styrene **Polystyrene**

$$n\,CH_2{=}CHCl \longrightarrow -\!\!\left[CH_2-CHCl\right]\!\!-$$

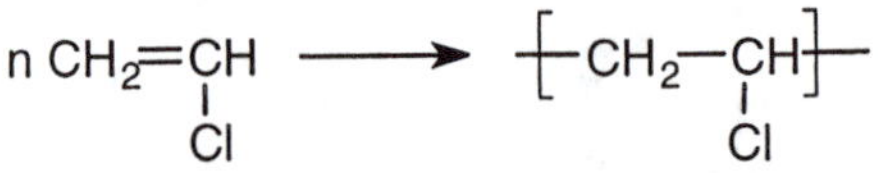

Vinyl chloride **Poly(vinyl chloride)**

Although it may appear that long-chain, linear polymers are always produced, this is not the case. Consider polyethylene for example. Depending on the initiation process used, and the reaction conditions employed, branching may be induced in the chain. This gives rise to the various forms of polyethylene shown in Figure 7.5. Branching usually results from an intramolecular transfer of the reactive radical center.

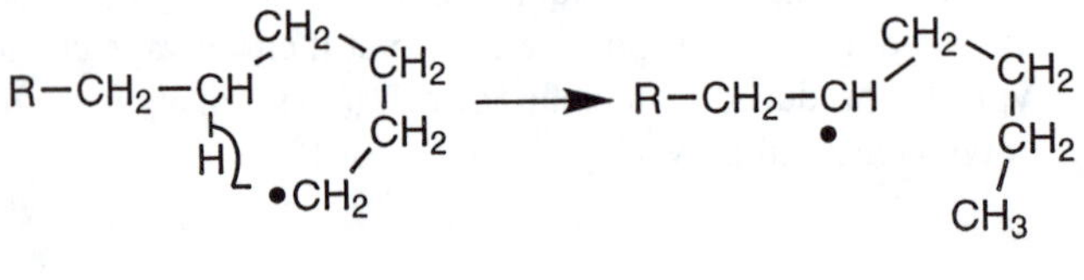

$$R-CH_2-CH(H)-CH_2-CH_2-CH_2-\bullet CH_2 \longrightarrow R-CH_2-\overset{\bullet}{C}H-CH_2-CH_2-CH_2-CH_3$$

LOW DENSITY POLY(ETHYLENE) or LDPE

Conditions: High Pressure (1000 to 3000 atm)
High Temperature (200 to 300°C)
FREE RADICAL INITIATOR

HIGH DENSITY POLY(ETHYLENE) or HDPE

Conditions: Low Pressure (3 to 30 atm)
Low Temperature (50 to 120°C)
TRANSITION METAL CATALYST

LINEAR LOW DENSITY POLY(ETHYLENE) or LLDPE

Conditions: As for HDPE
ETHYLENE COPOLYMERIZED WITH SMALL AMOUNTS
(UP TO 10%) OF A 1-ALKENE SUCH AS 1-BUTENE

FIGURE 7.5 The various forms of polyethylene and their modes of formation.

The new radical that is generated can then interact with ethylene monomer molecules in the usual way. It will now, however, have a saturated hydrocarbon side chain.

D. CROSS-LINKING

Bridges comprising covalent chemical bonds that occur between separate polymer chains are known as cross-links. The presence and frequency of occurrence of cross-links can have a significant influence on the physical and chemical properties of materials in which they exist. As mentioned previously, polymers not cross-linked are generally thermoplastic, which means that they will melt and flow at sufficiently high temperatures. In contrast, cross-linked polymers are almost invariably thermosets. They cannot melt because the cross-links restrict the motion and movement of individual chains. Cross-linking tends to impart rigidity to polymers. However, extensively cross-linked materials are usually very brittle. The dense three-dimensional network of covalent bonds allows little flexibility and hence no way for the material to take up stress. Eventually, the material fails, catastrophically.

Probably the best known example of cross-linking is the vulcanization of natural rubber. In its raw state, natural rubber is sticky when warm and prone to oxidative degradation. By heating with a few percent of elemental sulfur in the presence of catalysts, rubber molecules are linked by short chains of sulfur atoms. The vulcanized rubber produced is a useful elastomer, harder and more resistant to oxidation and attack by solvents than its natural counterpart. With about 30% sulfur, and heating to 150°C, ebonite is formed. This hard and rigid material has been used for making the casings of automobile batteries. Another commonly encountered example of cross-linking is the so-called "drying" of some paints, which involves cross-linking by molecular oxygen from the atmosphere, resulting in a tough resistant film.

Cross-linking also affects the solubility of polymers. A linear polymer that is not cross-linked will usually dissolve in an appropriate solvent (although it may take some time). Cross-linked polymers will not dissolve. They may, however, swell as solvent molecules diffuse into the network. Polystyrene, for instance, is soluble in many solvents such as toluene, benzene and carbon tetrachloride. Cross-linking by copolymerization of styrene with only a small amount of divinylbenzene produces a material that does not dissolve in these solvents, but only swells.

V. ENVIRONMENTAL CHARACTERISTICS

A. COMBUSTION

Polymers are organic materials, and like most organic materials almost all of them will undergo combustion if the temperature is high enough. Combustion represents a form of degradation for polymers. When they burn, polymers, even those comprising mainly carbon and hydrogen, give off significant amounts of smoke, and in some case toxic gases (e.g., HCl, HCN, CO, HNO_3) as well as leaving a solid, carbonaceous residue.

Combustion of polymers occurs in two stages. In the first, a source of heat (e.g., a flame) raises the temperature of a polymer sample to such a level that bonds begin to break. This generates small, volatile, low molecular weight products. Thermoplastics melt, which assists these products in migrating through the polymer and out into the surrounding air where they burn, which is the second stage of the process. This burning produces heat, which is fed back into the polymer. For many thermoplastics, this process can become self-sustaining, and will continue until production of volatile materials from the polymer ceases. Thermosets tend to be relatively nonflammable. The presence of many cross-links between chains retards movement of the small volatile molecules. Heating may in fact cause more cross-linking to occur, which will hinder movement even further. The result is that the combustion zone surrounding the polymer is starved of organic fuel.

The small volatile molecules produced in the decomposition of thermoplastics are often the result of either depolymerization or elimination reactions. Depolymerization is essentially the reverse of polymerization, so that monomers are formed from the breakdown of polymer molecules. Combustion of poly(methylmethacrylate) or perspex, leads to the formation of methyl methacrylate in high yield, as shown below.

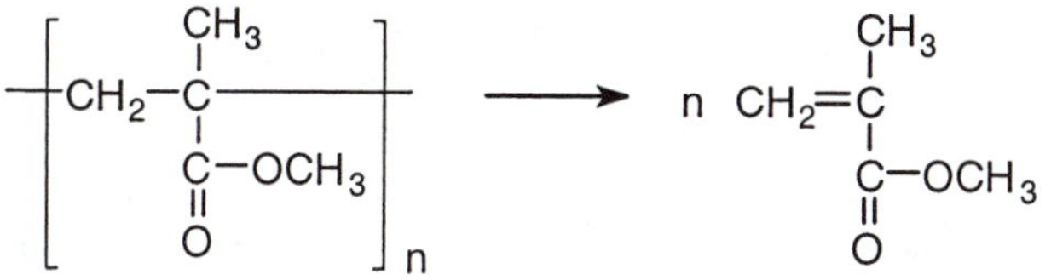

$$\left[-CH_2-\underset{\underset{O}{\overset{\|}{C}}-OCH_3}{\overset{CH_3}{\overset{|}{C}}}- \right]_n \longrightarrow n\ CH_2=\underset{\underset{O}{\overset{\|}{C}}-OCH_3}{\overset{CH_3}{\overset{|}{C}}}$$

Other polymers to undergo depolymerization include polystyrene and poly(oxymethylene).

$$\left[-CH_2O- \right]_n \longrightarrow n\ H_2C=O$$

Poly(oxmethylene) **Formaldehyde**

Poly(vinyl chloride) or PVC provides a good example of an elimination process. When PVC is heated, the polymer begins to discolor and evolve HCl. A carbon-carbon double bond is left in the chain when the HCl is eliminated. The HCl molecule can then catalyze loss of other HCl molecules. This produces conjugated sequences or stretches of alternating carbon-carbon double and single bonds, accounting for some of the color of thermally degraded PVC.

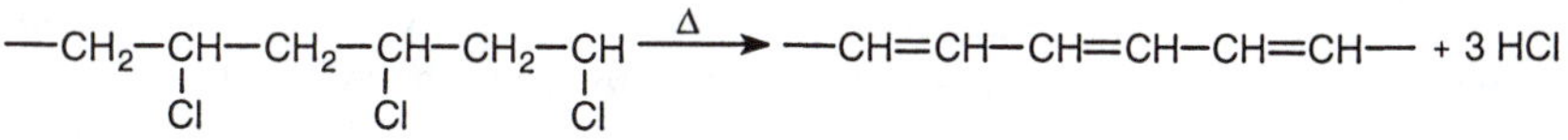

$$-CH_2-\underset{Cl}{CH}-CH_2-\underset{Cl}{CH}-CH_2-\underset{Cl}{CH} \xrightarrow{\Delta} -CH=CH-CH=CH-CH=CH- + 3\ HCl$$

Poly(vinyl chloride)

Polymers such as polyethylene and polypropylene degrade both by depolymerization and elimination, as well as more complex mechanisms.

B. BIODEGRADATION

Biodegradation of synthetic polymers involves breakdown, by microorganisms or higher organisms, using enzyme-catalyzed pathways. Most of the current applications of synthetic polymers, however, are based, at least partly, on their relative resistance to biodegradation as compared with natural macromolecular or polymeric substances such as starch and proteins. It would clearly be undesirable, for example, if plastic coating on telephone lines buried underground were to degrade, allowing corrosion of the metal fibers. Most of the synthetic polymers available today are bioresistant.

Biological systems tend to degrade natural polymers, e.g., starch by hydrolysis, then oxidation. It is not surprising then that most known biodegradable synthetic polymers have hydrolyzable functional groups in the main chain (e.g., ester, amide, urethane groups).

$$-\overset{\overset{\displaystyle O}{\|}}{C}-O- \qquad -\overset{\overset{\displaystyle O}{\|}}{C}-NH- \qquad -O-\overset{\overset{\displaystyle O}{\|}}{C}-NH-$$

Ester **Amide** **Urethane**

Aliphatic (i.e., non-aromatic) polyesters are among the most biodegradable. This is attributable not only to the possession of ester groups, but also the flexibility of the main chain. This flexibility facilitates the polymer chain binding to the active site of enzymes. The degradable and absorbable sutures employed in surgical procedures today are often made of polyglycolic acid, while another aliphatic polyester polycaprolactone is used as a matrix for the controlled release of drugs in the body.

$$n\ HO-CH_2-\overset{\overset{\displaystyle O}{\|}}{C}-OH \longrightarrow \left[O-CH_2-\overset{\overset{\displaystyle O}{\|}}{C}\right]_n$$

Glycolic acid **Poly(glycolic acid)**

$$n\ \text{Caprolactone} \longrightarrow \left[O-(CH_2)_5-\overset{\overset{\displaystyle O}{\|}}{C}\right]_n$$

Caprolactone **Polycaprolactone**

Thermoplastic materials made from polycaprolactone have been shown to almost completely degrade on burial in soil for 12 months. Here, microorganisms are presumably using the polymer as a growth substance, and converting it into microbial biomass, CO_2 and H_2O. Polyesters that contain aromatic rings in the main chain between the ester groups, e.g., PET, are not as degradable as their aliphatic counterparts. This is because the rigid aromatic ring restricts the flexibility of the main chain.

$$\left[-O-CH_2-CH_2-O-\overset{\overset{\displaystyle O}{\|}}{C}-C_6H_4-\overset{\overset{\displaystyle O}{\|}}{C}-\right]_n$$

PET or poly(ethylene) terephthalate

Lack of flexibility is also the reason why polyamides are less biodegradable than polyesters with analagous structures. It is well known that the internal rotation within an amide functional group is restricted compared to an ester group, because of considerable double bond character in the carbon-nitrogen bond.

$$\text{—C(=O)—NH—} \longleftrightarrow \text{—C(O}^-\text{)=N}^+\text{H—}$$

In many countries, the most widely used plastics are polyethylene, polypropylene, polyvinyl chloride and polystyrene. The polymers involved contain only carbon atoms in their main chain, with no hydrolyzable functional groups.

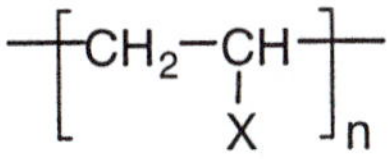

$$\left[CH_2-CH(X) \right]_n$$

$X = H$	**Polyethylene**
$X = CH_3$	**Polypropylene**
$X = Cl$	**Poly(vinyl chloride)**
$X = C_6H_5$	**Polystyrene**

These polymers are very bioresistant. Studies have shown that straight-chain alkanes are degraded by microorganisms such as fungi up until the alkane molecular weight reaches 500 (n-C_{35}). For larger alkanes, biodegradation becomes effectively zero. This is shown in Table 7.4, which also shows that chain branching inhibits biodegradation (a phenomenon also observed with surfactants). Given that polymers have a distribution of molecular weights, it would be expected that materials with low molecular weights (and hence a higher proportion of chain ≤35 carbons) and less branching would be more susceptible to biodegradation. It must be remembered though that chains greater than 35 carbons in length would be inert, and this is usually the majority of a polymer sample for materials such as polyethylene and poly(vinyl chloride). Cross-linking restricts access of enzymes to the polymer, and so most thermosets are not biodegradable. Most biodegradable plastics are technically unsuitable for uses such as packaging and thus cannot be readily substituted for the less biodegradable substances.

It is interesting to note that natural rubber has an all carbon backbone with branching off the main chain, yet is considered biodegradable. This may be due to the possession of the carbon-carbon double bond that could be the site of enzyme attack.

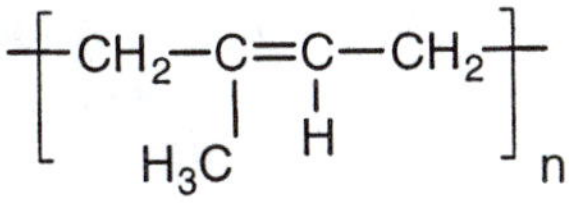

$$\left[CH_2-C(CH_3)=C(H)-CH_2 \right]_n$$

Rubber

There are a number of strategies that can be used to make synthetic polymers more biodegradable. One is to modify natural polymers, e.g., conversion of cellulose into cellulose nitrate and cellulose acetate. Cellulose nitrate has been used for movie film,

TABLE 7.4
Influence of Chain Length and Extent of Branching of Alkanes on Biodegradation by Fungi Mixture

Compound	MW	Relative biodegradability (O = Poor, 4 = Good)
Hexadecane ($C_{16}H_{34}$)	226	4
Octadecane ($C_{18}H_{38}$)	255	4
Tetracosane ($C_{24}H_{50}$)	339	4
Octacosane ($C_{28}H_{58}$)	395	4
Dotriacontane ($C_{32}H_{66}$)	451	4
Hexatriacontane ($C_{36}H_{74}$)	507	0
Tetracontane ($C_{40}H_{82}$)	563	0
2,6,11-Trimethyldodecane ($C_{15}H_{32}$)	212	0
2,6,11,15-Tetramethylhexadecane ($C_{20}H_{42}$)	283	0

From Potts, J.E. *et. al.*, The biodegradability of synthetic polymers in *Polymers and Ecological Problems*, J. Guillet, Ed., Plenum Press, New York , 1973. Copyright Plenum Press. With permission.

and old films using this material are often in a degraded state. Subjecting polymers to elevated temperatures or making them photodegradable can sometimes cause breaking of the main chain into fragments small enough for biodegradation to occur.

Biodegradable additives may also be included. Often, these additives serve as heat stabilizers or plasticizers, but also happen to be biodegradable. Occasionally, material such as starch is deliberately added to the synthetic polymer. It should be remembered that degradation of these additives cause the plastic to lose structural integrity and become brittle. The plastic eventually forms small pieces or even a powder, but the polymer itself does not degrade.

C. PHOTODEGRADATION

To undergo photodegradation in the environment, a polymer must contain a chromophore — a group capable of absorbing solar radiation. The lower wavelength limit of solar radiation reaching the surface of the earth is 290 nm. Radiation responsible for initiating photodegradation is in the 290 to 450 nm range, but particularly between 290 and 320 nm since this is of higher energy. Even if a polymer is potentially susceptible to photodegradation, this process cannot occur unless the material is actually exposed to sunlight. Therefore, plastics that are buried in a landfill, for example, cannot undergo photodegradation.

An approprate chomophore is the carbonyl structure, $>C=O$. When it occurs in the middle of a carbon chain, it is a ketone functional group and absorbance of radiation extends up to about 330 to 360 nm. A photodegradable polymer containing ketone groups can be made by copolymerizing ethylene (ethene) and carbon monoxide, as shown below.

$$CH_2{=}CH_2 + CO \longrightarrow -CH_2-\overset{\overset{\displaystyle O}{\|}}{C}-(CH_2)m-\overset{\overset{\displaystyle O}{\|}}{C}-(CH_2)n-\overset{\overset{\displaystyle O}{\|}}{C}-(CH_2)-$$

Upon absorption of solar radiation, the polymer chain is broken by either a Norrish Type 1 or Type 2 reaction.

$$R-CH_2CH_2\overset{\displaystyle O}{\overset{\|}{C}}CH_2CH_2-R' \xrightarrow{h\upsilon} R-CH_2CH_2\overset{\displaystyle O}{\overset{\|}{C}}{}^{\bullet} + {}^{\bullet}CH_2CH_2-R \longrightarrow R-CH_2CH_2 + CO$$

NORRISH TYPE 1

$$R-C(=O)-CH_2-CH_2-CH(H)-R \xrightarrow{h\upsilon} R-C(OH)=CH_2 + CH_2=CH-R \longrightarrow R-C(=O)-CH_3$$

NORRISH TYPE 2

The Type 1 reaction produces radicals that can undergo further reaction, and also evolves carbon monoxide. The Type 2 reaction yields a methyl ketone and an alkene as products. Interestingly, because of the particular range of wavelengths it absorbs, this copolymer will not photodegrade to any great extent indoors, but only outside in sunlight. Ordinary window glass absorbs most of the radiation less than about 330 nm. No radiation less than this is produced by incandescent light bulbs, and little by fluorescent lights. Therefore, a copolymer of this type should be useable behind windows under artificial light, but photodegradable outdoors in the environment.

The ester functional group contains a carbonyl structure; but for aliphatic polyesters, absorbance is only <250 nm, which is at a lower wavelength than for ketones. Aliphatic polyesters are therefore generally not susceptible to photodegradation. For aromatic polyesters, however, where the aromatic ring is in conjugation with the carbonyl, absorption up to and perhaps >300 nm can occur. Thus, PET, which is used for soft drink bottles, is potentially susceptible to photodegradation in the absence of any UV stabilizers.

$$\left[-O-CH_2-CH_2-O-\overset{\displaystyle O}{\overset{\|}{C}}-C_6H_4-\overset{\displaystyle O}{\overset{\|}{C}}-\right]_n$$

Chromophore

Poly(ethylene terephthalate) or PET

Unless stabilizers are incorporated in them, the four major thermoplastics (viz. polyethylene, polypropylene, poly(vinyl chloride) and polystyrene) degrade when exposed to the UV component of sunlight. This process is quite slow though, taking years in many cases, and so these materials can hardly be described as photodegradable. However, no photodegradation at all would be predicted by looking at the structure of these polymers, except perhaps for polystyrene. There is simply no suitable chromophore. The fact that commercial material does undergo photodegradation is due to branching and photosensitive impurities invariably introduced during polymerization and processing. Polypropylene gives a good example of why branching is important. Polypropylene has tertiary hydrogens that on removal, result in the formation of a relatively stable tertiary radical, as shown.

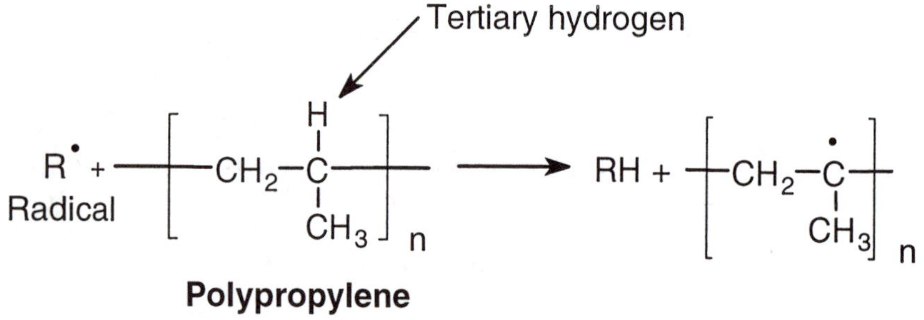

Tertiary hydrogen

$$\underset{\text{Radical}}{R^{\bullet}} + \left[CH_2 - \overset{\overset{H}{|}}{\underset{\underset{CH_3}{|}}{C}} \right]_n \longrightarrow RH + \left[CH_2 - \overset{\bullet}{\underset{\underset{CH_3}{|}}{C}} \right]_n$$

Polypropylene

Polypropylene is therefore more susceptible to photodegradation than polyethylene. Poly(vinyl chloride) degrades in sunlight by a mechanism much the same as that outlined earlier for thermal degradation. This occurs on the surface of the material only. The conjugated polyene structure that is formed on loss of HCl is a good absorber of UV radiation, and acts as a protective layer to the bulk poly(vinyl chloride) below, which is thus not directly exposed to sunlight.

In order to increase the photodegradability of polymers apart from introducing ketone groups into the main chain, introduction into branches off the main chain is sufficient. Thus, copolymers of styrene, methyl methacrylate and from 0.3 to 10% by mass of methyl vinyl ketone (or 3-buten-2-one, to be more correct) produces a polymer that degrades easily in sunlight.

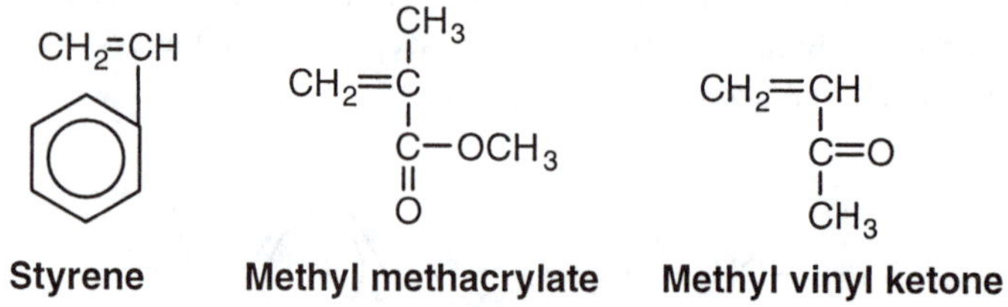

$CH_2{=}CH{-}C_6H_5$ **Styrene**

$CH_2{=}C(CH_3){-}C({=}O){-}OCH_3$ **Methyl methacrylate**

$CH_2{=}CH{-}C({=}O){-}CH_3$ **Methyl vinyl ketone**

Presumably, Norrish type reactions in the side chains can lead to the formation of radicals that can then attack the main chain of other polymers.

Various photosensitive additives can also be deliberately introduced to enhance photodegradation. Photosensitive additives are molecules, e.g., aromatic ketones or transition metal derivatives, that on absorption of sunlight form excited species which abstract hydrogen atoms from the main chain to form radicals and hence initiate polymer degradation. Note that photodegradation of polymers results in a weakening of the structure and a physical disintegration to small fragments. Chemically, the formation of smaller molecules means that the potential for biodegradation is increased.

D. RECYCLING

It would be desirable to recycle as much of the synthetic polymer waste as possible. Presently, some 30% of municipal solid wastes being placed in landfills is inorganics, e.g., glass, aluminium or plastics that are not readily biodegradable. Comparatively little of the energy required to produce a plastic is associated with processing, but the polymer itself represents a considerable investment of chemical energy. Some of this energy can be recovered by using the plastic as a fuel for incineration. This does not conserve as much energy as recycling though.

In theory, thermoplastic materials can be heated, softened and remolded repeatedly. In practice, while they are easily reprocessed, infinite recycling is impossible as degradation gradually occurs and contaminants are introduced. Ultimately, the recycled product acquires properties that render it unsuitable for its designed use. Thermosets cannot be heated and reformed into new products. However, they can be ground to a very fine powder, mixed with new materials and remolded. A high proportion of industrial and commercial plastic waste, perhaps greater than half, is recycled. This is because it is relatively clean, homogenous and in sufficient volumes to make it relatively economical.

Very little household plastic waste is recycled for exactly the opposite reasons. Plastics account for about 5% by mass, and 10 to 15% by volume of household garbage disposed of each year. In the U.S. in 1989, the overall recovery and reprocessing rate of plastics consumed by households was only 1 or 2%. The main plastics recycled were polyethylene (14% of quantity in household waste stream recovered and reprocessed) and PET (10%). Generally, however, it is not economical at present to collect, transport and segregate mixed plastic waste. Typical products made from reprocessed plastics are shown in Table 7.5. Recovering and reprocessing mixed plastic waste avoids the cost of segregating plastics, but the resulting product has variable properties, depending on the input material. Mixing incompatible polymers results in a lack of adhesion between the various phases, so that stresses cannot be transmitted. The products typically lack strength. Mixed reprocessed waste has not found significant applications and markets as yet.

TABLE 7.5
Typical Products Made from Reprocessed Polymers and Plastics

Polymer/plastic	Reprocessed products
Poly(ethylene terephthalate)	Carpet fiber, insulation, furniture stuffing
Polyethylene	Pipes, pot plant holders, garbage bags
Polyurethane foam	Carpet underlay, industrial paddings
Polyvinyl chloride	Garden hose, shoe soles
Mixed plastic waste	Replacement for timber, and other structural materials

VI. KEY POINTS

1. Synthetic polymers can be classified as plastics, elastomers and synthetic fibers and are used in many applications, including plumbing, textiles, paint, floor covering and clothing.

2. Packaging is the largest and fastest growing single market for synthetic polymers with a lifetime of less than a year before they are discarded as waste.

3. Synthetic polymers are generally resistant to environment transformation and degradation and may present a problem for the disposal of solid wastes and as litter.

4. Polymers consist of macromolecules built up from molecular units described as monomers. Synthetic polymers have molecular masses of 1000 or more, a degree of polymerization ($\overline{DP}$) of at least 100 and can have a linear, branched or cross-linked structure.

5. There are many natural polymers, including cellulose, DNA, RNA, proteins and rubber.

6. The physical and chemical properties of a synthetic polymer depend principally on
   - Molecular structure
   - Molecular weight
   - Component monomers

7. Some common synthetic polymers can be represented by the chemical structures below.

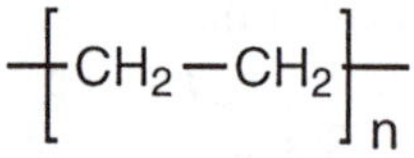

$$\left[CH_2 - CH_2 \right]_n$$ **Polyethylene**

$$\left[CH(CH_3) - CH_2 \right]_n$$ **Polypropylene**

$$\left[CH_2(C_6H_5) - CH_2 \right]_n$$ **Polystyrene**

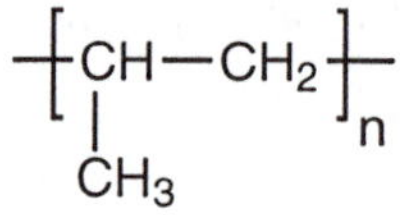

$$\left[CH - C(CH_3)(C{=}O)H_2 \right]_n$$ **Poly(methyl methacrylate) (Plexiglas, Lucite)**

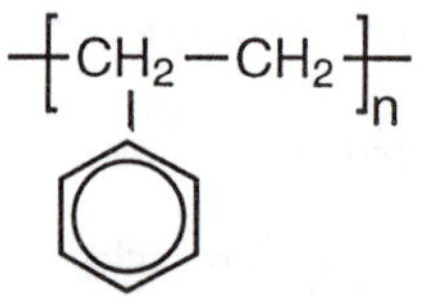

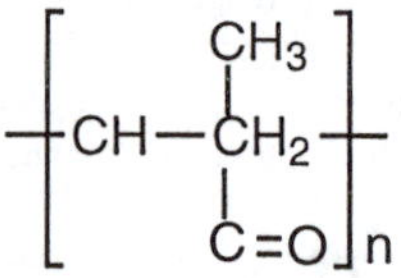

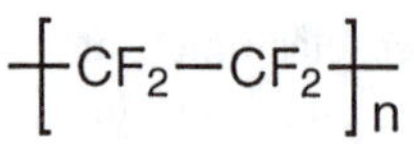

$$\left[CF_2 - CF_2 \right]_n$$ **Teflon**

8. Polymers can be grouped according to whether the polymerization occurs in a stepwise manner (step growth) or from a growing chain (chain growth).

9. Combustion of polymers occurs in two stages. First, a source of heat leads to the breakage of bonds and formation of small molecules, which burn in the second stage of the process. Depolymerization reactions can occur in which molecules of the monomer are produced and then burnt. Also, elimination reactions occur, with some polymers leading to the formation of small molecules such as HCl.

10. The commonly used synthetic polymers are generally resistant to biodegradation. Biodegradation capacity can be improved by:
    - Mixing with natural biodegradable substances such as starch or proteins
    - Incorporation of hydrolyzable functional groups such as esters, amides and urethanes
    - Ensuring the main chain is flexible and thus able to facilitate binding to active sites of enzymes

11. To undergo photodegradation, the polymer should contain a chromophore capable of absorbing solar radiation, particularly in the range 290 to 320 nm. This can be done by incorporating carbon monoxide into a polymer to yield a ketone group in the molecule. Aromatic rings conjugated with carbonyl groups, such as in poly(ethylene terephthalate) (PET), render the polymer potentially susceptible to photodegradation.

REFERENCES

Nicholsen, J.W., *The Chemistry of Polymers,* Royal Society of Chemistry, Cambridge, England, 1991.

Cowie, J.M.G., *Polymers: Chemistry and Physics of Modern Materials,* Intext Education, New York, 1973.

CHAPTER 7 PROBLEMS

1. A step polymerization is being carried out and the final product requires a degree of polymerization ($\overline{DP}$) of 100. How much of the monomer must react to give this product?

2. A polymer has the following chemical structure:

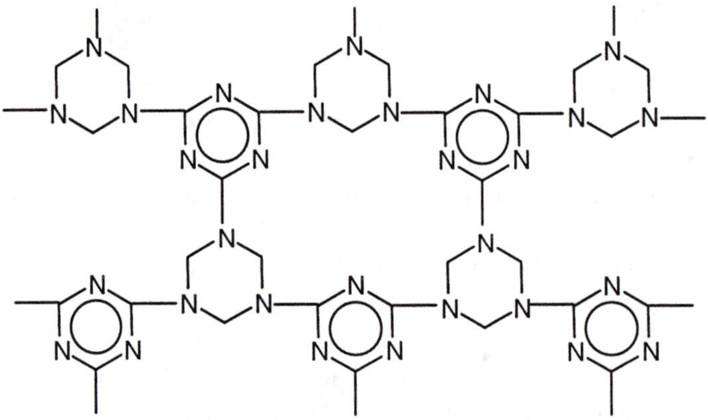

Based on this structure, briefly comment on the following properties:

   a. Physical properties such as melting point, etc.
   b. Solubility in solvents
   c. Combustibility
   d. Biodegradability in the environment

3. A polymer has the following structure:

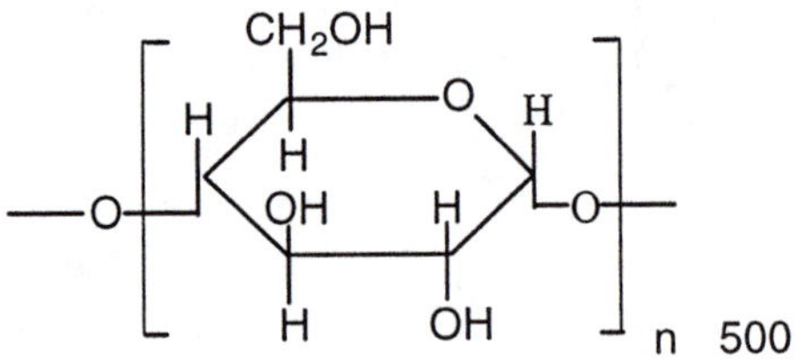

CH_2OH H H H OH H H OH n 500

Glucose

Based on this structure briefly comment on the following properties:

   a. Physical properties
   b. Solubility in solvents
   c. Combustibility
   d. Biodegradability

Answers on Page 181.

CHAPTER 7 SOLUTIONS

1. The degree of polymerization $\overline{DP}$ can be related to the proportion of monomer reacted as follows:

$$\overline{DP} = \frac{1}{1-p}$$

$$100 = \frac{1}{1-p}$$

$$p = 99\%$$

2. a. Due to the many cross-links, this polymer is not likely to possess a melting point and be brittle. It will tend to decompose rather than melt.
   b. Due to the extensive cross-linking, this polymer is unlikely to be soluble in any solvent.
   c. This polymer will be resistant to combustion and the formation of small molecules by heating.
   d. This polymer lacks any functional groups that are susceptible to hydrolysis or oxidation and is inflexible and thus likely to be resistant to biodegradation.

3. a. This polymer has limited cross-linking and consists essentially of long chains. It would be expected to be flexible.
   b. The polymer has many polar groups (–OH) and would be expected to exhibit some solubility in polar solvents, such as water, depending on chain length.
   c. This polymer would be expected to decompose on heating to yield small combustible molecules and thus burn fairly readily.
   d. The polymer contains many groups susceptible to chemical attack, such as hydroxyl and ether groups, and thus would be expected to biodegrade.

Chapter 8

PESTICIDES

I. INTRODUCTION

Pesticides are chemicals used to remove pests such as insects and weeds. Of all the environmental contaminants, pesticides have probably been the most widely criticized due to their direct use in natural systems. The nature of pesticide usage often requires broadcast distribution over large areas of crops. This wide treatment of the crop environment often results in treatment of adjacent areas as well. This in itself creates concern since people and natural organisms are exposed to these chemicals. The use of pesticides in agriculture and other areas is not recent. Prior to the 1940s, insecticides such as lime, sulfur, nicotine, pyrethrum, kerosene and rotenone were extensively used. This group of pesticides suffer from a number of deficiencies. In particular, they lack potency with a wide range of insects. In addition, they lack persistence in the environment so that repeated usage is frequently required. Finally, as a group, these substances are costly to produce and use.

A major development in insect control came in 1939 when the Swiss chemist Paul Müller working for the Geigy Company, patented DDT as an insecticide. DDT was not a new substance, but its insecticide activity was previously unknown. It was first prepared by Ziedler in Germany in 1874. It found extensive use during World War II due to its relatively long persistence, cheap cost and potency to a wide range of insect species. It achieved such a high level of success in helping control food pests and pests bearing human diseases that Müller was awarded the Nobel Prize in 1948 for its development.

Usage of DDT and related chlorohydrocarbon insecticides rapidly accelerated during the 1940s and subsequent decades, and the organophosphate pesticides became widely used as well. Little thought was given to ecological implications although there were a few reports of possible consequences due to the occasional "kills" of fish and other aquatic organisms associated with the use of these substances. In 1962, the book *Silent Spring* was published by Rachel Carson. This book raised many possible ecological problems that could be associated with the usage of DDT. It has had a major influence in that it initiated a large research program in the U.S. and other countries throughout the world. As a result, there have been many scientific and governmental inquiries into the usage of DDT and other chlorohydrocarbons. It is now clear that the use of DDT has caused a range of problems, including direct lethal effects and sublethal effects such as eggshell thinning in certain species of birds.

There are now severe restrictions on the usage of DDT and other chlorohydrocarbon compounds as pesticides, due to their adverse environmental effects. Thus, usage of these chemicals has declined and many have totally disappeared in a range of countries throughout the world. However, concern about public health and food production has led to manufacture and continued usage of DDT and related compounds in many tropical countries. The low cost and effectiveness of these substances in the tropical environment is a major factor in this continued usage.

The sale and use of a chemical in agriculture, and elsewhere, now usually requires registration with a governmental agency. Many steps are currently necessary to introduce a chemical into the market as a pesticide. These start with synthesis, screening and proceeds through trials, evaluation, and finally end with registration. Aside from the costs involved in the selection of a compound for effectiveness and economical manufacture, there is the testing for environmental effects now required as a major component of registration. Thus, it would be expected to take about 5 years, with costs up to many tens of million dollars to place a new pesticide on the market.

In a bid to find alternatives without adverse environmental effect, there has been extensive testing of natural pesticides known to occur in plants. Many potent pesticides have been isolated but of these, only **pyrethrum** has proven to be a commercial success, finding wide acceptance in the community. **Pyrethrum** is a natural pesticide prepared from the dried, powdered flowers of the chrysanthemum (*Chrysanthemum cinerariaefolium*). Synthetic compounds related to these natural compounds are also prepared. In addition, a variety of other techniques have been used, such as biological control, that have helped reduce chemical usage.

Pesticides can be conveniently divided into classes, depending on which particular pest they are directed toward. Thus, the main groups are the **herbicides, insecticides** and **fungicides**. The various chemical classes within these groups are shown in Table 8.1. Pesticides share only one common property: their toxicity to organisms regarded by humans as pests. Chemically, they are a very diverse group, ranging from metal compounds through to a range of diverse organic chemicals. The major groups of pesticides are considered below.

TABLE 8.1
Some Chemical Classes in the Various Groups of Pesticides

Herbicides
Carbamates, phenoxyacetic acids, triazines, phenylureas
Insecticides
Organophosphates, carbamates, organochlorines, pyrethrins, pyrethroids
Fungicides
Dithiocarbamates, copper, mercurials

II. THE CHLORINATED HYDROCARBON PESTICIDES

This group of substances is referred to as the **chlorinated hydrocarbons**, the **chlorohydrocarbons** or the **organochlorines**. Some caution should be adopted with the term "**organochlorines**" since this includes such pesticides as 2,4-D and 2,4,5-T, which have quite different properties. DDT is the most prominent member of the chlorohydrocarbon group, and its history and usage has had a major impact on the use of pesticides in particular and considerations of hazardous chemicals in the environment in general. After its introduction during World War II, its usage gradually expanded and extended into various parts of agriculture. It has been manufactured in many countries throughout the world in large quantities, with about 100,000 tons per year produced in the U.S. in the late 1950s. This declined to about 20,000 tons in 1971 due to concerns regarding its impact on the natural environment. The very properties that stimulated its usage had become a liability. Cheap production in large quantities and persistence in the environment reduced the need for retreatment, but resulted in many problems in environmental management. With the early success of DDT, a range of related chlorohydrocarbons were produced for particular applications. Most of these substances now face similar restrictions to DDT.

A. CHEMICAL STRUCTURE AND SYNTHESIS

The properties of the hydrocarbons were considered in Chapter 5. In general, these substances can be divided into three broad classes: **alkanes, alkenes** and **aromatic hydrocarbons.** Members of all these groups can form chlorinated derivatives. The C–Cl bond can be formed by substitution reactions of the various hydrocarbons with chlorine. For example, methane can undergo substitution reactions with chlorine to form carbon tetrachloride (tetrochloromethane) and chloroform (trichloromethane). In addition, benzene can form chlorobenzenes by similar substitution reactions. The chlorohydrocarbons can also be formed by addition reactions with alkenes.

DDT was first prepared in 1874 by condensation of chloral (trichloroacetaldehyde) with monochlorobenzene, and this has remained the basis for commercial synthesis since that time. An outline of the reaction is shown in Figure 8.1. It can be seen that two isomers are produced, with the more effective 4,4′-isomer produced in greater quantities. Usually, commercial DDT contains significant amounts of the 2,4′-isomer as well.

Commercial chlorohydrocarbon insecticides, apart from DDT, include heptachlor, lindane, dieldrin, chlordane, aldrin, endrin and mirex. Some chemical structures of these substances are shown in Figure 8.2. Within this group, there is a subgroup described as the **cyclodiene group** that has properties enabling a distinction to be made from other members. Heptachlor, dieldrin, chlordane and aldrin are usually included in this group. Members are produced by the Diels-Alder reaction of hexachlorocyclopentadiene with cyclopentadiene, as shown in Figure 8.3, or with other substances containing at least one carbon-carbon double bond. Another compound often included in this group is **endosulfan**. Strictly, endosulfan is not a chlorohydrocarbon and not a member of the cyclodiene group because, while it contains a part of the molecule that is chlorohydrocarbon in nature, its properties are strongly influenced by the presence of a sulfur-containing group. The structure of endosulfan is shown below,

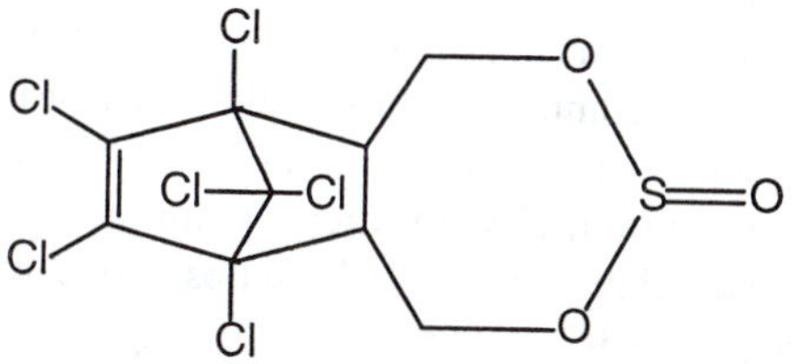

Endosulfan

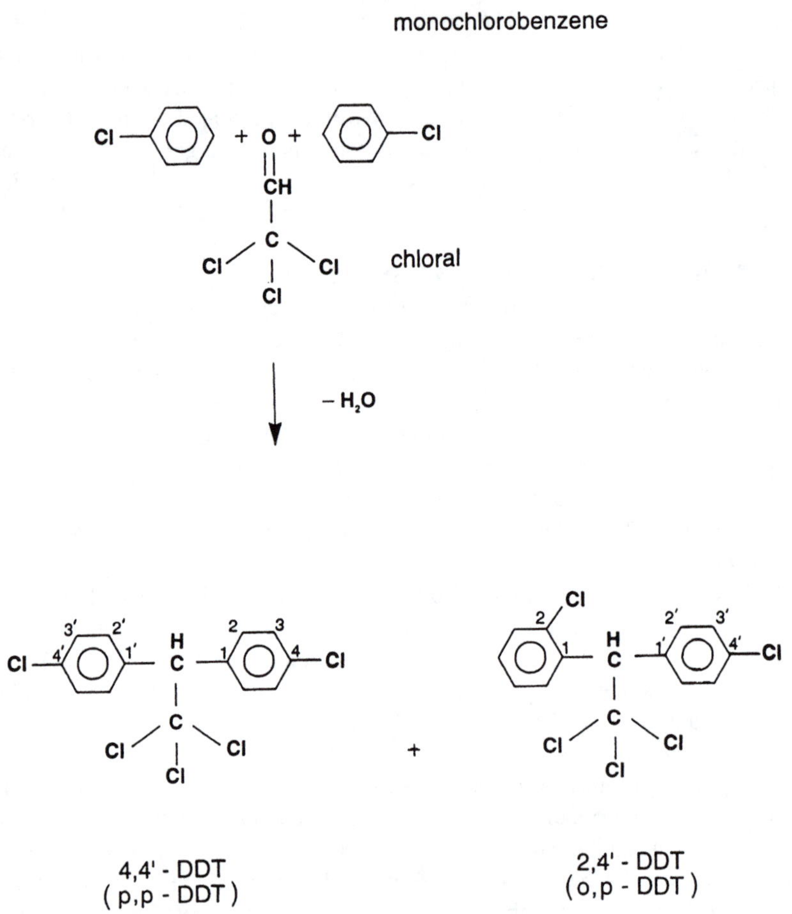

monochlorobenzene

chloral

$- H_2O$

4,4' - DDT
(p,p - DDT)

2,4' - DDT
(o,p - DDT)

FIGURE 8.1 Synthetic reaction process for the production of DDT.

B. PHYSICAL CHEMICAL PROPERTIES

There are a limited range of bond types present in this group. These are the C---C (aromatic), C=C, C–H, C–Cl with lesser numbers of C–C. The symmetrical bonds, C---C (aromatic), C–C and C=C have dipole moments close to zero. Only C–H and C–Cl bonds have dipole moments of 0.4 and 0.5 Debyes, respectively, though these are relatively low. This means the compounds in this group tend to have low polarity and dipole moments. These properties result in compounds in this group being **fat soluble** or **lipophilic** and having a low solubility in water. This is illustrated by some typical examples shown in Table 8.2. While the solubility in lipid of all of these compounds lies on the order of hundreds to thousands of grams per litre, the aqueous solubility ranges from only a few $\mu g\ L^{-1}$ to several hundred $\mu g\ L^{-1}$. The lipophilicity of these compounds is indicated by the octanol/water partition coefficient (K_{OW}), which lies between 470 ($\log K_{OW} = 2.67$) and 2,300,000 ($\log K_{OW} = 6.36$).

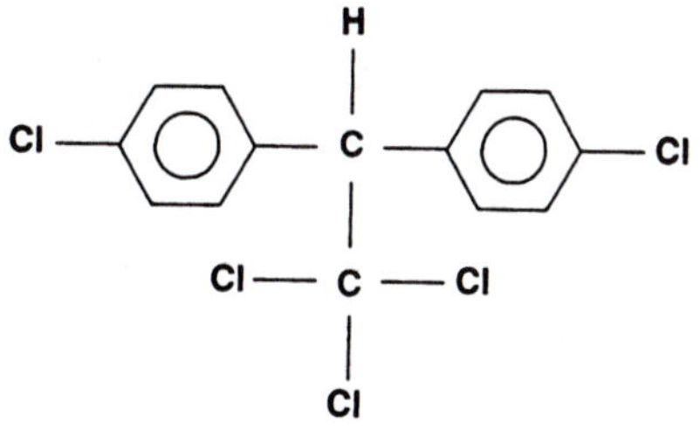

DDT
(dichlorodiphenyl trichloroethane or
1,1 - bis(4 - chlorophenyl) -2,2,2-
trichloroethane)

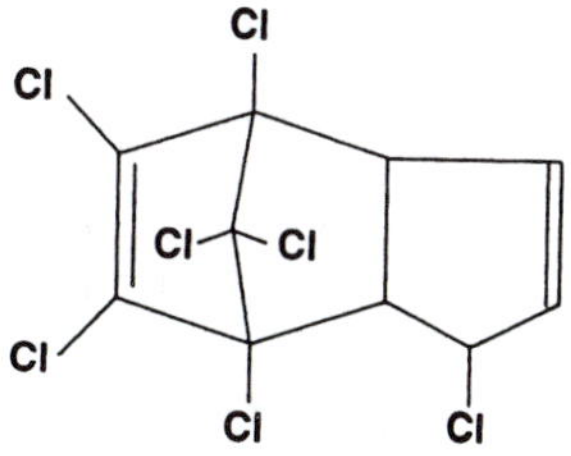

heptachlor

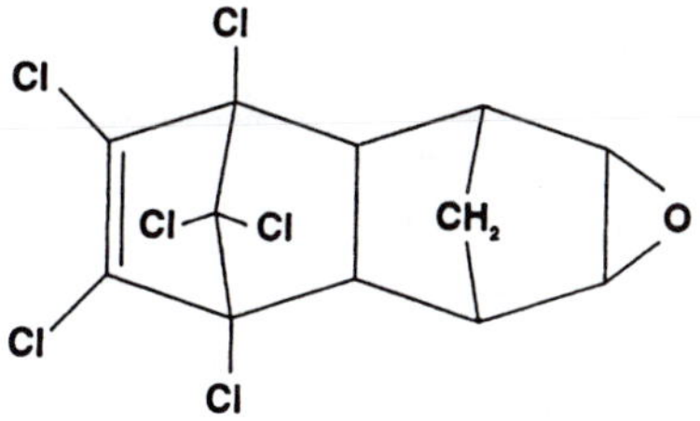

dieldrin

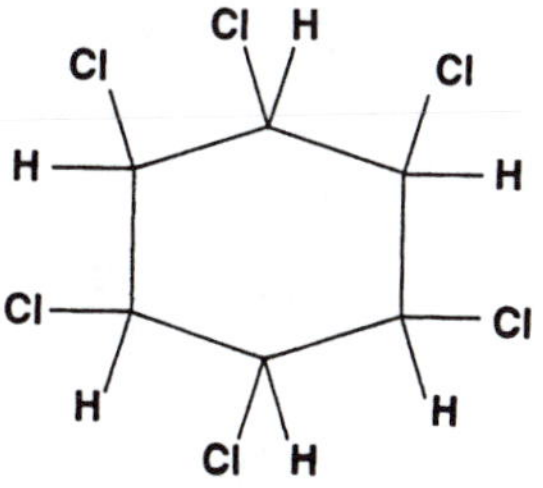

lindane (γ isomer)
(1,2,3,4,5,6 - hexachloro
cyclohexane also known as
benzene hexachloride)

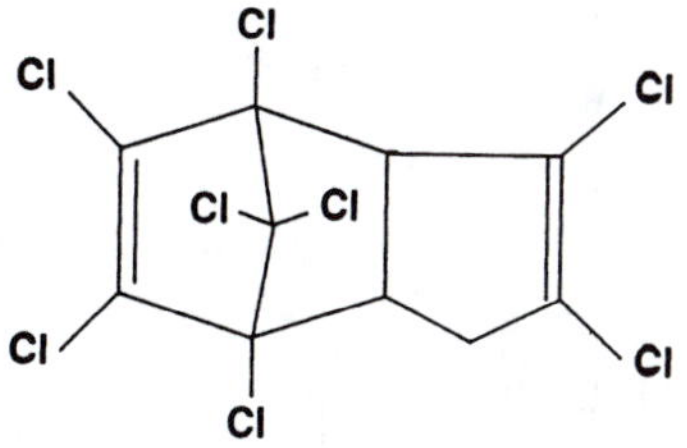

chlordane

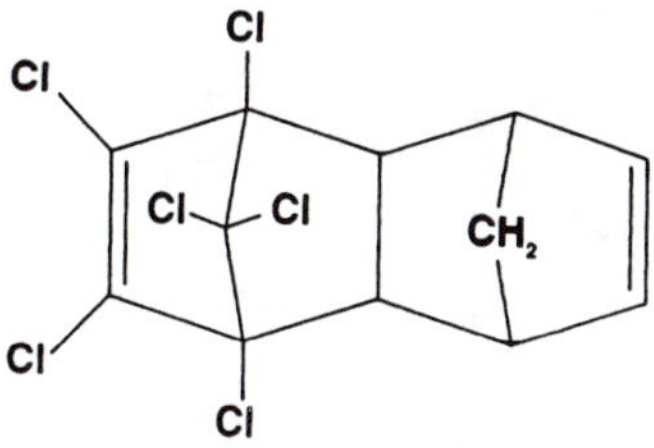

aldrin

FIGURE 8.2 Examples of the chemical structures of some insecticides included in the chlorohydrocarbon group.

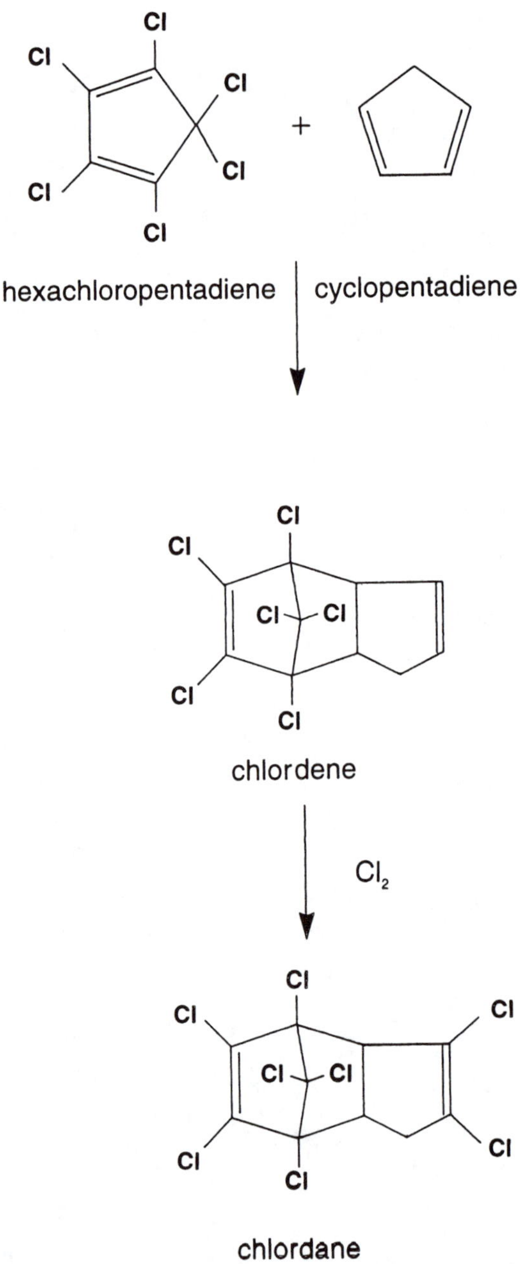

FIGURE 8.3 Synthesis of chlordane.

C. ENVIRONMENTAL PROPERTIES

The limited range of bond types present in the chlorohydrocarbon pesticides are generally relatively resistant to attack by abiotic or biotic agents in the environment. As a result, environmental degradation proceeds at a relatively slow rate. Most compounds in this group persist for long periods in soil and often exhibit half-lives

TABLE 8.2
Physicochemical Properties of Some Typical Chlorohydrocarbon Pesticides

Compounds	Solubility in lipid ($g\ L^{-1}$)	Aqueous solubility ($mg\ L^{-1}$)	Log K_{OW}
4,4′-DDT	330	3.36	6.36
Heptachlor	1000	50	4.11
Dieldrin	3700	200	3.88
Lindane	800	130	2.67

of many years, as illustrated by the data in Table 8.3. The half-lives can range from approximately 0.04 to 15.6 years for the compounds considered. The variability is due to the range of different conditions that can occur in soil. For example, soil moisture and temperature can vary considerably, affecting the microbial population and its growth rate and resulting in a considerable impact on the degradation rate.

TABLE 8.3
Toxicity and Persistence of Various Chlorohydrocarbon Pesticides

Compound	LD_{50} ($mg\ kg^{-1}$ body rats)	LC_{50} (estuarine fish $\mu g\ L^{-1}$; 96 hr)	EC_{50}[a] (Daphnia, $\mu g\ L^{-1}$; 48 hr)	Half-life range (soil, years)
4,4′-DDT	115	0.4–89	0.36	2.0–15.6
Dieldrin	50	0.9–34	250	0.5–3.0
Lindane	125	9–66	460	0.04–0.7
Aldrin	50	5–100	28	0.06–1.6

[a] Concentration that induces immobilization rather than lethality.

There are a range of possible environmental degradation patterns for the chlorohydrocarbon pesticides. A pathway for the biodegradation of DDT is shown in Figure 8.4. All pathways with all compounds would be expected to involve hydrolysis and oxidation at various stages. The ultimate products of degradation would be expected to be carbon dioxide, water and other substances.

D. BIOCONCENTRATION PROPERTIES

Bioconcentration in aquatic organisms occurs as a result of partitioning between the organism lipid and the surrounding water. If a substance is **lipophilic**, then equilibrium occurs with a relatively high concentration in the biota lipid as compared to water. The **Bioconcentration Factor (K_B)**, or the organism/water partition coefficient, is defined as follows:

$$K_B = \frac{C_{organisms}}{C_{water}}$$

The bioconcentration factor is a characteristic of a particular chemical and lipid content if it is on a whole-weight basis, just as its solubility in water, vapor pressure, melting point are also characteristics. To exhibit high bioconcentration, a compound

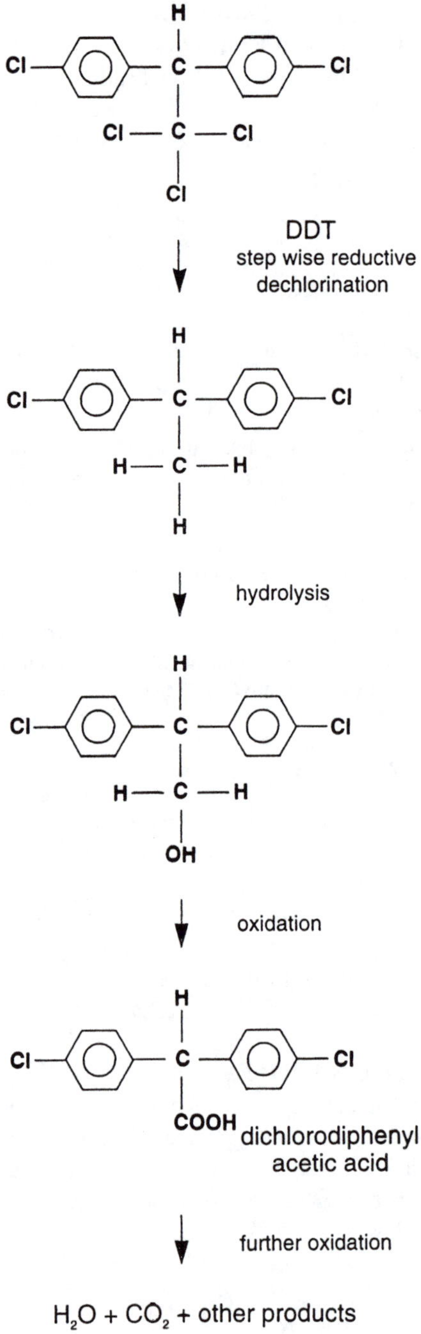

DDT

step wise reductive dechlorination

hydrolysis

oxidation

dichlorodiphenyl acetic acid

further oxidation

$H_2O + CO_2$ + other products

FIGURE 8.4 A pathway in the environmental degradation of DDT.

must also persist in the organism rather than be biodegraded and removed, resulting in lower concentrations in the organism. The chlorohydrocarbons have the properties of high lipophilicity, indicated by log K_{OW} values usually between 2 and 6.5, as well as persistence, and thus would be expected to exhibit a strong bioconcentration capacity. For example, the K_B value of 4,4′–DDT is 52,500 (log K_B = 4.72); heptachlor is 12,900 (log K_B = 4.11); dieldrin is 7600 (log K_B 3.88); and lindane is 470 (log K_B = 2.67). The K_{OW} value and lipophilicity of compounds are described in Chapter 2.

E. TOXICITY

The chlorohydrocarbons as a group tend to have a wide range of activity with different insects and related organisms. Also, the members of the group are usually powerful fish toxicants, despite relatively low aqueous solubility, as illustrated by the data in Table 8.3. All compounds listed are highly toxic to a range of species of fish at concentrations much less than 1 mg L^{-1}. Also, these compounds are generally quite toxic to other aquatic species such as Daphnia (see Table 8.3). The cyclodiene subgroup, e.g., dieldrin and aldrin, tend to have higher mammalian toxicity than other members of the chlorohydrocarbon family (see Table 8.3).

The mode of action of the compounds in this group is not fully clear. They are neurotoxins since they act on the nervous system, producing tremors followed by loss of movement, convulsions and death. DDT appears to act on the nerve axon. All members are strong inducers of mixed-function oxidase (MFO) in exposed organisms, which was considered in Chapter 3. Resistance to the chlorohydrocarbons has been observed in pesticide populations that have been repeatedly treated with a particular pesticide. With DDT, its often due to a dehydrochlorinase removing HCl and producing DDE, so resistance is concerned with the development of this enzymatic capacity. Resistance usually takes several generations to develop and may be specific for a particular pesticide.

F. ECOLOGICAL EFFECTS

The major ecological effect observed has been the reduction in the reproductive success of carnivorous birds, such as the Peregrine falcon with DDT. This has been shown to be due to interference with an enzyme system, causing reduction in the development of the eggshell in eggs produced. As a general rule, a 20% reduction in shell thickness leads to a reduction in the population due to the damage and destruction of eggs during the brooding process. A wide range of other ecological effects have been observed in aquatic and terrestrial ecosystems.

III. THE ORGANOPHOSPHATE INSECTICIDES

This group of substances came under intensive investigation during World War II for use as military gases. Initially, these compounds were considered quite unsuitable for agricultural use due to their high mammalian toxicity. This was particularly so since the chlorohydrocarbon pesticides were available which generally have relatively low mammalian toxicity; but with the various environmental problems that have become apparent with the chlorohydrocarbon pesticides, a great deal of attention has been focused onto the organophosphate group for development as commercial pesticides. In recent years, a wide range of organophosphate insecticides have been developed that are acceptable for agricultural use.

A. CHEMICAL FORM AND SYNTHESIS

The organophosphate pesticides have the following general formula:

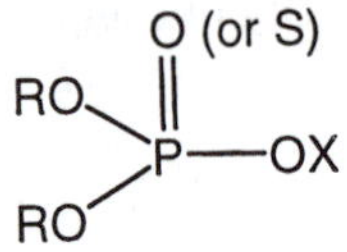

O (or S)
RO
P—OX
RO

The two R groups are usually methyl or ethyl groups. The oxygen atom in the OX group can be replaced by S with some compounds. Some examples of the chemical structures of this group are shown in Figure 8.5. Although the group has a common core structure, there is still considerable diversity due to variations in the attached chemical groupings.

B. PHYSICAL CHEMICAL PROPERTIES

The defining chemical structure of the organophosphate pesticides contains one P=O and three P–O bonds. From Table 2.3, the oxygen atom has an electronegativity of 3.5 and phosphorus of 2.1. The phosphorus electronegativity is comparable with hydrogen, which is also 2.1. Thus, the O–P bond would be expected to have similar polarity to the O–H bond and thus to be polar. At the same time, the molecule usually contains a range of other bond types, including O–alkyl which is of relatively low polarity. These compounds would generally be expected to have greater water solubility than the chlorohydrocarbons and lower lipophilicity. For example, the compounds in Table 8.4 have water solubilities in the range from 25 to 10,000 mg L^{-1}, whereas the chlorohydrocarbons (see Table 8.2) range from 6 to 200 µg L^{-1}. The K_{OW} values lie in the range from 2.71 to 3.81 (see Table 8.4), which is in the lower range of the lipophilic compounds. Thus, it would be expected that, depending on the identity of the R and particularly the X groups in the molecule, the organophosphate pesticides can have a range of physical chemical properties.

TABLE 8.4
Physicochemical Properties of Various Organophosphate Pesticides

Compound	Water solubility (mg L^{-1})	Log K_{OW}	Half-life range (soil, days)	LC_{50} (mg/kg body weight)	LC_{50} (Estuarine fish, mg L^{-1}; 96 hr)	EC_{50} (Daphnia, mg L^{-1}; 48 hr)
Malathion	145	2.89	3–7	2,500	27–3250	1.8
Parathion	24	3.81	7–10	—	—	0.60
Dichlorvos	10,000 (approx)	—	—	63	0.05–3.1	0.07
Dimethoate	25	2.71	11–37	500	—	—

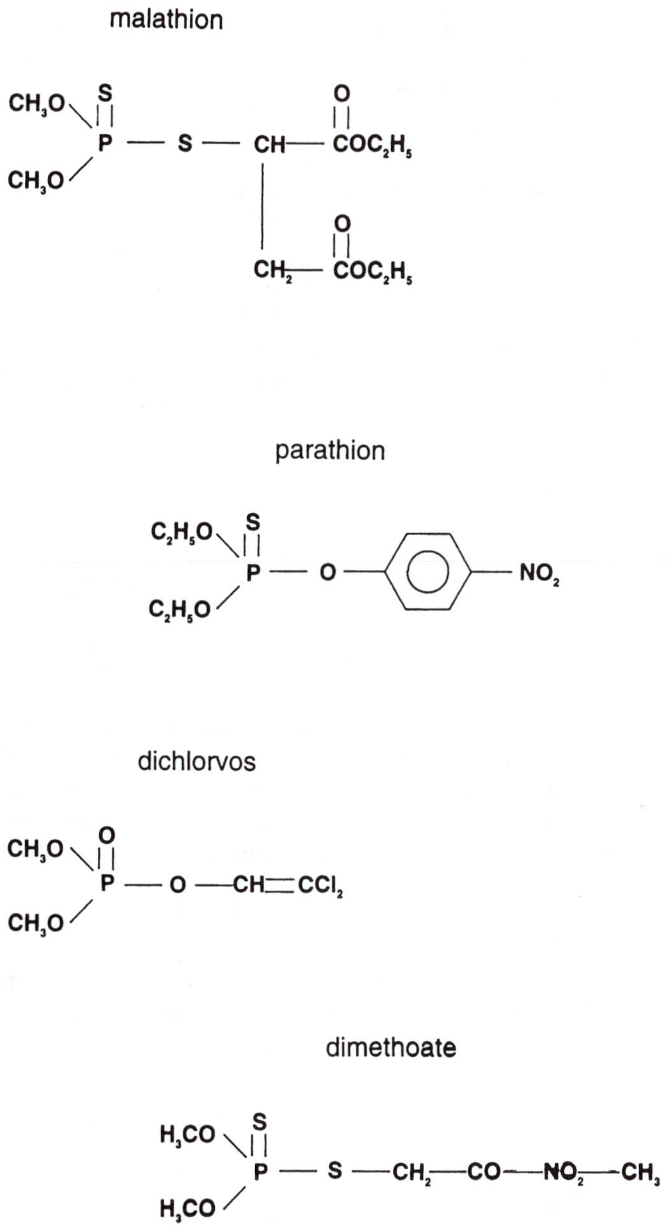

malathion

$$(CH_3O)_2P(=S)-S-CH(COOC_2H_5)-CH_2-COOC_2H_5$$

parathion

$$(C_2H_5O)_2P(=S)-O-C_6H_4-NO_2$$

dichlorvos

$$(CH_3O)_2P(=O)-O-CH=CCl_2$$

dimethoate

$$(H_3CO)_2P(=S)-S-CH_2-CO-NO_2-CH_3$$

FIGURE 8.5 Chemical structures of some common organophosphate pesticides.

C. ENVIRONMENTAL PROPERTIES

The organophosphate pesticides are a chemically reactive group of compounds. For example, they are susceptible to hydrolysis when they come into contact with water. This reaction is dependent on the pH with higher pH values, giving more rapid rates of reaction. The half-lives in soil are considerably less than the chlorohydrocarbons and usually range up to about 40 days, as illustrated by the data in Table 8.4. Their lack of persistence in soil indicates a similar lack of persistence in biota and this, together with their moderate water solubility and relatively low lipophilicity, leads to a lack of bioaccumulation capacity.

D. TOXIC ACTIVITY

The organophosphate pesticides exhibit strong toxic activity with a wide range of biota. The biota affected range from mammals through to insects. All chemicals in this group act by inhibiting the action of several ester-splitting enzymes present in living organisms, and they are particularly active in inhibiting acetylcholinesterase (ACh). Cholinesterase is an important enzyme that facilitates the transmission of nerve impulses. It operates by hydrolyzing the substance acetylcholine, which is generated in the transmission of nerve impulses. Acetylcholine contains an ester grouping that is the focus of the action of the enzyme. The structure of this substance is shown below.

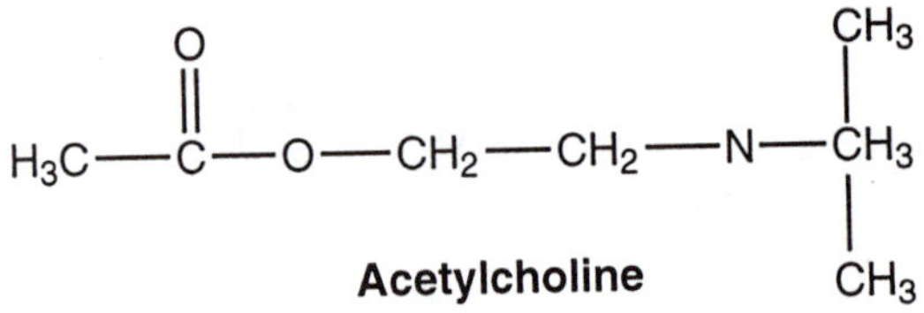

$$H_3C-\overset{O}{\overset{\|}{C}}-O-CH_2-CH_2-N\overset{CH_3}{\underset{CH_3}{<}}CH_3$$

Acetylcholine

Acetylcholine normally has a fraction of a second to interact before it is hydrolyzed by the acetylcholineesterase. The receptor site is thus cleared for the next incoming signal. If this substance is not removed, the acetylcholine accumulates and interferes with the coordination of muscle response. This interference with muscular function of vital organs can produce serious symptoms and eventually death.

IV. THE CARBAMATES

Carbamate insecticides are frequently employed in situations where insects do not respond to the organophosphate compounds. They tend to be more expensive to produce than the organophosphate compounds. The **carbamate** pesticides have the following general formula:

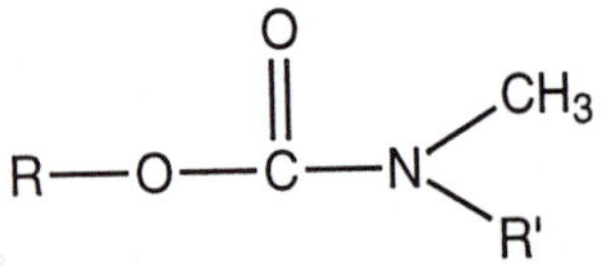

$$R-O-\overset{O}{\overset{\|}{C}}-N\begin{matrix}CH_3\\ R'\end{matrix}$$

where R can be a variety of groups often containing aromatic rings, and R′ can be hydrogen or other groups.

Examples of carbamate insecticides include Propoxur, also known as Baygon, and Carbaryl, the structures of which are illustrated below.

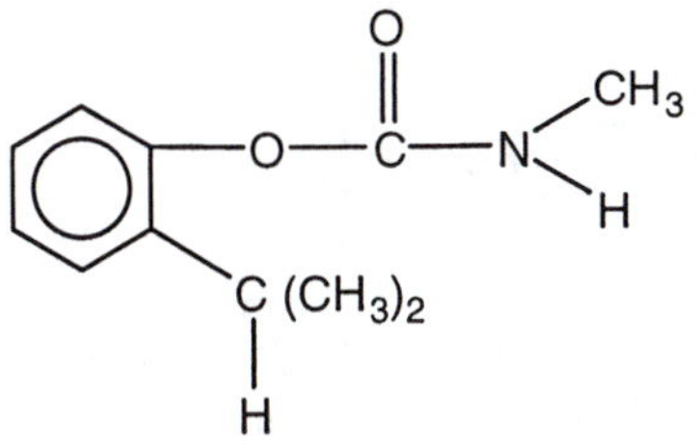

Propoxur 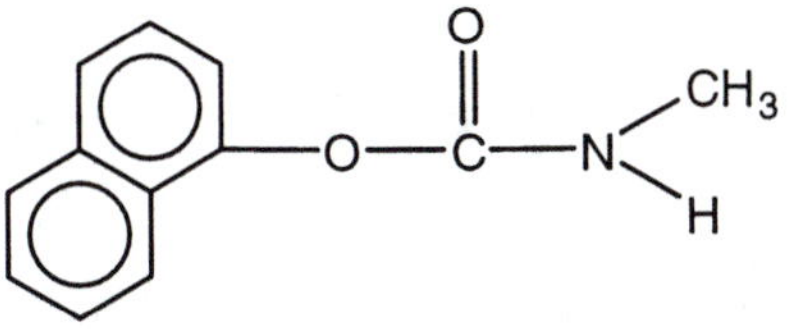 **Carbaryl**

A. ENVIRONMENTAL PROPERTIES

The general properties of the carbamates are somewhat similar to the organophosphate compounds and are summarized below:

1. The carbamate functional group can be regarded as a hybrid between ester and amide groups, both of which are polar (see Table 2.4), leading to a reasonably polar molecule.
2. These substances are relatively water soluble in comparison with the chlorohydrocarbons. For example, Propoxur dissolves to the extent of about 2 g L^{-1} and Carbaryl to the extent of about 120 mg L^{-1}.
3. They have limited persistence in the environment, which is reflected by the presence of reactive groups in the molecule. For example, the ester group is easily hydrolyzed when this substance comes in contact with water in the environment. The group, in general, is extremely susceptible to hydrolysis, particularly in alkaline water. This means that they will exhibit low persistence in the aquatic environment, which has been found to range from about 1.6 days to 4 weeks. The hydrolysis reaction is illustrated below.

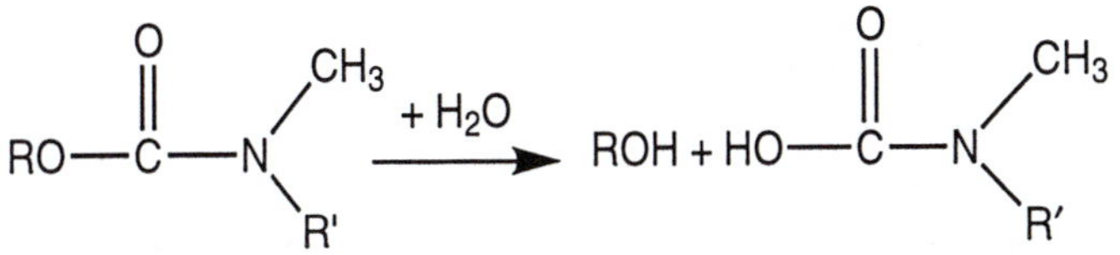

$$RO{-}C({=}O){-}N(CH_3)R' \xrightarrow{+\,H_2O} ROH + HO{-}C({=}O){-}N(CH_3)R'$$

   The carbamate insecticides also have generally short lives in soils, with similar half-lives to those observed in the aquatic environment.
4. Some are extremely toxic to mammals and are readily absorbed through the skin, but tend not to bioaccumulate, as would be expected due to their relatively low persistence and high solubility in water.
5. The principal biodegradation pathways are hydrolysis and oxidation.

B. MECHANISM OF TOXIC ACTION

Carbamates act in a manner analogous to the organophosphate compounds. This means that they are anticholinesterase compounds that inhibit the removal or breakdown of acetylcholine. An important distinction from the organophosphate compounds is that the enzyme cholinesterase is not attached or deactivated for such a long period. This means there is little delayed or long-term effect on the nervous system with this group of compounds.

V. PYRETHRINS AND PYRETHROIDS

Pyrethrum is a natural insecticide found in the flowers of certain plants belonging to the genus **Chrysanthemum** (*C. cinerariaefolium* and *C. coccineum*). This group of substances forms another major group of insecticides in wide usage. The Chrysanthemum genus are a daisy-like plant originating from the area now known as Iran. Today, pyrethrin flowers mainly come from the highland areas of Kenya where they contain up to 3% pyrethrins. There are about six principal active components in the pyrethrin flowers. These components have the general structure shown in Figure 8.6. These substances are esters and the constituent acids and alcohols are practically inactive toward insects.

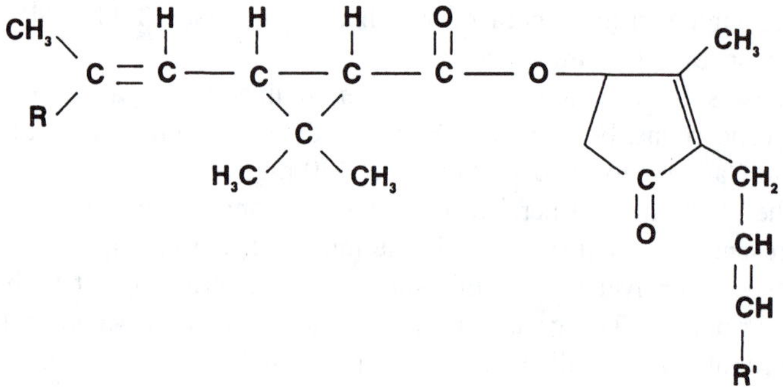

COMPOUND	R	R'
Pyrithrin I	CH_3	$CH_2 = CH$ -
Pyrithrin II	- $COOCH_3$	H_3C -
Cinerin II	- $COOCH_3$	$H_3C - CH_2$ -

FIGURE 8.6 Chemical structures of some active pyrethrum constituents.

A. PHYSICAL PROPERTIES

The chemical structures of these substances contain some polar groups, such as C=O, but have large nonpolar groups present. As a result, they are oily liquids that are soluble in alcohol and acetone but poorly soluble in water. In this respect, they resemble the chlorohydrocarbon insecticides, differing from most of the organophosphate and carbamate insecticides, and could be classified as **lipophilic**. Most **pyrethroids** are relatively high boiling, viscous liquids with low vapor pressure due to their relatively high molecular weight. Only a few, for example allethrin (see Figure 8.7), are sufficiently volatile to be useful constituents of mosquito coils. These coils are lit and work on the principle that the pesticide can be vaporized into the atmosphere, thereby repelling insects. Natural pyrethroids are unstable toward moisture, largely due to the presence of the ester group, which is particularly susceptible to hydrolysis. In water, natural pyrethroids are hydrolyzed, with the reaction being both acid and base catalyzed. The structure also contains a conjugated group, i.e., C=C–C=O, which strongly absorbs the high-energy ultraviolet radiation leading to light-induced photochemical reactions. These factors lead to about a 20% loss of insecticidal activity per year, even within the dried flower heads.

allethrin

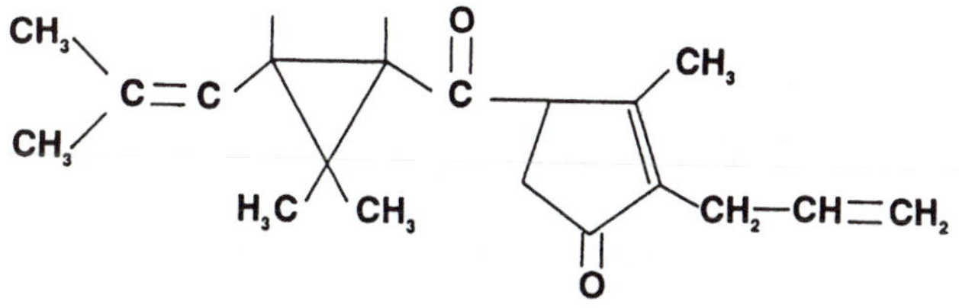

CH_3 CH_3 C=C H_3C CH_3 O C CH_3 CH_2–CH=CH_2 O

fenvalerate

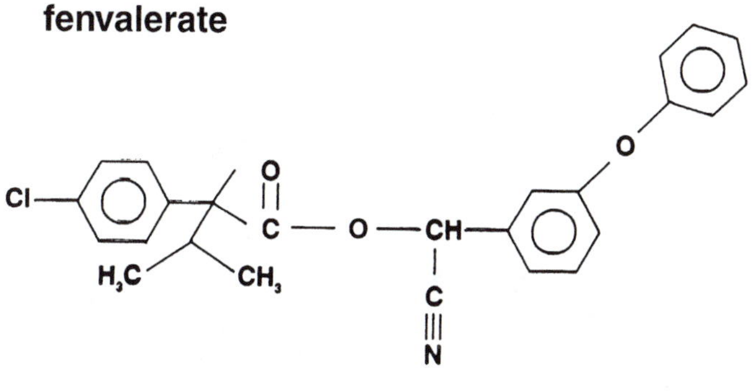

Cl H_3C CH_3 O C — O — CH C N O

FIGURE 8.7 Chemical structures of some synthetic pyrethroids.

B. MECHANISM OF ACTION AND TOXICOLOGY

Pyrethroids are neurotoxic, which means that they are poisons that attack the nerve ends. They are also contact insecticides as opposed to systemic insecticides and fumigants. This means that they enter the insect body by absorption through the cuticle following direct contact with the insecticide through droplets, dust or contaminated surfaces. This mechanism of absorption operates because of their lipophilic properties. The symptoms of pyrethrin poisoning follow the typical pattern of nerve poisoning. Firstly, excitation of the organism occurs, followed by convulsions, after which paralysis and death occur. These substances do not interact with acetylcholinesterase as do the organophosphate and carbamate insecticides. The toxic action is due to a loose chemical binding of the pyrethrin to a neural receptor that alters the sodium and potassium ion conductance.

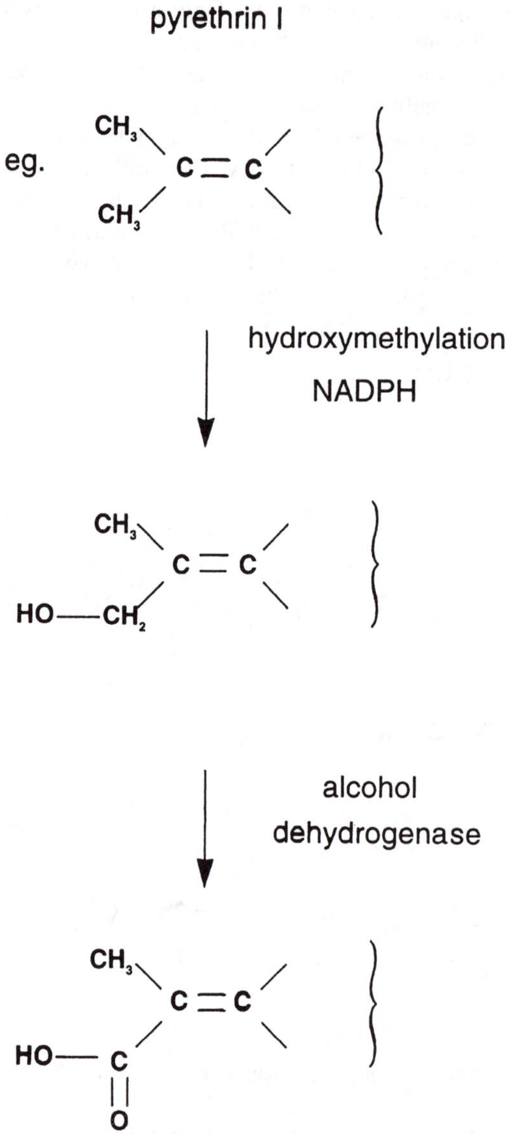

FIGURE 8.8 Oxidative degradation of Pyrethrin I.

Natural pyrethrins are rapidly degraded by mammals and also insects, and usually have low mammalian toxicity. The ester linkage of natural pyrethrins is unstable when exposed to enzymes in the insect gut. Once pyrethrins have gained access to the tissues, they appear to be oxidized by a system involving MFO induction, as indicated in Figure 8.8.

In relatively small doses, the pyrethrins typically exhibit a **knockdown effect**, which is the induction of temporary paralysis. There is normally a large difference in the dosage required to cause death over that required to cause a knockdown effect. The knockdown and lethal effects of pyrethrins may be due to two different mechanisms or two different sites of action. This is suggested by the observation that knockdown tends to be associated with the more polar pyrethroids, while lethality seems related to the more lipophilic pyrethroids. As a general rule, the pyrethrins are quite toxic to fish, despite their ready hydrolysis, and also to bees.

In domestic formulations, the pyrethrins are usually used in conjunction with a synergist. The ability of insects to degrade and detoxify pyrethrins is largely due to mixed-function oxidases (MFO), and the synergistic compound typically inhibits the MFO induction. Thus, while the synergistic compound alone is relatively innocuous in combination with the pyrethrin, a relatively toxic mixture is formed. Synergistic compounds include piperonyl butoxide and several other substances. The ratio of synergistic compound to pyrethrin is usually about 10:1.

C. SYNTHETIC PYRETHROIDS

The elucidation of the structure of the natural pyrethrins has made possible the synthesis of related compounds, synthetic pyrethroids, possessing similar insecticidal activity but being more stable to moisture and light. Allethrin (see Figure 8.7) was the first synthetic pyrethroid, developed in 1949. The synthesis of pyrethroids generally involves the maintenance or enhancement of the insecticidal activity while increasing the environmental stability.

Synthesis has been based on the principle of copying the molecular geometry of the natural compounds rather than to mimic their structural chemistry as illustrated by fenvalerate in Fig. 8.7. The low mammalian toxicity associated with the natural pyrethrins extends to some, but not all, of the synthetic pyrethroids. As a general rule, the synthetic compounds have high toxicity to fish.

VI. PHENOXYACETIC ACID HERBICIDES

The **phenoxy acetic acid herbicides** were developed during World War II, based on the structure of the natural plant hormones, the **auxins**. An example of an auxin, indole-3-acetic acid, is shown in Figure 8.9, and as a result, these substances are often described as the "**hormone**" weed killers. Examples of common phenoxy acetic acid herbicides are shown in Figure 8.9. These substances can be synthesized by the route shown previously in Chapter 6, IIIB. This involves the use of the sodium salt of 2,4,5-trichlorophenol and chloracetic acid.

These substances are polar and exhibit relatively high solubility in water. As a rule, they have comparatively low toxicity to aquatic animals, with LC_{50} values ranging from several hundred to several thousand parts per million. Persistence in the environment is also comparatively low. A particular environmental problem is concerned with trace contaminants, the **dioxins**, which are produced during the manufacturing process and are considered in detail in Chapter 6, IIIB. The dioxins produced include the particularly toxic compound, tetrachlorodibenzodioxin (TCDD). Possible mechanisms of formation of TCDD are shown in Figure 8.10. If the conditions of temperature and pressure are not carefully controlled during the manufacturing process, the formation of TCDD is enhanced, as is the production of other dioxins. It is believed that TCDD can possibly be formed from the phenoxy acetic acid herbicides after application by processes in the environment. TCDD is one of the most toxic substances ever made synthetically, and usually concentrations

indoyl - 3 - acetic acid(IAA) 24D 245T MCPA

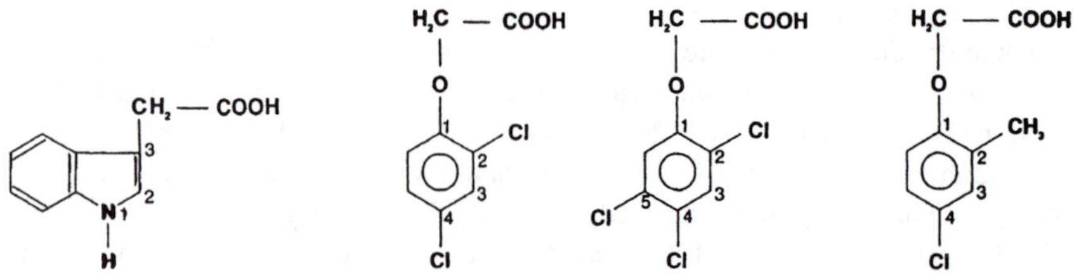

FIGURE 8.9 The chemical structures of some commonly used herbicides and IAA.

2.4.5 - trichlorophenol (or its sodium salt) TCDD

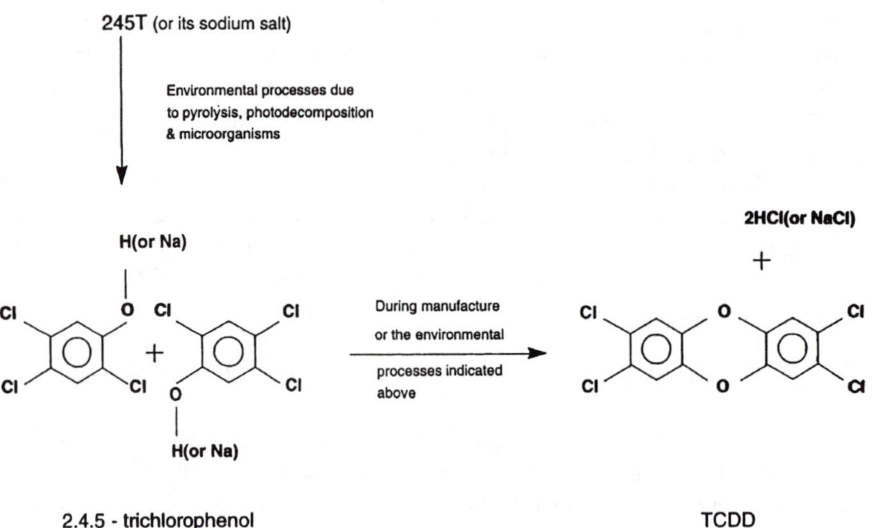

FIGURE 8.10. Possible mechanisms for the formation of TCDD.

have been restricted to below 1 part per million in the final product. TCDD is moderately persistent, having a half-life in soil of about 1 year. This substance has been implicated in a range of environmental problems, and as a result, the use of the phenoxy acetic acid herbicides is banned in many countries.

VII. KEY POINTS

1. In chemical terms, the pesticides are an extremely diverse group of substances, having only toxicity to pests as a common characteristic. They can be classified chemically as chlorohydrocarbons, carbamates, phenoxy acetic acids, organophosphates, organochlorines and so on. They are often classified according to function as herbicides, insecticides and fungicides, as well as other names according to the pest targeted.

2. Most pesticides are synthetic chemical compounds that are widely used since costs are relatively low. However, the pyrethrins, natural pesticides isolated from the Chrysanthemum species, have achieved commercial success.

3. The chlorinated hydrocarbon insecticides achieved outstanding commercial success during the 1960s and 1970s, but have been banned in many countries due to their persistence and other unsatisfactory environmental effects. This group is typified by DDT and dieldrin.

4. The chlorinated hydrocarbons contain a limited range of bond types, principally C---C (aromatic), C=C, C–H, C–Cl, with some C–C. The bonds have low dipole moments and are relatively resistant to attack by chemical agents in the environment. Thus, the members are lipophilic, with log K_{OW} values ranging from 2.67 to 6.36 and correspondingly low aqueous solubility.

5. The chlorohydrocarbons are strongly bioconcentrated with 4,4′-DDT, exhibiting a log K_{OW} value of 6.36. They are toxic to a wide range of biota, particularly aquatic biota and insects. The cyclodiene subgroup tend to have higher mammalian toxicity than the other members. The major ecological effect has been observed with DDT as a reduction of reproductive success with some bird species.

6. The organophosphate pesticides have the following general structure:

$(RO)_2P(=O\text{ (or S)})\text{—}OX$

The two R groups are usually methyl or ethyl groups, the O in the OX group is replaced with S in some compounds, while the X group can take a wide diversity of forms. This general structure indicates polar molecules with greater water solubility than the chlorohydrocarbons but lower lipid solubility. The group is reactive and generally susceptible to hydrolysis in the environment.

7. The organophosphate pesticides are toxic to a wide range of biota and act by inhibiting acetylcholinesterase. Their lack of environmental persistence leads to a lack of bioaccumulation capacity.

8. The carbamate pesticides have the following general structure:

$R\text{—}O\text{—}C(=O)\text{—}N(CH_3)(R')$

R and R′ can be a variety of different groups. The compounds in this group are relatively water soluble, with limited environmental persistence due to their ready hydrolysis. They are toxic to a wide range of biota and act by inhibiting cholinesterase.

9. The pyrethrins are a group of naturally occurring pesticides present in certain Chrysanthemum species. Their chemical structure is complex but contains the following grouping:

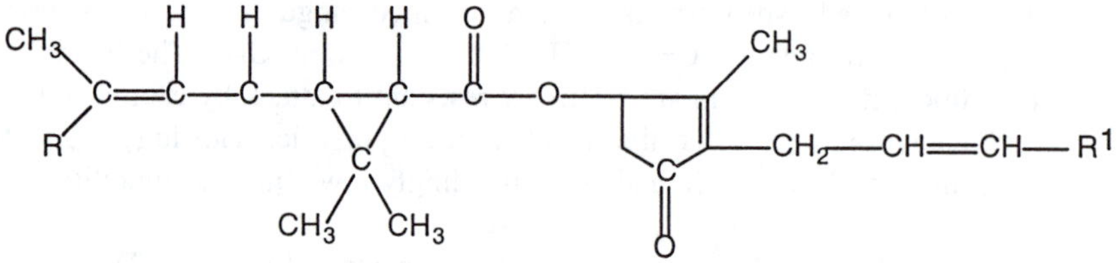

H H H H O CH$_3$ CH$_3$ C═C—C—C—C—C—O— CH$_2$—CH═CH—R^1 R C CH$_3$ CH$_3$ C O

where R and R^1 are ester and other groups. In addition, there are a group of structurally related synthetic pyrethroids. The molecules are relatively large and contain large hydrocarbon groupings with a limited number of polar groups. Thus, they are poorly soluble in water and tend to be lipophilic. They are readily degraded in the environment by hydrolysis of the ester linkage to inactive products. They do not interact with acetylcholinesterase, but by attachment to a receptor that alters sodium and potassium ion conductance. The group has low mammalian toxicity.

10. The phenoxyacetic acid herbicides are manufactured synthetically, and during this process the polychlorodibenzodioxins, particularly tetrachlorodibenzodioxin, TCDD, can be formed in trace amounts. TCDD is persistent, highly toxic and bioaccumulative.

REFERENCES

Connell, D.W. and Miller, G.J., *Chemistry and Ecotoxicology of Pollution,* John Wiley & Sons, New York, 1984.

Green, M.B., Hartley, G.S., and West, T.F., *Chemicals for Crop Improvement and Pest Management,* 3rd ed., Pergamon Press, Oxford, 1989.

Howard, H.P., Boethling, R.S., Jarvis, W.F., Heylan, W.M., and Michalenko, E.M., *Handbook of Environmental Degradation Rates,* Lewis, Boca Raton, FL, 1991.

CHAPTER 8 PROBLEMS

1. DDT has been banned in many areas and, in some cases, replaced with endosulfan. Compare the chemical structures and related physical chemical properties, as well as expected bioaccumulation and persistence of these two substances.

2. Malathion (see Figure 8.5) is a typical organophosphate pesticide and is not very persistent in the environment. Write equations for an initial set of reactions that this compound could undergo in the environment.

3. The pyrethrins are lipophilic compounds. Explain, in terms of their chemical structure and physicochemical properties, why they do not accumulate in soils or biota.

4. The phenoxyacetic acid herbicides contain an acidic group (–COOH) that can be converted into the sodium salt. Compare and contrast the chemical structure, physicochemical properties and expected environmental behavior of 2,4–dichlorophenoxyacetic acid (2,4-D) and its sodium salt.

CHAPTER 8 SOLUTIONS

1.

Chemical structure and property	DDT	Endosulphan
Chemical structure	Chlorohydrocarbon with only C---C (aromatic), C–H, C–Cl, C–C bonds present	Similar bonds to DDT but with a S-containing group
Physical chemical properties	A relatively nonpolar compound since nonpolar bonds are present; this means the log K_{OW} would be high and water solubility low	The S-containing group is polar and would be expected to make endosulphan much more water soluble with a much lower log K_{OW} than DDT
Persistence	High due to unreactive bonds present	Much less than DDT due to S containing group, which would be expected to be susceptible to hydrolysis
Bioaccumulation	Persistent and lipophilic and thus highly bioaccumulative	Much less bioaccumulative than DDT due to its relatively low persistence and log K_{OW} value

2. The following hydrolysis reactions could occur with malathion at the indicated bonds.

$$(H_3C{-}O)_2P({=}S){-}S{-}CH(C({=}O){-}O{-}C_2H_5){-}CH_2{-}C({=}O){-}O{-}C_2H_5$$

$$\downarrow$$

$$2H_3COH + (HO)_2P({=}S){-}OH + HS{-}CH(C({=}O){-}OH){-}CH_2{-}C({=}O){-}OH + HOC_2H_5$$

3. The chemical structures contain a high proportion of nonpolar bonds such as C–C, C–H, C=C that are nonpolar in character, making the compounds lipophilic. But the structures also contain the ester group, which is susceptible to hydrolysis. In addition, the structures contain a conjugated carbonyl group (C=C–C=O) that absorbs strongly in the UV range of the spectrum. This leads to activation of the molecule and degradation. So, while the compounds are lipophilic and thus could accumulate in biota lipid, they lack sufficient persistence to bioaccumulate.

4.

Chemical structure or property	2,4-D	Na salt OF 2,4-D
Chemical structure	Weakly ionic	More strongly ionic
Physicochemical properties	Weakly ionic so solubility in water would be low and vapor pressure low	Greater ionic strength would give greater water solubility but vapor pressure lower due to greater ionic character
Environmental behavior	Moderate accumulation in soil; low evaporation into the atmosphere	Poor accumulation in soil due to high water solubility, very low volatilization in the atmosphere

Chapter 9

POLYCYCLIC AROMATIC HYDROCARBONS (PAHs)

I. INTRODUCTION

The polycyclic aromatic hydrocarbons have been contaminants of the human environment ever since human life first evolved because they are ubiquitous contaminants of the natural environment. But the growing industrialization of human society has involved an increase in environmental management issues. Increasing environmental pollution is one of these issues and important among the environmental pollutants are the **polycyclic aromatic hydrocarbons (PAHs).**

The scientific investigation of the PAHs and their effects started in 1775 with Sir Percival Pott who attributed scrotum cancer in chimney sweeps in London exposed to soot and ash. Later investigations strongly suggest that the causative agents present in the soot and were the PAHs. Direct evidence of the involvement of PAHs as agents of cancer was produced during the 1930s. During this decade, some PAHs were shown to be powerful carcinogens; since that time, many other PAHs have been shown to possess similar properties.

The PAH family of hydrocarbons consists of molecules containing two or more fused six carbon atom aromatic rings. Two common members of the group are naphthalene and benzo(a)pyrene with two and five fused rings, respectively. In fact, benzo(a)pyrene is a powerful carcinogen, and often the occurrence of PAHs are reported in terms of this substance. In addition, the members share common properties of relatively low water solubility, with the most important members being lipophilic.

Their ubiquitous environmental occurrence stems from their many sources, both natural and anthropogenic. As a general rule, the PAHs are produced by combustion, which can be natural (e.g., forest fires) or anthropogenic (e.g., combustion in automobiles). These substances have been shown to be widely distributed in aquatic sediments, water, air, plants and animals. The interest in this group derives from their wide occurrence and the possible induction of cancer in organisms as a result.

II. CHEMICAL NATURE OF PAHs

The PAHs comprise a large family of hydrocarbons and thus contain carbon and hydrogen only. Each member consists of a number of benzene rings fused together through two or more carbon atoms. The possible number of different PAHs is enormous but there are common PAHs that occur throughout the environment, some of which are listed in Table 9.1. Naphthalene is considered to be the simplest member of the family and its structure is among those shown in Figure 9.1. On the other hand, coronene is considered to be the highest molecular weight PAH of environmental significance and its structure is also shown in Figure 9.1.

While PAHs usually contain fused benzene rings, there are some that can contain five-membered rings as well; for example, acenaphthene and fluoranthene, as shown

TABLE 9.1
Some Physical Properties of PAH Compounds of Environmental Interest

PAH	Molecular formula	Molecular weight	Melting point (°C)	Boiling point (°C)	Vapor pressure (kPa)	Aqueous solubility (mol/L)	log K_{OW}
Naphthalene	$C_{10}H_8$	128.2	81	218	1.09×10^{-2}	2.48×10^{-4}	3.36
Acenaphthene	$C_{12}H_{10}$	154.2	93	279	5.96×10^{-4}	2.55×10^{-5}	3.92
Phenanthrene	$C_{14}H_{10}$	178.2	100	340	2.67×10^{-5}	7.25×10^{-6}	4.57
Anthracene	$C_{14}H_{10}$	178.2	218	342	1.44×10^{-6}	4.10×10^{-7}	4.54
Fluoranthene	$C_{16}H_{10}$	202.3	107	384	2.54×10^{-4}	1.29×10^{-6}	4.90
Pyrene	$C_{16}H_{10}$	202.3	149	404	8.86×10^{-7}	6.68×10^{-7}	5.18
Benzo(a)-anthracene	$C_{18}H_{12}$	228.3	157	438	—	6.14×10^{-8}	5.61
Benzo(a)-pyrene	$C_{20}H_{12}$	252.3	178	495	6.67×10^{-13}	1.51×10^{-8}	6.04
Perylene	$C_{20}H_{12}$	253.3	277	503	—	1.59×10^{-9}	6.04
Coronene	$CH_{24}H_{12}$	300.4	438	590	—	4.67×10^{-10}	6.90

in Figure 9.1. It is also worthwhile keeping in mind that there are other PAHs based on these structures as a parent structure and that contain attached alkyl and other groups. In addition, there can be closely related compounds that contain oxygen, nitrogen and sulfur atoms, as well as fused benzene rings similar to the PAHs.

Naphthalene has a molecular formula of C_{10} H_8 and a molecular weight of 128, as shown in Table 9.1. Also, it has a carbon content of 94% and a hydrogen content of 6%. As the PAHs become larger and more complex (moving down in Table 9.1), the percentage of carbon tends to increase and that of hydrogen to decrease. For example, pyrene (see Figure 9.1 and Table 9.1) has a molecular formula of $C_{20}H_{12}$ and therefore contains 95% carbon and 5% hydrogen. While coronene is considered to be the highest molecular weight PAH of environmental interest, the ultimate PAH may be considered to be graphite. Graphite consists of layers of fused benzene rings in which the different layers are held together by comparatively weak forces so that they can slide over one another. Because of this property, graphite is a valuable lubricant.

III. ENVIRONMENTAL PROPERTIES

The PAHs are usually solids with naphthalene, the lowest molecular weight member, having a melting point of 81°C. The melting point increases with molecular weight to coronene, with a melting point of 400°C. Similarly, the boiling point increases from naphthalene at 200°C to coronene at 590°C (see Table 9.1). As expected with increasing molecular weight and molecular size, the aqueous solubility and vapor pressure decline. Both aqueous solubility and vapor pressure are comparatively low, even for low molecular weight compounds, but very low for the high molecular weight compounds (see Table 9.1). This suggests that aqueous solubility and vapor pressure may be important factors influencing environmental behavior for the lower molecular weight PAHs, but that the influence of these properties will decline with increasing molecular weight.

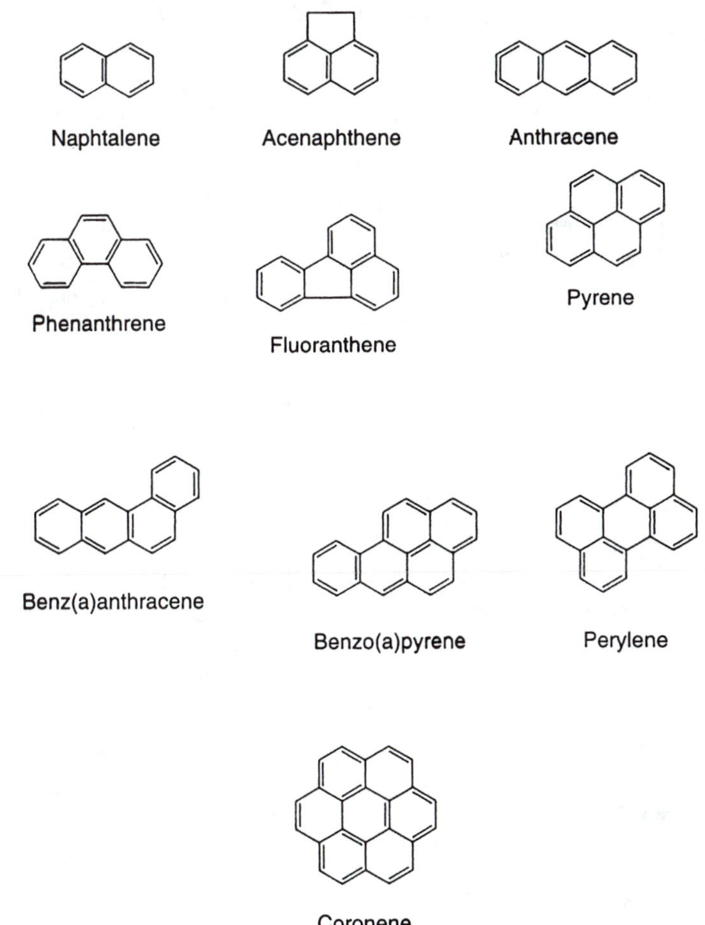

FIGURE 9.1 Chemical structures of some typical PAHs frequently encountered in the environment.

The octanol/water partition coefficient (K_{OW}) values are a measure of the tendency of a compound to dissolve in biota fat (see also Chapter 2) and other lipoidal substances such as humic acid in soil and sediments (see Chapter 17). The PAHs significant from an environmental perspective are listed in Table 9.1 and, except for coronene, all lie in the range of log K_{OW} 2 to 6.5, which is generally considered the range for lipophilic compounds. Thus, the PAHs would be expected to bioaccumulate and concentrate in sediments and soils in the environment to an extent depending on their persistence in these media as well.

IV. FORMATION OF PAHs

A. MECHANISMS OF FORMATION

Complete combustion of hydrocarbons in oxygen results in the complete oxidation of the carbon and hydrogen present to carbon dioxide and water, as shown below with naphthalene.

$$C_{10}H_8 + 12O_2 \rightarrow 10\ CO_2 + 4\ H_2O$$
Naphthalene

Combustion is not as simple as expressed in this equation, particularly with the

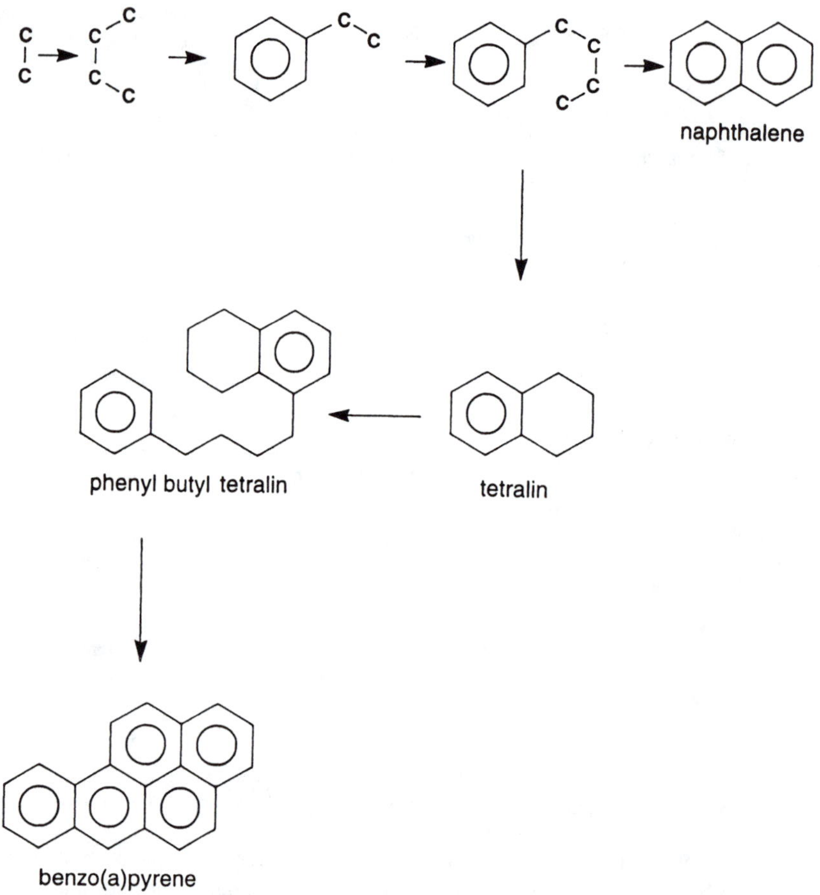

naphthalene

phenyl butyl tetralin

tetralin

benzo(a)pyrene

FIGURE 9.2 Possible mechanisms for the formation of PAHs during combustion.

combustion of organic substances in the natural environment. At the high temperatures of a flame, greater than 500°C, some of the C–C, C–H and other bonds are broken to ultimately form free radicals. Depending on the abundance of oxygen present, many of these fragments will react with oxygen to form carbon dioxide and water vapor. But usually, oxygen is not sufficiently well dispersed and mixed with the fragments to be able to react efficiently to form carbon dioxide and water. As a result, many organic fragments will react with other fragments close to them that may be other free radicals formed from the initial hydrocarbon. As the mixture cools it forms more complex fragments, often leading to PAHs, as shown in Figure 9.2. In this way, a variety of PAHs can be formed, depending on the conditions that exist at the time. For example, the formation of PAHs will tend to be more prevalent in an atmosphere where insufficient oxygen is available for complete combustion. The amount of PAHs formed also depends on the combustion temperature and the nature of the organic material combusted. However, irrespective of the type of material burned, which could be coal, cellulose, tobacco, polyethylene and other polymeric materials, similar ratios of PAHs are formed at a defined temperature. The PAHs listed in Table 9.1 are among the most common PAHs formed through the combustion process. It should be remembered that alkyl-substituted PAHs will be formed also. As a general rule, the higher the combustion temperature, the less alkyl-substituted PAHs will be produced.

During the geological formation of fossil fuels, PAHs can be formed by somewhat different processes. In the formation of coal and petroleum, biological material is broken down by pressure and modest temperatures (less than 200°C). Under these conditions, PAHs can be formed by mechanisms similar to those involved in incomplete combustion. Because of the lower temperatures involved, the transformations occur at a much slower rate. In addition, there may be differences in the types of PAHs which are formed in coal and petroleum. Relatively large amounts of alkyl-substituted PAHs are present in crude oil and coal-derived material compared with the amounts formed in combustion processes. On combustion of petroleum and coal, some PAHs are released unchanged in emissions and some are transformed into other PAHs. Of course the combustion process itself will also produce a range of PAHs as would normally be expected in the incomplete combustion process.

There have been suggestions that there are biological pathways for the formation of PAHs. This is still the subject of considerable scientific debate. In any case, the amount formed in biosynthesis is considered to be relatively low compared to the amounts formed abiotically.

V. SOURCES OF THE PAHs

There are a wide range of primarily natural sources of PAHs in the environment. Foremost among these are forest fires and volcanic activity. The actual quantities involved are variable, depending on the sporadic nature of these events. The data on production of benzo(a)pyrene shown in Table 9.2 indicates that open burning can produce gaseous emissions that have extremely high concentrations of PAHs. Significant proportions of a range of PAHs, including those in Table 9.1, are produced, although Table 9.2 only reports the PAHs as benzo(a)pyrene, which is considered to be indicative of this group. A large proportion of the PAHs released into the environment arise from anthropogenic sources such as coal-fired electricity power plants, incinerators, open burning and motor vehicle exhausts, as indicated in Table 9.2. The relative importance of the anthropogenic sources is indicated in Table 9.3. Industrial processes, residential combustion for heating, mobile transport emissions and incineration are the most significant sources of PAH production.

Considering the data in Table 9.2 and Table 9.3 it could be concluded that tobacco smoking is relatively insignificant as a source of PAHs. This would be true in overall quantitative terms but importantly this source results in direct exposure of humans and in many cases close exposure to others who are non-smokers. Its interesting to

TABLE 9.2
Benzo(a)pyrene Production from Different Processes

Process	Production
Coal-fired power plant	30–930 ng m^{-3} emission
Municipal incinerators	17–2700 ng m^{-3} emission
Open burning	2800–173,000 ng m^{-3} emission
Motor vehicle gasoline combustion	25–700 μg L^{-1} gasoline
Tobacco combustion	0.8–2.0 μg g^{-1} tobacco

TABLE 9.3
Estimated Total PAH Emissions from Different Sources in the United States, Sweden and Norway

Source	Quantity[a] (%)		
	United States	Sweden	Norway
Industrial processes	3497 (41)	312.3 (62)	202.7 (67)
Aluminium production			
Iron and steel works			
Coke manufacturing			
Ferro-alloy industry			
Asphalt production			
Carbon black			
Petroleum cracking			
Residential combustion	1380 (16)	132 (26)	62.5 (21)
Wood			
Coal			
Oil			
Gas			
Mobile sources	2170 (25)	47 (9)	20.1 (7)
Gasoline automobiles			
Diesel automobiles			
Air traffic			
Incineration	1150 (13)	3.5 (<1)	13.7 (5)
Municipal			
Open burning			
Forest fires			
Agricultural burning			
Power generation	401 (5)	13 (3)	1.3 (<1)
Coal and oil fired			
Peat, wood, straw			
Industrial boilers			
Total	8598	507.8	300.3

[a] Metric tons per year.

note that the cancer inducing effects of tobacco smoking are not due to exposure to nicotine, although nicotine is toxic, but exposure to PAHs produced by the combustion of tobacco. The range of PAHs identified as being produced by smoking is considerable and includes several established carcinogens including benzo(a)pyrene. The PAHs identified in ambient air in which cigarettes are being smoked are shown

TABLE 9.4
Concentrations of PAHs in the Ambient Atmosphere Produced by Cigarette Smoking

PAH	Ambient air concentration ($ng\ m^{-3}$)
Flouranthene	99
Pyrene	66
Benzo(a)anthracene	100
Benzo(b,j,k)flouranthene	35
Benzo(a)pyrene	22
Perylene	11

TABLE 9.5
Typical Examples of Some PAHs in the Environment

PAH	Combustion particulates ($mg\ kg^{-1}$)	Air (typical city locations) ($ng\ m^{-3}$)	Smoked Fish ($\mu g\ kg^{-1}$)	Sewage sludge ($\mu g\ kg^{-1}$, dry weight)
Fluoranthene	4–400	1–15	300–3000	2–7
Benzo(a)anthracene	2–160	0.1–20	20–200	1–4
Perylene	0.1–138	1	1–4	0.1–2
Benzo(a)pyrene	0.2–64	1–1,000	4–16	0.5–3
Coronene	0.1–40	2	1–10	0.1–2

in Table 9.4. It can be seen that this includes the well known carcinogen benzo(a)pyrene.

VI. OCCURRENCE AND BEHAVIOR OF PAHs IN THE ENVIRONMENT

In accord with the many and varied human and natural sources of PAHs these substances are very widespread in the environment. The concentrations which occur are highly variable ranging from very high concentrations in combustion particulates to the very low amounts which generally occur in fresh and sea water. Some typical examples of the occurrence of PAHs are shown in Table 9.5. The PAHs as a group share the common property that they generally consist of linked six membered aromatic rings. The aromatic bonding within the rings tends to make aromatic compounds resistant to attack by chemical and other agents. So it would be expected that the PAHs would tend to be stabilized by the presence of aromatic rings in conjugated form. On the other hand the compounds have strong ultraviolet and visible radiation absorption. This means that there is an uptake of energy which can be used to chemically modify and transform these substances. So this characteristic would tend to make the compounds less stable in the environment. In addition, the extensive occurrence of these substances in the natural environment would tend to develop populations of organisms that would have the capacity to degrade them.

The major degradation pathways involve chemical, photolytic or metabolic processes associated with microorganisms. In many situations, all of these processes may occur together; whereas, in some circumstances, one or various combinations of them may be in operation. These processes are also strongly influenced by environmental conditions such as temperature, availability of oxygen, populations of microorganisms present and so on.

The chlorination and ozonation of treated water from wastewater treatment plants can result in significant degradation of any PAHs present. Laboratory experiments with these processes have indicated that many PAHs exhibit very low persistence on the order of minutes or hours under these conditions. Somewhat similarly, PAHs in water where oxygen is available dissolved within the water mass when exposed to sunlight can exhibit rapid rates of degradation. The half-lives for PAHs in clear water exposed to strong sunlight and oxygen within the water mass can be less than 1 hour, as indicated by the data in Table 9.6. In turbid waters, however, this photolysis would be greatly slowed through the diminution of sunlight and also the partitioning of the PAHs onto particulate matter present and onto bottom sediments. Once sorbed to particulates and bottom sediments, the rate of degradation would be expected to be substantially reduced. The metabolic degradation of PAHs by microorganisms usually occurs through the co-metabolism of the PAHs with normal organic food material. The process usually involves the oxidation of the PAHs to produce oxidized materials that are subsequently further degraded to simple products. The biodegradation of PAHs by bacterial populations in aquatic systems has been shown to proceed at different rates for different compounds. The half-lives measured in a sediment water system are shown in Table 9.6. This indicates that, as a general rule, the persistence of the PAHs increases with increasing size of the molecule. The higher molecular weight PAHs such as benzo(a)pyrene show a half-life of the order of up to about 1 year in this situation.

TABLE 9.6
Persistence of PAHs in the Environment

	Half-life	
PAH	**Clear water exposed to sunlight (hr)**	**Sediment/water microcosm (weeks)**
Naphthalene	—	2.4–4.4
Benzo(a)anthracene	0.54	—
Phenanthrene	—	4–18
Pyrene	0.75	34–>90
Benzo(a)pyrene	0.034	200–>300

The PAHs of environmental significance (see Table 9.1) generally lie in the log K_{OW} range from 2 to 6.5, and thus would therefore be classified as lipophilic and would have the potential to bioaccumulate. The relatively low persistence of naphthalene and other low molecular weight PAHs indicate that these would have a limited capacity to exhibit bioaccumulation. On the other hand, the high molecular weight compounds, for example benzo(a)pyrene, are persistent and lipophilic and therefore would be expected to bioaccumulate.

VII. CARCINOGENICITY AND TOXICITY OF THE PAHs

The major area of environmental concern with the PAHs is their ability to produce cancers in exposed organisms. A range of PAHs has been found to have strong carcinogenic activity with animals, particularly benzo(a)anthracene and benzo(a)pyrene with humans, as listed in Table 9.7. With many PAHs, evidence establishing carcinogenicity is not available and they are not classified one way or the other. But importantly, almost all PAHs are suspected to be carcinogens to some degree or another, although this level could be very low.

TABLE 9.7
Carcinogenic Activity of Some PAHs

Compound	Overall Evaluation[a]
Phenanthrene	3
Anthracene	3
Fluoranthene	3
Pyrene	3
Benzo(a)anthracene	2A
Benzo(a)pyrene	2A

[a] I, carcinogenic to humans; 2A, probably carcinogenic to humans; 2B, possibly carcinogenic to humans; 3, not classified; 4, probably noncarcinogenic to humans.

To understand the process of cancer induction, it is necessary to examine the metabolic fate of the PAHs within organisms. Mammals for example have evolved a group of enzymes to convert xenobiotic lipophilic compounds, including PAHs, to polar water-soluble products. The enzymes involved are the mixed function oxidases (MFO), which consist of enzymes in the cytochrome P_{450} group (also discussed in Chapter 3). This enzyme system is actually stimulated within an organism by exposure to persistent lipophilic compounds. Repeated exposure results in the induction of increased quantities of the P_{450} enzyme. The capacity to induce these enzymes by organisms is variable, and depends on the organism group involved. For example, mammals have quite strong inductive capacity and as a result have a good ability to degrade persistent lipophilic compounds. Other organism groups that have much more limited capacity to induce MFO, for example fish, have a relatively limited capacity.

The initial products of the metabolic degradation of the PAHs involve the insertion of an oxygen atom into the PAH structure to form an epoxide. Hydrolysis, followed by further epoxidation, leads to the formation of dihydrodiol epoxide. This process is illustrated by the reaction sequence in Figure 9.3. These substances have the capacity to bind to centers in biological molecules such as DNA and hemoglobin. Certain of these complex substances formed by this process have the capacity to cause tumor formation. However, this later formation can often take a considerable period of time and is not particularly well understood at present.

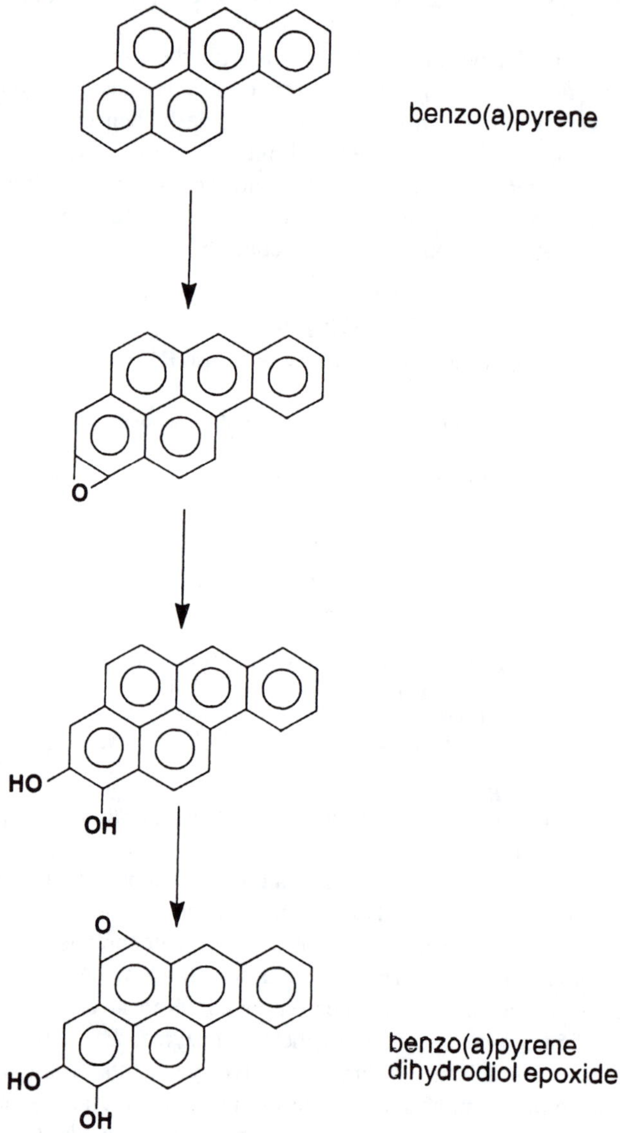

benzo(a)pyrene

HO OH

O

benzo(a)pyrene dihydrodiol epoxide

FIGURE 9.3 Metabolic oxidation of benzo(a)pyrene to the dihydrodiol epoxide by the P_{450} enzyme system.

The PAHs have acute toxicity to aquatic organisms, which can be measured as the LC_{50} value shown in Table 9.8. As a general rule, toxicity has been found to increase with molecular weight and log K_{OW} value, and the PAHs are quite toxic to aquatic organisms. Its interesting to note that solar radiation has been found to significantly increase the toxicity of PAHs to aquatic organisms. It is believed that this occurs from photoactivation of the PAH molecules present on or within the organisms and not metabolism or formation of other degradation products.

TABLE 9.8
Toxicity of PAHs to Various Aquatic Organisms

PAH	Organism	LC_{50} (96 hr) (mg L^{-1})
Naphthalene	Fish	0.1–8.0
	Crustacea	1.0–2.4
Acenaphthene	Fish	0.6–3.0
Phenanthrene	Fish	0.04–0.6

TABLE 9.9
Cancer Mortality Associated with Environmental Exposure in the United States

Factor	Percentage of total cancer	Year estimated
Tobacco	30 (mortality)	1977
	76 (mortality)	1980
Diet	35 (mortality)	1977
Air pollution	2 (mortality)	Future

VIII. EFFECTS ON HUMAN HEALTH AND THE NATURAL ENVIRONMENT

There is a range of information available on the chronic effects of PAHs on aquatic organisms in natural systems as well as in laboratory experiments. Generally, exposure to sublethal levels of PAHs, in the lipophilic range of log K_{OW} values from 2 to 6.5, results in morphological, physiological and developmental abnormalities in fish. Low concentrations have also been found to cause significant reduction in the hatchability of eggs as well as larval length and weight. Of particular concern is the occurrence of mutagenic and carcinogenic effects. Benzo(a)pyrene has been shown to induce chromosome aberrations in fish. In addition, the PAHs have been found to cause epidermal hyperplasia and neoplasia. Some species of fish have been shown to develop liver cancer as a result of exposure to PAHs.

Tobacco smoking results in a high mortality in human populations (see Table 9.9) due to the occurrence of cancer and other effects, and the PAHs must be considered to be implicated in this situation. Diet is particularly noteworthy and there is a high incidence of mortality associated with agents present in food. Smoked fish, which are relatively high in PAHs, may be among a range of causative factors in some individuals and populations. Also, the incidence of mortality due to cancer associated with air pollution may be related to the effects of PAHs.

IX. KEY POINTS

1. The PAHs are produced by a wide range of combustion processes used in human society, as well as natural combustion and geological processes.

2. A range of PAHs, including naphthalene, phenanthrene, anthracene, pyrene, benzo(a)pyrene, perylene and coronene, commonly occur in air, soil and biota in the natural environment.

3. The PAHs, which commonly occur in the environment, range in molecular weight from naphthalene at 128 to coronene at 300, and have log K_{OW} values ranging from 3.36 to about 6.90, respectively. This means that the common PAHs are lipophilic compounds.

4. The low molecular weight PAHs have limited persistence in the environment, but the higher molecular weight compounds are more persistent; for example, benzo(a)pyrene persists in aquatic systems for up to about 300 weeks.

5. The higher molecular weight PAHs in the lipophilic group, with log K_{OW} values from about 2 to about 6.5, are persistent in the environment and are thus capable of being bioaccumulated.

6. The common group of environmental PAHs are relatively toxic to aquatic organisms and have LC_{50} values to fish in the range 0.1 to 8 mg L^{-1}.

7. Many PAHs are carcinogenic, with human exposure occurring through tobacco smoking as well as compounds in food and the atmosphere.

REFERENCES

Neff, J.M., *Polycyclic Aromatic Hydrocarbons in the Aquatic Environment,* Applied Science, London, 1979.

Grimmer, G., *Environmental Carcinogens: Polycyclic Aromatic Hydrocarbons,* CRC Press, Boca Raton, FL, 1983.

Bjorseth, A., *Handbook of Polycyclic Aromatic Hydrocarbons,* Marcel Dekker, New York, 1983.

CHAPTER 9 PROBLEMS

1. Using the data in Table 9.1, make estimates of the log K_{OW} of the following compounds:

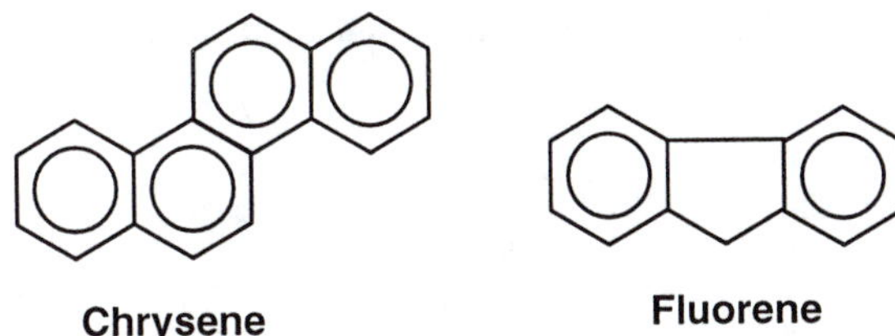

2. PAHs can be formed during pyrolysis by a variety of processes. Suggest possible fragmentation combination patterns, starting with a two-carbon fragment, for anthracene, acenaphthene and pyrene.

3. An environmental investigation of PAHs has commenced. Data on the persistence in natural aquatic systems is required for fluoranthene and perylene. Make estimates of these characteristics using information available in this chapter.

CHAPTER 9 SOLUTIONS

1. (a) The molecular weights can be calculated from the structural formula: Chrysene, $C_{18}H_{12}$, 228; fluorene, $C_{13}H_{10}$, 166.
   (b) Using the data in Table 9.1, plots can be made of MW against log K_{OW}, as shown in Figure 9.4. Projections can be made to give the following log K_{OW} values:

   Chrysene5.53
   Fluorene4.24.

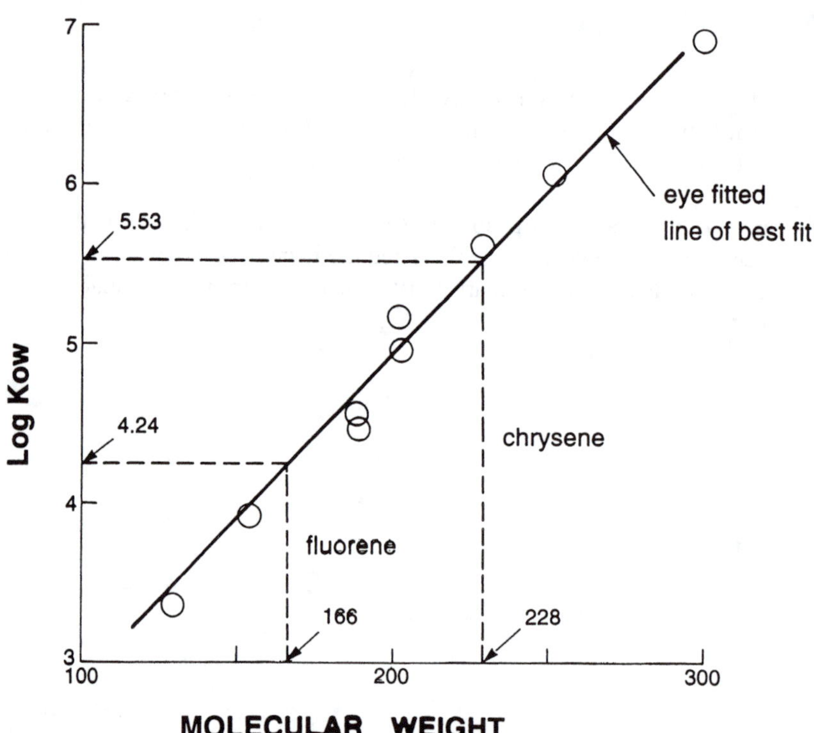

FIGURE 9.4 Plot of the log K_{OW} against molecular weight using the data in Table 9.1 with projections to obtain log K_{OW} for chrysene and fluorene (answer to Question 1).

2. Possible combination pathways are shown in Figure 9.5.

3. The half-life data for the PAHs in Table 9.6 can be plotted against log K_{OW} or molecular weight, as shown in Figure 9.6. The maximum and minimum half-lives can be joined by an eye-fitted line of best fit. Fluoranthene and perylene have molecular weights of 202 and 252, respectively, and log K_{OW} values of 4.96 and 6.04, respectively. Using these values, the following half-lives can be obtained by interpolation.

	MW estimate (weeks)	Log K_{OW} estimate (weeks)
Fluoranthene	30–90	20–50
Perylene	200–300	200–300

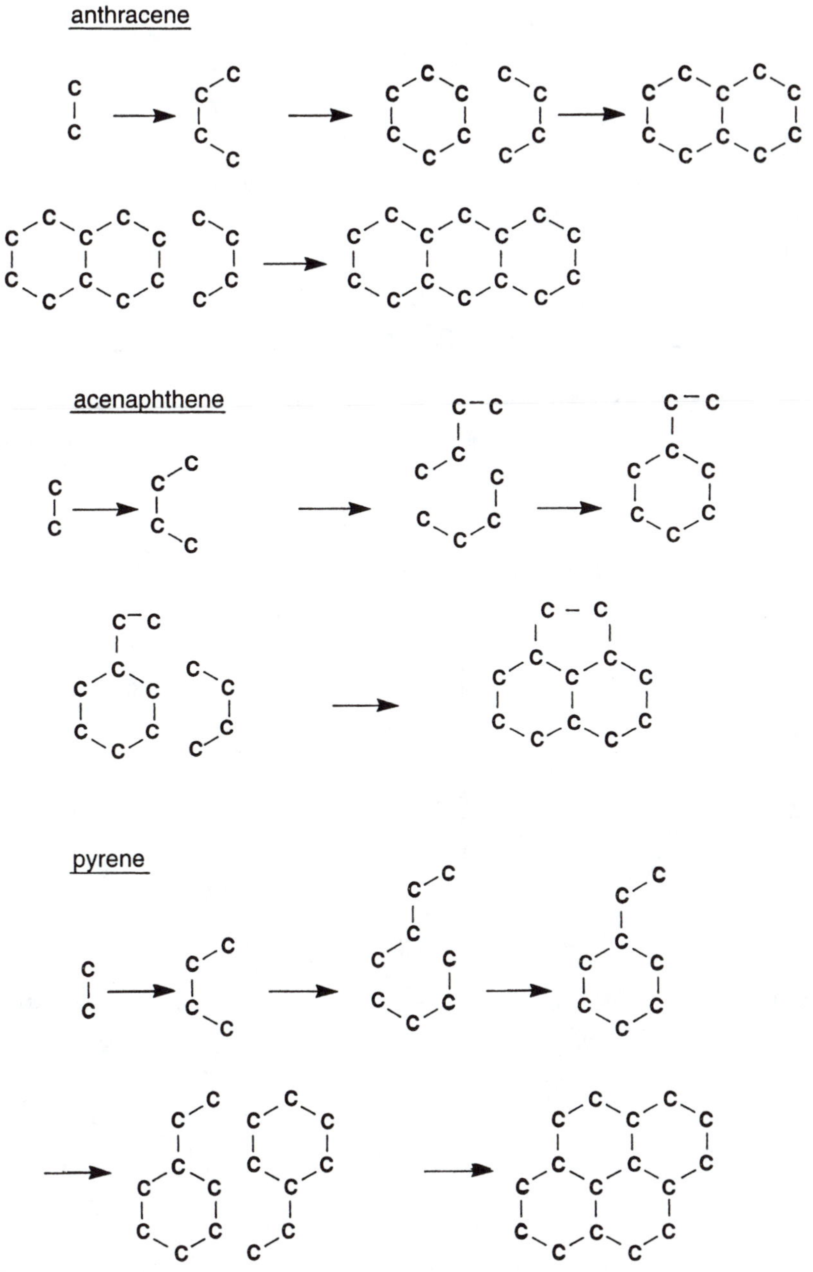

anthracene

acenaphthene

pyrene

FIGURE 9.5 Possible two-carbon fragment combination processes that can lead to the formation of anthracene, acenaphthene and pyrene (answer to Question 2).

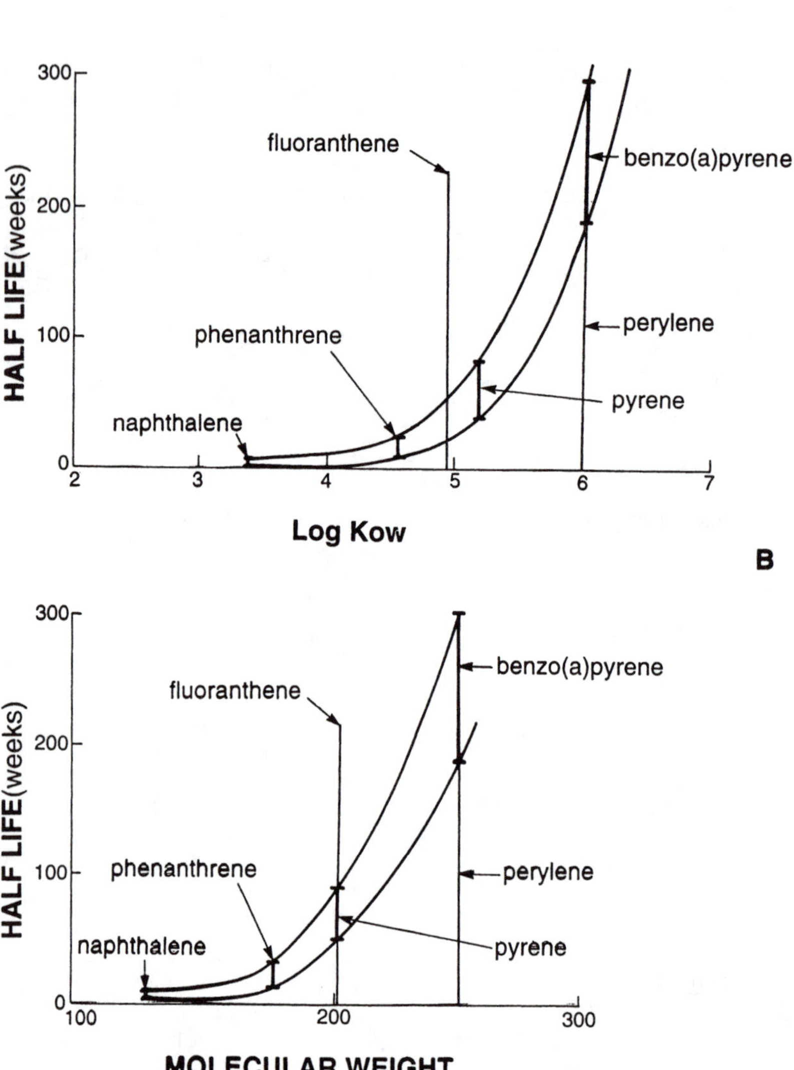

FIGURE 9.6 A. Plot of half-life against log K_{OW}. B. Plot of half-life against molecular weight using the data in Table 9.6 (answer to Question 3).

Chapter 10

SOAPS AND DETERGENTS

I. INTRODUCTION

Cleaning ourselves, our clothing, our homes, our eating utensils and so on has been a concern of human beings since the beginning of human society. This serves an esthetic purpose but is also important in the control of disease and maintenance of good health. The cleaning process almost always involves the use of water as a solvent and a carrier of the removed contaminants. This wastewater is often discarded into waterways where it can cause adverse effects.

Soap is the oldest and best known chemical agent to assist cleaning. Soap has been prepared since ancient times by recipes handed down from generation to generation. Little was known in the past regarding the chemistry of its preparation and mode of action. Now we have a detailed knowledge of the chemistry of soap, which has been used to make soap more effective and to develop cleaning agents commonly called **detergents**.

The development of new cleaning agents started in Germany during World War I when blockades prevented the importation of raw materials for soap manufacture. This steadily increased in both range of types and overall quantities until today the annual worldwide production exceeds 15 million metric tons and grows about 3% per year. About half of this is soap, but this proportion is declining. In 1994, consumption in the U.S. totalled roughly 4×10^9 kg, only about 5% being soap. Some of the broad categories are shown in Figure 10.1. The quantities are roughly split between domestic and industrial use. Domestic laundry detergents alone account for one quarter of the total amount of surfactants made.

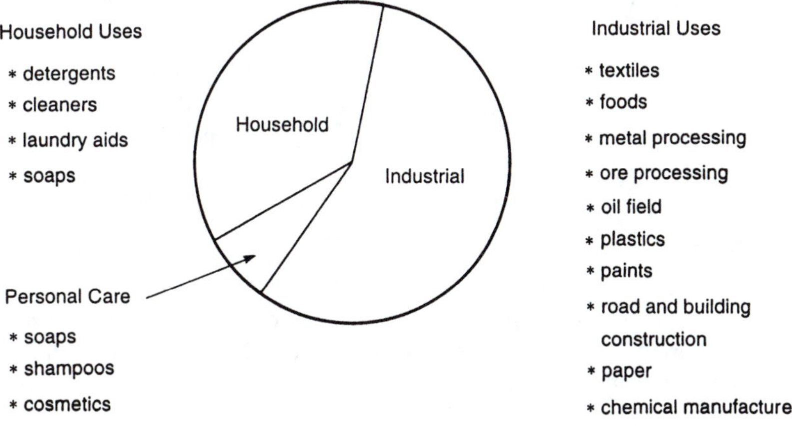

FIGURE 10.1. Surfactant use categories with approximate proportions.

During the 1950s and 1960s, large masses of foam could be seen on treatment ponds in sewage plants and many rivers and streams. This was caused by detergents in wastewaters and aroused concern regarding possible harmful effects on the natural environment. The occurrence of detergent components in natural water bodies has been reduced in recent decades by the use of substances that are more readily degraded. Soap has limited persistence in waterways and has not generally caused environmental problems.

II. SURFACTANTS — THE ACTIVE CLEANING AGENTS

Soaps and detergents contain substances described as **surfactants (surface active agents).** They act at the surface, or interface, between polar and nonpolar phases, to modify the properties of the phases. The surface of water, for example, can be seen as the interface between the air and water phases, with water having polar properties and air having nonpolar properties. The presence of surfactants modify the surface properties of both phases. They have a key molecular property, illustrated in Figure 10.2 by sodium myristate, a component of soap. The molecule has two parts with very different characteristics. The long hydrocarbon chain forms a nonpolar "tail," and the carboxylate group forms a polar "head." The head group is hydrophilic and imparts water solubility, but the hydrophobic tail would prefer to dissolve in nonpolar materials such as grease. Surfactants are often termed **amphiphiles** (from the Greek that means having a love of two kinds). For this reason, surfactants have low solubilities in water and other solvents.

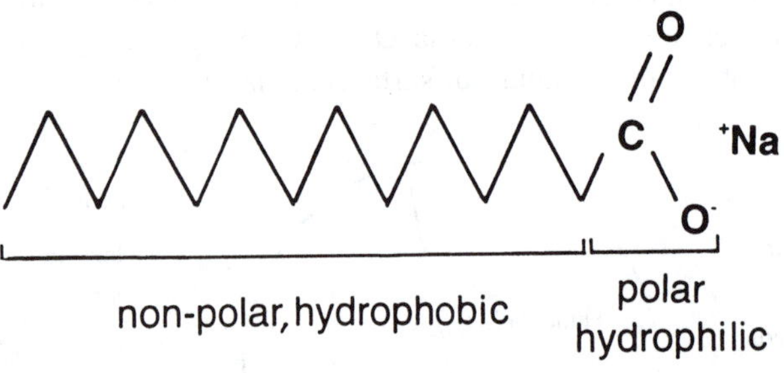

FIGURE 10.2. Structure of a soap component, sodium myristate, $CH_3(CH_2)_{12}COO^-Na^+$; the hydrophobic alkyl "tail" is flexible.

The surfactant molecules are most stable at interfaces such as air-water or oil-water surfaces (Figure 10.3). They gather together, orienting themselves with the hydrophilic part in the water and the hydrophobic tail projecting into the air or grease. This aggregation is so efficient that the surfaces can be completely covered with a surfactant **monolayer**, even though the surfactant concentration in the bulk water may be as low as 10^{-4} molar. As a result the surfactant monolayers greatly reduce the **surface tension**. Surface tension is a measure of the energy required to form a surface (about 74 and 50 mJ m^{-2} for water-air and water-hydrocarbon surfaces). Surfactants can reduce the water-air tension to about 25% of its usual value, and the effect is much more dramatic at the water-grease surfaces, where the tension can fall more than 10^5 fold. This effect promotes the formation of structures with very large surface areas, such as foams and emulsions. Soap bubbles are short-lived since they are both kinetically and thermodynamically unstable. On the other hand, oil and water and a little surfactant can form stable, permanent emulsions, largely due to the extremely low tensions produced.

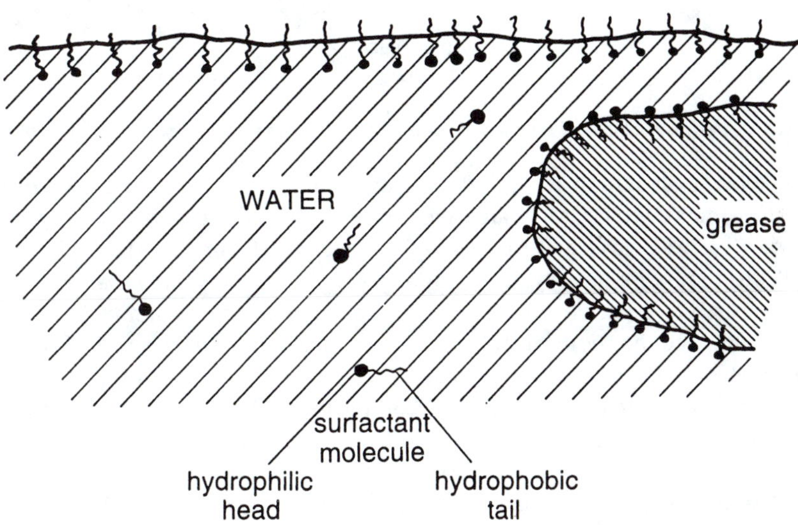

FIGURE 10.3. Surfactant molecules aggregating at water/air and water/grease interfaces. The polar "heads" are shown as filled circles.

When surfactants reach their solubility limit, they often do not precipitate as do conventional solutes, but display another effect. The molecules begin to clump together, forming **micelles**, tiny aggregates containing about 50 to 100 individuals. More surfactant can seem to dissolve by forming more micelles, while the actual concentration of single molecules remains about the same. Quite concentrated apparent "solutions" can result. The thick, clear dishwashing liquids that we use daily are not true solutions, but are instead concentrated suspensions of invisibly small surfactant micelles.

Micelles adopt various shapes and sizes. They may be spheres, disks, rods, sheets or vesicules (hollow spheres), depending on the type of surfactant, concentration, and other conditions such as temperature. Sphere and sheet-like micelles are shown in Figure 10.4. The polar heads face outward toward the water, while nonpolar tails form a loosely tangled core. The hydrophobic interior of the micelles behaves much like a tiny drop of oil — a good medium for similar hydrophobic compounds such as petroleum or PCBs. These normally water-insoluble compounds apparently can be **solubilized** in surfactant solutions by partitioning into the oil-like interior of micelles.

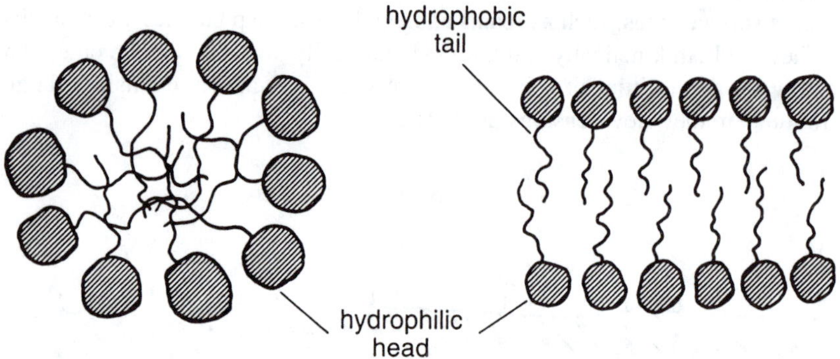

FIGURE 10.4. Cross-sections of spherical and bilayer (sheet-like) micelles of surfactant molecules.

Micelles begin to form at a well-defined concentration, called the **critical micelle concentration** (CMC). CMCs are detected by rather subtle changes in solution properties such as surface tension or light scattering (Figure 10.5). CMCs range from 0.1 to 10 m*M* for most synthetic surfactants. These concentrations are relatively high compared to expected environmental concentrations or toxic limits (by at least tenfold) and, accordingly, the micellization phenomenon does not normally influence the environmental behavior of surfactants.

Surfactants are broadly categorized according to the charged nature of the **hydrophilic** part of the molecule:

- **Anionic**: negatively charged
- **Cationic**: positively charged
- **Nonionic**: neutral, though highly polar
- **Amphoteric**: a "zwitterion" containing positive and negative charges

Some examples of surfactants in each class are illustrated in Table 10.1 and the broad proportions used shown in Figure 10.6. The hydrophobic fragments (R groups) are organic groups, usually alkyl chains.

III. SYNTHESIS

A. SOAPS

Soaps are salts of long-chain fatty acids. Traditionally, soaps are made by the saponification of natural fats or oils, which are triesters of fatty acids and glycerol.

$$\begin{array}{l} R{-}CO{-}O{-}CH_2 \\ \quad\quad\quad\quad\quad\;\; | \\ R{-}CO{-}O{-}CH \\ \quad\quad\quad\quad\quad\;\; | \\ R{-}CO{-}O{-}CH_2 \end{array} \xrightarrow[H_2O]{NaOH} 3R{-}\overset{\overset{\displaystyle O}{\|}}{C}{-}O^- \; Na^+ + \begin{array}{l} HO{-}CH_2 \\ \quad\quad\;\; | \\ HO{-}CH \\ \quad\quad\;\; | \\ HO{-}CH_2 \end{array}$$

a triglyceride ester **soap** **glycerol**

Table 10.1
Some Synthetic Surfactants in the Four Surfactant Classes

Common name	Structure	
Anionics		
Carboxylates (soaps)	$R—CH_2—COO^-Na^+$	$R = C_{10-18}$
Linear alkylbenzene sulfonates (LAS)	R–(benzene ring)–SO_3^- Na^+	$R = C_{10-13}$
Alcohol sulfates	$R—CH_2—OSO_3^-Na^+$	$R = C_{11-17}$
Alcohol ethersulfates	$R—O\ (CH_2CH_2O)_nSO_3^-Na^+$	
Nonionics		
Fatty alcohol ethoxylates	$R—(O\ CH_2CH_2)_nOH$	$R = C_{8-18}$, n = 9–70
Alkyl glucosides	(glucoside ring: H_2COH, OH, OH, HO, R—O)	$R = C_{12-14}$, n = 1–4
Cationics		
A quaternary ammonium chloride	$R_2N^+(CH_3)_2\ Cl^-$	$R = C_{16-18}$
Amphoterics		
A sulfobetaine	$R—N^+(CH_3)_2—(CH_2)_3—SO_3^-$	$R = C_{12-18}$

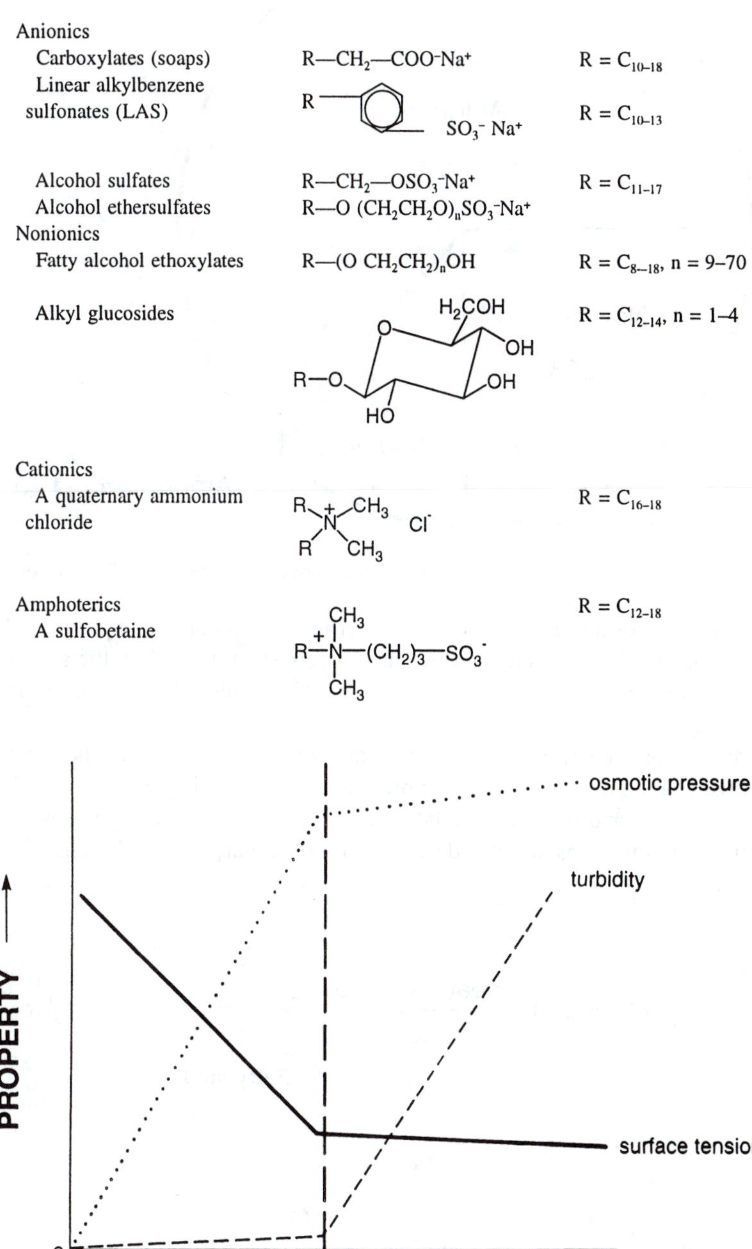

FIGURE 10.5. Changes in solution properties near the critical micelle concentration value.

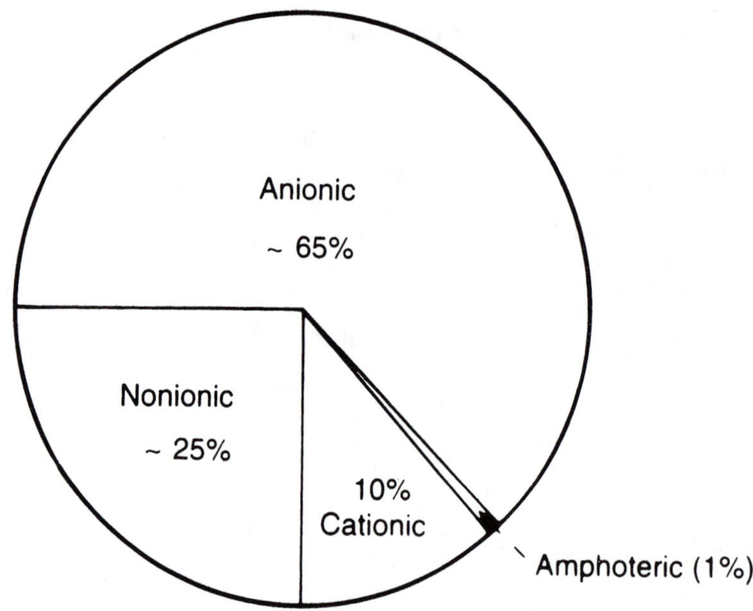

FIGURE 10.6. Broad categories of consumption of surfactants in the United States in 1994.

The fat sources are mainly tallow (animal fat) and coconut and palm seed oils. The R groups are linear, with variable chain lengths that depend on the source fats. Coconut and palm oil soaps are mainly C_{12} and C_{14}, while tallow soaps are mainly C_{16} and C_{18}.

Today, soaps are usually synthesized industrially by reacting together vegetable oil and water at high temperatures and pressures (where the ingredients are soluble), splitting the ester and forming free fatty acids. The fatty acids can be subsequently neutralized to form soaps, or refined and used for other purposes, including surfactant synthesis.

$$\text{oil or fat (triglyceride)} \xrightarrow[H_2O]{\text{heat pressure}} \underset{\textbf{Fatty acid}}{R-\overset{O}{\overset{\|}{C}}-OH} + \text{glycerol}$$

$$\underset{\textbf{Fatty acid}}{R-\overset{O}{\overset{\|}{C}}-OH} \xrightarrow{NaOH} \underset{\textbf{Soap}}{R-\overset{O}{\overset{\|}{C}}-\bar{O}\ Na^+} + H_2O$$

B. ALKYLBENZENE SULFONATES AND ALKYL SULFATES

The synthetic paths for most other surfactants are varied, but basically involve chemically linking hydrophobic and hydrophilic groups. The raw materials for the hydrophobes can be sourced from natural fats, but most come from the petrochemical industry (~70% worldwide). Today, surfactants consume about 0.4% of the world's petroleum supply.

Alkanes are too unreactive for direct surfactant synthesis. Intermediate compounds containing active centers such as double bonds, benzene rings or hydroxyl groups are used. **Linear alkylbenzene sulfonates** (LAS) are usually made by attaching a benzene ring to a long-chain olefin, an unsaturated hydrocarbon, followed by sulfonation, which provides the hydrophile. Neutralization with sodium hydroxide finishes the sequence, yielding the surfactant.

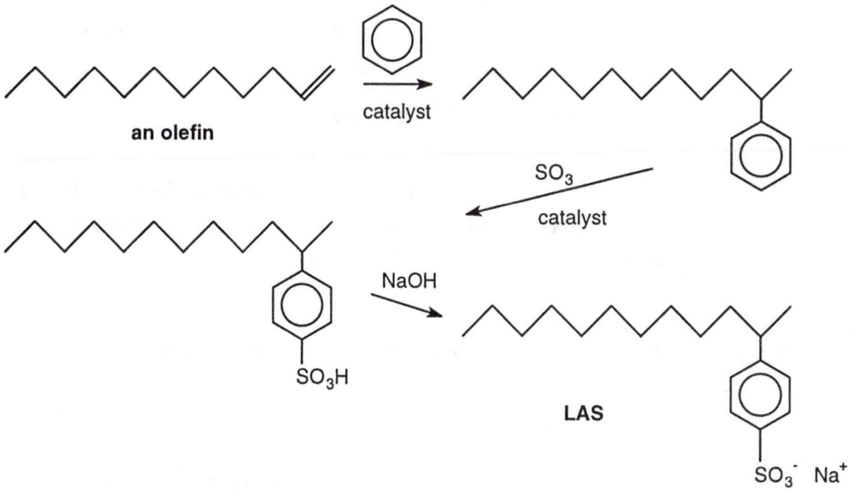

catalyst
an olefin
SO_3
catalyst
NaOH
SO_3H
LAS
SO_3^- Na^+

The starting alkene (olefin) usually has a range of chain lengths, and the intermediate alkylbenzene can have the phenyl group attached at any position except the terminal carbon. Accordingly, the LAS surfactant synthesized is a **complex mixture** of homologs (different chain lengths) and isomers (different phenyl group attachment). Only the C_{12} homolog and 2-isomer are shown above.

Sulfur trioxide (a product of the sulfuric acid industry) is used in the manufacture of several types of anionic surfactants. **Alkyl sulfates**, for example, are formed by the sulfation of fatty alcohols.

$$\text{R—OH} + SO_3 \xrightarrow{\text{catalyst}} \text{R—O—}SO_3H$$

Fatty alcohol

$$\downarrow \text{NaOH}$$

$$\text{R—O}SO_3^- \ Na^+$$ **Alkyl sulphate**

Sulfates differ from sulfonates in having an extra oxygen, contained in the C–O–S link.

C. ALKYL ETHOXYLATES

The nonionic **alkyl ethoxylates** are formed by reacting ethylene oxide with a long-chain fatty alcohol. The ethylene oxide molecules react with the terminal OH in a short polymerization sequence, building up the water-soluble portion of the molecule:

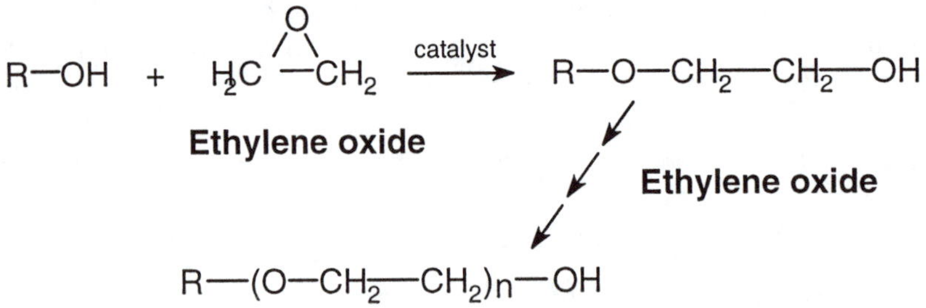

$$R{-}OH + H_2C{-}CH_2 \text{ (O bridge)} \xrightarrow{\text{catalyst}} R{-}O{-}CH_2{-}CH_2{-}OH$$

Ethylene oxide

Ethylene oxide

$$R{-}(O{-}CH_2{-}CH_2)n{-}OH$$

The extent of polymerization can be controlled, depending on the end use of the surfactant. Detergent-grade ethoxylates average about 15 ethylene oxide units per molecule, but the mixture can have a range spanning n = 10 to 20.

D. ALKYL GLUCOSIDES

Alkyl glucosides are formed by **condensing** fatty alcohols with glucose.

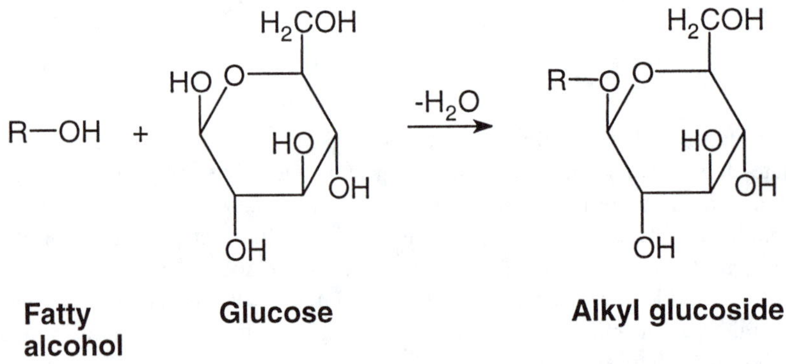

$$R{-}OH + \text{Glucose} \xrightarrow{-H_2O} \text{Alkyl glucoside}$$

H$_2$COH, HO, O, HO, OH, OH — H$_2$COH, R–O, O, HO, OH, OH

Fatty alcohol **Glucose** **Alkyl glucoside**

Usually, more than one glucose can join the molecule (the reaction is quite complex). These surfactants are novel in that both parts of the molecule are drawn from "renewable" resources.

The production of commercial surfactants as **mixtures** is an advantage, since the different components can have synergistic effects. The disadvantage is in assessing environmental effects, since the different components can vary in their toxicity and biodegradation.

IV. DETERGENTS

A. SURFACTANT COMPONENTS

The major household use of surfactants is in commercial mixtures such as cleaners or detergents. Laundry detergents are about one third surfactants, and a breakdown of the main ingredients of a modern compact detergent is given in Table 10.2.

TABLE 10.2
Typical Formulation of a Compact Powder Laundry Detergent

Ingredient	%
Surfactants	30–40
Anionic	(30)
Nonionic	(5)
Soap	(2)
Builders	~35
Polyphosphate/zeolite	
Sodium carbonate/silicate	
Polycarboxylate	
Enzymes	0–1.5
Bleaching agents	0–25
Auxiliaries	~5
Stabilizers	
Brighteners, etc.	

Detergent action is quite complex, but basically involves the dislodgement of dirt particles and their suspension in water (Figure 10.7). Dirt particles (food, grease, soil stains) are relatively hydrophobic and firmly attach to weakly polar fabrics. Surfactant molecules attach to the dirt and substrate surfaces, reducing the interfacial tension and promoting wetting. Heat and agitation tend to tear the dirt particles apart and from the substrate. More surfactant absorbs to and stabilizes the freshly exposed surfaces. Micelles play an essential role here in providing a large reservoir of surfactant molecules. Eventually, the dirt particles break free and are dispersed as an emulsion. The new particles have hydrophilic surfaces that have the same charge, which prevents recombination or redeposition by mutual repulsion.

The surfactant components of detergents are anionic, nonionic or, more commonly, mixtures of the two. The mixtures exploit the synergistic properties of the two types. Cloth fibers, soil and organic particles usually carry **negative surface charges**. Sorbed anionic surfactant enhances this negative charge, increasing mutual repulsions and assisting the detersive action. Nonionics promote surface activity at lower surfactant concentrations, since the absence of a head charge permits easier packing into surface regions and micelles (they have CMCs 10 to 100 times lower than those of anionics). Anionics are better at cleaning cottons, and nonionics are better on synthetics.

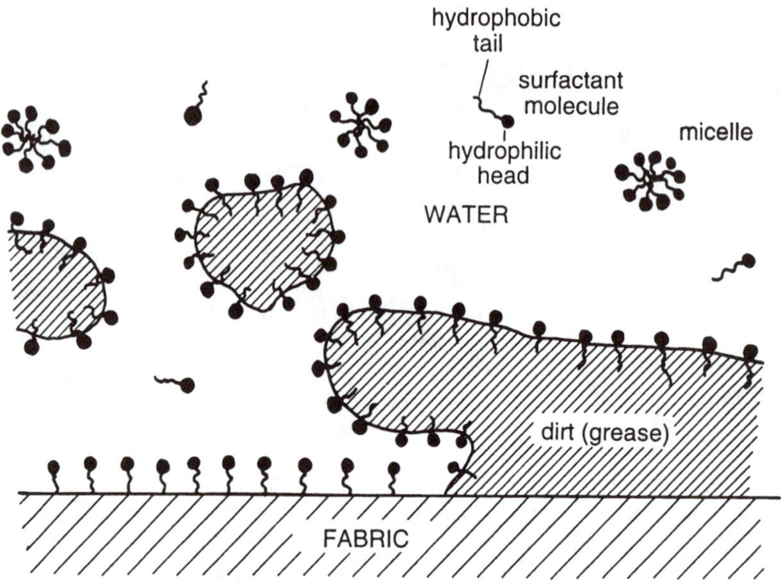

FIGURE 10.7. Detergent action of a surfactant. The cleaned surface fabric is on the left.

Soaps now have declined in their role as detersive agents. A major factor in this lack of use is that soaps are sensitive to hard water, forming very water-insoluble precipitates with polyvalent ions such as calcium and magnesium.

$$2\,R{-}\overset{\overset{\displaystyle O}{\|}}{C}{-}O^-\,Na^+\,(aq) \;+\; Ca^{2+}\,(aq) \longrightarrow Ca\,(R{-}\overset{\overset{\displaystyle O}{\|}}{C}{-}O)_2\,(s) \;+\; 2\,Na^+\,(aq)$$

These greasy solids form the familiar bathtub "rings." They can irreparably stain and give clothing a bad odor.

The surfactants do not suffer these precipitation reactions, but their detersive action is still more or less hardness impaired. Not surprisingly, the nonionic surfactants are much less hardness affected than their anionic counterparts.

Small quantities of soap are included in laundry detergents to **retard foaming**, by actually exploiting the hard water precipitation reaction. The precipitated soap is still strongly surface active, and spreads rapidly over the bubble film, displacing other surfactant molecules. But the solid lowers the elasticity of the foam, making it more prone to rupture and collapse (bubbles, like rubber balloons, need to be elastic to resist damage). Foams do not contribute to washing performance and are tolerated to appease consumer preference.

B. BUILDERS

The second major component of detergents are the **builders**, so-called since they increase the detersive action of the surfactants. The builders have several modes of action, including water softening, maintaining an alkaline pH and improving dirt dispersion.

The major builder in use worldwide is **sodium tripolyphosphate**, $Na_5P_3O_{10}$. The anion of this water-soluble salt is a powerful sequestrant, binding firmly to hardness-causing ions such as Ca^{2+}, Mg^{2+} and other polyvalent ions.

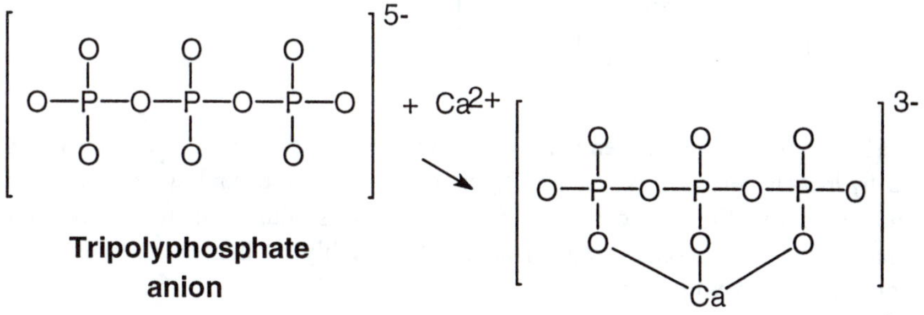

Tripolyphosphate anion

The sequestered metal ion still remains in solution, but in a form that is not available to cause hardness interference with surfactants. Tripolyphosphate is also alkaline and binds to soil particles. Both effects serve to increase the negative surface charges, improving detersive performance.

Tripolyphosphate, however, is a ready source of nutrient phosphorus, and its discharge in large tonnages to waterways is believed to contribute to excessive algal growth or eutrophication. For this reason, many government authorities have restricted its use in detergents.

Zeolites have been used as a phosphate builder substitute for 20 years. They behave as a water softener, removing hardness-causing ions by ion exchange. Zeolites are insoluble solids called *aluminosilicates.* They occur naturally as minor rock-forming minerals. Detergent zeolite is a synthetic zeolite made from sand, sodium chloride and bauxite, and has the formula $Na_{12}(AlO_2)_{12}(SiO_2)_{12}{\cdot}27H_2O$. The anion is a polymer made up of alternating AlO_4 and SiO_4 tetrahedra.

This "giant" anion is porous, with water molecules and sodium ions occupying the pore spaces. Calcium ions from hard water enter the pores, displacing the sodium and preferentially binding to the surrounding anion.

Zeolites are environmentally benign substances. They have a very low toxicity to mammals and aquatic organisms. Being mineral solids, zeolites deposit as sediments in sewage treatment plants and waterways.

C. ENZYMES AND BLEACHES

Enzymes and **bleaches** are two ingredients that augment the cleaning process. Protease and lipase enzyme additives catalyze the hydrolysis of protein and lipid stains, forming hydrophilic products more readily removed by the detergent. Recently, cellulase enzymes have been developed that have a more novel action — they hydrolyze cellulose in the **underlying** cotton fibers. The cellulases, however, selectively attack damaged, nonstructural parts of the fiber, parts that hold stains and cause the cotton to look prematurely worn and jaded. Cellulase washes apparently give a cleaner, newer-looking cloth!

The development of bleaching agents has involved some novel surfactant technology. Bleaches compatible with enzymes are based on peroxygen compounds such as sodium perborate, which contains the active –O–O– group

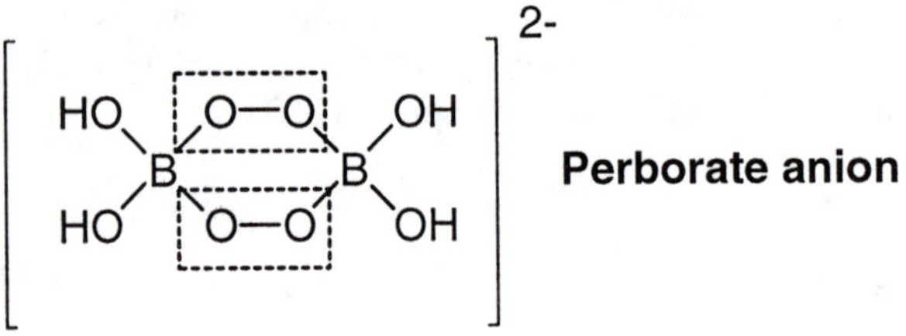

Perborate anion

Perborate releases hydrogen peroxide, H_2O_2, when dissolved in water. However, dilute hydrogen peroxide is not an effective bleach in the cool wash conditions popular currently. Bleach boosters are needed, such as sodium nonanoyloxybenzene sulfonate (SNOBS). SNOBS is a surfactant that readily exchanges its hydrophilic portion for the peroxo group:

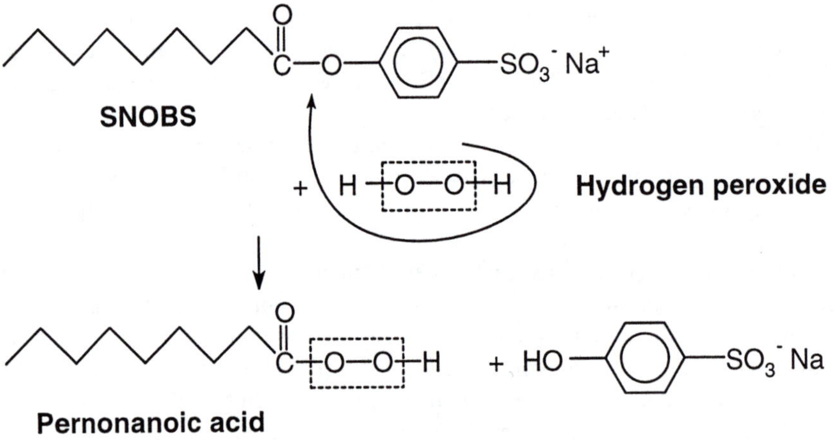

SNOBS

Hydrogen peroxide

Pernonanoic acid

The end result is an in-the-wash synthesis of a surface-active bleach (pernonanoic acid) that has a much higher affinity (and bleaching ability) for the stained fabric.

V. CATIONIC SURFACTANTS

Cationic surfactants are generally ineffective as detergents. Their major household use is as fabric and hair conditioners, and as disinfectants. Because most natural organic and mineral particles usually carry a negative surface charge, cationic surfactants adopt a "head down" position on the surface.

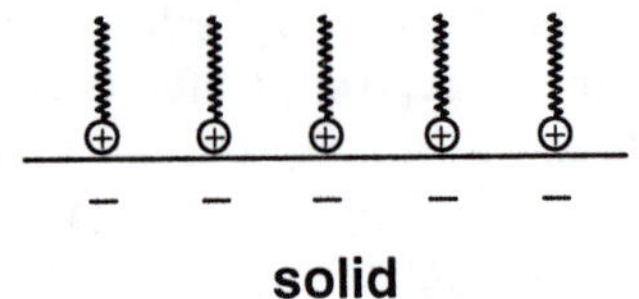

This orientation tends to neutralize the electric charge and offers a hydrophobic surface to the water, both effects discouraging detergency. Just this alignment of surfactant molecules, however, is exploited in fabric and hair conditioners. The fiber surfaces acquire an oil-like film that separates and lubricates the strands, giving a softer feel and reducing static electricity.

Most cationic surfactants behave as **germicides**. Molecules with one long alkyl group are most effective, such as cetylpyridinium chloride, an ingredient of popular mouthwashes.

Cetylpyridinium chloride

Cationic and nonionic surfactants are compatible, but cationics and anionics generally are not. The positive and negative molecules combine to form **neutral 1:1 complexes** that are very insoluble and have much less surface activity. For this reason, fabric conditioners are usually added after the wash cycle, either in the rinse or (as solid sheet formulations) in the drier.

VI. TOXICITY

The oral toxicity of surfactants to mammals is low, ranging from about the same as sodium chloride to exhibiting no lethal effect at all (1000 to >10,000 mg kg^{-1} body weight). Many surfactants are classed as "edible" and used in foods. Sodium dodecylsulfate [$CH_3(CH_2)_{11}OSO_3^-Na^+$], for example, is a detergent-grade surfactant used also as an emulsifier and whipping agent. Surfactants are not accumulated by mammals but are very rapidly metabolized and excreted as shorter molecules.

Aquatic organisms are much more sensitive to surfactants than mammals, having LC_{50} values in the milligram-per-litre range (Table 10.3). Cationics, in particular, are potent algicides and bacteriocides, with responses below 1 mg L^{-1}. The mode of toxicity is not clear, but surfactants seem to interfere with sensitive **external tissues** such as gill membranes.

Toxicities increase with increasing hydrophobic character of the molecule (Figure 10.8). As the chain length in the linear alkyl benzene sulfates increases the hydrophobic nature of the compound increases thus homologues that have longer alkyl chains become more toxic. The same trend is observed by shortening the polyether chain in ethoxylates (essentially making the hydrophile fragment less effective). This pattern of toxicity fits surfactants into the **general narcosis model** where partition into a lipid phase is a prime requirement. Surfactants have toxicities not unlike nonsurfactant organic compounds with similar K_{OW} values and nonspecific modes of action.

The toxicity of cationics is mitigated by the presence of particles or anionic surfactants. The toxicity of the clarified river water without particles is higher (lower L_{50}) than the unclarified river water (see Table 10.3). Sorption to the negative surfaces

TABLE 10.3
Surfactant Toxicity to Some Aquatic Organisms

Surfactant	Fish LC_{50} (mg L^{-1})	*Daphnia* LC_{50} (mg L^{-1})	Algae growth inhibition EC_{50} (mg L^{-1})
Anionic			
Linear alkylbenzene sulfonates	3–10	1–10	1–100
C_{12} fatty alcohol sulfates	3–20	5–70	~60
Nonionic			
Fatty alcohol ethoxylates	1–10	1–10[a]	4–50
Alkylglucosides	3.7	—	6.5
Cationic			
Distearyldimethyl	0.62–3[b]	0.16–0.48[b]	0.1–10
Ammonium chloride	10–24[c]	3.1[c]	

[a] Various invertebrates
[b] Clarified water
[c] River water

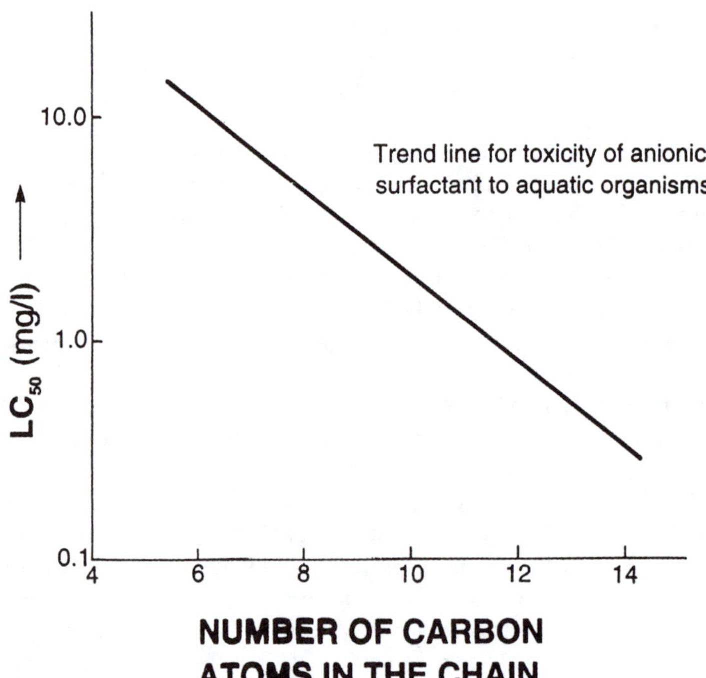

FIGURE 10.8. Variation in surfactant toxicity to aquatic organisms with change in the nonpolar chain length.

or the formation of neutral complexes probably reduce the effective numbers of toxic molecules in the water.

Commercial mixtures of surfactants decline in toxicity as they degrade forming intermediates that have lost their surface activity. An exception are the **alkylphenol**

ethoxylates. Enzymes degrade this surfactant by "snipping" segments off the polyether chain. The intermediates are still surface active, and with a shrunken "head" become more hydrophobic and more toxic. For this reason, the use of alkylphenol ethoxylates is restricted in the U.S. and some other countries.

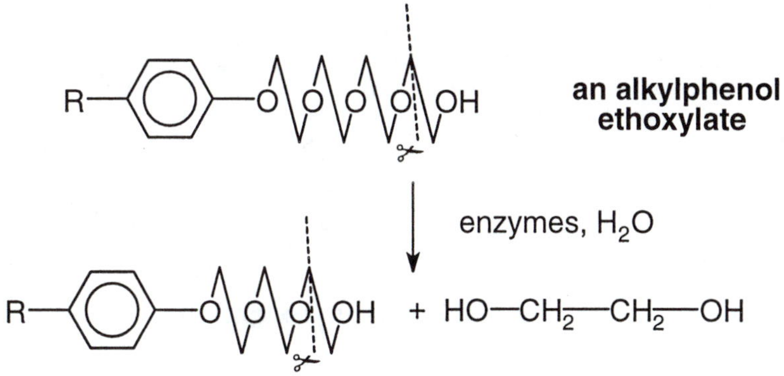

an alkylphenol ethoxylate

enzymes, H_2O

+ $HO-CH_2-CH_2-OH$

VII. BIODEGRADATION

Surfactant biodegradation first became an issue in the 1950s, when alkylbenzene sulfonates (ABS) largely replaced soaps as mentioned in the Introduction. Surfactant residues accumulated in sewage treatment plants and waterways, causing spectacular foaming episodes, tastes in recycled water, and reaching levels potentially toxic to aquatic life.

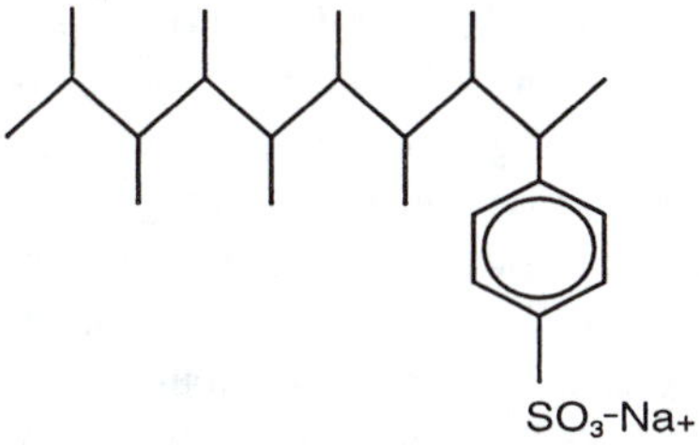

$SO_3^-Na^+$

A branched alkylbenzene sulfonate

A branched alkylbenzene sulfonate

The early surfactants contained **highly branched alkyl hydrophobes** that were resistant to biodegradation. Today, these recalcitrant surfactants are largely obsolete, having been replaced by the linear alkylbenzene sulfonates (LAS) and other biodegradable surfactants.

The variety of surfactant types, coupled with the great diversity of bacterial populations, means that no one degradation process dominates. Enzymatic attack can occur at the hydrophobe, the hydrophile, the connecting link, or all three. Linear alkylbenzene sulfonates degrade via **oxidation** of the hydrophobic chain. Oxidation begins at the terminal carbon, forming a carboxyl group; this is the slow step and causes the molecule to lose its surfactant properties. Successive oxidations shorten the chain two carbons at a time (a process that mirrors the way all living organisms metabolize the fatty acids in their cells). Ring opening and mineralization complete the reaction sequence as shown below.

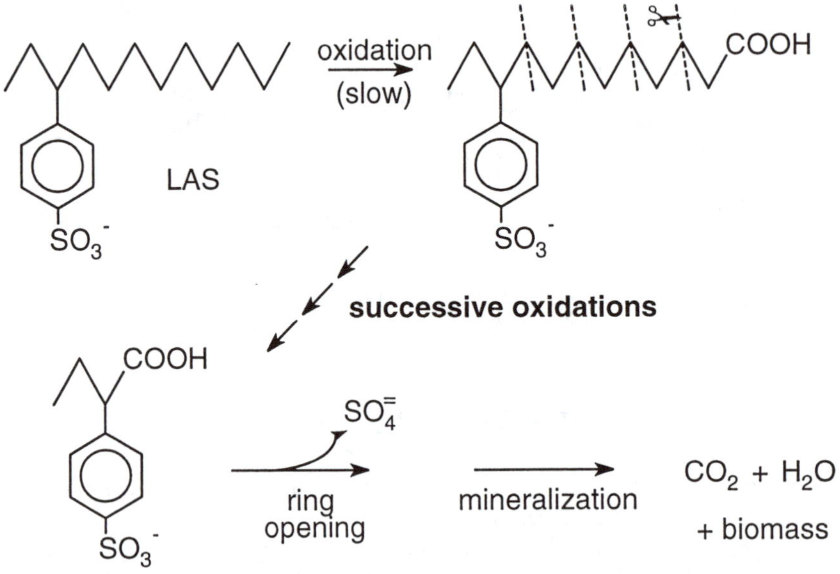

LAS $\xrightarrow[\text{(slow)}]{\text{oxidation}}$ COOH

SO_3^- SO_3^-

successive oxidations

COOH

SO_3^-

$SO_4^=$

ring opening

mineralization

$CO_2 + H_2O$ + biomass

The remainder of the anionic and nonionic surfactants listed in Table 10.1 are regarded as highly biodegradable, often exceeding 98% breakdown when the test criteria demand a minimum of 80 or 90%. Most of the molecules contain chemical groups that exist in natural materials. These groups are prone to attack by hydrolytic enzymes that are common in bacterial assemblages — etherases, sulfatases, esterases, etc.

Cationic surfactants are biodegradable, as long as toxicity thresholds are not exceeded. At concentrations below 1 mg/L, breakdown is rapid and virtually complete, but much higher; and their biocidal effect can slow or stop bacterial action. Cationic surfactant biodegradation is improved in the presence of the anionic variety, probably a consequence of the "neutralization" effect, which reduces toxicity.

VIII. SORPTION AND BIOACCUMULATION

Cationic surfactants bind strongly to soils and sediments, sorbing to the particles and complexing with humic matter. The organic soil matter, humic and fatty acids, behave as a negative surfactant, forming micellular aggregates. Sediment-water partition coefficients (concentration in soil/concentration in water) can exceed 10^4 for cationics, but are orders of magnitude lower for anionic and nonionic surfactants.

Surfactants show little facility for bioaccumulation, with bioconcentration factors spanning about 3 to 500. The high rates of metabolism tend to limit biological concentrations. Cationics are only slowly taken up by fish since they cling to the gill membranes but are not transported across this tissue.

IX. NATURAL SURFACTANTS

Surfactants occur widely in plants and animals, performing many essential functions (Table 10.4). Lecithins, for example, are phospholipids that occur in the walls of living cells. Lecithins have long been commercially extracted from plants (e.g., soya bean) and used in foods. Some bacteria, yeasts and fungi are prolific surfactant producers, apparently to act as extracellular emulsifiers of water-insoluble substances such as hydrocarbons. Surfactant biosynthesis is an emerging technology to provide surfactants designed for specific purposes.

X. KEY POINTS

1. During the 1950s and 1960s, large masses of foam in sewage plants and waterways indicated the presence of alkylbenzene sulfonate (ABS) surfactants, which were resistant to degradation. These were later replaced with the linear alkylbenzene sulfonate (LAS) surfactants with less persistence in water.

2. Soaps and detergents contain surfactants that assemble at the interfaces (e.g., air/water) and modify the properties of the phases.

3. The molecules of surface-active agents (surfactants) usually contain two parts with different characteristics. A hydrophilic part (a polar head) which orients at the two-phase interface to the polar phase, and a hydrophobic part (a nonpolar tail) which orients to the nonpolar phase. When this occurs, a monolayer forms at the interface with like charges in proximity. The repulsion of these groups tends to reduce the surface tension. This allows the phases to form mixtures.

4. Surfactants are broadly categorized according to the charged nature of the hydrophilic part of the molecule:

   - **anionic**: negatively charged
   - **cationic**: positively charged
   - **nonionic**: neutral, though highly polar
   - **amphoteric**: a "zwitterion" containing positive and negative charges

5. Soaps are the metal salts of long chain fatty acids and traditionally are made by saponification of natural fats or oils which are triesters of fatty acids and glycerol.

$$\begin{array}{l} R{-}CO{-}O{-}CH_2 \\ R{-}CO{-}O{-}CH_2 \\ R{-}CO{-}O{-}CH_2 \end{array} \xrightarrow[H_2O]{NaOH} 3R{-}\overset{O}{\overset{\|}{C}}{-}O^- Na^+ + \begin{array}{l} HO{-}CH_2 \\ HO{-}CH \\ HO{-}CH_2 \end{array}$$

Fat **Soap** **Glycerol**

TABLE 10.4
Some Natural Surfactants

a bile salt (intestinal emulsifier)

a lecithin, a phospholipid (cell walls, lung surfactant)

Surfactin, a lipopeptide
(bacteria - extracellular emulsifier, antibiotic)

a rhamnolipid (bacteria)

6. The linear alkyl benzene sulfonates are a major group of synthetic anionic surfactants and are synthesized by the following reaction sequence:

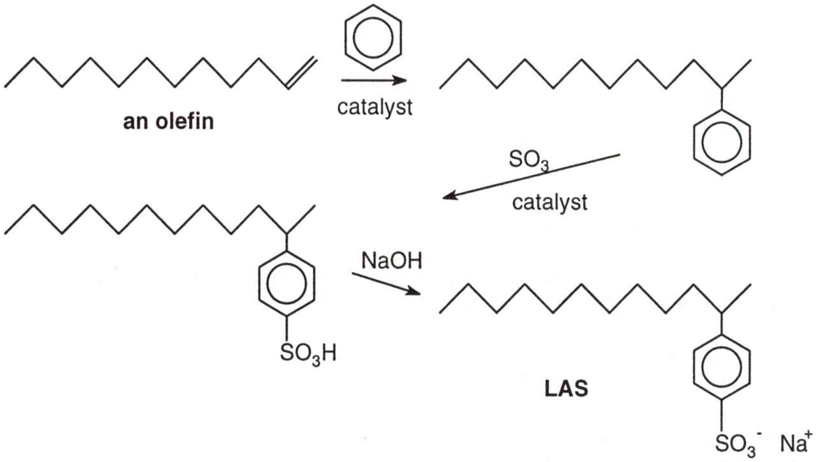

an olefin

catalyst

SO_3

catalyst

SO_3H

NaOH

LAS

SO_3^- Na^+

Linear alkylbenzene sulphonate

7. Washing detergents contain surfactants (30%-40%), builders to enhance the detersive action (about 35%), enzymes to assist in removing stains etc. from fabrics, bleaching agents (0.25%) auxiliaries (about 5%).

8. Cationic surfactants are used as fabric and hair conditioners and as disinfectants. The surfactant molecules tend to adopt a head-down (positive charge) position on the surface, which usually contains a negative charge. The fabric and hair surface acquire a neutral oil-like film that lubricates movements and gives a soft feeling.

9. The toxicity of surfactants to mammals is generally low and many are used in foods. However, toxicity to aquatic organisms is relatively high ($LC_{50} \cong 0.1$ to 70 mg L^{-1}). Toxicity increases with carbon chain length, which increases hydrophobicity.

10. The ABS surfactants used in the 1950s and 1960s had highly branched chains, as shown below.

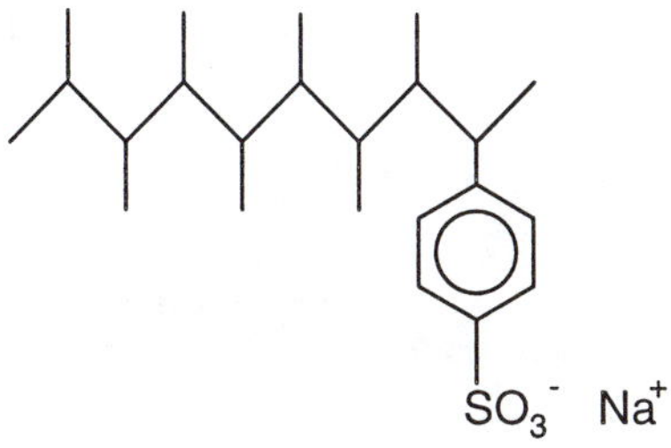

SO_3^- Na^+

This branching causes resistance to degradation. The LAS surfactants do not have this branched structure and are relatively rapidly degraded in the environment.

REFERENCES

Hutzinger, O., Ed., Detergents. *The Handbook of Environmental Chemistry,* Vol. 3, Part F. Springer-Verlag, Berlin, 1992.

Falbes, J., Ed., *Surfactants in Consumer Products,* Falbe, J., Ed., Springer-Verlag, Berlin, 1987.

Clint, J.H., *Surfactant Aggregation,* Chapman and Hall, New York, 1992.

CHAPTER 10 PROBLEMS

1. Which of the following compounds are likely to be surface active?

   (a) 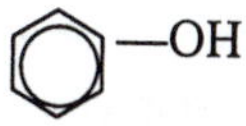 C_6H_5—OH
   (b) $CH_3(CH_2)_{10}Cl$
   (c) Sucrose
   (d) $CH_3(CH_2)_8SO_3^-\ Na^+$
   (e) $CH_3{}^+NH_3\ Br^-$
   (f) $CH_3(CH_2)_{10}(OCH_2CH_2)_6OH$

2. Classify (a) lecithin and (b) the rhamnolipid (Table 10.4) according to their surfactant type.

3. Suggest reasons why the surface tension (a) drops as the CMC is approached, and then (b) levels off after the CMC is passed (Figure 10.5).

4. Compare the toxicity and biodegradability that which would be expected with the following two surfactants:

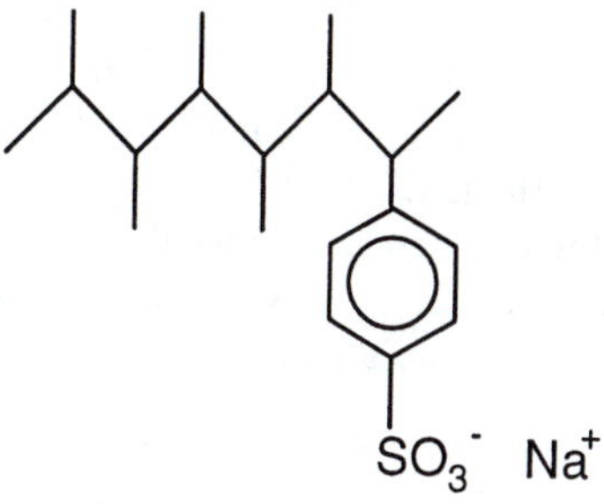

SO_3^- Na^+

Compound (a)

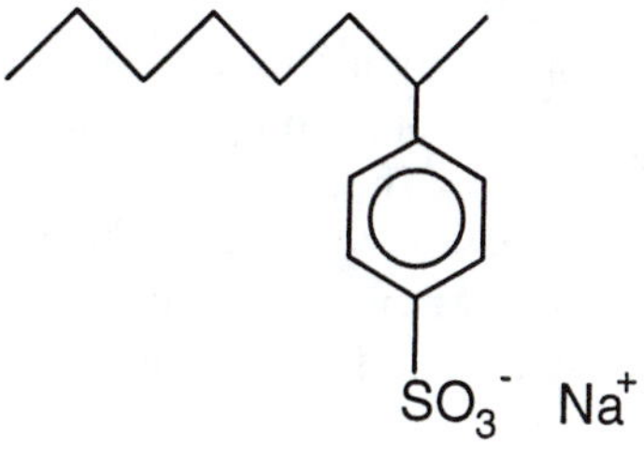

SO_3^- Na^+

Compound (b)

CHAPTER 10 SOLUTIONS

1. For a compound to be a surfactant, it must contain within the molecule a clearly differentiated highly polar group and a highly nonpolar group. Compound (a) (phenol) has some of these characteristics, with the weakly polar benzene ring and the polar hydroxyl group, but these characteristics are not strong enough to produce a surfactant. Compound (b) does not have the required groups. Compound (c) (sucrose) has some polar groups but it doesn't have a nonpolar group. Compound (d) has a nonpolar group, $CH_3(CH_2)_8^-$, and a polar group (SO_3^-) that are well differentiated and is be expected to be a surfactant. Compound (e) lacks a significant nonpolar part of the molecule and so does not have the required groups. Compound (f) is a nonionic fatty acid ethoxylate with a nonpolar group, $(CH_3(CH_2)_{10}^-$, and a polar group, $(OCH_2CH_2)_6OH$, which give it surfactant properties.
   This means that Compounds (d) and (f) would be expected to be surfactants.

2. Lecithin contains both positive and negative charges within the molecule and cannot form free ions. It would be classified as **amphoteric**.
   Rhamnolipid contains nonpolar and polar groups within the molecule but does not form ions. The acid groups would be expected to have a very low ionization capacity. It would be classified as **nonionic**.

3. **Part (a).** More surfactant molecules are available to occupy the interface as concentration increases. This leads to more intermolecular repulsion at the interface and a drop in surface tension as the CMC is approached.
   Part (b). The interface is now saturated with surfactant molecules, and so the number of molecules at the interface and the surface tension reaches a steady value. At increased concentrations, the excess surfactant molecules would be expected to form micelles.

4.

Characteristic	Compound (a)	Compound (b)
Surfactant classification	Anionic	Anionic
Surfactant type	Alkylbenzene sulfonate (ABS)	Linear alkylbenzene sulfonate (LAS)
Toxicity[a]	Mammals: low Aquatic organisms >Compound (b)	Mammals: low Aquatic organisms: <Compound (a)
Biodegradability[b]	<Compound (b)	>Compound (a)

[a] Toxicity increases with the size of the carbon chain, and Compound (a) has 11 carbons and Compound (b) has 7 carbons.

[b] Biodegradability increases with branching, and Compound (a) has more branching than Compound (b).

Chapter 11

ORGANOMETALLIC COMPOUNDS

I. INTRODUCTION

There is an important group of substances, from an environmental prospective, that consists of metals in combination with organic groups, commonly referred to as **organometallic compounds.** These substances include environmental contaminants as well as chemical transformation products that result from environmental processes. A well-known example of this group is the gasoline additive tetraethyllead [$(CH_3CH_2)_4Pb$], which is now being phased out of use in many areas due to environmental contamination.

Many metals, after discharge into the environment, have organic groups attached to them by environmental processes, converting them into organometallic compounds. Mercury, in an inorganic form, can be an important environmental contaminant but on discharge, most mercury within organisms is converted into dimethylmercury ($H_3C–Hg–CH_3$). The properties of this substance are quite different, in most respects, from the original mercury. It is more volatile and more toxic than metallic mercury and its distribution in the environment is different.

The organometallic compounds share many properties with organic compounds in general. So, organometallic compounds of environmental significance are considered in this chapter.

II. THE NATURE OF ORGANOMETALLIC COMPOUNDS

Organometallic compounds are commonly defined as compounds having metal-carbon bonds, where the carbon atom is part of an organic group. An example of an organometallic compound is dimethylmercury ($H_3C–Hg–CH_3$). Most metals are capable of forming organometallic derivatives. Metalloids are elements that in elemental form exhibit one or more characteristics of metals, but not all. In the Periodic Table, metalloids are found on the boundary between metals and non-metals. Metalloids are boron (B), silicon (Si), germanium arsenic (As), antimony (Sb) and tellurium (Te), with the metals to the left and the nonmetals to the right (see Figure 11.1). Metalloids form bonds to carbon; but for simplicity, organometalloids will be referred to as organometallics in this chapter. In some organometallics, the entire bonding capacity of the metal is taken up by bonds to carbon [e.g., $(CH_3)_2Hg$ and $(CH_3)_4$ Pb]. Such compounds are named by first taking the names of the organic groups, in alphabetical order, followed by the name of the metal. The prefixes di-, tri- and tetra- are used when the same organic group occurs more than once in a molecule. The name is written as one word. This is the system adopted in this chapter. Examples, according to this nomenclature system, are:

$H_3C{-}Sn(CH_2CH_3)_2{-}CH_3$

Diethyldimethyltin

$H_3C{-}As(CH_2CH_3)(CH_2)_3CH_3$

Butylethylmethylarsenic

IA	IIA											IIIA	IVA	VA	VIA	
H 2.1																
Li 1.0	Be 1.5											B 2.0	C 2.5	N 3.0	O 3.5	F 4.0
Na 0.9	Mg 1.2	Transition Metals										Al 1.5	Si 1.8	P 2.1	S 2.5	Cl 3.0
K 0.8	Ca 1.0	Sc 1.3	Ti 1.5	V 1.6	Cr 1.6	Mn 1.5	Fe 1.8	Co 1.9	Ni 1.9	Cu 1.9	Zn 1.6	Ga 1.6	Ge 1.8	As 2.0	Se 2.4	Br 2.8
Rb 0.8	Sr 1.0	Y 1.2	Zr 1.4	Nb 1.6	Mo 1.8	Tc 1.9	Ru 2.2	Rh 2.2	Pd 2.2	Ag 1.9	Cd 1.7	In 1.7	Sn 1.8	Sb 1.9	Te 2.1	I 2.5
Cs 0.7	Ba 0.9	La 1.1	Hf 1.3	Ta 1.5	W 1.7	Re 1.9	Os 2.2	Ir 2.2	Pt 2.2	Au 2.4	Hg 1.9	Tl 1.8	Pb 1.9	Bi 1.9	Po 2.0	At 2.2

FIGURE 11.1 The Periodic Table with the electronegativities of the elements.

In other organometallics, the metal forms some bonds to carbon, and some to other elements [e.g., methylmercury chloride CH_3HgCl and trimethyllead chloride $(CH_3)_3PbCl$].

The first recorded preparation of an organometallic (an organoarsenic compound) occurred over 200 years ago. Today, organometallic compounds are produced in ever increasing amounts for a wide range of purposes. These substances are used as catalysts (e.g., organoaluminium compounds), pesticides (e.g., organoarsenic compounds), gasoline additives (organolead compounds) and polymers (organosilicon compounds). Recent production figures (shown in Table 11.1) indicate the scale on which these substances are produced. With reduction in the lead content of gasoline, amounts of organolead compounds manufactured may decline in the future. However, production of organic derivatives of many other metals can be expected to increase.

Organometallic compounds often exhibit properties (e.g., toxicity, solubility and volatility) that are different from those of the metal itself and inorganic derivatives of the metal. The relatively toxic nature of some organometallics was first demonstrated over a century ago. A number of poisoning incidents were reported that were ultimately found to be due to a mold producing the volatile trimethylarsenic (Me_3As) from nonvolatile arsenic-containing dyes in wallpaper. This also showed that organometallic species can be formed in the environment from inorganic starting mate-

TABLE 11.1
Production Figures for Some Organometallics

Organometallic	Production (tons per year)
Silicones	700,000
Organolead	600,000
Organoaluminum	50,000
Organotin	35,000
Organolithium	900

From Elschenbroich, C. and Salzer, A., *Organometallics — A Concise Introduction*, 1989. Copyright VCH Verlagsgessellschalft. With permission.

rials. The different properties of organometallics compared to those of the metal element and inorganic derivatives influences the environmental mobility and distribution of that element. Those metals that can form environmentally stable organometallic derivatives have new and added pathways in their biogeochemical cycles.

III. STABILITY AND SOURCES

A. STABILITY

1. Hydrolysis

Most elements of the Periodic Table are metals, and practically all of them can form organometallic derivatives. Comparatively few, however, form organometallic compounds that are relatively stable in the environment. Essentially all the metals that carbon bonds to in organometallic compounds are more electropositive (or less electronegative) than carbon. Carbon-metal bonds are thus polarized in the direction $C^{\delta-}$–$M^{\delta+}$. The electronegativities of some metals and carbon are shown in Periodic Table format in Figure 11.1. The difference in electronegativity between carbon and a metal can give some indication of the polarity of a C-M bond. Clearly, the nature of the organic group to which the carbon belongs also plays a role. Ionic bonds have a larger electonegativity difference, a more complete charge transferral, and greater distortion of the electron cloud than polar covalent bonds, but there is no sharp distinction between them. It has been suggested that ionic bonds are considered to be those with an electronegativity difference of $\geq$1.5. On this basis, organic derivatives of most Group IA and Group IIA metals (e.g., Na, K, Ca) can be thought of as possessing ionic bonds. This is borne out by their salt-like properties. They are generally colorless, high melting point solids with low volatilities and low solubilities in nonpolar solvents such as pentane or hexane. The crystal structure of methylpotassium is in fact similar to that of sodium chloride; each potassium ion is surrounded by six methyl anions and vice versa. Methylpotassium could well be written Me^-K^+.

Under environmental conditions, organometallics are exposed to moisture or water. The greater the polarity of the carbon-metal bond, the greater the rate of hydrolysis (reaction with water). It has been observed that for metals with an electronegativity of about 1.7 or less, their organometallic derivatives react relatively rapidly with water to give a hydrocarbon and the metal hydroxide. For example,

$$\begin{array}{l} H_3C^- \; K+ \\ \quad\quad + \\ H^{\delta+} -^{\delta-} OH \rightarrow CH_4 + KOH \end{array}$$

$$(CH_3)_3Al + 3H_2O \rightarrow 3CH_4 + Al(OH)_3$$

In some cases, the reaction is explosively violent!

Because electronegativity increases as you move across the Periodic Table, the only metals capable of forming hydrolysis-resistant organometallic compounds are essentially those on the right in Groups IVA (Si, Ge, Sn, Pb) and VA (As, Sb, Bi), selenium, tellurium, mercury and thallium, together with some transition metals. If an organometallic molecule possesses only weakly polar carbon-metal bonds (e.g., Me_2Hg, Me_4Sn), then the compound will be weakly polar and organic-like in its properties. Typically, for such compounds with small organic groups such as methyl, they are low melting, volatile materials poorly soluble in water and soluble in nonpolar solvents. The boiling points of methyl derivatives of Group IVA and VA metals are shown in Table 11.2.

TABLE 11.2
Boiling Points (°C) of Methyl Derivatives of Some Group IVA and V Metals

Group IVA		Group V	
Me_4C	10	Me_3As	52
Me_4Ge	43	Me_3Bi	110
Me_4Sn	77		
Me_4Pb	110		

From Wilkinson, G., Stone, F.G.A., and Abel, E.W., Eds., *Comprehensive Organometallic Chemistry,* Vol. 1, p.3, 1982. Copyright Pergamon Press. With permission.

2. Thermal Stability

Another factor to consider in the environmental occurrence of organometallics is thermal stability. Will the compound be stable under ambient conditions or when elevated temperatures are encountered? Typically, when alkyl derivatives of metals decompose thermally, the products are hydrocarbons, the elemental metal and perhaps some H_2. For example, Me_4Pb has a number of pathways by which it undergoes thermal decomposition, as shown below.

$$\mathrm{Me_4Pb} \xrightarrow{\Delta} \mathrm{Pb + 2CH_3{-}CH_3}$$

$$\mathrm{Pb + 2CH_4 + H_2C{=}CH_2}$$

$$\mathrm{Pb + 2H_2 + 2H_2C{=}CH_2}$$

For some organometallic compounds such as Me_4Si, the organometallic is more stable than the products of thermal decomposition. However, for many other compounds (e.g., Me_2Cd, Me_3In, Me_4Pb), the reverse is true. In these cases, the organometallic compound is thermodynamically unstable with respect to thermal decomposition. Some of these compounds do exist in the environment though because the *rate* of thermal decomposition is small.

Factors influencing the rate include the strength of the carbon-metal bond and the presence of empty orbitals that are not too much higher in energy than the filled ones. The first, and rate-determining step in thermal decomposition is probably homolysis of a carbon-metal bond. (Homolysis is the breaking of a covalent bond such that each fragment retains one of the two electrons from the bond). This is shown for tetramethyllead.

$$\mathrm{Me_4Pb \rightarrow Me_3Pb^{\bullet} + {}^{\bullet}CH_3}$$

Carbon-metal bonds tend to become longer and weaker as you go down a group of the Periodic Table (see Figure 11.1). This suggests that organometallic derivatives of the lower members of each group would tend to be most susceptible to thermal decomposition. For example, tetramethyllead, Me_4Pb, decomposes above about 265°C, while Me_4Si is stable at over 500°C. This is the reason that tetralkyllead compounds have been added to gasoline. At temperatures achieved in internal combustion engines, the organolead molecules break down into radicals,

$$\mathrm{(CH_3CH_2)_4Pb \xrightarrow{\Delta} 4(CH_3CH_2^{\bullet}) + Pb}$$

which act to promote smoother combustion of gasoline.

The accessibility of low-energy, vacant orbitals enhances thermal decomposition by accommodating electrons promoted from metal-carbon bonding orbitals. This weakens the metal-carbon bond. This effect is seen in organometallic derivatives of transition metals. Consider the case of titanium, which has four valence electrons and an outer electron configuration of $3d^2 4s^2$. There are unfilled $3d$ orbitals on the titanium atom. On forming covalent bonds to four methyl groups, the carbon atoms in each methyl group acquire a filled shell of electrons. In the Me_4Ti molecule, there are accessible empty orbitals derived from the original vacant d-orbitals of the titanium atom. This predisposes the molecule to thermal decomposition. It is found that this compound is unknown in the environment, decomposing at temperatures above –78°C.

There is another mechanism by which organometallics decompose thermally. It is called a β-elimination mechanism, and requires the presence of a hydrogen atom on the β-carbon atom (the second carbon from the metal atom). The organic group thus needs to contain at least two carbon atoms.

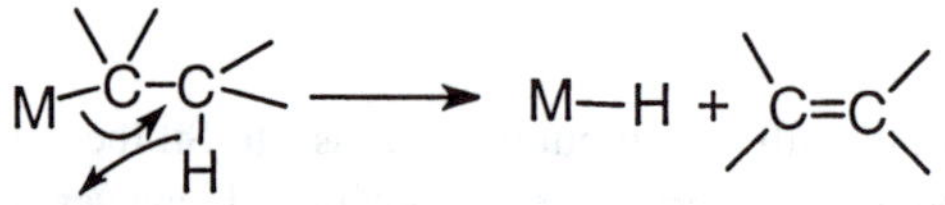

$$M-C-C(H) \longrightarrow M-H + C{=}C$$

The products include alkenes. This process is also aided by the presence of a vacant, accessible orbital on the metal to interact with the electron pair of the β-carbon-hydrogen bond. Tetraethyltitanium $(CH_3CH_2)_4Ti$ is capable of β-elimination and has never been found in the environment. However, a compound like $(C_6H_5CH_2)_4Ti$ does not have this capacity, and is stable at room temperature.

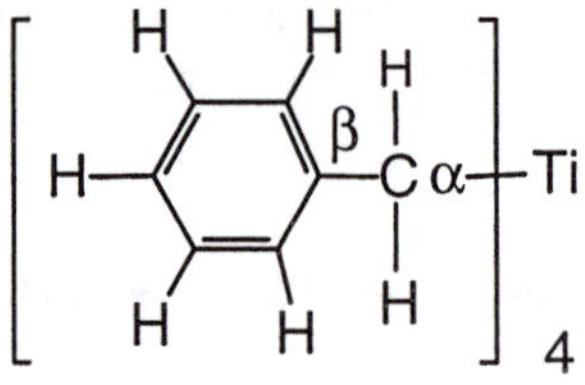

$$[C_6H_5{-}\overset{\beta}{}CH_2(\alpha){-}]_4Ti$$

3. Photolysis

In areas of the environment to which light penetrates, organometallics may be subject to photolysis (light-induced cleaving of bonds). The lower wavelength limit of solar radiation reaching the surface of the Earth is about 290 nm. The energy contained in this radiation can be calculated from Planck's law:

$$\text{Energy of a photon: } E = h\nu \text{ (Planck's law) and } c = \lambda\nu$$

where h= Planck's constant 6.626×10^{-34} J
c= velocity of light 3×10^{8} ms^{-1}
λ= wavelength of light 290 nm = 290×10^{-9} m

$$E = \frac{hc}{\lambda} = 6.85 \times 10^{-19} \text{ J photon}^{-1}$$

Since Avagadro's Number (molecules in a mole) is 6.02×10^{23}, then

$$\begin{aligned}\text{Energy of a mole of photons} &= 6.85 \times 10^{-19} \times 6.02 \times 10^{23} \text{ J mole}^{-1} \\ &= 413\,000 \text{ J mole}^{-1} \\ &= 413 \text{ kJ mole}^{-1}\end{aligned}$$

One mole of photons at 290 nm contains 413 kJ energy. Higher wavelengths toward the visible region would contain less energy. Metal-carbon bond strengths cover a wide range. Table 11.3 contains mean bond enthalpies, or the average energy required to break the metal-carbon bonds, in gaseous methyl derivatives of metals (Me_nM). These bond enthalpies can be used as a measure of bond strength. They show that light encountered at the Earth's surface *potentially* has sufficient energy to break some metal-carbon bonds. To undergo photolysis, the molecule actually has to absorb this radiation. Whether or not this occurs depends on the energy level differences within the molecule. The absorption spectrum of gaseous dimethylmercury is shown in Figure 11.2. Maximum absorbance occurs at just over 200 nm, and there is practically no absorbance at wavelengths of 290 nm and above. Dimethylcadmium (Me_2Cd) has an absorbance maximum near 220 nm, as does tetraethyltin [$(CH_3CH_2)_4Sn$]. This indicates that alkyl derivatives of metals are stable toward photolysis under environmental conditions encountered near the surface of the Earth. Degradation is more likely if the organic group itself contains a suitable chromophore. Tetraphenyltin for example has an absorbance band at 245 to 270 nm.

TABLE 11.3
Mean Metal-Carbon Bond Enthalpies (kJ mol^{-1}) for Gaseous Methyl Derivatives of Metals

Group IIB	Group IIIA	Group IVA	Group V
—	Al, 274	Si, 311	—
Zn, 177	Ga, 247	Ge, 249	As, 229
Cd, 139	In, 160	Sn, 217	Sb, 214
Hg, 121	—	Pb, 152	Bi, 141

From Wilkinson, G., Stone, F.G.A., and Abel, E.W., Eds., *Comprehensive Organometallic Chemistry*, Vol. 1, p.5, 1982. Copyright Pergamon Press. With permission.

4. Oxidation

Complete oxidation of organic compounds produces CO_2 and H_2O. Oxidation of most organometallic compounds produces the metal oxide, CO_2 and H_2O. For example,

$$(CH_3)_2Zn(g) + 4O_2(g) \rightarrow ZnO(s) + 2CO_2(g) + 3H_2O(g) \rightarrow \Delta H = -1918 \text{ kJ mole}^{-1}$$

$$(CH_3)_4Sn(g) + 8O_2(g) \rightarrow SnO_2(s) + 4CO_2(g) + 6H_2O(g) \rightarrow \Delta H = -3591 \text{ kJ mole}^{-1}$$

These reactions are potentially exothermic, and the equilibrium position is heavily in favor of products. For many organometallics, this process occurs rapidly, for example, with alkyl derivatives of transition metals, Groups IA, IIA and IIIA, as well as zinc (see Figure 11.1). Some are so reactive they spontaneously burst into flames in air! Rapid reaction with oxygen is associated with the presence of empty, accessible orbitals, but also the presence of electron lone pairs on the metal. It may seem strange to talk about metals having lone pairs, but remember that elements such as arsenic and antimony are in the same group (Group VA) of the Periodic Table as nitrogen, which has one lone pair. A compound such as trimethylarsenic could well be written as $:AsMe_3$. These metallic elements commonly exhibit valen-

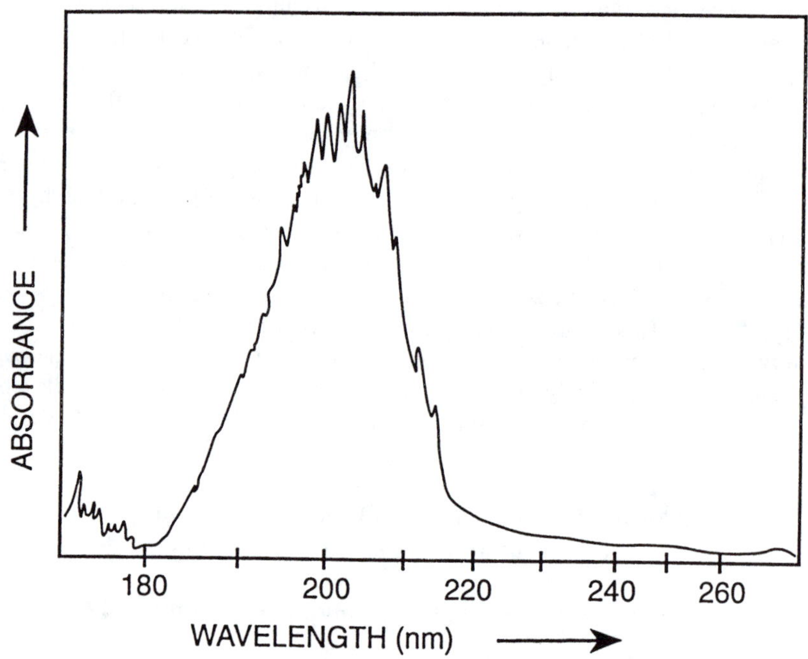

FIGURE 11.2 Absorption spectrum of gaseous dimethylmercury.

cies of three and five. On exposure to oxygen, trimethylarsenic forms $Me_3As{=}O$, in which arsenic has a valency of five. Metal-like elements of Group IVA (selenium and tellurium) also possess lone pairs and are reactive toward oxygen.

Despite exothermicity and an equilibrium position in favor of oxidation, some organometallics will withstand exposure to oxygen. This is because the rate of oxidation is relatively slow, usually due to the absence of suitable empty orbitals or lone pairs. The Group IVA organometallics (like Me_4Sn) are good examples of this.

5. Biotransformation

Organometallics may also be transformed and degraded in the environment by bacteria and higher organisms. In aquatic or sedimentary environments, biotransformation may be a major degradation pathway. Biotransformation occurs at two sites within an organometallic molecule: the metal-carbon bond and on the organic group. Both bacteria and higher organisms possess the ability to cleave metal-carbon bonds. Bacteria for example can convert Me_2Hg to methane (CH_4) and elemental mercury. In humans, tetraalkyllead compounds are converted to trialkyllead salts in the liver, as shown.

$$R_4Pb \rightarrow R_3PbX$$

In organisms that possess the MFO system, hydroxylation of organic groups attached to the metal occurs. For example, with tributyltin salts such as tributyltin chloride, the following reaction is seen. The hydroxylation usually takes place on the β-carbon. For larger alkyl groups, it may be that this is the first step in a multistep pathway leading to cleavage of the organic group from the metal.

$$(CH_3CH_2CH_2CH_2)_3\,SnCl \xrightarrow{MFO} (CH_3CH_2CH_2CH_2)_2\,Sn(Cl)\,CH_2\,CH(OH)CH_2CH_3$$

6. Summary

Overall, environmental stability of organometallics is associated with

- Strong metal-carbon bonds of low polarity
- Lack of electron lone pairs on the metal atom
- Lack of energetically low-lying empty orbitals on the metal atom
- Organic groups with no hydrogens attached to the β-carbon, and no C-C unsaturated bonds

Also, we have largely focused on compounds where the metal is only bonded to carbon. Generally, compounds in which the metal is bonded not only to carbon, but to other non-metals (e.g., Me_3PbCl, $MeAsBr_2$) are more stable.

B. SOURCES

1. Formation in the Environment

Those organometallic compounds found in the environment arise from two sources: discharge of synthetic organometallics either deliberately or inadvertently, and formation in the environment through the actions of organisms or by abiotic chemical reactions. Organisms, primarily microorganisms, or compounds produced by them are known to be able to transfer alkyl groups to metal ions. The most common alkyl group transferred is the methyl group. The principal naturally occurring methylating agents are methylcobalamin (a form of Vitamin B_{12}), S-adenosyl methionine and methyl iodide (CH_3I). Methyl iodide is probably formed from the methylation of the iodide anion by S-adenosyl methionine. These methylation reactions take place within organisms, but can also occur in solution if these methylating agents are released in some way.

Within the environment, it also appears that methyl derivatives of metals may be formed in abiotic transalkylation reactions. This type of reaction is well known in the laboratory and involves transfer of a methyl group from one methylmetal species to another, as shown below.

$$Me_3SnCl + HgCl_2 \rightarrow Me_2SnCl_2 + MeHgCl$$

It is suspected that this process may also occur in the environment, but it is difficult to prove conclusively. For example, microorganisms acting directly on the metal ion may be responsible.

2. Environmental Occurrence

Generally speaking, the metals capable of forming organometallic compounds with sufficient stability to be detected in the environment are silicon, germanium, tin, lead, mercury, arsenic, antimony and selenium. Silicon, germanium, tin and lead are in the same group of the Periodic Table as carbon and, to that extent, may be considered to form organometallic derivatives that are similar to organic compounds (see Figure 11.3). However, there are some differences in the environmental chemistry of these elements compared with carbon. Particularly in aqueous solution, they can utilize outer, unoccupied *d*-orbitals to form species with five or six groups coordinated or bound to the central metal atom. Most organometallic derivatives of these Group IVA elements found in the environment have alkyl groups (e.g., methyl, ethyl, butyl) attached to the metal. Methyl derivatives of germanium have been found in the oceans, while organometallic forms of tin and lead have significant anthropogenic sources in the environment. Although the mercury-carbon bond is relatively weak, it is not particularly susceptible to oxidation or hydrolysis, and so organomercury compounds are also observed in the environment.

													IVA			
1 H																
3 Li	4 Be											5 B	6 C	7 N	8 O	9 F
11 Na	12 Mg											13 Al	14 Si	15 P	16 S	17 Cl
19 K	20 Ca	21 Sc	22 Ti	23 V	24 Cr	25 Mn	26 Fe	27 Co	28 Ni	29 Cu	30 Zn	31 Ga	32 Ge	33 As	34 Se	35 Br
37 Rb	38 Sr	39 Y	40 Zr	41 Nb	42 Mo	43 Tc	44 Ru	45 Rh	46 Pd	47 Ag	48 Cd	49 In	50 Sn	51 Sb	52 Te	53 I
55 Cs	56 Ba	57 La	72 Hf	73 Ta	74 W	75 Re	76 Os	77 Ir	78 Pt	79 Au	80 Hg	81 Tl	82 Pb	83 Bi	84 Po	85 At

FIGURE 11.3 Position of elements in the Periodic Table that form stable organometallic compounds in the environment.

Organoarsenic compounds are relatively widespread in the environment, as a result of both formation in the natural environment and from anthropogenic inputs. The ability to transform inorganic arsenic into organoarsenic compounds is observed with both microorganisms and higher organisms. Apart from simple methylarsenic derivatives, relatively complex molecules such as arsenic-containing lipids and sugars are produced, typically by aquatic organisms. Antimony is in the same Periodic Table Group as arsenic, but few organoantimony compounds have been found in the environment; those that have been, tend to have a valency of five (e.g., $CH_3(OH)_2Sb{=}O$). Methylantimony compounds such as this probably don't have any anthropogenic sources but can be formed with methyl transfer agents such as methylcobalamin.

The remaining element forming organometallic compounds encountered in the environment is selenium, which is in the same group of the Periodic Table as sulfur. Just as sulfur occurs in some amino acids, selenium-containing amino acids are also known.

Several microorganisms that can convert inorganic [e.g., $O{=}Se(OH)_2$] or other organic forms of selenium to methylselenium compounds have been found.

IV. BEHAVIOR

A. AIR

For a given metal, boiling points of organometallic derivatives tend to decrease as the size of the organic substituents become smaller, and as the number of metal-carbon bonds increase, as shown below.

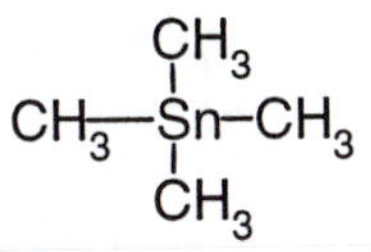

Tetramethyltin BP=78 °C 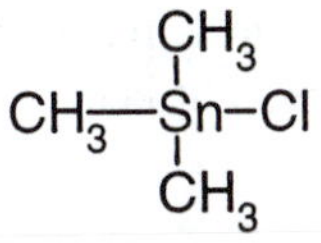 **Trimethyltin chloride BP=115 °C**

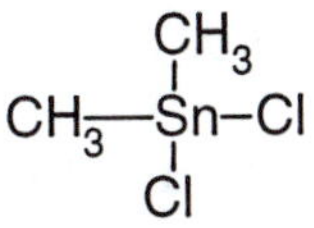

Dimethyltin dichloride BP=187 °C

Metals completely substituted with methyl groups (e.g., Me_2Hg, Me_4Pb, Me_3As) are sometimes called permethyl derivatives. These generally have the highest vapor pressure and volatility among organometallic derivatives of a metal, and would be expected to be encountered in the atmosphere. Volatile molecules such as Me_2Hg or Me_4Pb may travel long distances before undergoing degradation. Removal by rain is likely to be unimportant for these permethyl derivatives because they are relatively insoluble in water.

In the atmosphere, the fate of organic molecules includes photolysis, reaction with ozone (O_3), hydroxyl radicals ($OH^{\bullet}$) and nitrate radicals ($NO_3^{\bullet}$), or sorption to particulate matter. For organometallic species, photolysis and reaction with O_3 are generally of minor importance, unless the organic substituent contains carbon-carbon multiple bonds or aromatic rings. Hydroxyl radicals react with organometallic molecules by abstraction of a hydrogen atom from the organic group, addition to the metal, or displacement of an organic group from the metal. These mechanisms are illustrated in Figure 11.4.

Reaction of $NO_3^{\bullet}$ with organometallics generally proceeds similarly to the scheme outlined for $OH^{\bullet}$ in Figure 11.5. For some molecules such as Me_2Se, reaction

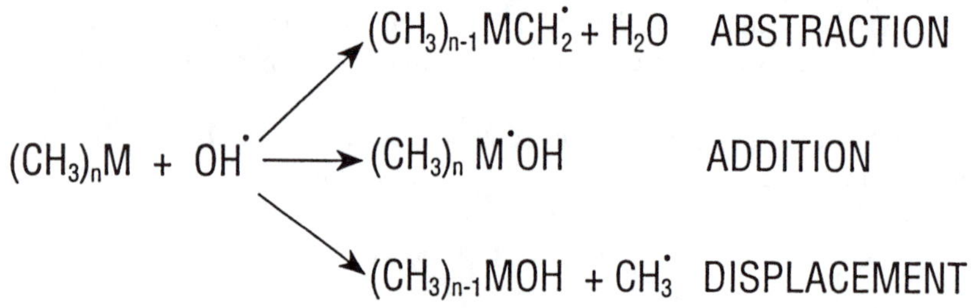

$$(CH_3)_nM + OH^{\bullet} \rightarrow \begin{cases} (CH_3)_{n-1}MCH_2^{\bullet} + H_2O & \text{ABSTRACTION} \\ (CH_3)_n M^{\bullet}OH & \text{ADDITION} \\ (CH_3)_{n-1}MOH + CH_3^{\bullet} & \text{DISPLACEMENT} \end{cases}$$

FIGURE 11.4 Mechanisms of reaction of a hydroxyl radical with organometallic compounds.

with $NO_3^{\bullet}$ is particularly important. Given average values for $OH^{\bullet}$, $NO_3^{\bullet}$ and O_3 concentrations in the atmosphere, half-lives for permthylmetal compounds range from days for Me_4Si to minutes for Me_2Se (at night). With compounds such as Me_3PbCl or Me_3SnCl, reaction with $OH^{\bullet}$ is slower than that with the permethyl derivative and washout becomes a more significant removal process.

B. WATER

Organometallics may exist in the aquatic environment in a truly dissolved state. The aqueous solubilities of methyl derivatives of some metals, compared with $(CH_3)_4C$ are shown in Table 11.4. Generally, these molecules resist hydrolysis. For compounds in which one or more of the organic groups have been replaced by a more polar substituent (e.g., Cl) ionization may occur in solution; for example,

$$(CH_3)_3SnCl \rightarrow (CH_3)_3Sn^+ + Cl^- \text{ (aqueous solution)}$$

Furthermore, ligands can bind or coordinate to the central metal atom. Examples of ligand binding and consequent molecular alterations of organometallics in solution, as compared to the vapor phase, are shown in Figure 11.5. There are many different types of potential ligands in natural waters, including H_2O, Cl^-, NH_3 and humic substances. Coordination of ligands occupies the previously vacant orbitals on the metal atom that promote instability toward some decomposition processes. Coordination in solution thus stabilizes some organometallics.

Organometallics with weakly polar carbon-metal bonds are often hydrophobic. In addition to dissolving in water to a small extent, they are readily sorbed onto particulates and sediments. They also tend to concentrate in a narrow (20 nm to 200 µm) microlayer consisting of microorganisms, organic compounds and other hydrophobic molecules that exists on the surface of most natural waters. Concentrations of hydrophobic organometallics may be thousands of times greater in this surface microlayer than in the underlying water.

C. SOILS AND SEDIMENTS

Methylation of metal ions in soils and sediments by agents such as methylcobalamin and S-adenosylmethionine is a widespread phenomenon. Organometallics may also accumulate in these phases by sorption in aqueous systems. For weakly polar, hydrophobic organic molecules, such as chlorinated hydrocarbons, the extent

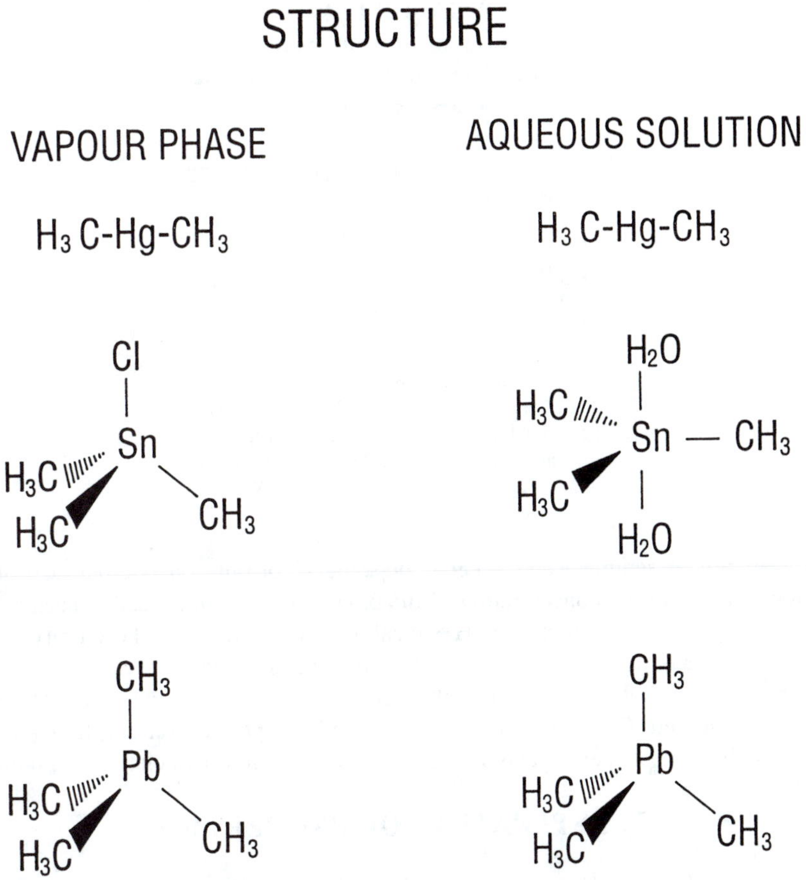

STRUCTURE

VAPOUR PHASE	AQUEOUS SOLUTION
$H_3C\text{-}Hg\text{-}CH_3$	$H_3C\text{-}Hg\text{-}CH_3$
$(CH_3)_3SnCl$	$(CH_3)_3Sn(H_2O)_2$
$(CH_3)_4Pb$	$(CH_3)_4Pb$

FIGURE 11.5 A comparison of the molecular forms of various organometallic compounds in the aqueous and vapor phase that can differ due to ligand binding to the metal.

of sorption, to a good approximation, is dependent not only on the nature of the compound itself, but also on the organic matter or organic carbon content of the soil. The ratio of sorbed and solution concentrations at equilibrium (K_D) is given by expression:

$$K_D = xK_{OW} F_{OC}$$

where x is a constant and F_{OC} the organic carbon mass fraction in the soil or sediment. The octanol/water partition coefficient (K_{OW}) is a physicochemical characteristic of a compound, and is a measure of its hydrophobicity. Expressions such as this suggest that for a given soil, the more hydrophobic a compound (is as evidenced by its K_{OW} value), the greater the sorption. This principle probably holds true for organometallics that tend not to ionize and subsequently form a number of ligand-coordinated species in solution. Organometallics in this category include Me_2Hg and Me_4Sn. For organometallics capable of ionization and speciation depending on the ligands present, the situation is more complex. The partition coefficient

TABLE 11.4
Logarithm of Aqueous Solubility of Methyl Derivatives of Some Metals Compared with $(CH_3)_4C$

Compound	Logarithm of aqueous solubility (mol L^{-1})
$(CH_3)_4C$	–2.90
$(CH_3)_2Hg$	–1.88
$(CH_3)_4Sn$	–3.64
$(CH_3)_3Sb$	–3.13

From Brickman, F.E., Olsen, G.J., and Iverson, W.P., The production and fate of volatile molecular species in the environment: metals and metalloids, in *Atmospheric Chemistry*, Goldberg, E.D., Ed., p. 243, 1982. Copyright Springer-Verlag. With permission.

between soil or sediment and water is dependent not only on f_{OC}, but also other factors, including the concentration of the organometallic entity and concentration of ligands present, as well as the pH and salinity of the water. As the salinity of the water is increased, it might be expected that the aqueous concentration of the organometallic would decrease and values of K_D would increase. In fact, the opposite trend is observed. One reason may be increased competition for sorption sites of soil or sediment between organometallic cations and cations associated with salinity.

V. ESTIMATION OF PROPERTIES

For organic compounds, a relatively large database of environmentally relevant physicochemical properties (e.g., aqueous solubility, K_{OW}, vapor pressure) exists. Where values are unknown, they can usually be estimated from molecular characteristics such as surface area, volume or molecular weight. Knowledge of these physicochemical properties provides some idea about the environmental distribution and behavior of the compound. Properties such as K_{OW}, for example, can be used to predict toxicity, bioconcentration factors and soil/water partition coefficients.

For organometallic compounds, comparatively little physical chemical data exist. If an organometallic does not ionize, but forms a single neutral molecular species on dissolving in water (e.g., peralkylmetals), then it may be possible to consider the compound as an organic chemical and employ estimation methods derived for such chemicals. For example, vapor pressures of both hydrocarbons and tetraalkyltin compounds can be derived from a single relationship involving molecular surface area.

Organometallics where the metal is bonded to carbon, as well as other non-metals (e.g., MeHgOH, Me_3SnCl) may ionize and form a variety of species in solution. This complexity makes it difficult to measure parameters such as K_{OW} and even molecular surface area since this requires some knowledge of the processes occurring in solution. Relationships do exist between toxicity, for example, and the surface area for ionizable organometallics. However, they are usually specific for a particular type of compound, such as diorganotins or triorganotins.

VI. ORGANOMETALLICS OF ENVIRONMENTAL IMPORTANCE

A. ORGANOMERCURY COMPOUNDS

Organomercury compounds have been used for a variety of purposes such as slimicides, seed dressings (fungicides) and antiseptics. Most of these compounds are monoorganomercury compounds with the general formula RHgX, where R is an organic group and X is an anion. Usage is now declining, as are quantities being released into the environment.

Organomercury compounds have been responsible for a number of poisoning incidents around the world, through direct ingestion of contaminated foodstuffs. Mercury has no known metabolic function. Organometallic species such as RHg^+ are toxic, relatively lipid soluble and tend to bioaccumulate within organisms. Their uptake from the gastrointestinal tract is greater than 90% (compared with less than 10% for most inorganic mercury compounds). Toxicity is due to a high affinity for –SH groups on proteins and enzymes. Within cells, protein and RNA syntheses are inhibited. Methylmercury ($MeHg^+$) has an exceptional ability to cross the blood/brain barrier and bind, via –SH groups, to various parts of the brain. Slow release of Hg^{2+} ions to binding sites in the central nervous system can then occur, resulting in secondary toxic effects. The relatively high vapor pressure of small alkylmercury compounds, and also of Me_2Hg, together with their lipid solubilities means that uptake, via a pulmonary route, can also take place.

Most mercury introduced into the environment is actually in inorganic or elemental form. Environmental methylation of inorganic or elemental mercury leads to low levels of methylmercury ($MeHg^+$) compounds in sediments and water. This is likely to be most prevalent where mercury-polluted waterways flow into bacteria-rich areas. Inorganic and elemental mercury is primarily used in batteries, electric and scientific apparatus, and in the chloralkali process. This process involves the electrolysis of NaCl to produce molecular chlorine (Cl_2) and sodium hydroxide. Elemental mercury is used as the cathode. Here, Na^+ is reduced to sodium metal, which forms an amalgam with mercury. Treating the sodium amalgam with water releases the mercury, as shown.

$$2\ Na/Hg + 2\ H_2O(l) \rightarrow 2\ NaOH(aq) + H_2(g) + 2\ Hg$$

Although the mercury is recycled, some is often discharged into the environment, resulting in mercury contamination and possible methylmercury formation. The presence of sulfide ions (S^{2-}) in natural systems promotes the formation of mercury sulfide (HgS), which is extremely insoluble in water and relatively unreactive toward methylation.

In natural waters, most organomercury compounds are found sorbed to suspended particulates. Bioaccumulation by plankton, algae and fish can occur. The rate of uptake of compounds such as methylmercury by fish is faster than that for Hg^{2+}, and its elimination is slower. Relatively high concentrations of methylmercury accumulate in muscle and tissue. In fact, most of the mercury in fish is in the form of methylmercury ($MeHg^+$). It is generally observed, for both marine and freshwater fish, that the higher the trophic level and age, the greater the methylmercury levels in tissue.

B. ORGANOTIN COMPOUNDS

Organotins are among the most rapidly expanding class of organometallic compounds in terms of amount produced. They find application in three broad areas: as stabilizers in polymers, as catalysts in polymer manufacture, and as biocides (e.g., marine antifouling materials, fungicides, insecticides). Over two thirds of the annual production of organotins is used for additives for the stabilization of poly(vinyl chloride) (PVC) polymers. There are a larger number of organometallic derivatives of tin in use than any other metal. Most eventually find their way into the environment. With the exception of methyltin compounds, essentially all known organotin compounds are anthropogenically produced.

Organotin compounds are generally less toxic than similar organolead and organomercury compounds. Within each class, however, alkyl-containing organometallics are often more toxic than those with aromatic substituents. Maximal toxicity is usually observed for the species derived by the loss of one alkyl group from the peralkyl molecule, i.e., by R_3Sn^+, R_3Pb^+, or RHg^+. Among alkyltins, the order of toxicity is trialkyltin > dialkyltin > monoalkyltin. Tetraalkyltin compounds have few uses, but are important intermediates in the synthesis of other organotin molecules. They have been observed to exhibit a delayed toxicity in higher organisms, due to conversion of trialkyltin species, primarily in the liver. With trialkyltin compounds (R_3SnX), the nature of the organic group(s) is the most important determinant of biological activity. The anionic group (X) influences physical factors such as volatility and solubility. In the tri-n-alkyltin series, the trimethyl- and triethyltins are most toxic to mammals, while the tri-n-propyl and tri-n-butyltins are most toxic to bacteria and fungi. As the alkyl chain becomes larger, biological activity declines. The tri-n-octyltin compounds are effectively nontoxic to living organisms.

Much of the toxicity of triorganotin compounds is due to the interaction with mitochondria. Mitochondria are small, membrane-enclosed structures within cells. The citric or tricarboxylic acid cycle, and the electron transfer chain of respiration occur within mitochondria. In humans, some 90% of the body's supply of ATP is produced in these structures. Triorganotins inhibit the conversion of ADP to ATP, interfere with ion transport through the mitochondrial membrane, and also cause the membrane to swell.

Because organotin compounds are so widely used industrially, they are likely to be found in waste-waters. Here, they are likely to be associated with suspended solids. Sorbed materials are generally considered to be not bioavailable for microorganisms. In sewage treatment plants with secondary treatment, microbial degradation of organotins is relatively unimportant. Most of these compounds are transferred from wastewater to sewage sludge.

C. ORGANOLEAD COMPOUNDS

The largest, most well-known and possibly most controversial use of organolead compounds is as additives in gasoline. This commenced in the United States in 1923, but today is declining around the world. Typically, mixtures of tetramethyllead (Me_4Pb), tetraethyllead (Et_4Pb) and the intermediate ethylmethyl-substituted compounds ($EtMe_3Pb$, Et_2Me_2Pb, Et_3MePb) are added to gasoline in order to prevent detonation or uncontrolled combustion of the gasoline/air mixture which is known as *knocking*.

Most of the organolead compounds are converted (with the aid of other additives) to inorganic lead salts (e.g., $PbCl_2$, PbBrCl) that are emitted into the atmosphere. Processes such as incomplete combustion, spillage and evaporation of gasoline result in about 2% of the original alkyllead content entering the atmosphere. There are a relatively large number of organolead compounds that have been prepared in the laboratory. The environmental chemistry of organoleads however is focused on ethyl- and methyl-containing tetraalkyllead compounds, and related tri-, di- and monoalkyllead salts. This is because the predominant anthropogenic use is as a gasoline additive, and the other possible source is environmental methylation of lead ions.

The tetraalkylleads are colorless liquids. Some physical properties of tetramethyl- and tetraethyllead are shown in Table 11.5, together with those of the analogous hydrocarbons 2,2-dimethylpropane (Me_4C) and 3,3-diethylpentane (Et_4C). The alkylleads have higher densities and boiling points than their carbon analogs. Under ambient conditions, they are relatively stable toward oxygen, water and light. At elevated temperatures, they decompose thermally to elemental lead, alkanes, alkenes and hydrogen (H_2). Organolead salts (e.g., Me_3PbCl, Me_2PbCl_2) are colorless, crystalline solids at room temperature, and are relatively soluble in water.

TABLE 11.5
Physical Properties of Tetramethyllead and Tetraethyllead Together with Those of Analogous Carbon-Containing Compounds

Compound	Structure	Density (g cm^{-3})	Boiling point (°C)	Melting point (°C)
Tetramethyllead	$(CH_3)_4Pb$	1.995	110	−30.2
Tetraethyllead	$(C_2H_5)_4Pb$	1.650	200	−130.0
2,2-Dimethylpropane	$(CH_3)_4C$	0.614	9.5	−16.6
2,2-Diethylpentane	$(C_2H_5)_4C$	0.754	146	−33.1

In the atmosphere, tetraalkyllead compounds decompose primarily via reaction with OH$^{\bullet}$ radicals. Eventually, R_3Pb^+ and R_2Pb^{2+} species are formed, existing either in the vapor phase or sorbed to particulate material. Tri- and diorganolead compounds, in particular, may be removed by rain, or on becoming associated with particles, undergo dry deposition.

Organolead compounds in the environment have caused concern because of their toxicity. Alkyllead compounds are quite toxic. The toxic effects of tetraalkyllead compounds on higher organisms are due to conversion to trialkyllead species in the liver:

$$R_4Pb \rightarrow R_3Pb^+$$

In cases of acute poisoning, highest levels of lead are found in the liver. Anemia is often the first symptom of chronic exposure to both inorganic and organic forms of lead. After that, patterns differ. The critical organ for organoleads appears to be the brain. Various symptoms such as convulsions and coma are observed.

D. ORGANOARSENIC COMPOUNDS

Arsenic, together with nitrogen, phosphorus, antimony and bismuth, comprise Group VA of the Periodic Table. Unlike the previous organometallics we have been discussing, organoarsenic compounds found in the environment can contain arsenic in two valency states: three and five. The most common organoarsenic compounds encountered in the environment are:

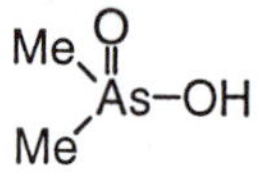

$Me_2As(=O)OH$

Dimethylarsinic acid
(also known as cacodylic acid)

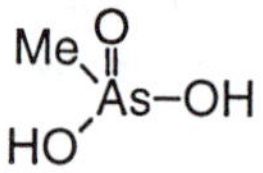

$Me(HO)As(=O)OH$

Methylarsonic acid
(also known as dihydroxymethylarsine oxide)

In both these compounds, arsenic has a valency of five. Use and release of organoarsenic compounds into the environment is diminishing. They have been used primarily as pesticides and herbicides but have also found application in such diverse roles as an animal feed additive, as a remedy for syphillis and as a component of Lewisite ($ClCH_2=CH\text{-}AsCl_2$), which was developed for poison gas warfare. Organoarsenic compounds have a variety of physical forms under ambient conditions. Trimethylarsenic (Me_3As) for example is a low boiling point liquid that oxidizes in air, and is poorly soluble in water. Methylarsonic acid [or dihydroxymethylarsine oxide $Me(OH)_2As=O$] is a colorless solid with a high melting point and soluble in water.

Where organoarsenic compounds are found in the environment, they may arise from anthropogenic sources, or natural processes. The ability to convert inorganic arsenic species into organic derivatives is reasonably widespread in the environment. Microorganisms have been shown to be able to do this.

In the terrestrial environment and in the atmosphere, organoarsenic compounds are mainly methyl derivatives. In aquatic systems, more structurally complex organoarsenic molecules may be formed as well. Arsenic has no recognized role in organisms, but its presence can potentially interfere with biochemical systems involving phosphorus. While many organoarsenic compounds are toxic, they are often less toxic than inorganic species (in contrast to mercury for example). The widespread capacity of organisms to produce organic derivatives of arsenic may be a strategy to reduce the toxicity of ingested arsenic. It is interesting to note that the capacity to metabolize organoarsenic compounds is relatively poorly developed. Arsenic occurs in natural waters in both inorganic and organic forms. A significant proportion of the arsenic in aquatic organisms is organic however.

The mechanism of toxicity for organoarsenic compounds differs with the valency of arsenic. Trivalent organoarsenic compounds have a high affinity for –SH group on protein or enzymes. The toxic mechanism of pentavalent organoarsenic compounds is less well understood.

VII. KEY POINTS

1. Organometallic compounds have metal-carbon bonds where the carbon is part of an organic group. Organometalloids, comprising elements such as silicon and arsenic, share many characteristics with organometallic compounds and can be considered in this group.

2. Organometallic compounds are being produced in increasing quantities for use as catalysts, polymers and other purposes.

3. Organometallic compounds have properties such as toxicity, solubility and volatility that are quite different from the component elemental metals.

4. The carbon-metal bonds in organometallic compounds of environmental importance are polarized in the direction $C^{\delta-}–M^{\delta+}$. The greater the polarity, the greater the rate of hydrolysis by water in the environment.

5. Organometallic compounds can be decomposed thermally to yield hydrocarbons, the elemental metal and possibly some hydrogen.

6. Organometallic compounds can occur in the environment as a result of anthropogenic discharges, but also as a result of environmental chemical processes acting on naturally occurring metals.

7. Organic compounds in which a nonpolar organic group has been replaced by a polar group may ionize in aqueous solution. For example,

$$(CH_3)_3SnCl \rightarrow (CH_3)_3Sn^+ + Cl^-$$

8. Organomercury, organotin, organolead and organoarsenic compounds are the most important organometallic environmental contaminants.

REFERENCES

Wilkinson, G., Stone, F.G.A., and Abel, E.W., Eds., *Comprehensive Organometallic Chemistry — The Synthesis, Reactions and Structures of Organometallic Compounds*, Pergamon Press, Oxford, 1982.

Craig, P.J., Ed., *Organometallic Compounds in the Environment — Principles and Reactions*, Longman, Harlow, Essex, 1986.

Bodek, I., Ehntholt, D.J., Glazer, A.E., Loretti, C.P., and Lyman, W.J., Organometallics, in *Environmental Inorganic Chemistry — Properties, Processes and Estimation Methods*, Bodek, I., Lyman, W.J., Reehl, W.F., and Rosenblatt, D.H., Eds., Pergamon Press, New York, chap. 10, 1988.

Brinckman, F.E. and Bellama, J.M., Eds., *Organometals and Organometalloids — Occurrence and Fate in the Environment*, American Chemical Society, Washington, D.C., 1978.

CHAPTER 11 PROBLEMS

1. Place the following set of metals in a list of decreasing order of susceptibility to reactions with water in the environment, starting with the highest: Al, As, Cd, Hg, Na, Pb, Se. Describe the type of reaction that would be expected to occur and write a general equation for reaction with M–R.

2. Estimate the boiling points of Me_4Si and Me_3Sb using only the data available in this chapter.

Answers on Page 264.

CHAPTER 11 SOLUTIONS

1. Using the data in Figure 11.1, the differences in electronegativity can be determined and the metals placed in order of susceptibility to reaction as: Na (1.6), Al (1.0), Cd (0.8), Pb (0.6), Hg (0.6), As (0.5), Se (0.1). Thus, organo-sodium compounds would be expected to be very susceptible to reaction with water and organoarsenic compounds being relatively stable. The reaction with water is hydrolysis, which would be expected to occur as follows:

$$\begin{matrix} \delta^{-}R\text{—}M\delta^{+} \\ \delta^{+}H\text{—}O\delta^{-} \\ \quad\;\; | \\ \quad\;\; H \end{matrix} \longrightarrow \begin{matrix} R \\ | \\ H \end{matrix} \; + \; \begin{matrix} M \\ | \\ O \\ | \\ H \end{matrix}$$

Hydrocarbon

Metal hydroxide

2. Boiling point is often related to molecular weight with groups of related organic compounds. The atomic weights of the various metals are not available in this chapter, so molecular weight cannot be calculated; however, the Atomic Number of various elements is shown in Figure 11.3.
   A plot can be made of boiling point from Table 11.2 against Atomic Number, as shown in Figure 11.6. An eye-fitted line of best fit can be shown to give a line with a slight curvature. Projections can be made indicating the following:

Compound	Estimated bp (°C)	Measured bp (°C)
Me_4Si	25	27
Me_3Sb	77	79

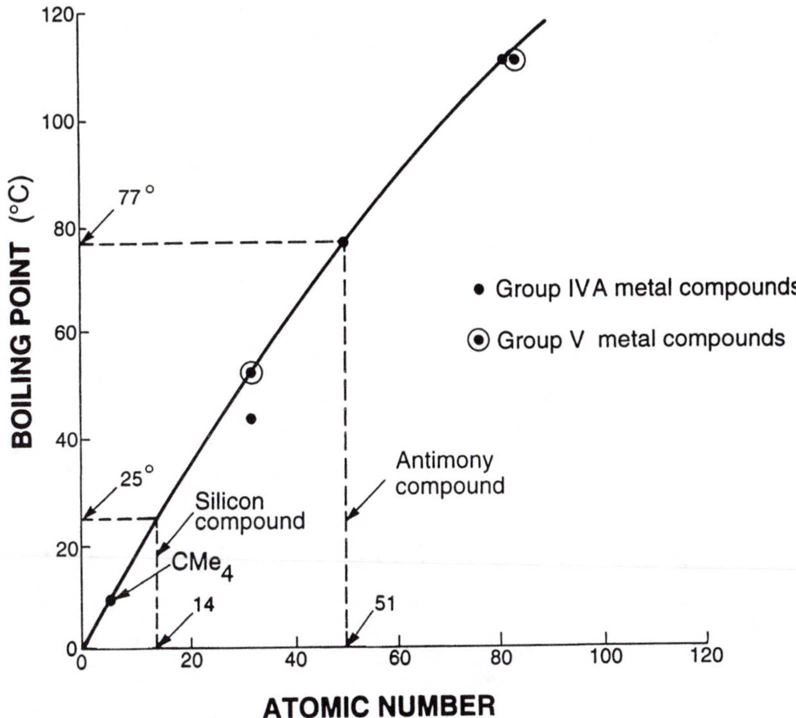

Figure 11.6 Plot of the boiling points of tetramethyl metals (Group IVA), tetramethyl carbon and trimethyl metals (Group V) against boiling point, with projections to estimate MeSi and Me_3Sb (answer to Question 2).

Processes in the Natural Environment

Chapter 12

PHOTOSYNTHESIS AND RESPIRATION — KEY GLOBAL ENVIRONMENTAL PROCESSES

I. INTRODUCTION

Plants and some bacteria are **autotrophs**, which means they synthesize all their organic requirements from inorganic components. Such organisms convert solar energy to forms of energy that can be stored and be made available to other organisms. The chemical reactions that allow this creation of bioavailable energy are called **photosynthesis**. The basic importance of plants and autotrophic bacteria to the global ecosystem is evident from the fact that all multicellular organisms utilize energy, in the form of organic compounds, that are derived originally from photosynthesis. With a few minor exceptions, all living systems are dependent on energy produced by photosynthesis.

In the oceans, the floating plants described as phytoplankton provide food for zooplankton, small animals, which are in turn consumed by fish. In terrestrial systems, the grasses provide food for herbivores, which are subsequently consumed by carnivores, including human beings.

The functioning of all plants is the result of an extremely complex array of chemical reactions; of which photosynthesis, photorespiration and respiration are but three. Photosynthesis is the sunlight-driven conversion of carbon dioxide gas to carbohydrates such as sugars and the release of oxygen into the atmosphere. Respiration is the reverse of photosynthesis; it is the conversion of carbohydrates back to carbon dioxide with the subsequent production of utilizable energy. Photorespiration is a light-dependent type of respiration that causes a net loss of energy and reduces the overall efficiency of plants. As photosynthesis and respiration reactions affect the composition of the atmosphere, it is logical that they must occur at roughly the same rate; otherwise, the carbon dioxide and oxygen contents of the atmosphere would change. Such changes have occurred in the past and may continue in the future, with significant implications for living systems and climate stability.

This chapter will briefly discuss aspects of plant physiology and chemistry relevant to these vital processes. The environmental significance of the development of photosynthesis and respiration and the role these processes have in carbon cycling and climate stabilization will also be discussed.

II. OVERVIEW OF PHOTOSYNTHESIS AND RESPIRATION

Photosynthesis is the conversion of energy from sunlight to energy stored in chemicals (carbohydrates) that is biologically available. It requires water, sunlight and carbon dioxide. The conversion of carbon dioxide to carbohydrate is an endothermic reaction that can be summarized by Equation 1.

$$CO_2 + H_2O + \text{Energy} \rightarrow CH_2O + O_2 \tag{1}$$

Photosynthesis can be conveniently divided into two sets of reactions; the **light** and **dark** reactions. The **light reactions**, as indicated by the name, require sunlight in order to proceed. The light reactions produce two compounds; adenosine triphosphate (ATP) and a reduced form of nicotinamide adenine dinucleotide (NADPH + H^+), which are a means of transferring energy and reducing power to the dark reactions. The **dark reactions** do not require light; however, they can proceed in its presence. Here, the ATP and NADPH + H^+, synthesized in the light reactions, are used to drive the conversion of carbon dioxide to carbohydrates. The series of reactions that form carbohydrates is called the Calvin-Benson cycle.

Respiration is the reverse reaction to photosynthesis (Equation 1). It oxidizes the carbohydrates, formed in photosynthesis, back to carbon dioxide and water, and releases the energy that was stored in the chemical bonds. This released energy is stored by synthesizing adenosine triphosphate (ATP), which can then be utilized to drive other chemical reactions necessary for normal plant metabolism. Respiration occurs in three consecutive stages: glycolysis, the tricarboxylic acid cycle and oxidative phosphorylation.

III. PLANT PHYSIOLOGY

It is important to know some basic plant physiology in order to better understand photosynthesis and respiration. Plants can be visualized as being composed of three separate sections: the leaves, the root system, and the stem or trunk. The leaves are the site of photosynthesis and gaseous exchange, the roots assimilate water and nutrients, and the stem is a means of distributing the products from the leaves and root system to all plant cells and transporting the waste products of metabolism to the sites of excretion.

The leaf has two primary roles to perform: photosynthesis and the exchange of gases (carbon dioxide is absorbed and oxygen released during photosynthesis and vice versa during respiration). In order to maximize photosynthesis, leaves are thin with large surface areas; whereas, to maximize gaseous exchange, leaves have evolved a series of openings called *stomata* (derived from the Greek word for mouth), which can be opened and closed. Directly beneath stomates are large voids or air spaces (Figure 12.1). This structure greatly decreases the number of cells that gases must diffuse through in order to either enter or leave the leaf. A leaf without stomates, that had to rely solely of gaseous diffusion, would not be able to maintain high rates of photosynthesis due to a lack of carbon dioxide.

Photosynthesis in plants occurs within discrete subcellular organelles called chloroplasts, which are located in mesophyll cells of leaves (Figure 12.1). The chloroplasts are separated from the rest of the cell by a lipid bilayer membrane identical to the membranes that separate animal cells. Within the chloroplasts there are other lipid bilayers that are either arranged in parallel rows or folded into a series of interconnected stacks resembling a pile of plates. Each unit within a stack is called a thylakoid. The remainder of a chloroplast is called the stroma.

The light reactions of photosynthesis occur on the inside of the lipid bilayers, forming the thylakoids, while the dark reactions occur in the stroma. Thus, transfer of compounds across the lipid bilayers within the chloroplasts must occur.

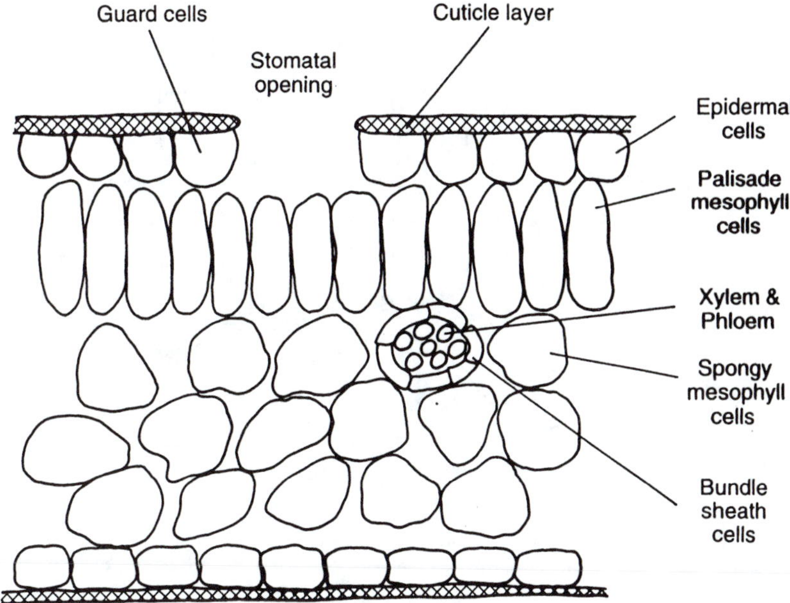

FIGURE 12.1 Internal structure of leaves.

IV. PHOTOSYNTHESIS

A. ABSORPTION OF SOLAR ENERGY

The energy in sunlight is captured by a series of pigments located in the chloroplasts. All pigments contain one or more aromatic rings attached to a long hydrophobic hydrocarbon chain (Figure 12.2). Only the conjugated double bonds in the aromatic rings of pigments, the chromophore, can absorb energy from sunlight.

Photosynthetic pigments of plants can be subdivided into three classes:

Chlorophylls
Carotenoids
Phycobilins

Each type of pigment has its own unique way of absorbing visible light; that is revealed in absorption spectra. For example, chlorophyll *a*, the most common pigment in green plants, absorbs light maximally at 420 and 660 nm, while β-carotene absorbs maximally at 425, 450 and 480 nm. The presence of many different pigments in chloroplasts increases the range of wavelengths of light absorbed and thus maximizes the energy absorbed from sunlight (Figure 12.3).

Photosynthetic pigments are arranged together in groups called *photosynthetic units*. In green plants, there are two different photosynthetic units named photosystem I and photosystem II. These are often abbreviated as P700 (Photosystem I or PSI) and P680 (Photosystem II or PSII), where the P stands for pigment and the

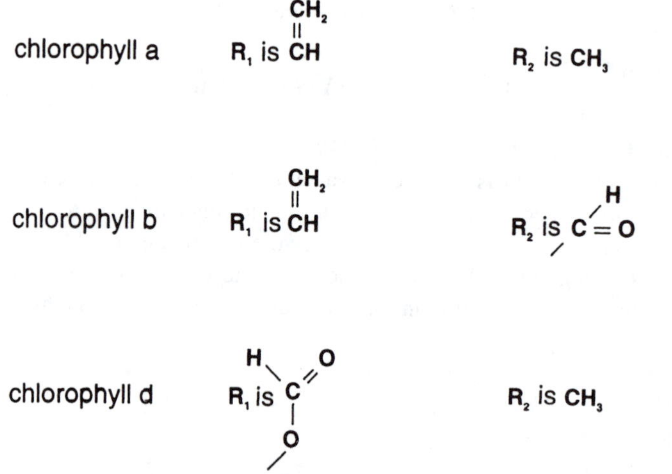

FIGURE 12.2 Chemical structures of some chlorophyll pigments present in plants.

number indicates the wavelength of light that is optimally absorbed by the photosystem.

A typical photosynthetic unit has a central chlorophyll to which proteins are attached (i.e., P680 or P700). The type of protein attached to the chlorophyll *a* molecule affects the absorption spectra (i.e., the wavelengths at which light is absorbed) and causes a shift in absorption maxima. The reaction center is closely surrounded by numerous pigment molecules termed the *antennae.* For example, the antenna for Photosystem I is thought to contain approximately 200 molecules of chlorophyll *a*, 50 molecules of chlorophyll *b,* and 50 to 200 carotenoid molecules.

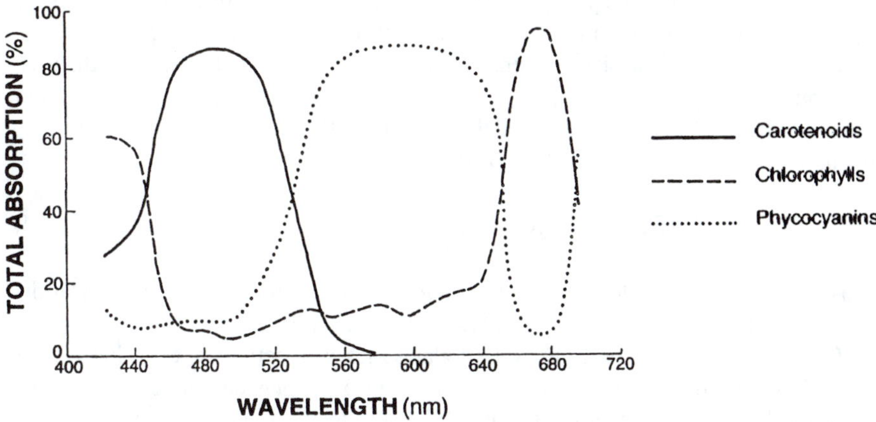

FIGURE 12.3 Visible absorption spectra of some plant pigments.

Solar energy that strikes any antennae pigment molecule is passed via other antennae molecules to the reaction center. What the energy transferred to the reaction center is used for depends on the reaction center; Photosystem I uses the energy to split water into oxygen, hydrogen ions (H^+) and electrons while Photosystem II transfers the energy to other compounds used in converting CO_2 to carbohydrates.

B. LIGHT REACTIONS

There are two separate pathways within the light reactions: cyclic phosphorylation and noncyclic phosphorylation, both of which involve the transfer of electrons and hydronium ions (H_30^+). Phosphorylation is the process of adding phosphorus to atoms or molecules. In fact, the sole products of the light reactions are two phosphorylated compounds, ATP and NADPH.

In noncyclic phosphorylation, the absorption of solar energy by PSII leads to the hydrolysis of water in the following manner:

$$2H_2O + \text{Energy} \rightarrow 4H^+ + O_2 + 4e^- \qquad \Delta H = -286\text{kJ mole}^{-1} \qquad (2)$$

The oxygen produced is released into the atmosphere. The four protons (H^+) are converted to hydronium ions (H_3O^+) by reacting with water molecules. The four electrons are transferred along PSII to quinone and then to plastoquinone, which releases them into the stroma. Then, in a complex series of reactions called the Q cycle, eight H_3O^+ ions and the four electrons are transferred into the thylakoid and into the thlakoid membrane, respectively.

These eight H_3O^+ ions combine with the four H_3O^+ ions produced by the hydrolysis of water and diffuse toward a trans-membrane protein enzyme complex described as ATP-ase or sometimes ATP-synthase. When the 12 H_3O^+ ions pass through the complex, to the stroma, they catalyze the conversion of four molecules of ADP to four molecules of ATP:

$$4\ \text{ADP} + 4\ \text{P}_i \rightarrow 4\ \text{ATP} \qquad (3)$$

where P_i is inorganic phosphorus; eight of the twelve H_3O^+ ions then become the H_3O^+ ions that reacted with the four electrons from PSII in the Q cycle.

Meanwhile, the four electrons that were released into the stroma have diffused and bound to PSI, where they react with NADP and four H_3O^+ ions (the four remaining from the twelve that passed through the ATP-ase complex) in the following manner.

$$2\ NADP^+ + 4\ H_3O^+ \rightarrow 2\ (NADPH + H^+) + 4\ H_2O \qquad (4)$$

Thus, for every two molecules of water hydrolyzed in noncyclic phosphorylation, four molecules of ATP and two of NADPH + H^+ are formed.

Cyclic phosphorylation follows a similar series of reactions except that it does not include any of the reactions that occurred prior to plastoquinone in noncyclic phosphorylation. Solar energy striking PSI leads to the release of two electrons, which react with plastoquinone and four H_3O^+ ions via the Q cycle. This leads to the electrons being transferred along the thylakoid membrane back to PSI and the release of the H_3O^+ ions into the thylakoid. The four H_3O^+ ions move through the ATP-ase complex into the stroma and are then available to react with two new electrons from PSI in the Q cycle. On passing through the ATP-ase complex, the H_3O^+ ions catalyze the formation of approximately 1.3 ATP molecules. Thus, cyclic phosphorylation is not as efficient as noncyclic phosphorylation, producing less ATP and no NADPH + H^+.

C. DARK REACTIONS

The dark reactions of photosynthesis occur in the stroma of chloroplasts and utilize the products from the light reactions to convert carbon dioxide to glucose, which can in turn be converted to all the other carbohydrates required by plants. Plants can be subdivided into three types based on the mechanism used in the dark reactions. However, we will only examine the dark reactions that occur in most temperate plants, called C3 plants because the first product of dark reactions is a molecule containing three carbon atoms.

The series of reactions used by C3 plants to fix carbon dioxide are called the Calvin-Benson cycle after the scientists who first proposed the cyclical nature of the dark reactions. All carbon fixation processes are endothermic and utilize the chemical forms of energy (NADPH + H^+ and ATP) synthesized in the light reactions to drive the reaction forward.

The general equation for the fixation of six molecules of carbon dioxide is endothermic, as shown below.

$$6\ [CO_2] + 6\ [H_2O] \rightarrow 6\ [CH_2O] + 6\ O_2 \qquad \Delta G = +288 \times 10^4\ J \quad (5)$$

It requires 288×10^4 J of energy. Eighteen molecules of ATP and twelve molecules of NADPH + H^+ are required in order to synthesize one glucose molecule. The equations for the dephosphorylation of these molecules are:

$$18\ ATP \rightarrow 18\ ADP + 18\ P_i \qquad \Delta G = -5.52 \times 10^5\ J \qquad (6)$$

and

$$12\ NADPH + H^+ + 6\ O_2 \rightarrow 12\ NADP^+ + 12\ H_2O \qquad \Delta G = -2.64 \times 10^6\ J \quad (7)$$

The total energy released by the conversion of these 30 molecules is 3.19×10^6 J compared with the 2.88×10^6 J needed for carbon dioxide fixation. Thus, there is an excess of 310 kJ and the C3 mechanism of carbon fixation is spontaneous.

D. OVERALL PHOTOSYNTHETIC PROCESS

The overall photosynthetic process is shown in Figure 12.4. The products from the light reactions are energy-rich ATP and NADPH + H^+, which are fed into the dark reactions with CO_2 to produce glucose, $NADP^+$, ADP and P_i. The glucose is utilized to maintain normal metabolic processes. The ADP, P_i and $NADP^+$ are recycled through the light reaction to reform ATP, NADPH and H^+ and facilitate the production of further glucose. It can be seen that the inorganic phosphorus is essential in the formation ADP and ATP, which are the main energy transfer compounds; thus, phosphorus is an essential trace element for plant growth.

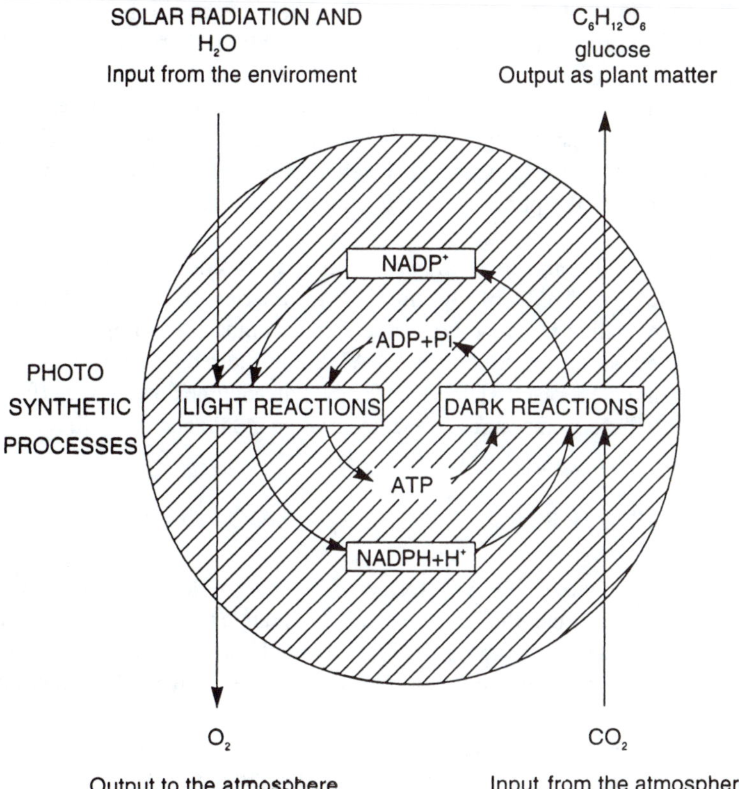

FIGURE 12.4 The overall photosynthetic process, indicating the cycling of products from the light and dark reactions.

V. PHOTORESPIRATION

As stated previously, photorespiration is a mechanism by which previously fixed carbon is lost from the plant and returned to the atmosphere as CO_2. This form of respiration, unlike normal respiration, does not lead to a net gain of energy in the form of ATP or NADPH + H^+; in fact, it causes a net loss of energy. The process is light dependent, as the energy used in the process comes from the light reactions of photosynthesis. Photorespiration is a complex pathway that involves many enzymes located in three different subcellular organelles: the chloroplast, peroxisome and the mitochondria.

VI. RESPIRATION

While photosynthesis is a process that can only be carried out by plants and some bacteria, the respiration process occurs in all organisms that use atmospheric oxygen. The carbohydrates produced by photosynthesis provide food for many other organisms that consume plant carbohydrates as their source of energy.

Respiration is a series of processes: glycolysis, the tricarboxylic acid cycle (TCA) and oxidative phosphorylation (Figure 12.5). These processes release the energy stored in carbohydrate for use by oxidation using atmospheric oxygen. In fact, a wide range of biological compounds are suitable as reactants for respiration. Glycolysis will accept any carbohydrate that can be converted to either glucose or sucrose. Glycolysis will also accept glycerol, which is a breakdown product of lipids. The TCA cycle accepts all proteins and fatty acids (the other breakdown product from lipids). However, for the purposes of this chapter, only the fate of polysaccharides will be discussed.

Under ideal conditions (i.e., sufficient oxygen), respiration leads to carbohydrates being completely oxidized to CO_2 and H_2O, in a series of reactions that releases energy, much of which is temporarily stored in a number of biologically utilizable compounds [i.e., adenosine triphosphate (ATP) and the reduced forms of nicotinamide adenine dinucleotide (NADPH + H^+) and flavin adenine dinucleotide ($FADH_2$)]. Ultimately, all these compounds are converted to ATP. The overall reaction for respiration is:

$$C_6H_{12}O_6 + 6\ O_2 + 32.5\ ADP + 32.5\ P_i \rightarrow 6\ CO_2 + 6\ H_2O + 32.5\ ATP \quad (8)$$

There are 32.5 molecules ATP formed for every molecule of glucose that is completely respired. When the ATP is used in the plant's metabolism, it is converted back to ADP for reuse in the above reaction. It should be noted that Equation 8 only occurs under the most favorable conditions; under less favorable conditions, the energy yield may be lower.

An overall view of respiration and how the various processes contribute to the total energy output is shown in Figure 12.5. Glycolysis is the first process; its products enter the ATP pool, the TCA cycle and the oxidative phosphorylation process. The products of the TCA cycle move to the ATP pool and the oxidative phosphorylation process. All the products of oxidative phosphorylation enter the

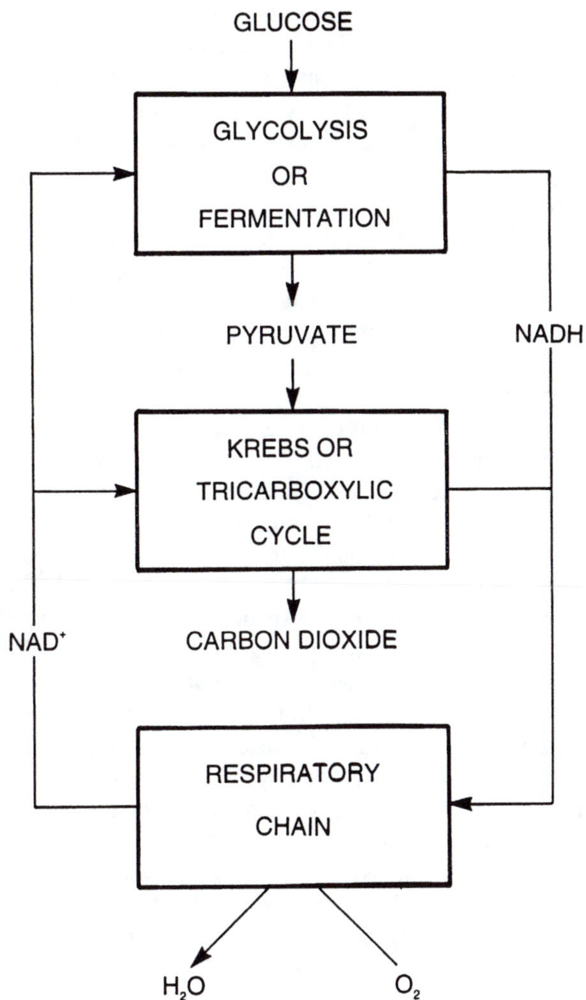

FIGURE 12.5 The respiration of one molecule of glucose with the quantity and pathway of the products from glycolysis, the TCA cycle and oxidative phosphorylation.

ATP pool for use in the metabolic processes (e.g., protein, lipid synthesis and cell replication) and catabolic processes (e.g., respiration) of the plant.

The three processes of respiration occur in different portions of plant cells. Glycolysis occurs in the cytoplasm of cells, the cytoplasm being all the contents of a cell apart from the subcellular organelles. The TCA cycle and oxidative phosphorylation occur within the mitochondria.

The mitochondria is a rod-shaped, subcellular organelle, approximately 0.5 μm wide and 5 to 10 μm long, that occurs in all cells of multicellular organisms (eukaryotes). It consists of an outer lipid bilayer membrane separated by a small fluid-filled space from a second, inner lipid bilayer membrane. The inner membrane has numerous invaginations, called cristae, that protrude into the mitochondria, like teeth on a comb. The inner membrane has imbedded in it or attached to it, all the proteins and enzymes involved in oxidative phosphorylation. Contained within the inner membrane is the mitochondrial matrix. The matrix contains, in a dissolved form, all the enzymes necessary for the TCA cycle except one: succinate dehydrogenase, which is imbedded in the mitochondrial inner membrane.

A. GLYCOLYSIS

The term "glycolysis" comes from the Greek *glykos* meaning sweet and *lysis* meaning losing. It is a sequence of reactions in which sugar molecules are split to yield energy. Glycolysis is believed to be the most primitive form of energy-yielding reactions, as it occurs almost universally. The other energy-producing mechanisms, which require oxygen, probably evolved as plants commenced photosynthesising and producing oxygen.

Any polysaccharide that contains either glucose or fructose, or can be converted to these, can be used as starting inputs for glycolysis. For example, starch is a polymer of glucose, sucrose consists of a glucose and a fructose molecule, and lactose is a glucose molecule attached to galactose that in turn can be converted to glucose. The point at which compounds enter glycolysis depends on the type of sugar formed.

The overall energy yield from glycolysis depends on the sugars that enter glycolysis. Every molecule of glucose will yield 2 molecules each of ATP and NADPH + H^+; whereas, a fructose molecule will yield 3 molecules of ATP and 2 molecules of NADH + H^+. Thus, the energy yield from complex carbohydrates will also vary according to the number of glucose or fructose molecules they yield.

The fate of the products of glycolysis is varied. The ATP becomes part of the ATP pool used for all metabolic processes and some catabolic processes (reactions that break down compounds to yield energy, e.g., glycolysis). The fate of NADH + H^+ and pyruvate are intertwined and depend on the presence or absence of oxygen. If anaerobic conditions prevail, both pyruvate and NADH + H^+ enter a series of reactions called *fermentation.* Whereas, if aerobic conditions prevail, they enter a different series of reactions termed the "TCA cycle."

B. FERMENTATION

Fermentation occurs in plants, anaerobic bacteria, cells subject to intense metabolic activity and yeasts. There are two different forms of fermentation: lactic acid and alcohol fermentation. In lactic acid fermentation, pyruvate and NADPH + H^+ from glycolysis are converted to lactate by the lactate dehydrogenase enzyme in the following manner:

$$\text{2 Pyruvate} + 2\ (\text{NADH} + H^+) \rightarrow \text{2 Lactate} + 2\ \text{NAD}^+ + 4\ H^+ \qquad (9)$$

Alcoholic fermentation occurs in two stages. Firstly, pyruvate is converted to acetaldehyde by the enzyme pyruvate decarboxylase as shown in the following equation;

$$\text{2 Pyruvate} \Leftrightarrow \text{2 Acetaldehyde} + 2\ CO_2 \qquad (10)$$

The acetaldehyde is then reduced by the enzyme ethanol dehydrogenase to ethanol, as indicated below:

$$2\ CH_3CHO + 2\ (NADH + H^+) \Leftrightarrow 2\ CH_3CH_2OH + 2\ NAD^+ + 2\ H^+ \quad (11)$$

Acetaldehyde Ethanol

In both fermentation reactions, the source of the NADH + H^+ molecules that reduce the pyruvate is glycolysis. Thus, if fermentation is the sole use of pyruvate, the total energy due to the respiration of one molecule of glucose is 2 ATP. This is approximately 6.5% of the energy yield when oxygen is present, and the pyruvate enters the TCA cycle and the NADH + H^+ enters oxidative phosphorylation.

C. THE TRICARBOXYLIC ACID CYCLE

As glycolysis and the TCA cycle are linked, with the products of glycolysis forming the reactants of the TCA cycle, the products of glycolysis (pyruvate and NADH + H^+) must cross the mitochondrial membrane in order to enter the mitochondria. The mitochondrial membranes are basically impervious to these compounds and cross-membrane transport is mediated by a number of transport proteins. Once inside the mitochondria and assuming there is sufficient oxygen, the pyruvate enters the TCA cycle while the NADPH + H^+ enters the oxidative phosphorylation process.

The TCA cycle is alternatively called the Krebs cycle after its discoverer, or the citric acid cycle, as the first product of the cycle proper is citric acid (citrate). In the TCA cycle, pyruvate molecules are completely oxidized to carbon dioxide. Each oxidation reaction is associated with a decrease in free energy, some of which is captured by the formation of the energy transfer compounds NADH + H^+, the reduced form of flavin adenine dinucleotide ($FADH_2$) and ATP.

On average, every pyruvate molecule must pass through the TCA cycle more than twice before all its carbon atoms have been oxidized to CO_2 and all hydrogen atoms have reacted with NAD^+ or FAD^+.

The carbon dioxide formed by the TCA cycle diffuses out of the mitochondria and eventually is released to the atmosphere, ATP moves to the ATP pool, while NADH + H^+ and $FADH_2$ move to the inner membrane of the mitochondria where oxidative phosphorylation occurs.

D. OXIDATIVE PHOSPHORYLATION

Oxidative phosphorylation occurs due to the passage of electrons along a series of compounds that are collectively termed the "electron transport chain" or alternately the "respiratory chain." There are six compounds involved in the process: four large enzymes and two smaller molecules imbedded in or on the surface of the inner membrane of the mitochondria.

The NADH + H^+ produced by glycolysis, which occurs in the cytoplasm of the cell, must enter the mitochondria. However, the membrane is impermeable and thus the NADH + H^+ cannot simply diffuse into the mitochondria. This is overcome by transferring the electrons associated with the NADH + H^+ across the membrane. Two possible transfer mechanisms are available: the glycerol phosphate shuttle and the malate shuttle. The amount of ATP formed by the subsequent oxidative phosphorylation of these shuttle products differs. Each molecule of nicotinamide adenine dineculeotide ($NADH + H^+$) transferred by the malate shuttle yields 1.5 molecules of ATP, while each molecule transferred by the glycerol phosphate shuttle results in 2.25 molecules of ATP. In comparison, each molecule of $NADH + H^+$ and $FADH_2$, produced by the TCA cycle in the mitochondria and therefore don't need to be transported by either shuttle mechanism, yields 2.5 and 2 molecules of ATP, respectively.

The complete respiration of one molecule of glucose leads to the formation of different quantities of ATP depending on whether the glycerol phosphate or malate shuttle pathways are used to transfer the $NADH + H^+$ produced by glycolysis into the mitochondria. If the malate shuttle path is used, 32.5 molecules ATP are produced per molecule of glucose whereas, the glycerol phosphate shuttle leads to the synthesis of 31 ATP molecules. The contribution of each of the processes in respiration to the total ATP production is shown in Figure 12.5.

E. OVERALL RESPIRATION PROCESS

The overall respiration process, which is comprised of glycolysis, the tricarboxylic acid cycle and oxidative phosphorylation, is shown in Figure 12.5. This process is used as the means of providing energy in most forms of life. The products of photosynthesis (i.e., carbohydrates) and further metabolism (lipids, proteins, etc.) are broken down via a series of oxidation reactions to ultimately yield energy in the form of ATP. This ATP becomes part of the ATP pool used for all metabolism (e.g., protein and enzyme synthesis) as well as some of the catabolic processes that may require energy (e.g., the early stages of glycolysis).

VII. THE ENVIRONMENTAL SIGNIFICANCE OF PHOTOSYNTHESIS AND RESPIRATION

A. PAST EFFECTS

The evolution of the linked processes of photosynthesis and respiration has had and continues to have a highly significant effect on the chemical composition of the atmosphere and environmental conditions on earth. From the available geochemical evidence for the earth and other planets, it seems that the very earliest atmosphere of the earth contained essentially no oxygen. The principal components were nitrogen, carbon dioxide and water vapor, with trace amounts of nitrogen monoxide, hydrochloric acid, hydrogen and carbon monoxide. With such a composition, the conditions were mildly reducing.

The only possible sources for the production of oxygen gas were the photolysis of water molecules and carbon dioxide by sunlight. The reactions for water are outlined below.

$$H_2O + \text{Light} \rightarrow OH^- + H^+ \quad (12)$$

$$OH^- + OH^- \rightarrow O^{2-} + H_2O \quad (13)$$

$$O^{2-} + OH^- \rightarrow O_2 + H^+ \quad (14)$$

For carbon dioxide,

$$CO_2 + \text{Light} \rightarrow CO + O^{2-} \tag{15}$$

$$O^{2-} + O^{2-} \rightarrow O_2 \tag{16}$$

However, unless either the hydrogen or carbon atoms are removed from the atmosphere, they will eventually react with the released O_2, reforming CO_2 and H_2O, and leading to no net gain of oxygen. Hydrogen atoms have a very small mass (1.66×10^{-27} kg) and are therefore very hard to retain in the atmosphere. They simply cannot be held by gravity and are lost into space. Carbon was removed from the atmosphere by forming carbonate minerals. These two processes could have led to the concentration of atmospheric oxygen gradually increasing; however, there were a number of substances such as reduced forms of iron (Fe^{2+} and Fe^{3+}) and manganese (Mn^{7+}, Mn^{4+}, Mn^{2+}) that would readily remove any oxygen in the atmosphere. This anoxic period, termed the Hadean Period, extended from the formation of the earth approximately 4.6 billion years ago (Bya) until 3.7 Bya.

The Archean Period extended from 3.7 to 2.5 Bya. It was during this period that life first evolved. The earliest forms of life would have been simple single-celled bacteria. These would not have had a nucleus or any other subcellular organelles and are termed "prokaryotes." Two main mechanisms for obtaining energy were developed (Figure 12.6). One group of bacteria, chemotrophs, catalyzed the reduction of inorganic atoms such as iron and sulfur and in this manner gained sufficient energy to oxidize organic carbon and form carbohydrates (Figure 12.6). The second group, autotrophs, used energy from the sun (photosynthesis) or from the oxidation of mineral matter (chemosynthesis) to convert CO_2 to carbohydrates (Figure 12.6).

The oldest preserved evidence of life on earth comes from rocks found in a location called North Pole in Western Australia. There, in rocks approximately 3.5 million years old, are the characteristically column or mushroom-shaped calcium carbonate deposits called **stromatolites**. These are deposited by colonial forms of *cyanobacteria* (also called, although incorrectly, blue- green algae). These bacteria were able to photosynthesize and the deposits are a direct result of the photosynthesis. Despite being among the earliest forms of life to appear, stromatolites still occur in several locations throughout the world, perhaps the largest and most famous being those at Shark Bay in Western Australia and in the Bahamas, although examples also occur in glacial lakes in Antartica and volcanic springs in Yellowstone National Park. The number and activity of these photosynthetic organisms must have been low as there is no evidence to support a marked increase in the levels of atmospheric oxygen, which was probably stable at approximately 1 part per million during the Archean Period.

The period from 2.5 to 0.7 Bya, the Proterozoic, was marked by great changes in the chemistry of the earth. The number and activity of photosynthetic organisms increased markedly and resulted in the conversion of the atmosphere from mildly reducing to oxidizing. One of the key pieces of evidence to support the claim of increased oxygen production during the Proterozoic is the change in abundance of

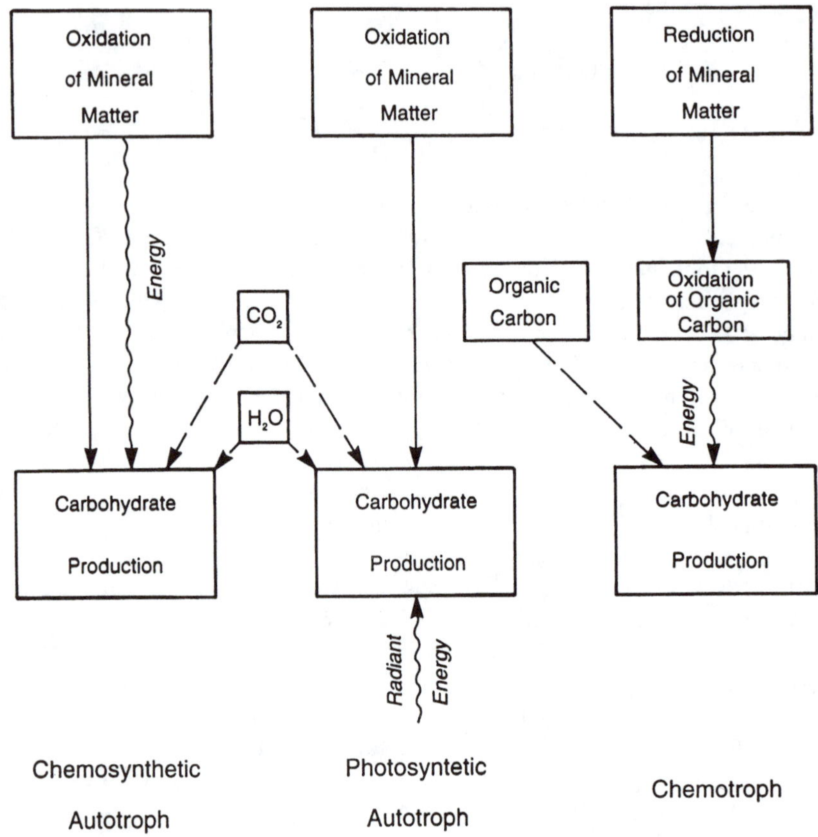

FIGURE 12.6 Mechanisms for the release of energy for growth by bacteria.

banded iron formations over time. As stated previously, reduced metals (particularly iron and manganese) readily absorb any free oxygen (Equations 17–20).

$$Fe^{2+} + O_2 \rightarrow Fe_2O_2 \tag{17}$$

$$4\ Fe^{3+} + 3\ O_2 \rightarrow 2\ Fe_2O_3 \tag{18}$$

$$4\ Mn^{7+} + 7\ O_2 \rightarrow \rightarrow 2\ Mn_2O_7 \tag{19}$$

$$2\ Mn^{4+} + 2\ O_2 \rightarrow Mn_2O_4 \tag{20}$$

When reduced forms of iron and manganese ions are oxidized, their aqueous solubility decreases substantially and they precipitate out of solution, forming the iron oxide deposits. The increase in oxygen production over time is evident by the parallel increase in the amount of these iron ore deposits. Known reserves deposited during the Hadean and Archean Periods total 0.8×10^{14} tons, between 2.5 and 2.0 Bya, the mass increased to 6.4×10^{14} tons, while between 2.0 Bya and the present,

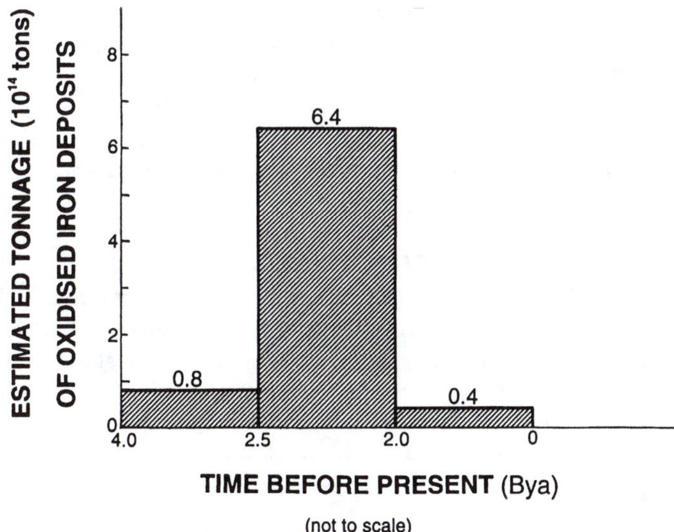

FIGURE 12.7 Tonnages of oxidized iron deposited over geological time to the present.

there was approximately 0.4×10^{14} tons (Figure 12.7). The marked decrease in tonnage of iron ore deposits after 2 Bya is not due to a decrease of oxygen, but rather to a reduction in the amount of iron present in a reduced state.

The oxidation of iron and other reduced chemicals would have negated the increased synthesis of oxygen, and concentrations of atmospheric oxygen most probably did not increase until essentially all the reduced forms were oxidized. Estimates of the oxygen content about 2 Bya are typically 1% (10, 000 ppm), compared with the current content of about 21%.

It was also during the Proterozoic Period that **eukaryotic bacteria** developed. **Eukaryotes** are organisms that have subcellular organelles, including a nucleus in which the DNA is located. The development of eukaryotes was not possible before this period as the replication of eukarotic DNA requires actinomyosin, a protein that requires oxygen for its synthesis.

The Phanerozoic Period (0.7 Bya to the present) was marked by the evolution of multicelled organisms. For example, there are fossils of jellyfish that have been dated as being 670 million years old (0.67 billion years old). The presence of sufficient concentrations of oxygen (>1%) was also vital to the evolution of multicellular organisms for two reasons. First, relatively high oxygen concentrations are needed in order to supply cells within the organism, and therefore isolated from the atmosphere, with sufficient oxygen. Second, multicellular organisms require collagen, the synthesis of which requires oxygen.

It was only with the development of large photosynthesizing plants with solid cell walls that oxygen levels rose above 7 to 10% (70, 000 to 100, 000 ppm) to the current levels of approximately 21% (210, 000 ppm). This happened because the plants not only produced more oxygen, but also the amount of carbon that could be removed from the atmospheric cycle was greatly increased. Single-celled organisms with their soft cell walls are very rapidly broken down, releasing carbon back to the atmosphere; whereas, with trees, there is a much greater chance of the carbon stored in the plant being preserved in sediments with a subsequent reduction of carbon re-entering the atmosphere.

As can be seen, the role of photosynthesis and respiration has had a very significant effect on the environment. It led to the conversion of the atmosphere from a reducing environment to an oxidizing one, which has allowed the evolution of animals and life as we currently know it.

B. PRESENT AND FUTURE EFFECTS

Photosynthesizing plants, algae and bacteria form an integral role in the carbon cycle and the modulation of the Earth's climate. The link between plants and climate is **carbon dioxide**, a greenhouse gas and the gas used to create the sugars of all plants. The carbon cycle is illustrated in Figure 12.8.

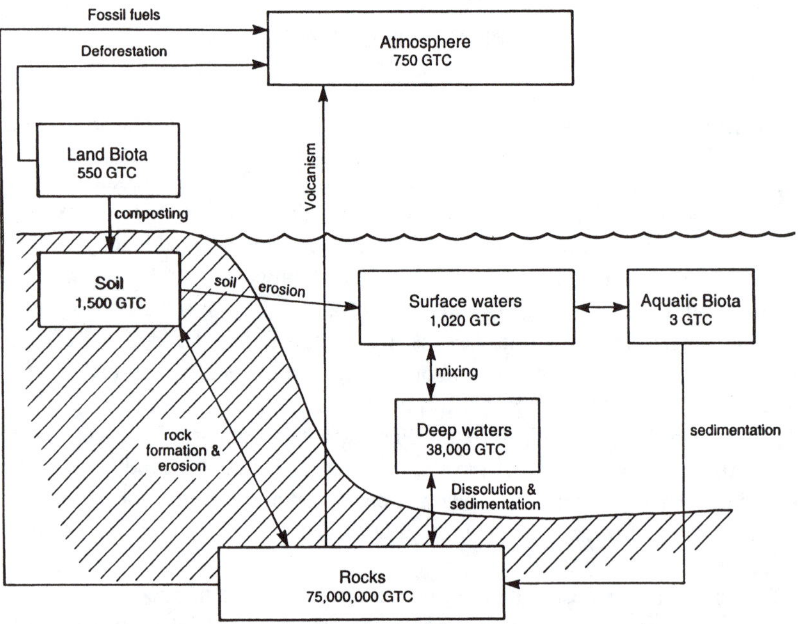

FIGURE 12.8 The carbon cycle, with the amounts of carbon in the various compartments indicated as tons of carbon (GTC: a giga ton of carbon is 10^8 tons).

The carbon cycle consists of a series of linked compartments through which carbon, in various forms, passes in various quantities and rates. It can be subdivided into three sections based on the average residence time of carbon. The biological portion consists of the atmosphere, land and aquatic biota, and surface waters. The exchange between the compartments of this portion of the cycle is very rapid, with an average residence time of between 1 and 5 years. The second portion, the soil compartment, retains carbon for a longer period; with typical residence times being 30 to 50 years. The third portion, the geological, consists of the deep water and rock compartments. Exchange of carbon between these compartments and the others is extremely slow. Carbon can reside in deep water or rocks for periods of approximately 1000 years and many millions of years, respectively.

As illustrated by Figure 12.8, all of the compartments are reversibly linked to each other, either directly or indirectly. Thus, in a sense, an equilibrium is established between the compartments. However, Figure 12.8 also illustrates that the combustion of fossil fuels and deforestation can short-circuit carbon from the cycle and increase the atmospheric concentration of carbon dioxide. It is these sources that have led to the observed increase of carbon dioxide levels and the associated problem termed the "Greenhouse effect." From 900AD to 1750, the atmospheric concentration of carbon dioxide was fairly steady at 280 parts per million (ppm) with maximum deviations of 10 ppm. Since 1800, when the industrial revolution began, carbon dioxide levels have increased up to the present value of 355 ppm, with the rate of change increasing with time.

There are a number of so called **greenhouse gases**: carbon dioxide, methane and water vapor are particularly important. Water is the most abundant of the greenhouse gases. However, as its concentrations are not rising, it is not a problem. In contrast, atmospheric concentrations of CO_2 and methane are increasing. At present, the main concern is with the levels of CO_2, although the methane concentrations are increasing more rapidly than CO_2, and each CH_4 molecule has the same effect as 30 CO_2 molecules.

The daily variation in the temperature of the earth is quite small, yet every day the earth absorbs more energy from the sun. The temperature is maintained at a relatively consistent level by the earth acting in a manner similar to a black body. In other words, it reradiates essentially the same amount of energy that it absorbs. Greenhouse gases do not interfere with the earth absorbing energy from the sun; however, they do reduce the amount of energy that can be lost by radiation. Therefore, it is expected that increases in greenhouse gases will lead to increases in the Earth's temperature, which could raise sea levels; decrease the temperature gradient between the equator and the poles; change weather patterns; and change the amount and distribution of rainfall. Such predicted changes could have catastrophic effects on human society and the environment.

Most of the models predict that the temperature should have risen by 1 or 2° since in the last 100 years and will rise by another 2 to 4° by 2100. It now appears that the models may well have overestimated the temperature increases. This may well reflect the fact that there are many gaps in our knowledge such as the role of the deep ocean in the carbon cycle and the effects that clouds have on the amount of solar radiation reaching the earth.

Analysis of gases trapped in ice from Greenland has revealed a similarity in the trends of the concentrations of CO_2 and temperature when viewed over the past 150,000 years. However, when viewed from a longer time scale, a different picture emerges. Since the formation of the earth, there has been an increase of 25 to 30% in the solar energy released daily by the sun and absorbed by the earth, yet mean air temperatures have been remarkably stable. For example, there is no evidence to indicate the temperature reached 100°C (water would boil) or 0°C (water would freeze). This maintenance of temperatures suitable for life has been maintained by the continual removal of CO_2 from the atmosphere with the ensuing increased levels of energy radiated into space. All the evidence indicates that the solar output will continue to rise. So in order to maintain temperatures, atmospheric CO_2 levels should decrease. Instead, due to the activities of humans, the reverse is happening.

VII. KEY POINTS

1. Photosynthesis by plants uses energy from sunlight to convert carbon dioxide and water to plant carbohydrate. This reaction can be summarized by the following equation:

$$6\ CO_2 + 6\ H_2O + \text{Energy} \rightarrow 6\ CH_2O + 6\ O_2 \rightarrow \text{Carbohydrate}$$

2. Solar energy is absorbed by pigments that contain a photoreactive center called the "chromophore." Pigments are arranged into photosynthetic units that consist of a central molecule, termed the "reaction center," usually a chlorophyll molecule surrounded in a radial fashion by other pigment molecules, termed the "antennae pigments." This structural arrangement maximizes the rate of activity of the photosynthetic units. Plants have two different photosynthetic units: Photosystem I (P700) and Photosystem II (P680).

3. Two separate sets of reactions occur with photosynthesis: the light and dark reactions. The light reactions require the energy from light to proceed and can be simply expressed by the following equations:

$$2\ H_2O + 2\ NADP \rightarrow O_2 + 2\ NADPH + 2\ H^+$$

$$ADP + P_i \rightarrow ATP$$

The dark reactions do not require darkness and can proceed in the presence of light and be simply expressed by the following equations:

$$CO_2 + 2\ NADPH + 2\ H^+ \rightarrow \underset{\text{Glucose}}{1/6\ C_6H_{12}O_6} + H_2O + 2\ NADP^+$$

$$ATP \rightarrow ADP + P_i$$

4. The plant matter produced by photosynthesis is utilized as food by animals and other organisms by the respiration process. This process is the reverse of photosynthesis and can be represented by the following equation:

$$\underset{\text{Carbohydrate}}{6\ CH_2O} + 6\ O_2 \rightarrow 6\ CO_2 + 6\ H_2O + \text{Energy}$$

5. The photosynthesis and respiration process must occur at approximately the same rate to maintain the O_2 and CO_2 content of the atmosphere.

6. Over geological times, the development of photosynthesis by larger plants has had an effect on the chemistry of the Earth. The production of oxygen gas by plants has changed the composition of the atmosphere and converted the chemical state of the Earth from reducing to oxidizing. Photosynthesis and respiration currently play a major role on Earth, as the key processes in the carbon cycle and maintaining the existing atmospheric levels of carbon dioxide.

REFERENCES

Galston, A.W., Davies, P.J., and Satler, R.L., *The Secret Life of Plants.* Prentice-Hall, Englewood Cliffs, NJ, 1980.

Hall, D.O. and Rao, K.K., *Photosynthesis.* 3rd ed., Studies in Biology No. 37. Edward Arnold, London, 1981, 84p.

James, D.C. and Mathews, G.S., *Understanding of Biochemistry of Respiration.* Cambridge University Press, Cambridge, 1991, 90p.

Lehninger, A.L., *Principles of Biochemistry.* Worth Publishers, New York, 1982, 1011p.

Purves, W.K., Orians, G.H., and Heller, H.G., *Life: The Science of Biology.* 3rd ed. W.H. Freeman & Co., Salt Lake City, 1992, 1145p.

Zubay, G., *Biochemistry.* 3rd ed. Wm.C. Brown Publishers, Dubuque Iowa, 1993, 1024p.

CHAPTER 12 PROBLEMS

1. The composition of the atmosphere has changed over geological time from predominantly carbon dioxide, nitrogen and water to predominantly nitrogen, oxygen and water with trace amounts of carbon dioxide. Use equations to explain how this has occurred and outline in broad terms which global components have changed in elemental composition as a result of this.

2. The chlorophylls strongly absorb visible light from solar radiation at the red end of the spectrum. At depths greater than about 10 meters in seawater, the only visible light present is in the blue-green part of the spectrum. Would chlorophylls be expected to be common marine plant pigments at depth in the sea? If not, what pigment group would be expected to predominate?

3. In 1750, the atmospheric concentration of carbon dioxide was about 280 ppm (volume/volume) compared with about 355 ppm (volume/volume) today. Calculate the proportion of carbon present in rocks which has been used as fuels. In carrying out this calculation, assume that air and carbon dioxide have the same density, the total mass of the atmosphere is 5.14×10^{15} tons, all of the carbon dioxide produced by burning fossil fuels still remains in the atmosphere, and burning fossil fuels has been the only process producing an increase in atmospheric carbon dioxide.

Answers on page 290.

CHAPTER 12 SOLUTIONS

1. The change in composition has occurred because after plants evolved on Earth, the global rates of photosynthesis and respiration were not the same:

   Photosynthesis:

$$6\ CO_2 + 6\ H_2O + \text{Energy} \rightarrow \underset{\text{Plant carbohydrate}}{6\ CH_2O} + O_2$$

   Respiration:

$$\underset{\text{Plant carbohydrate}}{6\ CH_2O} + O_2 \rightarrow 6\ CO_2 + 6\ H_2O + \text{Energy}$$

   Photosynthesis has resulted in the production of plant carbohydrate matter not fully consumed by organisms, such as animals and microorganisms. Thus, the rate of photosynthesis exceeded the rate of respiration. The excess plant matter was substantially incorporated into sedimentary strata over geological time. Thus, the upper compartments of the crust of the Earth became enriched with carbohydrate and, of particular importance, carbon. This material then formed large deposits of coal and other fossil fuels. On the other hand, the atmospheric compartment was steadily depleted in carbon dioxide, particularly carbon, and enriched with oxygen. The impact of these processes on nitrogen have not been as great and thus nitrogen has remained substantially unchanged.

2. The blue-green light at depth in seawater indicated that the red end of the solar spectrum has been absorbed. This means that the chlorophylls would not be expected in marine plants at depth since these pigments absorb at the red end of the spectrum and this has been removed by the seawater. Consideration of Figure 12.3 indicates that the carotenoids absorb strongly at the blue-green end of the spectrum and so would be able to utilize radiation at depth. Thus, these pigments would be expected in marine plants at depth.

3. The increase in concentration of carbon dioxide is 355 – 280 ppm (volume/volume); i.e., 75 ppm. If it is assumed that the atmosphere and carbon dioxide have the same density, then the mass of carbon dioxide produced is $(75 \times 10^{-6}) \times (5.14 \times 10^{15})$ tons CO_2.

$$\text{Tons } O_2 \text{ produced} = 38.55 \times 10^{10}$$

$$\text{Tons carbon produced} = 12/44\ (38.55 \times 10^{10})$$

$$\text{Therefore, carbon produced} = 10.48 \times 10^{10} \text{ tons}$$

Amount of carbon in rocks $= 7.5 \times 10^{15}$ tons
(from Figure 12.9)

Thus, the fraction of carbon $= (10.48 \times 10^{10}) / (7.5 \times 10^{15})$
in rocks consumed

$= 1.4 \times 10^{-5}$ or 0.0014%.

Chapter 13

CHEMICAL EVOLUTION

I. INTRODUCTION

The origin of life, and particularly the human species, is a topic that has occupied the minds of philosophers throughout the ages. Most concepts have focused on the spontaneous appearance of the human species or of primitive life forms; but current knowledge indicates a long history of gradual change and evolution over eons of time. Humans have been a fairly recent development at the end of this gradual evolutionary process. The actual start of the evolution of life is difficult to decide since a long period of **chemical evolution** preceded the evolution of life. Chemical evolution is the gradual development of complex chemicals starting from simple substances.

It is currently accepted that the universe, as we know it, began with a colossal explosion about 15 billion years ago. In the aftermath, material consisting largely of hydrogen and helium filled the expanding universe. Gravity began acting on regions of relatively high material density, eventually producing massive concentrations of matter in isolated areas of space. This process represents the beginning of galaxy and star formation. Most stars initially produce energy by fusion of hydrogen to form helium. If the star is sufficiently large, as reserves of hydrogen diminish, exothermic fusion of helium begins, and the star expands greatly, entering the so-called "red giant phase." During this period, elements such as carbon and oxygen are formed. For example, the formation of carbon can be represented as follows:

$$^{4}He + {}^{4}He \longrightarrow [^{8}Be]^{*} + \gamma$$

$$[^{8}Be]^{*} + {}^{4}He \longrightarrow {}^{12}C + \gamma$$

As the star ages, again depending on size, an explosion or supernova can result, scattering elements such as carbon and oxygen throughout nearby space. The process of star formation can then begin again. It is thought that this is the way in which our sun and solar system formed, about 5 billion years ago, because of evidence for the presence of many heavy elements in the sun. This suggests that the carbon that makes up the organic molecules we are familiar with today probably did not originate on the Earth.

Following the formation of the Earth around 4.6 billion years ago, a period of chemical evolution of organic biomolecules (molecules associated with living organisms) must presumably have occurred. Organic compounds, such as proteins, nucleic acids and carbohydrates, are major components of organisms, and it is unlikely that these relatively complex molecules have always existed since the formation of the Earth. In the chemical evolution process, organic molecules were formed from inorganic precursors, and eventually incorporated into and formed part of "living" matter. This most likely occurred via a series of steps as shown below.

1. Formation of small organic molecules (e.g., amino acids, monosaccharides and nucleotides) from inorganic precursors
2. Formation of macromolecules (e.g., proteins, polysaccharides and nucleic acids
3. Organization of macromolecules into primitive cell-like structures (a protocell)
4. Evolution into a contemporary cell comprising a defined nucleus, with a nucleic acid-based mechanism for protein synthesis

Following this period of abiotic chemical evolution, biological evolution (based on biological processes) would have commenced.

Of course, the details of chemical evolution cannot be known for certain, and are the subject of much speculation. However, important clues have been found and assembled from areas as diverse as the study of fossils (paleontology), geology, astronomy and environmental chemistry.

II. CONDITIONS ON PRIMITIVE EARTH

There is evidence that the Earth may have originally had an atmosphere comprised mainly of H_2 and He were lost at an early stage in its development. Planets such as Venus, Earth and Mars obtained secondary atmospheres, probably from release of volatile material incorporated within their interiors as they formed. These gases may have been physically trapped, or chemically bound to minerals in the form of water of crystallization, for example.

There is strong evidence that our secondary atmosphere originally comprised water vapor, CO_2 and N_2. Importantly, it is also clear that there was virtually no molecular oxygen (O_2) present. Some estimates have placed the level of O_2 at about 10^{-14} times that presently found in the atmosphere. Because of the distance of the Earth from the Sun, surface conditions would eventually have allowed the presence of liquid H_2O and the formation of oceans. Oceans must have existed by 3.8 billion years ago, since this is the age of the earliest known sedimentary rocks. These oceans would have contained dissolved salts; but because of the absence of O_2, ions such as SO_4^{2-} would be absent. Much of the water vapor originally present in the atmosphere condensed to form the oceans. The atmospheric CO_2 was soluble in H_2O, subsequently forming bicarbonate and carbonate ions susceptible to precipitation as carbonate-containing minerals. Today, the oceans on Earth contain about 50 times as much CO_2 as our atmosphere. Carbonates (e.g., $CaCO_3$) in sedimentary rocks are the major repository for CO_2, accounting for 100,000 times as much CO_2 as presently exists in the atmosphere. Interestingly, it is estimated that if all the CO_2 fixed as carbonate were released in gaseous form, our atmosphere would possess a total mass of CO_2 very similar to the Venusian atmosphere. It would have a total pressure of up to 60 atmospheres, with the N_2 content at only 1%, similar to that found on Venus and Mars.

The starting point for chemical evolution of organic compounds (and our present atmosphere) is likely to have been an atmosphere containing predominantly N_2, together with some water vapor and CO_2. Minor or trace constituents may have included NO, CO, HCl and/or H_2.

III. PREBIOTIC EVOLUTION OF ORGANIC COMPOUNDS

A. SIMPLE MOLECULES

Many organic compounds that we are familiar with today, such as petroleum hydrocarbons, amino acids, carbohydrates and lipids, have been biosynthesized by living organisms. Originally, however, there were no living organisms, and compounds such as these must have been formed abiotically. To try and establish whether this was possible, a number of laboratory experiments have attempted to simulate prebiotic primitive Earth conditions.

Among the early experiments were those of the American chemist Stanley Miller, using apparatus such as that shown in Figure 13.1. Water was alternately boiled and then condensed to promote circulation in a system where initially a mixture of NH_3, CH_4, H_2 and water vapor was employed as the "atmosphere" or vapor phase, and sparks passing between tungsten electrodes simulated lightning discharges. After a week of continuous operation, CO and N_2 were the main gaseous products; some CH_4 remained, but most of the NH_3 was consumed. From the original carbon source of CH_4, a number of organic molecules such as amino acids and simple carboxylic acids were found in the aqueous phase. Also formed was an insoluble polymeric material. As mentioned previously, however, recent research suggests that it was unlikely that significant levels of NH_3 and CH_4 were present in the prebiotic atmosphere. With NH_3 for example, it is subject to photolysis and is also quite soluble in water. Later experiments carried out using a gaseous mixture comprised of N_2, CO, CO_2, H_2 and water vapor were found to give essentially the same products provided that, initially, H_2:CO and H_2:CO_2 ratios were greater than 1 and 2, respectively. In these systems, which more closely resemble the likely prebiotic atmosphere of Earth, the carbon source for the organic products is CO_2 or CO.

Experiments such as these are not conclusive but do indicate that organic compounds necessary for the construction of living organisms could have been formed abiotically on Earth. Further evidence is that amino acids and other organic material have been found in meteorites, and molecules such as formaldehyde ($H_2C{=}O$), methanol, ethanol and methylamine (CH_3NH_2) have been detected spectroscopically in interstellar space.

Contemporary living organisms contain many different organic compounds, including macromolecules such as proteins, lipids, polysaccharides and nucleic acids. However, all the different types of protein (over 3000 in a single *Escherichia coli* cell) are constructed from about 20 different amino acids. The polysaccharides generally found in living organisms are made from only a few monosaccharides, such as glucose.

Several organic nitrogen-containing molecules called pyrimidines and purines, together with the monosaccharide ribose and phosphate ion, can be considered the precursors or building blocks of nucleic acids. In the same manner, glycerol, choline

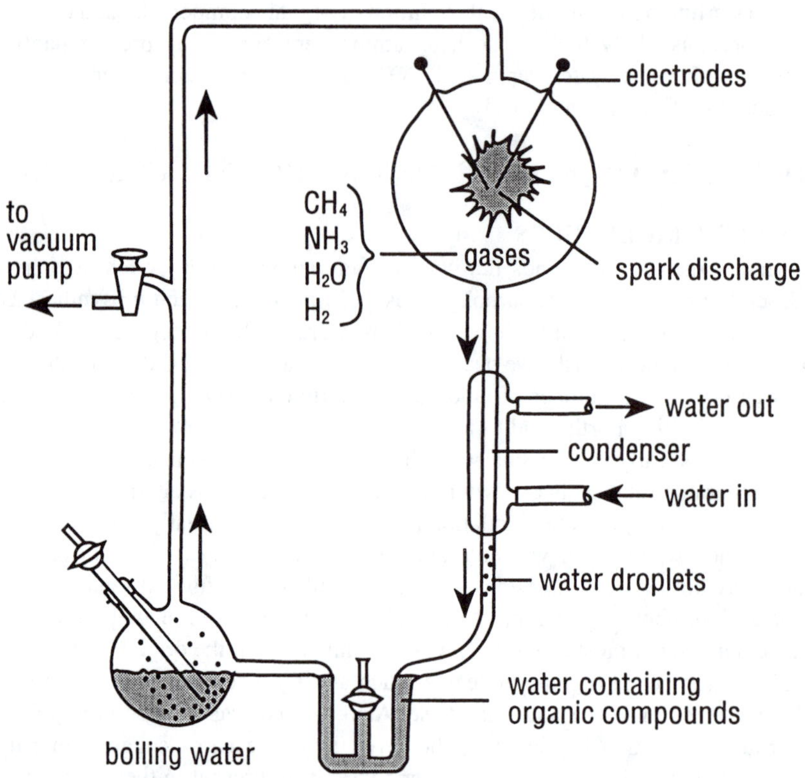

FIGURE 13.1 Apparatus used to simulate the composition and conditions in the prebiotic atmosphere.

[$(CH_3)_3N^+CH_2CH_2OH$], and fatty acids are the building blocks of many types of fats. The earliest life forms most probably comprised the same biomolecules as found in contemporary organisms and so these relatively few organic building block molecules, from which most others can be derived, must have been synthesized abiotically on Earth. A list of these molecules together with their structures is shown in Figure 13.2. This set of relatively simple organic molecules may be regarded as ancestors of the biomolecules found today; they have been used to construct biota both in the past and currently.

The prebiotic source of monosaccharides is generally believed to be formaldehyde. It was probably formed photochemically in the prevailing atmosphere of the time, from water vapor and CO_2. Although formaldehyde is itself susceptible to photolysis in the atmosphere,

$$H_2C{=}O \xrightarrow{h\nu\,\lambda<338\,\mathrm{nm}} H^{\bullet} + H\dot{C}O$$

Nevertheless, some would have dissolved in rainwater and deposited in the oceans. It is well known that in the laboratory, under basic (high pH) conditions, formaldehyde reacts to form mainly pentoses and hexoses, presumably on account of their relatively stable cyclic or ring structures. The first stage is this process may be:

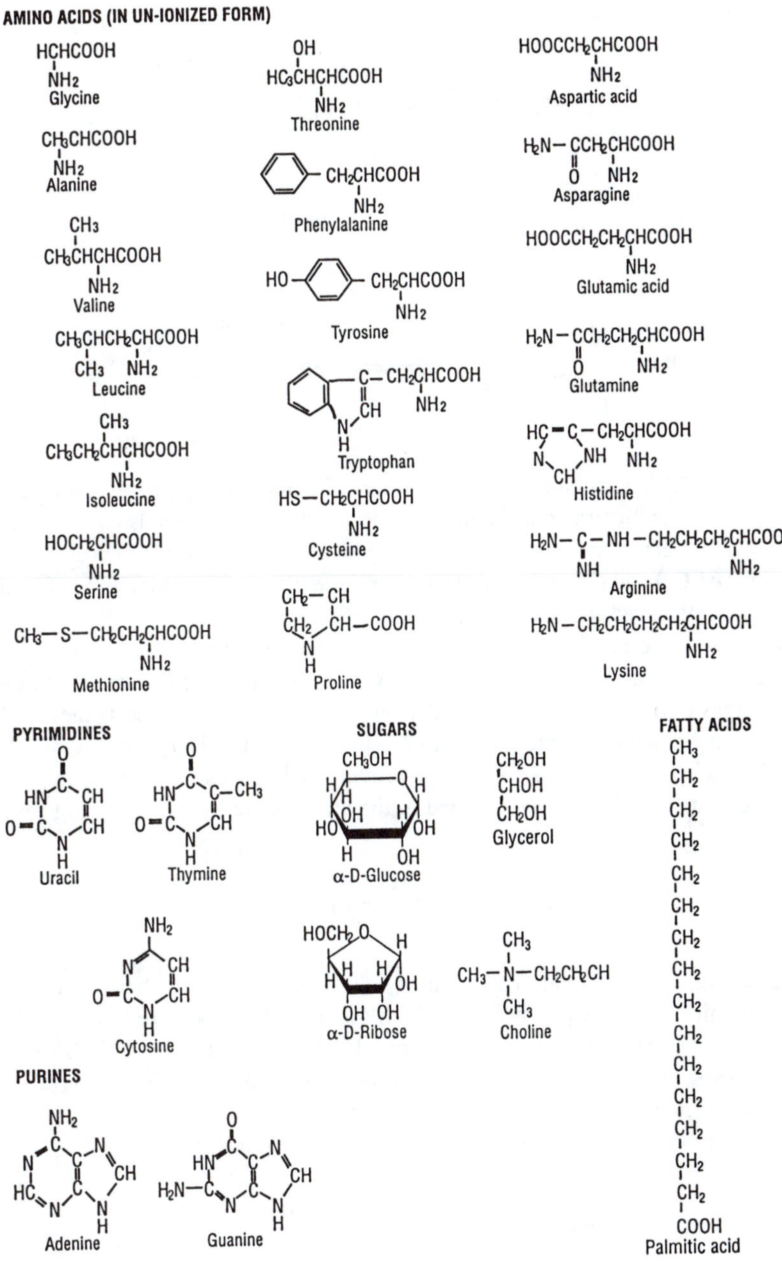

FIGURE 13.2 Chemical structures of simple organic compounds that could be precursors of the molecules existing today.

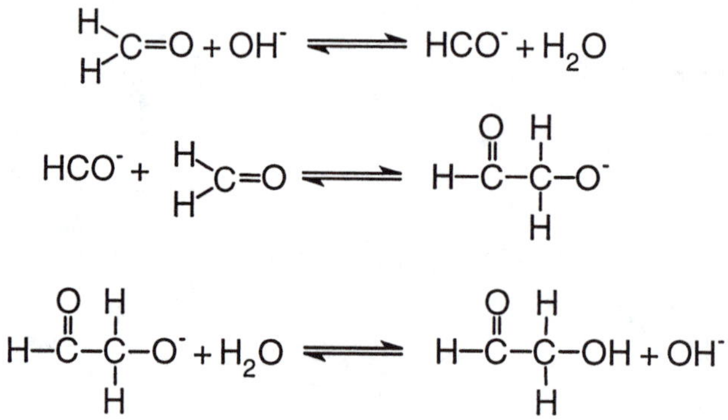

$$\mathrm{H_2C{=}O + OH^- \rightleftharpoons HCO^- + H_2O}$$

$$\mathrm{HCO^- + H_2C{=}O \rightleftharpoons H{-}C({=}O){-}CH_2{-}O^-}$$

$$\mathrm{H{-}C({=}O){-}CH_2{-}O^- + H_2O \rightleftharpoons H{-}C({=}O){-}CH_2{-}OH + OH^-}$$

A possible mechanism for the formation of pentoses (e.g., ribose) and hexoses (e.g., glucose) is illustrated in Figure 13.3. Recently, heterogeneous catalysts for this process have been shown to exist. They include carbonate minerals such as limestone, alumina (Al_2O_3) and clay minerals (e.g., kaolinite, illite). Monosaccharide formation could have occurred in the presence of these minerals.

A critical intermediate in the prebiotic synthesis of organic nitrogen-containing compounds is hydrogen cyanide (HCN). Recent research has shown that in certain desert areas rich in titanium dioxide (TiO_2) (e.g., the Imperial Sand Dunes, California), photochemical reduction of N_2 to NH_3 occurs on the TiO_2 surface, which acts as a catalyst. Such a process could have provided a local source of NH_3 on prebiotic Earth so that HCN could have formed in the atmosphere by the following reaction:

$$\mathrm{CO + NH_3 \xrightarrow{h\nu} HCN + H_2O}$$

Once formed, HCN can react further, in the presence of aldehydes and NH_3, to form α-amino acids. A pathway by which this may have occurred is illustrated in Figure 13.4 by the synthesis of glycine from HCN, formaldehyde ($H_2C{=}O$) and NH_3 in aqueous solution. This process is used today in laboratories for the synthesis of amino acids and is known as the Strecker synthesis.

Nucleic acids are macromolecules with the following structure:

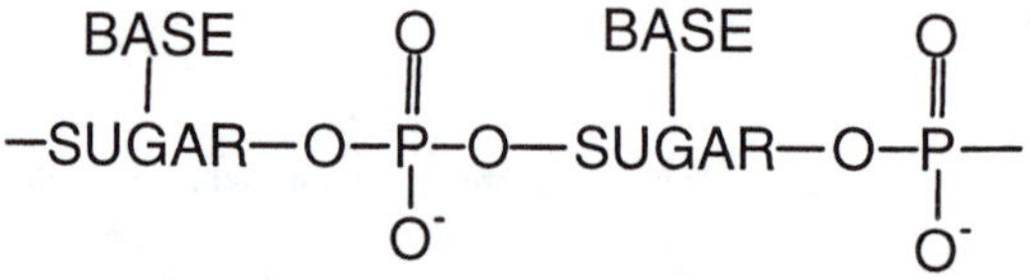

```
 BASE       O        BASE        O
  |         ||        |          ||
—SUGAR—O—P—O—SUGAR—O—P—
            |                    |
            O⁻                   O⁻
```

The sugar is ribose or the related 2-deoxyribose. A base-sugar unit is termed a nucleoside, while a base-sugar-phosphate unit is called a nucleotide. As shown in

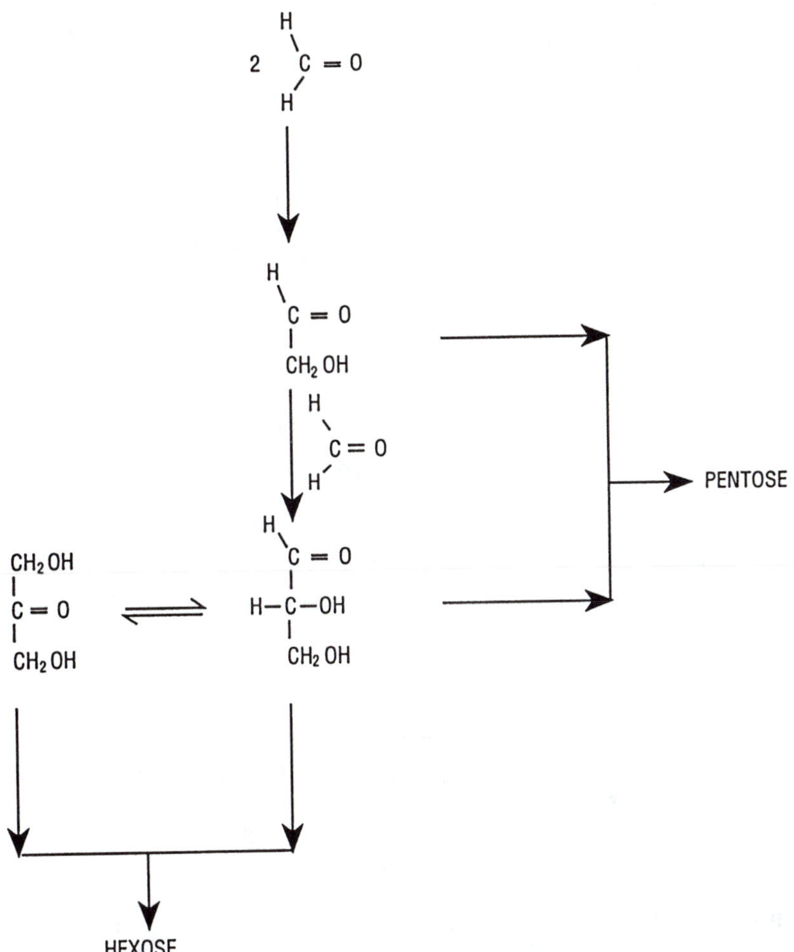

FIGURE 13.3 A possible mechanism for the formation of pentoses and hexoses from formaldehyde.

Figure 13.2, the bases are organic nitrogen-containing molecules. It has been known for more than 100 years that HCN polymerizes in aqueous solution. For example, a 0.1 *M* solution of HCN at pH 9.2 left for 6 months at room temperature in a bottle will result in about half the starting material being converted to a complex mixture of HCN oligomers (compounds comprising a few HCN units). Although yields are small, many of the nucleic acid bases could have been formed on prebiotic Earth by this means. As an illustration, the formation of adenine ($C_5H_5N_5$), which can be thought of as a pentamer of HCN, is shown in Figure 13.5. Phosphate ions (PO_4^{3-}) were probably present in the oceans of prebiotic Earth, as was ribose and 2-deoxyribose. Thus, all the raw materials necessary for the construction of nucleic acids were available.

It is striking that the monosaccharides, amino acids, nucleic acid bases and other compounds that are of biological importance today can be obtained abiotically, under

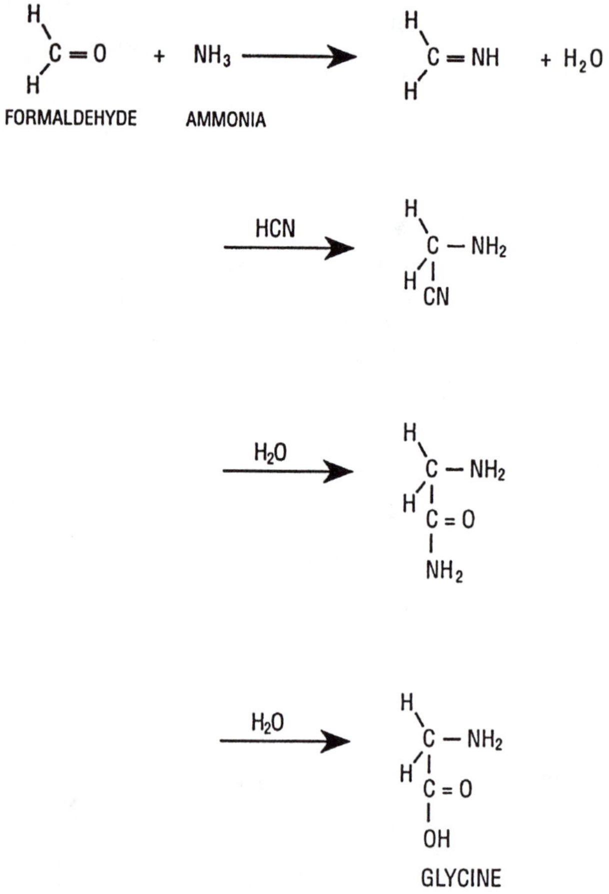

$$H_2C{=}O \; (\text{FORMALDEHYDE}) + NH_3 \; (\text{AMMONIA}) \longrightarrow H_2C{=}NH + H_2O$$

$$\xrightarrow{HCN} H_2C(NH_2)(CN)$$

$$\xrightarrow{H_2O} H_2C(NH_2)\text{-}C({=}O)\text{-}NH_2$$

$$\xrightarrow{H_2O} H_2C(NH_2)\text{-}C({=}O)\text{-}OH$$

GLYCINE

FIGURE 13.4 A possible pathway for the formation of an amino acid from simple compounds.

relatively mild conditions in aqueous solution, from the same set of small organic compounds that exists in interstellar space. Coincidence is unlikely. Why this particular set of compounds is important in the construction of all contemporary biota is unclear. For example, why are the proteins found in all living organisms composed of only about 20 α-amino acids and not 40? Perhaps these amino acids were prominent among those formed on prebiotic Earth, or perhaps some process of selection took place.

B. CONDENSATION AGENTS

With the formation of the monomeric raw materials needed to construct biopolymers such as protein and nucleic acids, the next stage of chemical evolution was the synthesis of the biopolymers themselves, and other complex molecules such as lipids. The covalent bonds between the monomer units of biopolymers are generally formed by condensation reactions. A simple definition of a condensation reaction is a reaction where two reactants combine to form a larger molecule, usually by eliminating a smaller molecule such as water. Unfortunately, the equilibrium for

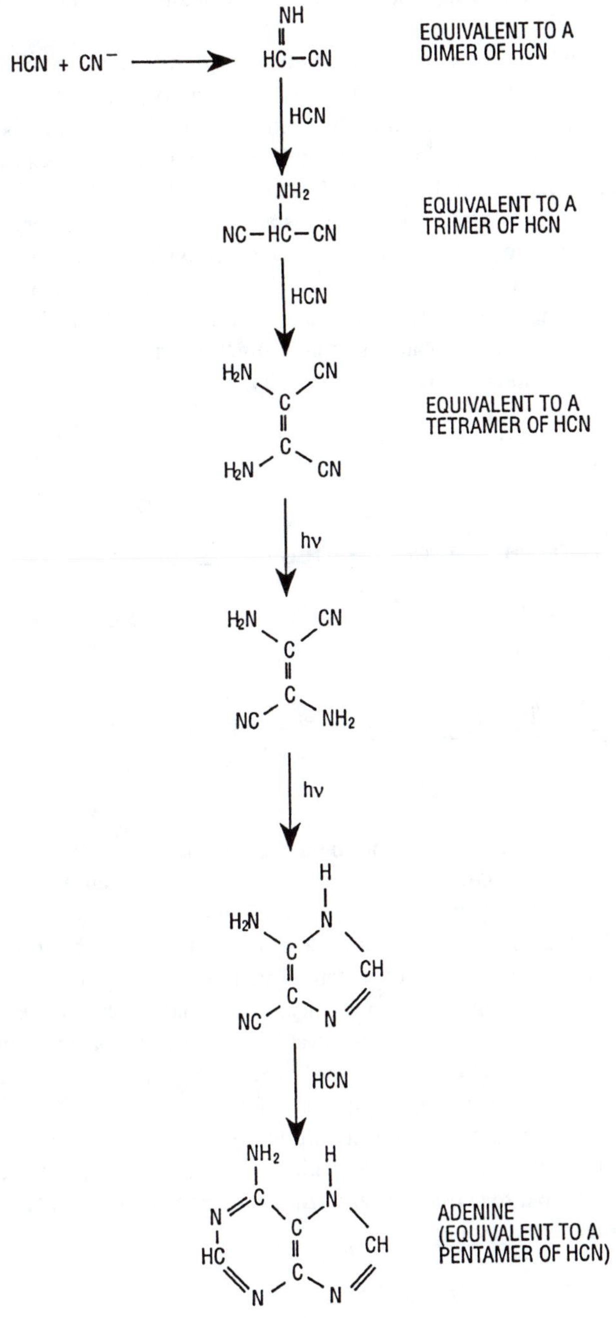

HCN + CN⁻ → HC(=NH)—CN — EQUIVALENT TO A DIMER OF HCN

↓ HCN

NC—HC(NH₂)—CN — EQUIVALENT TO A TRIMER OF HCN

↓ HCN

(H₂N)(NC)C=C(NH₂)(CN) — EQUIVALENT TO A TETRAMER OF HCN

↓ hv

(H₂N)(NC)C=C(CN)(NH₂)

↓ hv

H₂N, NC, C=C, N—H, CH, N

↓ HCN

NH₂, H, C, N, HC, CH — ADENINE (EQUIVALENT TO A PENTAMER OF HCN)

FIGURE 13.5 Adenine can be seen as a pentamer of HCN and could be formed by the sequence illustrated.

such processes in dilute aqueous solution is typically very much in favor of the reactants.

There are only a few ways in which condensation-type reactions can take place with reasonable yield. One is to carry out the reaction with condensation agents, compounds that appear to essentially chemically bind the water being released. Possible prebiotic condensation agents include cyanamide (H_2N–CN), dicyanamide (NC–NH–CN), carbodiimides (R–N=C=N–R) together with polyphosphates and polyphosphate esters. While carbodiimides are used in the laboratory for condensation of amino acids, recent work suggests that the polyphosphates and polyphosphate esters were the most likely. They would have been forerunners of ATP, the most important contemporary biological condensation agent. Consider the following example of how a polyphosphate ester may function as a condensation agent in the linking of two amino acids to from a peptide.

$$\underset{\text{Amino acid}}{NH_2\text{-}CH(R)\text{-}C(=O)\text{-}OH} + \underset{\text{Amino acid}}{NH_2\text{-}CH(R')\text{-}C(=O)\text{-}OH} \xrightarrow[\text{Condensation Agent (R''=Organic Group)}]{R''\text{-}O\text{-}P(=O)(O^-)\text{-}O\text{-}P(=O)(O^-)\text{-}O\text{-}P(=O)(O^-)\text{-}O^-}$$

$$NH_2\text{-}CH(R)\text{-}C(=O)\text{-}NH\text{-}CH(R')\text{-}C(=O)\text{-}OH + R''\text{-}O\text{-}P(=O)(O^-)\text{-}O\text{-}P(=O)(O^-)\text{-}OH + HO\text{-}P(=O)(O^-)\text{-}O^-$$

Few details regarding condensation on prebiotic Earth are known with certainty. Apart from condensation agents, satisfactory yields in condensation reactions may be obtained by carrying out the process under anhydrous (i.e., without water) or nearly anhydrous conditions. For example, heating a mixture of amino acids, without water, at 150 to 180°C affords a substantial yield of polypeptides. Anhydrous conditions could conceivably have arisen near lakes and pools on the edge of land masses. Another mechanism for concentrating smaller, monomeric compounds and bringing about condensation to form polymers is by adsorption to clay minerals. One type of clay, montmorillonite, has been found to promote the condensation of amino acids under relatively mild conditions, to form long polypeptides in high yield. There may also have been competition for adsorption by the various amino acids, amounting to a selection process. Another condensation agent could well have been ultraviolet light.

C. PREBIOTIC MACROMOLECULES

The oldest known fossils are believed to be about 3.5 billion years old. Given the age of the Earth, this means that lifeforms took up to 1.1 billion years to emerge. These early lifeforms were presumably composed of macromolecules such as proteins that had also formed, within this time span, from simpler compounds. The approximately 20 α-amino acids have greatly differing stabilities. The most stable are amino acids with aliphatic side chains (glycine, alanine, valine, isoleucine and leucine), which decompose irreversibly by decarboxylation.

$$R{-}\underset{\underset{{}^{+}NH_3}{|}}{C}H{-}CO_2^- \longrightarrow R{-}CH_2{-}NH_2 + CO_2$$

The half-life for alanine due to this process is about 3×10^9 years at 25°C. The remaining amino acids are less stable, and have other decomposition pathways in addition to decarboxylation. Serine and threonine, for example, have decomposition half-lives of less than a thousand years at 25°C.

$$R{-}\underset{\underset{OH}{|}}{C}H{-}\underset{\underset{{}^{+}NH_3}{|}}{C}H{-}CO_2^- \quad \left(\begin{array}{ll} R = H & \textbf{Serine} \\ R = CH_3 & \textbf{Threomine} \end{array}\right)$$

This relative instability means that amino acids would have to be synthesized continuously, and at such a rate that a sufficient concentration was achieved for protein synthesis. Once formed, proteins are unstable due to the peptide bonds linking the amino acid units being susceptible to hydrolysis in aqueous solutions. No single protein molecule would be expected to survive indefinitely. Perhaps sorption to clays and other minerals aided in replication of these molecules. It is also possible that proteins were not synthesized from amino acids but from other starting materials.

One of the characteristics of proteins found in contemporary living matter is that the constituent amino acids (except glycine) have only one of two possible stereochemical or three-dimensional configurations. If a carbon atom has four different substituents, then these substituent groups can occupy two different spatial arrangements around the carbon atom. This is illustrated for the amino acid alanine in Figure 13.6. The two different forms of alanine are mirror images of each other, but they are not superimposable and therefore not identical. No matter how one form is twisted or turned, it cannot be made to coincide with the other. (This can be confirmed by building molecular models.) This is an example of stereoisomerism or chirality (from the Greek for "hand"). Recall that isomers are compounds with the same elemental composition, but different structures. Alanine is called a chiral compound, and the two different forms are known as stereoisomers. Larger molecules can have more than one chiral center. A good analogy to simple chiral compounds are our hands. The right and left hand are mirror images of each other, but are not superimposable on one another. The two stereoisomers of alanine can be regarded as right-handed and left-handed forms of the molecule. The two forms of alanine separately have generally identical physical and chemical properties; one exception is when interacting with other chiral compounds. To continue the hand analogy, right hands of two people interact with each other via a handshake quite naturally, but the interaction of a right and left hand from different people is quite awkward. In this way, some enzymes can discriminate between stereoisomers. One form of alanine is found in proteins, and can serve as an energy source for organisms, while the other is not metabolized efficiently. They even taste differently because the sense organs of the human tongue are chiral agents that respond differently to the two stereoisomers. The chiral amino acids found in proteins of organisms then all have the same "handedness." Indeed, essentially all organic compounds found in living material and having one or more chiral centers occur in only one stereochemical configuration.

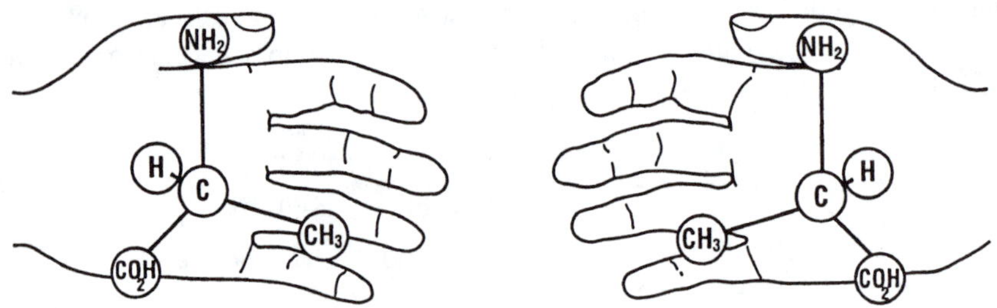

FIGURE 13.6 Stereoisomerism with alanine, indicating the two possible configurations.

The fact that proteins contain amino acids all of the same form is not unusual. For proteins that have a helical secondary structure, it is well known that a stable helix will not result from a random mixture of right- and left-handed amino acids. Hence, there is a stability advantage in having amino acids all of one form or the other. What is intriguing is why one form was selected over the other. Primitive earth simulation experiments (Figure 13.1) and samples from meteorites both produce essentially equal numbers of right-handed and left-handed forms. Presumably then, both forms were available on prebiotic Earth. It could be that it was purely by chance. If so, then this suggests that all living organisms are derived from a single ancestor, since the proteins of all organisms comprise amino acids of the same form. It also follows that if life evolved in a similar manner on other planets, lifeforms there would have an equal chance of comprising proteins with amino acids of the opposite configuration or "handedness."

An alternative explanation for this discrimination may again be sorption to clay minerals. Some experiments have shown preferential attraction of one stereochemical form of amino acid over the other; however, this effect was reported to be only slight and remains unconfirmed. Another interesting possibility comes from the area of quantum mechanics. This is a branch of science that focuses on subatomic particles and the forces governing their behavior. Chiral compounds are not equivalent from a quantum mechanical point of view. Although the difference in stability is not great, over the long periods of time associated with chemical evolution and under the conditions envisaged on prebiotic Earth, selection of the more stable stereoisomer is predicted. The particular amino acid and monosaccharide stereoisomers found today in macromolecules of living organisms are in fact those that are more stable, based on quantum mechanical considerations.

No completely satisfactory prebiotic synthesis of polynucleotides or nucleic acids has yet been proposed. The structure of nucleic acid chains e.g., RNA in contemporary organisms has the ribose or deoxyribose linked to phosphate groups at the 3′ and 5′ positions, and to the base at the 1′-position, as shown below:

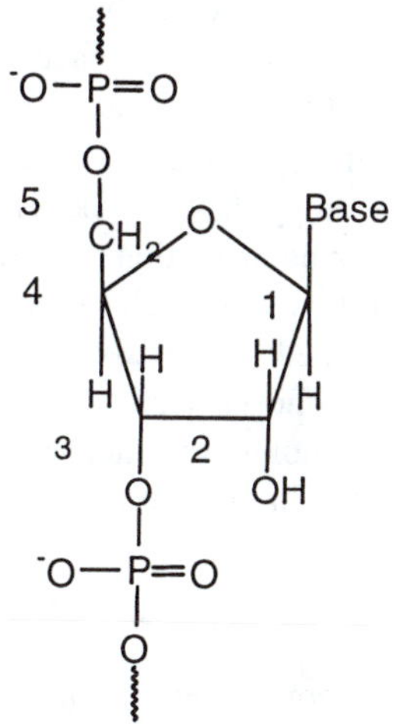

It has been demonstrated that nucleosides can be phosphorylated by tripolyphosphoric acid at temperatures in the range 0 to 22°C; however, the reaction is not specific. With adenosine as the starting material, for example, products are reported to include several mono-, di- and triphosphates. The next stage in the chemical evolution is the joining together or condensation of different nucleotides through the formation of internucleotide linkages. Condensation agents in laboratory experiments have included both radiation and tripolyphosphoric acid. In most cases, any oligo- or polynucleotides formed show linkages between the 2′- and 5′-positions of the sugar ring, as well as the "natural" 3′,5′-linkage that forms less readily. It has been determined however that the 2′,5′-linkage is less stable that the 3′,5′-linkage. An abiotic stability-based selection process may therefore have eventually favored the linkage found in living organisms today.

Monosaccharides react relatively rapidly in neutral or slightly basic solution to form various derivatives with other organic and inorganic compounds. It is possible that monosaccharides did not occur to any great extent as free compounds in the oceans of prebiotic Earth. Rather, they occurred in some combined form, including perhaps as oligo- and polysaccharides. Ultraviolet light has been used to polymerize monosaccharides in aqueous solution. This condensation is aided by clay minerals such as kaolin, the role of which is uncertain.

IV. PROTOCELLS

There are widely differing estimates of the amount of organic material that could have been formed on prebiotic Earth. Some calculations suggest that the maximum possible concentration of dissolved organic compounds in the oceans was extremely dilute (about 10^{-12} M), while others indicate that it could have been much more concentrated. Generally, it is felt that concentrations were low, particularly given that ultraviolet radiation reached the surface of the Earth since there was no ozone layer. Some prebiotic syntheses would probably have occurred in the oceans, but others would have been restricted to areas where the concentrations of organic compounds were higher (e.g., small ponds). The efficient synthesis of macromolecules, in particular, is not likely to have taken place in the open ocean.

In order to construct the first lifeform, prebiotic macromolecules would have to collect or assemble. This assembly may have occurred in a similar way to the formation of micelles. In an entropy-driven process, macromolecules could have clumped together so that their nonpolar sections could associate and avoid unfavorable interactions with water. In this way, a cluster with a hydrophilic exterior and nonpolar interior could have formed. Alternatively, assembly could have occurred on the surface of clay minerals. The later stages of chemical evolution may well have occurred on surfaces since polymeric materials are, in general, much more strongly absorbed than monomers. In fact, one current theory suggests that our first ancestors were literally made from clay!

It is difficult to define exactly what "life" is. It is unlikely, however, that the first lifeforms resembled relatively complex modern cells which for all organisms is the basic unit of organization that is considered living. Minimum requirements are that they had the capacity to self-replicate and also the potential to evolve into modern cells. The first lifeforms presumably arose from organized collections of macromolecules called "prototype cells" or "protocells." Because of our difficulty in defining the meaning of "life," it is uncertain as to precisely when, in the course of events leading from the formation of protocells, it arose.

The major classes of macromolecules serve similar functions in cells of all modern organisms. Nucleic acids store and transmit genetic information, while proteins function as enzymes or are structural form material which the organisms is constructed. Polysaccharides act as a storage form for cellular fuel, or serve as extracellular structural components. Lipids serve as sources of fuel, and are a major component of the membranes that surround all cells. One important point needs to be made here. It has been observed that nucleic acids and proteins are informational molecules, whereas polysaccharides and lipids are not. The specific nucleotide or amino acid sequence of nucleic acids and proteins represents a source of information, or an information-carrying capacity. Polysaccharides such as starch or cellulose are polymers of the same repeating unit, glucose, and thus lack the molecular complexity to carry information.

The polysaccharides and lipids may not have been necessary components of protocells and the first lifeforms. This means that protocells and the first lifeforms may have comprised or been based upon:

1. Proteins
2. Nucleic acids
3. Both proteins and nucleic acids

We shall now explore these possibilities in turn.

Much of the support for the view that protocells and early lifeforms were based on proteins comes from the relative difficulty of prebiotic syntheses of nucleic acids compared with proteins, and the properties of proteinoid microspheres. These microspheres are remarkable cell-like structures that may be prepared in the laboratory. Heating an anhydrous mixture of α-amino acids for 2 to 5 hours at 180°C produces an amino acid polymer. The mixture usually contains a high proportion of the acidic amino acids aspartic acid and glutamic acid. Because of the possibility of unusual bonding patterns such as cross-linking between different protein chains, the polymeric products are referred to as proteinoids. On adding hot saline water, and allowing the resultant solution to cool, proteinoid microspheres are formed. This method of preparation is not inconsistent with prebiotic Earth conditions. Microspheres tend to be spherical and uniform in diameter. The size of the microspheres are generally about 2.0 μm, and falling within the range 0.5 to 7.0 μm. This uniformity in microsphere size is said to represent one similarity to living cells.

Several other cell-like properties are also observed. Microspheres are relatively stable since they can be centrifuged without disruption. In addition, when composed of acidic amino acids, increasing the solution pH sees the formation of a crude double layer, resembling a natural lipid bilayer membrane, around the microsphere. Cleavage can also be induced by pH changes, or by exposure to $MgCl_2$. If suspensions of microspheres are allowed to stand for 1 to 2 weeks, buds appear on them spontaneously and can be detached by mild warming or agitation. On transfer to hypertonic (greater salinity) or hypotonic solutions, proteinoid microspheres shrink or swell, respectively, demonstrating an osmotic-like function. This suggests they have a semipermeable membrane surrounding an inner compartment containing salts and dissolved proteinoids and other organic material. This membrane also shows some selectivity in the compounds that are able to pass through it. Polysaccharides are retained within proteinoid microspheres under conditions in which monosaccharides diffuse out freely. Microspheres can also selectively recognize polynucleotides. This selective recognition could have been the beginning of a genetic code. This is important because while proteinoid microspheres show cell-like properties and are possible models for protocells, evolution could not proceed very far without a genetic system to record information on cellular characteristics and control replication. A mechanism involving complementary replication of proteinoids analogous to polynucleotides has been proposed. A proteinoid consisting of a sequence of amino acids with positively, negatively and neutral side chains could replicate, with positive and negative sidechains pairing, and neutral amino acids going together, as shown in Figure 13.7. In this way, protocells function in the absence of a nucleic acid based genetic system, which they presumably acquired later. According to this scenario, protocells came before enzymes, which came before genes.

Another view suggests that nucleic acids formed the basis of protocells, and that they provided the mechanism for the evolution of proteins. Viruses have been referred to as structures at the threshold of life, and may represent precursors of the first lifeforms. It has been speculated that the largest known viruses are in fact very small cells. The simplest and smallest viruses are little more than a single nucleic acid molecule containing only a few genes, with a protein coating serving as a protective sheath. A nucleic acid molecule can code for proteins, undergo self-replication, and undergo mutation, which is important for evolutionary modification of the molecule. The acquisition of a surrounding membrane and the development of enzymes may be regarded as later evolutionary events.

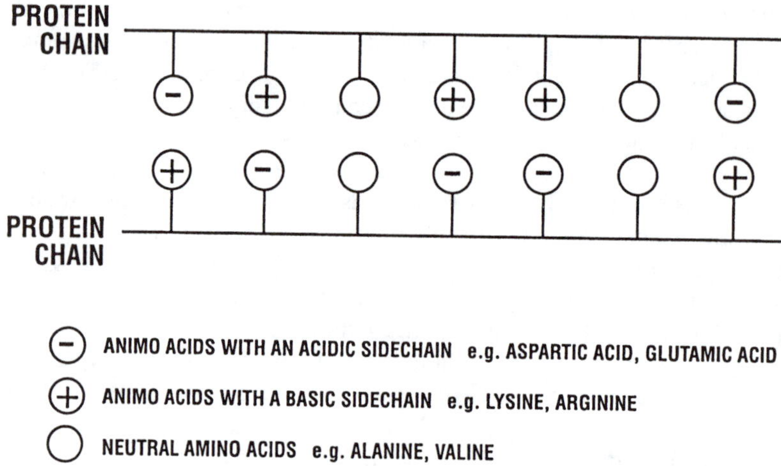

FIGURE 13.7 Possible replication mechanism in proteins with pairing in a sequence of amino acids.

A central role for nucleic acids in the origin of life is also suggested by the large number of functions that nucleotides or their derivatives perform in contemporary cells. Nucleotides serve as the monomeric units or building blocks of DNA and the three major types of RNA. Certain nucleotides or their derivatives also act as energy carriers (e.g., ATP, ADP), as hydrogen or electron carriers, as sugar carriers and as transporters of lipid components. Nucleotides are important in metabolism and energy transfer as well as in replication. Their widespread use today suggests a fundamental metabolic and genetic role in protocells and the first lifeforms.

We have seen that whichever of proteins or nucleic acid were considered to form the basis of protocells, the other information-carrying macromolecule was eventually acquired. It has also been argued that both primitive nucleic acids and proteins were required simultaneously, to form protocells and ultimately a living cell. According to this model, protocells and lifeforms without nucleic acids would lack a means of achieving some genetic continuity in "offspring" or successor structures. Protocells and lifeforms without the versatile proteins would be severely restricted in their ability to use the chemicals of their environment. Certainly, it is possible to prepare proteinoid microspheres containing polynucleotides that have some stability.

V. LATER DEVELOPMENTS

Presumably, in the early stages of biological evolution, the chemical evolution process would also have continued simultaneously for some time. The first lifeforms continued to select from the environment those organic compounds that were beneficial to their survival. The earliest lifeforms probably used energy-rich organic molecules formed abiotically in their environment to carry out protein synthesis and nucleic acid replication. The first living cells were probably prokaryotes (i.e., cells that do not have a nucleus or other compartments surrounded by a membrane). They could be described as anaerobic heterotrophic organisms because they used abiotically built organic compounds for energy, and to get carbon for growth. Moreover, they did not use molecular oxygen in these processes.

Fermentation is loosely defined as the conversion of glucose, either to lactic acid or to ethanol and CO_2, with the release of energy in the form of energy-rich molecules such as ATP. Despite the relative inefficiency of energy production in fermentation compared to say aerobic respiration, it is widespread among contemporary organisms, suggesting it is a very ancient process. Perhaps this is the way in which the first living organisms obtained the energy necessary to drive cellular activities. Eventually, a few essential organic molecules would have become depleted in the surrounding water. Perhaps the amino acid glycine became scarce for example, but glycolic acid was still relatively abundant, since it had never been uptaken and incorporated into organisms. An organism could have developed the capacity to convert glycolic acid to glycine, and thus gained an important advantage over competitors.

$$\underset{\textbf{Glycolic Acid}}{H-\overset{H}{\underset{OH}{C}}-CO_2H} \longrightarrow \underset{\textbf{Glycine}}{H-\overset{H}{\underset{NH_2}{C}}-CO_2H}$$

A significant and largely untapped energy source for organisms was the sun. It has been suggested that some cells acquired the capacity to synthesize ATP photochemically in a process called photophosphorylation (this comprises part of the so-called "light" phase of photosynthesis described in Chapter 12):

$$ADP + PO_4^{3-} \xrightarrow{h\nu} ATP + H_2O$$

Energy in the form of ATP would have permitted the transformation of relatively simple organic compounds, e.g., acetic and formic acids, into the more complex organic molecules on which heterotrophic lifeforms were reliant, as shown below:

$$\underset{\textbf{Formic Acid}}{12H\overset{O}{\overset{\|}{C}}-OH} + \text{energy} \longrightarrow \underset{\textbf{Glucose}}{C_6H_{12}O_6} + 6CO_2 + 6H_2O$$

Photosynthesis as we know it involves not only photophosphorylation, but also reduction whereby NADPH and ATP formed in the "light" phase are used to bring about reduction of CO_2 to glucose. As organisms carrying out phosphorylation evolved, they presumably developed a reduction capability, and in this way the first photosynthetic organisms developed.

In photosynthetic cells of higher green plants, water is "split," producing H^+ ions, electrons and O_2 as byproducts (see Chapter 12). We say that H_2O is an electron donor. This process requires substantial amounts of energy and is relatively complex. It is likely that the first photosynthetic organisms did not utilize H_2O and evolve O_2. Anaerobic photosynthesis, as seen today in certain bacteria, is simpler than the oxygen-producing form, and probably evolved first. These bacteria use H_2S, or organic compounds such as isopropanol, as hydrogen donors in the reduction of CO_2 represented by the equations below:

$$2H_2S + CO_2 \longrightarrow (CH_2O) + 2S + H_2O$$

$$2CH_3{-}\underset{\text{Isopropanol}}{\overset{\displaystyle OH}{\overset{|}{C}}H}{-}CH_3 + CO_2 \longrightarrow (CH_2O) + 2CH_3{-}\underset{\text{Acetone}}{\overset{\displaystyle O}{\overset{\|}{C}}}{-}CH_3 + H_2O$$

It is likely that reactions similar to these were carried out by the early photosynthetic lifeforms. The vast deposits of sulfur found in Louisiana and Texas may be the products of these primitive organisms.

When all relatively abundant electron donors other than water had been depleted, organisms would have had to use water of necessity, and oxygen evolution would have commenced:

$$H_2O + CO_2 \rightarrow CH_2O + O_2$$

Water would have also provided some protection against ultraviolet radiation. Earth's ozone layer is derived from O_2 and it is generally accepted that the O_2 in our atmosphere came from photosynthesis. A summary of the occurrence of oxygen in the atmosphere and evolutionary characteristics in the environment is shown in Table 13.1.

Somewhere between 1.4 and 2 billion years ago, eukaryotic cells appeared. These cells possess a nucleus and have their genetic material enclosed within a membrane. A size increase in some fossils around 1.4 billion years ago suggests the appearance of eukaryotes at this time. The importance of this development is that eukaryotic cells require large amounts of O_2 to function. Calculations indicate the atmospheric O_2 level had reached 1% of present levels. The accumulation of oxygen also meant that if all life on Earth ceased, it is unlikely that it could begin again in the same manner. Abiotic synthesis of organic compounds is more difficult in the presence of O_2. An outline of these processes is contained in Chapter 12 on photosysnthesis and respiration.

It seems that atmospheric O_2 reached present levels some 350 to 400 million years ago. Eventually, through the process of biological evolution, mammals and humans appeared. With or without human influence, the atmosphere and our environment that we know today will continue to evolve and change.

TABLE 13.1
Oxygen Levels in the Atmosphere and Accompanying Environmental Characteristics and Development

Years before present	Oxygen level in atmosphere as a fraction of present level	Environmental characteristics and developments
$>2 \times 10^9$	10^{-9}–10^{-14}	Abiotic synthesis of organic molecules and macromolecules
		Formation of protocells
		Earliest lifeforms appear.
2×10^9	10^{-3}	Oxygen evolving photosynthesis developed
		Significant accumulation of O_2 in atmosphere begins
1.4×10^9	10^{-1}	Appearance of eukaryotes
6.7×10^8	0.07	Appearance of soft-bodied multicellular organisms
5.0×10^8	0.1	Multicellular organisms with exterior skeletons
		Protective ozone layer develops.
$3.5–4 \times 10^8$	1.0	Atmospheric O_2 reaches contemporary levels
		Animals and plant life on land.

VI. KEY POINTS

1. The period of biological evolution on Earth was preceded by a period of chemical evolution. In this period, complex molecules used in biological evolution and the life processes of living organisms developed over long periods of time from simple substances.

2. Most stars of sufficient size produce carbon from the following reactions:

$$^4He + {}^4He \rightarrow [^8Be]^* + \gamma$$

$$[^8Be]^* + {}^4He \rightarrow {}^{12}C + \gamma$$

   Oxygen is also formed at later stages in these processes.

3. The Earth formed about 4.6 billion years ago and the following steps in the chemical evolution process, preceding biological evolution, commenced:
   (a) Formation of small organic molecules, e.g., amino acids, monosaccharides and nucleotides, from inorganic precursors.
   (b) Formation of macromolecules, e.g., proteins, polysaccharides and nucleic acids.
   (c) Organization of macromolecules into primitive cell-like structures (a protocell).
   (d) Evolution into a contemporary cell comprising a defined nucleus, with a nucleic acid-based mechanism for protein synthesis.

4. The atmosphere of the ancient earth comprised water vapor, CO_2 and N_2, with virtually no molecular oxygen present.

5. All contemporary biota are constructed from a common limited array of simple molecules assembled into larger biomolecules that form living tissue and conduct life processes. These simple molecules include amino acids (about 20 different compounds such as glycine, alanine, threonine and aspartic acid), sugars (e.g., glucose and ribose), pyrimidines (e.g., uracil and thymine), purines (e.g., adenine and guanine) and fatty acids (e.g., palmitic acid).

6. Monosaccharides are believed to have formed from the polymerization of formaldehyde formed photochemically from CO_2 and H_2O. Ammonia can be formed by photochemical reduction of atmospheric N_2 to NH_3, which then further reacts with carbon monoxide to form HCN. This HCN and formaldehyde can undergo a reaction sequence to form the amino acid glycine, and other amino acids can be formed by similar processes. The HCN can also form adenine, which together with sugars such as ribose and phosphate ions are the raw materials necessary for nucleic acid synthesis.

7. Prebiotic organic molecules can undergo condensation reaction sequences perhaps facilitated by sorption to clay minerals and other materials to give proteins, carbohydrates, etc.

8. Contemporary living material contains constituent amino acids that all have the same stereochemistry. This also suggests a common evolutionary pathway from a limited set of simple molecules.

9. The formation of protocells from the available biomolecules could have occurred through the formation of micelle-like clusters of molecules. These are generated by the molecular properties of the constituent biomolecules oriented to form a hydrophilic exterior and a nonpolar interior, thereby avoiding unfavorable interactions with the polar water.

10. In contrast to carbohydrates, the proteins and nucleic acids can contain different sequences of constituent units that represent an information source. These sequences can contain information that will allow replication of biomolecules.

11. Chemical evolution made limited use of the available solar energy, but some cells developed the capacity to carry out photophosphorylation. This comprises part of the "light" phase of photosynthesis and results in the formation of ATP from ADP. It is probable that this process was utilized in anaerobic photosynthesis (without oxygen), but later the currently common photosynthetic process in green plants evolved.

12. Oxygen-evolving photosynthesis probably evolved between 1.4 and 2 billion years ago, making significant contributions to the atmospheric oxygen content.

REFERENCES

Mason, S.F., *Chemical Evolution — Origin of the Elements, Molecules and Living Systems,* Clarendon Press, Oxford, 1991.

Cairns-Smith, A.G., *Genetic Takeover and the Mineral Origins of Life,* Cambridge University Press, Cambridge, 1982.

Wayne, R.P., *Chemistry of Atmospheres,* Clarendon Press, Oxford, 1991.

Lehninger, A.L., *Biochemistry,* 2nd ed., Worth Publishers, New York, 1975, Chap. 1 and 37

Fox, S.W. and Dose, K., *Molecular Evolution and the Origin of Life,* Marcel Dekker, New York, 1974.

Miller, S.L. and Orgel, L.E., *The Origins of Life on the Earth,* Prentice-Hall, Englewood Cliffs, NJ, 1974.

CHAPTER 13 PROBLEMS

1. Many of the reactions involved in chemical evolution require the molecules involved to be in specific configurations for reactions to occur. This could have been facilitated by various materials in the environment, such as clay minerals. Suggest the orientation of the two molecules involved and show the overall electron transfers that could occur in the following reactions, keeping in mind these electron transfers probably involve a sequence of steps.

A. $$2\ H_2C{=}O \longrightarrow H{-}C({=}O){-}CH_2{-}OH$$

B. $$H_2C{=}O + NH_3 \longrightarrow H_2C{=}NH$$

2. When ideas were first developing regarding chemical evolution on Earth, it was deduced that Earth surface conditions would have been reducing (little oxygen and high hydrogen concentrations). As a result of this, it was suggested the Earth's primitive atmosphere would consist mainly of the reduced forms of carbon and nitrogen i.e., CH_4 and NH_3, respectively. However, since essentially no oxygen was available in the atmosphere, there was also no ozone layer to filter out short wavelength solar radiation. If CH_4 and NH_3 were present in the Earth's atmosphere, how would this radiation affect these compounds and the composition of the atmosphere? (Keep in mind that high-energy radiation will cause dissociation of some molecules as well as water.)

CHAPTER 13 SOLUTIONS

1.A • Formaldehyde has the following molecular charge distribution, with all atoms lying in a plane:

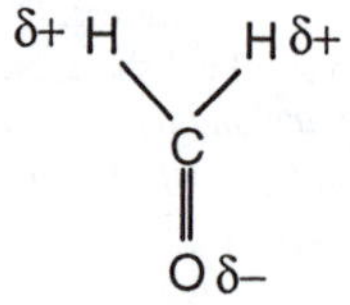

δ+ H H δ+
C
O δ–

• The two molecules would be expected to orient in the following manner.

• The ammonia molecule has the following charge distribution, with the nitrogen atom at the apex of a pyramid and the hydrogen atoms at the corners:

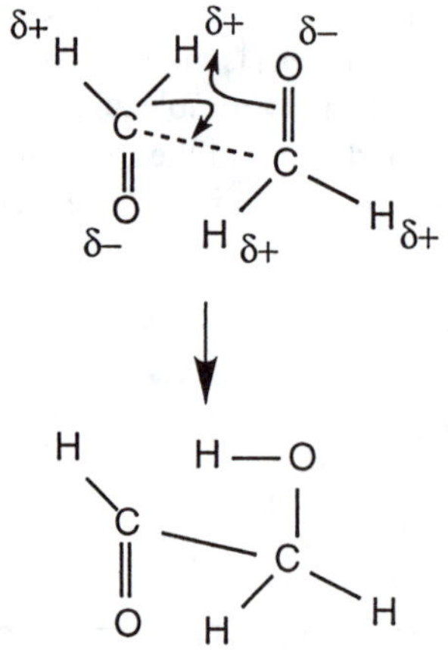

1.B • The ammonia molecule has the following charge distribution, with the nitrogen atom at the apex of a pyramid and the hydrogen atoms at the corners:

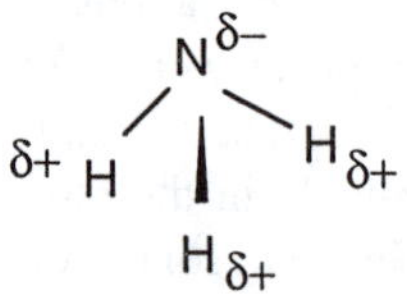

• The two molecules would be expected to orient in the following manner and the following depicts an overall summary of a reaction that probably occurs in a sequence of steps.

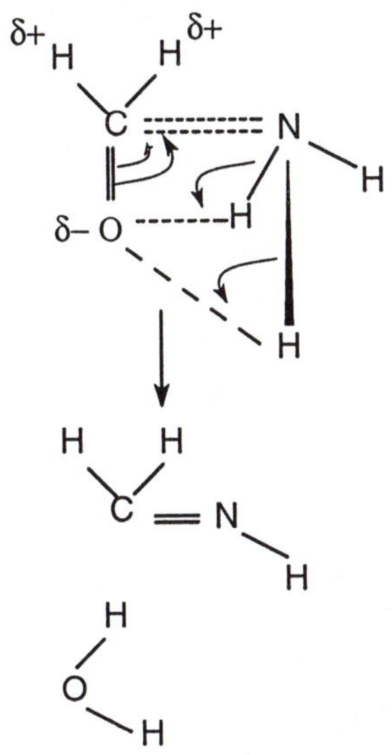

2. **Ammonia (NH_3):** Ammonia is susceptible to dissociation by radiation to give N_2 and H_2 by a series of reactions and thus would not be expected to be present to any significant extent in the primitive atmosphere.

   Methane: Methane is a relatively stable molecule and would not be expected to be directly dissociated by radiation. However, water is susceptible to dissociation and the products can react with methane by the following reactions:

$$H_2O \xrightarrow{h\nu} HO\bullet + H\bullet$$

$$HO\bullet + CH_4 \xrightarrow{h\nu} \bullet CH_3 + H_2O$$

   Overall: Owing to the intense solar radiation effects, neither CH_4 nor NH_3 would be expected to be present to any appreciable extent in the Earth's primitive atmosphere.

Chapter 14

CHEMISTRY OF NATURAL WATERS

I. INTRODUCTION

Water covers about 71% of the global surface; thus, the water environment comprises a major part of the human environment. About 97% of this water environment is in the oceans and the remainder in lakes, rivers and so on. The study of water bodies has been a major area of scientific activity since the beginning of scientific endeavors.

Limnology is the study of fresh waters. This branch of science is concerned with the physics, chemistry and biology of lakes, rivers and other freshwater bodies. In addition, limnologists have always had a strong interest in the effects of pollution on natural systems, particularly sewage. In fact, in the early part of this century, some of the basic principles of the pollution ecology of streams were established. This emphasis on the human interaction with freshwater systems continues to the present day.

Oceanography has a similar long history of activity. Some say that the voyages of Captain James Cook that started in 1768 with the *Endeavour* marks the start of scientific oceanography. In Cook's voyages, observations were made of the temperature and depth of the oceans as well as biological phenomena. Many oceanographic cruises have been undertaken since the time of Cook and these continue to the present day. Later expeditions began the study of the ocean currents and winds, principally due to their importance in the movement of sailing ships carrying cargoes for trading purposes.

The chemistry of natural water bodies has developed enormously, in parallel with the development of the science of chemistry in general. Currently, we are able to characterize and quantify chemical components of water bodies and evaluate the dynamics of their transformations. This capacity was not available to early investigators.

In recent years, with the development of governmental bodies and agencies for aquatic research, the water-based sciences have received a boost in activity. At the present time considerable resources and effort are employed in the development of an understanding of the functions of natural water bodies as a basis for conservation and management of these important natural environments.

II. PHYSICAL CHEMICAL PROPERTIES OF WATER

To appreciate the functions of natural water bodies, we must start at the molecular level. The water molecule H_2O is planar, with an angle of 104.5° between the O-H bonds, as shown in Figure 14.1. Each O-H bond is polar and has a dipole moment of 1.5 Debyes, with the positive charge directed toward the hydrogen atom. This is in agreement with the difference in electronegativity between oxygen (3.5) and hydrogen (2.1) of 1.4. Due to its polarity (1.84 Debyes), water has a high boiling point and vapor pressure compared with other compounds of similar molecular weight. Thus, water would be expected to be a good solvent for ionic salts, such as sodium chloride. This characteristic is reflected in the high salt content that is dissolved in seawater. On the other hand, nonpolar substances, such as animal and plant fats present in biotic cells, do not dissolve in water to any significant extent. If this were not so, water would penetrate living cells freely and thereby threaten their survival.

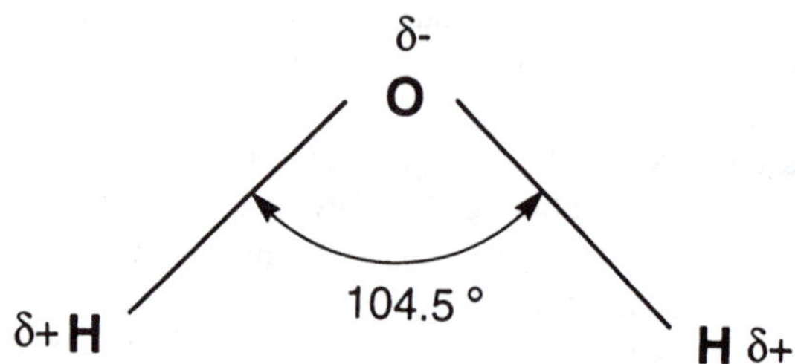

FIGURE 14.1. Diagrammatic representation of the water molecule.

Water has a number of remarkable properties. Perhaps one of the most noteworthy is its freezing behavior. Although it freezes at 0°C, it exhibits a maximum density at about 4°C. This means that when ice forms, instead of falling to the bottom, it floats to the surface. In this way, the ice forms a layer on the surface, insulating the lower waters and preventing them from freezing.

Most large water bodies, both natural and artificial, exhibit a vertical stratification into layers at different temperatures. For example, in the oceans, the surface temperatures show considerable variation; but at the lowest depths, the water temperature is usually close to 4°C at all latitudes. This is the temperature at which water exhibits its maximum density. This suggests that whenever water at 4°C is formed in the ocean, it falls to the bottom and accumulates where it is substantially insulated from thermal changes by the overlying water and the sea floor. In lakes and dams, as well as the oceans, solar radiation can heat the upper layers of the water body, producing lower densities. The heating produced by solar radiation penetrates to the depth permitted by the clarity of the upper layers of the water. The upper layer, which is often at a higher temperature and lower density, is referred as the **epilimnion** and the lower layer, having higher density and lower temperatures, is referred as the **hypolimnion**. A distinctive temperature profile is observed, with a thermocline separating the two layers as shown in Figure 14.2.

Gases from the atmosphere dissolve in natural water bodies and have an important influence on chemical and biological systems. Gases such as oxygen, nitrogen and methane are all symmetrical molecules and would be classified as nonpolar compounds. Thus, it would be expected that these substances would have limited solubility in water which is polar. This is reflected in the low aqueous solubility values shown in Table 14.1. Carbon dioxide also has low solubility in water, but the solubility of this gas is more complex than the others since it dissolves in water and also dissociates to form ions.

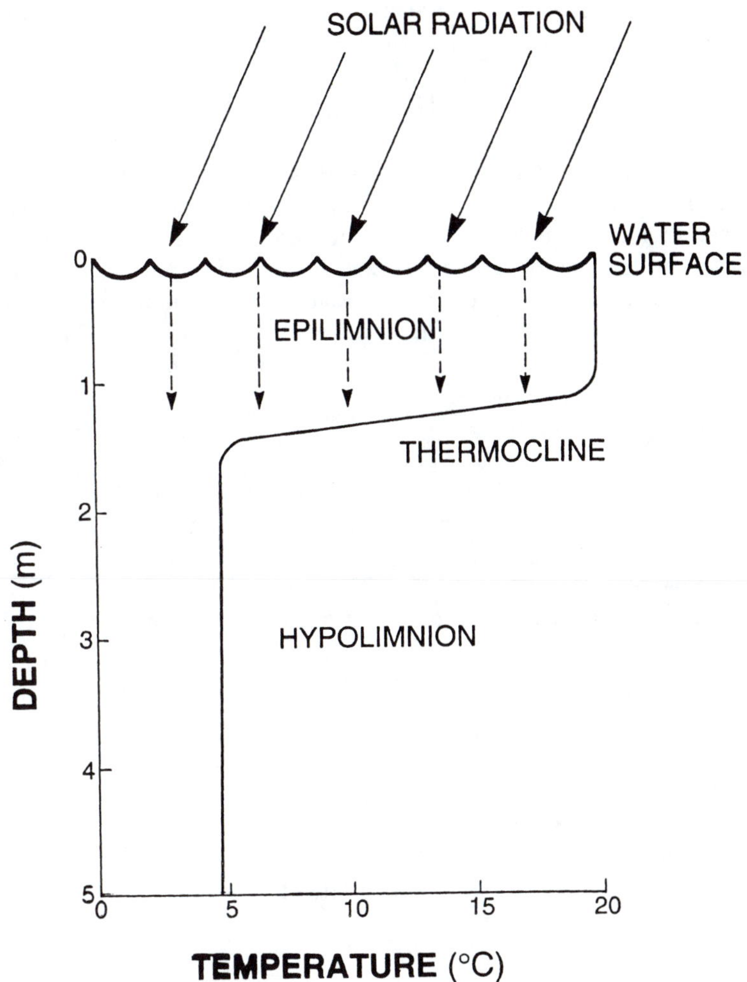

FIGURE 14.2. Stratification of a water body produced by solar radiation.

TABLE 14.1
Henry's Law Constants for Some Atmospheric Gases (1 atm pressure and 25°C)

Gas	Aqueous solubility (mg L^{-1})	Henry's law constant, H (atm mole L^{-1})
O_2	8.11	773
CO_2	0.55	29
N_2	13.4	1610
CH_4	24	746

A convenient measure of the environmental behavior of gases is the Henry's law constant (H). The Henry's law constant is defined as follows:

$$H = P/C_W$$

where P is the partial pressure of the substance in the atmosphere above the water and C_W is the corresponding concentration of the substance in water. Some Henry's law constants for atmospheric gases are shown in Table 14.1. It should be noted that the Henry's law constant for carbon dioxide may not be constant, as the other gases are, since carbon dioxide undergoes dissociation in water, as previously mentioned.

The Henry's law constant is useful for calculating concentrations that would be expected in water and the atmosphere. For example, the solubility of oxygen in pure water can be calculated as follows:

1. Calculation of partial pressure of oxygen in the atmosphere: The total pressure of the gases in the atmosphere is one atmosphere. But the atmosphere also contains 0.03 atmospheres of water vapor at 25° as well as 20.95% of oxygen as a permanent atmospheric gas. This means that the partial pressure of oxygen in the atmosphere is 0.2095/(1.00-0.03) which is 0.0216 atmospheres.
2. Calculation of the concentration of oxygen in water: From Henry's Law

$$S = P/H$$

thus,

$$S = \frac{0.216 \text{ atm}}{773 \text{ atm mol}^{-1} \text{ L}^{-1}}$$

$$= 2.79 \times 10^{-4} \text{ mol L}^{-1}$$

$$= 8.9 \text{ mg L}^{-1}$$

This is in reasonable agreement with the directly measured value reported in Table 14.1 at 8.11 mg L^{-1}.

The oxygen dissolved within natural waters has a critical role since it is needed for biotic respiration. Without this oxygen supply, biota would not be able to survive within the water mass. Although there is a relatively large supply of oxygen in the atmosphere, which contains about 21% oxygen, this substance has only limited solubility in water, as reflected in the data in Table 14.1. It is also interesting to note that the solubility of oxygen in water varies with a variety of factors. Of particular importance is the physical factor of temperature, with increases resulting in a major decline in solubility in the range from 0 to 35°C, as shown in Table 14.2. The salinity of seawater and estuarine waters also causes a decline in the solubility of oxygen in water.

Pure water weakly dissociates into hydrogen ions and hydroxyl ions with a dissociation constant of 10^{-14}, producing 10^{-7} g ions L^{-1} of each ion. The pH is defined as:

TABLE 14.2
Variation of Solubility of Oxygen in Water with Temperature (at 760 mm Pressure)

Temperature	Oxygen (mg L^{-1})
0	14.16
5	12.06
10	10.92
15	9.76
20	8.84
25	8.11
30	7.53
35	7.04

$$pH = -\log [H^+]$$

where p designates the power of the hydrogen ion activity.

Thus, the pH of pure water is 7. In natural waters, the pH is governed by a variety of factors. Of particular importance is the solubilization of carbon dioxide from the atmosphere to produce carbonic acid, which dissociates into ions as shown in Figure 14.3. The solubility of carbon dioxide in natural waters varies with temperature, and the dissociation into ions is strongly influenced by salts and other substances present. The pH of natural freshwaters is usually in the range from 6 to 9.

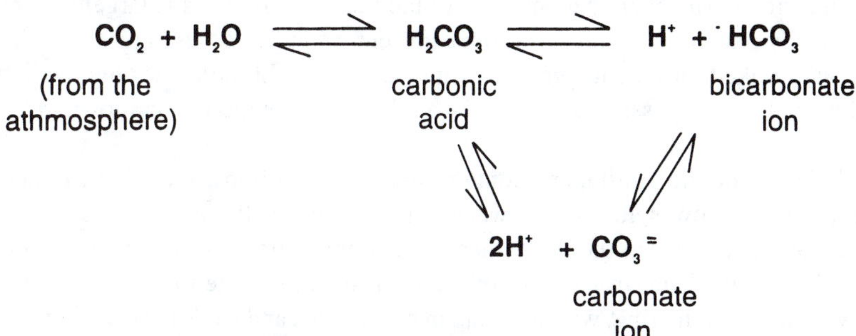

FIGURE 14.3. Ionization patterns of carbon dioxide dissolved in water.

III. ORGANIC COMPONENTS IN NATURAL WATERS

There is a wide variety of organic compounds in natural waters that can range in type from mountain lakes to the oceans. Most are the degradation products of animals and plants following death. There can be low concentrations of fatty acids, fats, hydrocarbons and many other substances. Generally, low molecular weight compounds, such as amino acids and monosaccharides, do not occur in any significant quantities. This is due to the ready degradation of these substances upon formation from more complex substances. This is facilitated by a large and varied population of microorganisms, together with many nutrient substances present in natural water bodies. For example, glucose has a half-life of about 10 hr in natural waters at temperatures of about 20°C.

Polysaccharides, such as cellulose, are relatively persistent in natural waters and therefore are present for long periods in relatively large amounts resulting from the death of plants in particular. Cellulose is usually present in plant detritus present in water bodies. Similarly, there can be relatively high amounts of proteins and polypeptides that are resistant to degradation.

In many coastal and freshwater areas, there are large zones of swamps and marshlands that are periodically or permanently inundated with water. These areas generally have a high plant biomass, which leads to large amounts of dead plant matter being formed within these zones. This dead plant material is subject to degradation, particularly by microorganisms. After the organic matter has been attacked by microorganisms and other processes residues remain that are resistant to further degradation. These residues are mainly derived from plant lignins, which are high molecular weight polyphenolic substances in plants. The same processes can occur within soil, producing similar substances resistant to biodegradation. As a group, these substances are referred to as **humic substances** and they have a brown, or dark, color with molecular weights ranging from 100s to 1000s.

This complex group of substances can be fractionated into crude components by acid and base solubility, as shown in Figure 14.4. This allows the isolation of substances described as **fulvic** and **humic acids** and **humin**. In fact, these fractions are complex mixtures with properties covering a range of characteristics rather than being specific. The fractions all share somewhat similar properties and chemical structures. Fulvic acid is the lowest molecular weight substance and has the highest oxygen content. The chemical behavior of the humic substances is largely governed by the many phenolic and related groups present, as indicated in Figure 14.5. These groups are linked into the larger, higher molecular weight molecules by a variety of bonding patterns. A possible structure for a fulvic acid component is shown in Figure 14.6.

While the humic substances contain many polar groups, their high molecular weight leads to low solubility in water. This is due to the strong intermolecular forces that exist between the molecules of the humic substances related to the high molecular weight. The strong color of these substances is due to extended conjugation with the phenolic ring, which occurs in many cases and leads to strong ultraviolet and visible absorption as shown in Figure 14.7.

An important environmental property of the humic substances is their ability to chelate, thereby strongly binding metal ions into their structure. This can occur with the phenolic hydroxyl group and the acid groups, as indicated in Figure 14.8. The ions of iron and aluminium are strongly bonded in this system, whereas magnesium is relatively weak, with nickel, lead and calcium being intermediate in their bonding strength. This may make essential trace metals unavailable to biota in aquatic areas with high concentrations of humic substances.

The humic substances and other organic matter, as well as inorganic matter, may not dissolve in water, but form small particulates that do not settle out over time. These are usually referred as **colloids**. The presence of colloids in an aquatic area alters the behavior of metals in solution in a somewhat similar manner to the humic

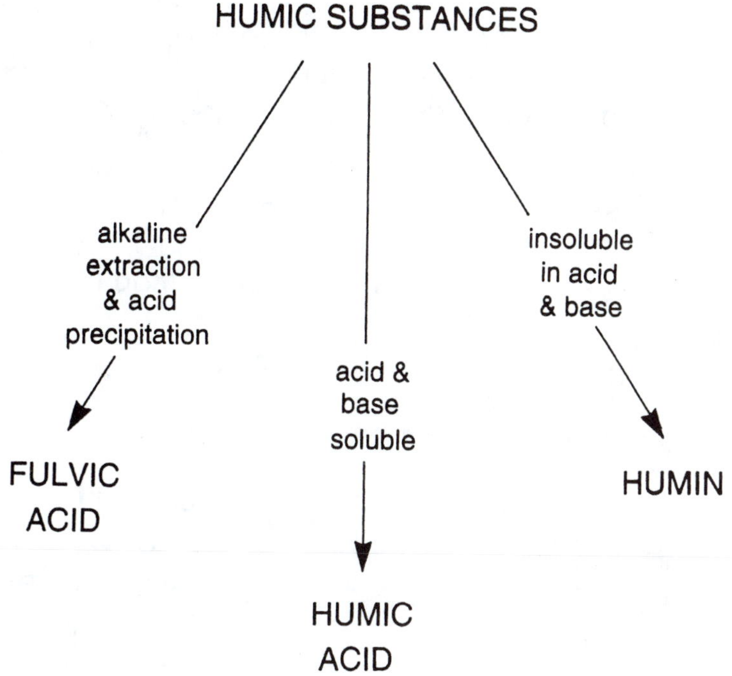

FIGURE 14.4. Fractionation of humic substances into broad component classes.

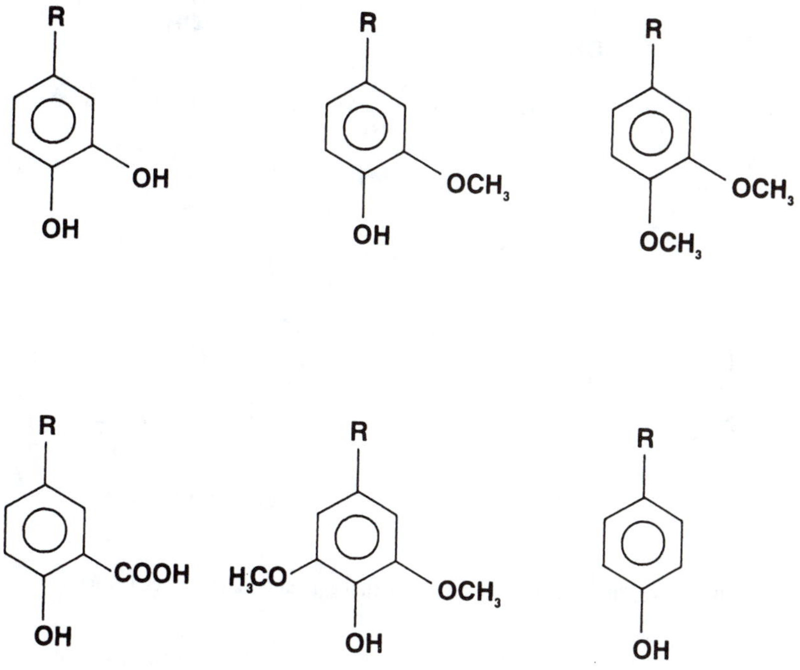

FIGURE 14.5. Phenolic groups present in the structure of humic substances.

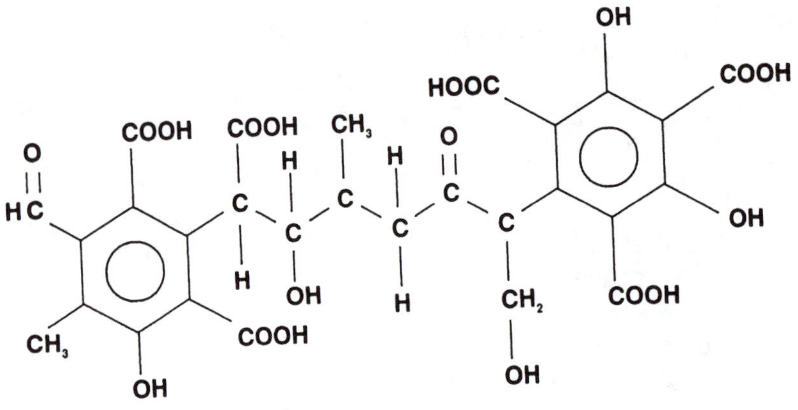

FIGURE 14.6. A possible structure for a fulvic acid component.

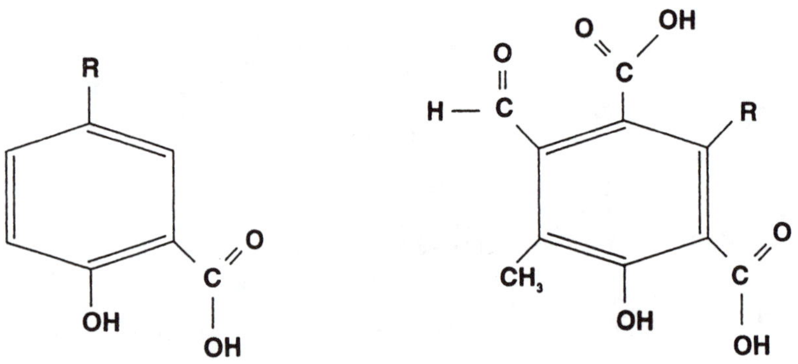

FIGURE 14.7. Examples of extended conjugation in groups present in humic substances.

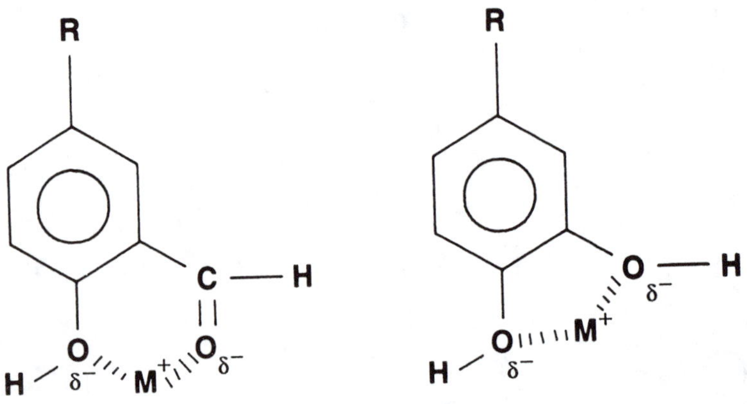

FIGURE 14.8. Chelation of metal ions by phenolic groups present in humic substances.

substances due to charged surfaces that may exchange and bind metals and other ions. In addition, colloids with lipoid characteristics may alter the behavior of lipophilic substances in water bodies.

IV. CARBON AND NITROGEN TRANSFORMATIONS IN NATURAL WATER BODIES

A. THE CARBON CYCLE

The fundamental process that feeds carbon and energy into aquatic systems is photosynthesis. In this process, carbon dioxide from the atmosphere is dissolved and dissociates into the various forms previously discussed in water bodies. Plants perform photosynthesis by converting carbon dioxide, water, energy and various trace components into plant matter. In addition to photosynthesis, the plants consume some of the energy produced by using it in respiration and so release some carbon dioxide back to the water mass. Some plant material is consumed by aquatic animals that utilize the energy in the food by respiration, which results in the release of carbon dioxide to the water mass, as shown in Figure 14.9.

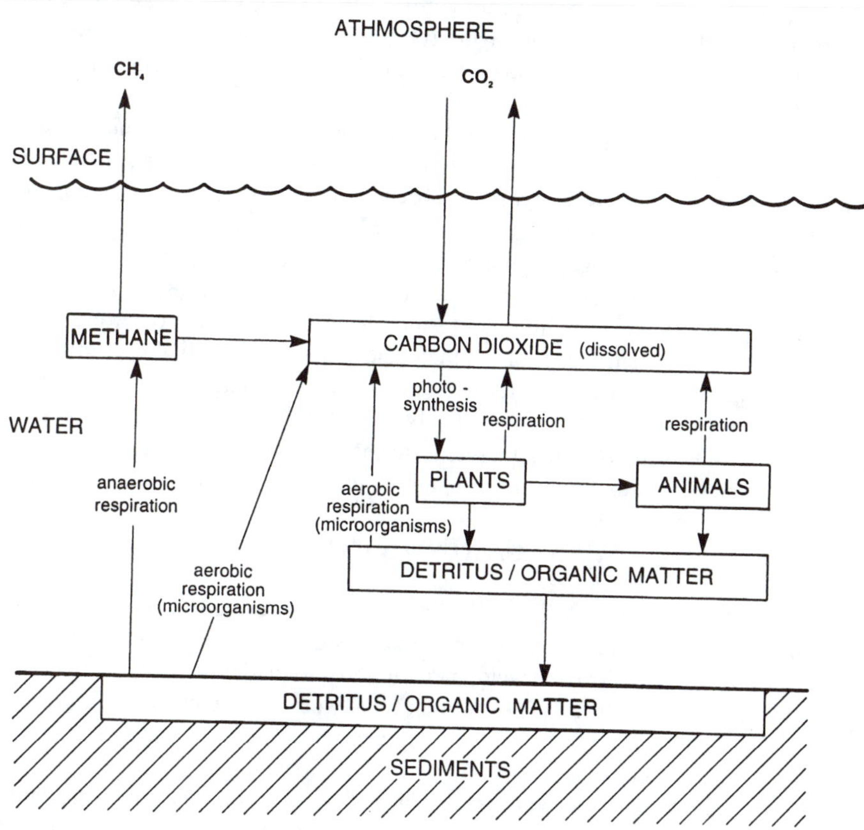

FIGURE 14.9. Transformations of carbon in aquatic systems.

On death, both animals and plants produce biotic fragments referred to as **detritus**. Detritus is rich in carbon and energy in the form of carbohydrate, protein and fat. Detritus enriches sediments with organic matter where it is acted on by the population of microorganisms present. Some of the resultant respiration occurs by **aerobic** processes and produces carbon dioxide, but often sediments are depleted in oxygen and **anaerobic** processes occur, resulting in the production of carbon dioxide and methane. Both methane and carbon dioxide produced from all respiration processes are released into the water mass and can escape to the atmosphere. These processes can be seen as a cycle of carbon from the atmosphere through the aquatic ecosystem and back to the atmosphere again. (See Figure 14.9).

B. THE NITROGEN CYCLE

Nitrogen exists in aquatic systems in a variety of chemical forms, including both inorganic and organic. The principle inorganic forms all occur in different proportions, depending on the conditions in the water mass. Nitrate (NO_3^-) is the principal inorganic nitrogen salt in well-aerated water bodies where oxygen is in plentiful supply. Nitrite (NO_2^-) occurs under certain specific conditions, but ammonia (NH_3) is the principal inorganic form of nitrogen under anaerobic conditions. Ammonia dissolves in water to form ammonium hydroxide (NH_4OH), which dissociates into the ammonium (NH_4^+) ion and the hydroxyl ion (HO^-). Oxidation can convert all the inorganic nitrogen forms into the nitrate ion, while reduction converts them all into ammonia.

The oxidation of inorganic nitrogen to nitrate is referred to as *nitrification,* as shown in Figure 14.10. *Denitrification* is the process whereby nitrogen is converted into N_2 through the nitrite ion and then subsequently can be released to the atmosphere as N_2 or N_2O (nitrous oxide). On the other hand, *nitrogen fixation* is a process whereby atmospheric nitrogen is fixed through ammonia into biological systems. This process requires a substantial amount of energy to convert atmospheric nitrogen into ammonia; for this reason, nitrogen fixation is less favored than other processes by biota. Proteins in detritus, from animals and plants, can be hydrolyzed to the component amino acids and then further degraded to ammonia and the other inorganic nitrogen forms, as shown in Figure 14.10. These processes can also be seen as a cycle of nitrogen from the inorganic forms in water through the biological ecosystem, finally resulting in the reformation of the inorganic nitrogen forms again.

V. ESTUARINE SYSTEMS

Many of the great cities of the world are situated on estuaries. This is because the estuaries form the junction between the oceans and freshwater, and often form natural harbors suitable for the development of ports. Estuaries are difficult to define and there are a range of different water bodies that would be included in this general category. It is generally accepted that the following definition of estuaries forms a reasonable description.

Estuaries are semi-enclosed coastal bodies of water that have a free connection with the open sea and have waters measurably diluted with freshwater from land drainage. This means that the salinity of the water in estuaries is between that of the ocean and freshwater. Oceanic salinity is usually somewhere around 35 parts

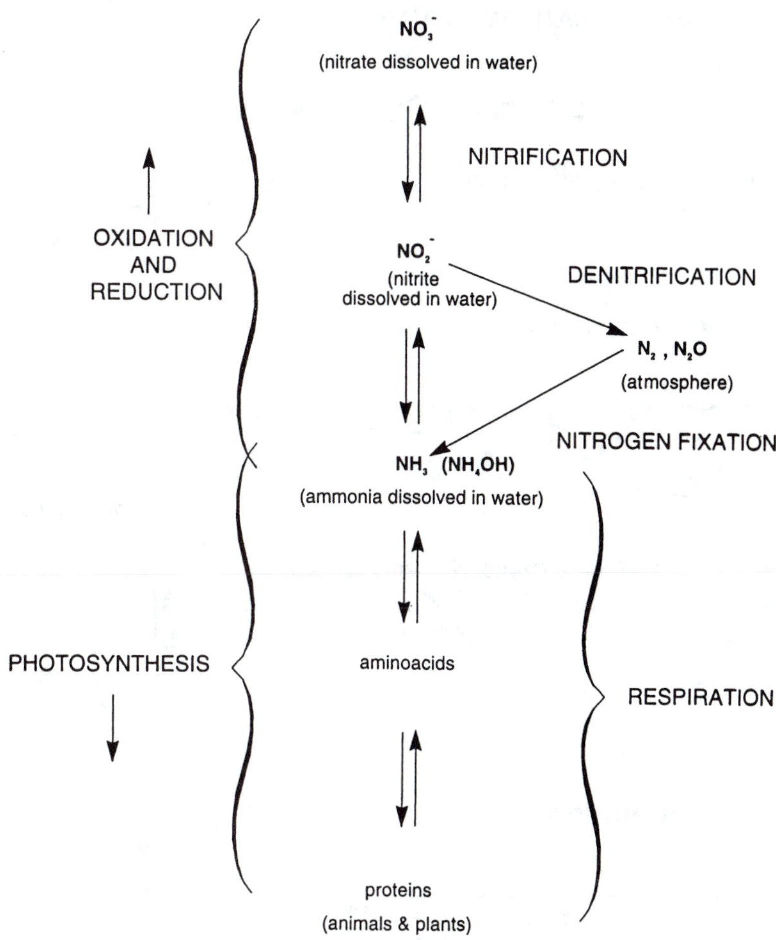

FIGURE 14.10. Transformation of nitrogen in an aquatic system.

per thousand (3.5% or 35,000 parts per million) and freshwater is usually less than 3000 parts per million. Estuaries are dynamic areas where sedimentation and erosion are occurring.

A. PHYSICAL INTERACTION OF FRESH- AND SEAWATER IN ESTUARIES

Estuaries are mixing zones for freshwater runoff with seawater. This can occur in a number of different ways. The density of freshwater is about 1.00, while the density of seawater is about 1.03. Since freshwater is slightly less dense that seawater, freshwater tends to float on top of seawater when it flows down from the land and enters an estuary. If the river flows are strong and dominant over the tidal flows, there will be little mixing caused by the tides or by turbulent flow resulting from an irregular bottom profile. In this situation, the estuary is normally classified as **highly stratified**; see Figure 14.11. If a vertical profile is taken at any point in the estuary, it exhibits a distinctive pattern of a homogeneous upper layer, a sharp rise

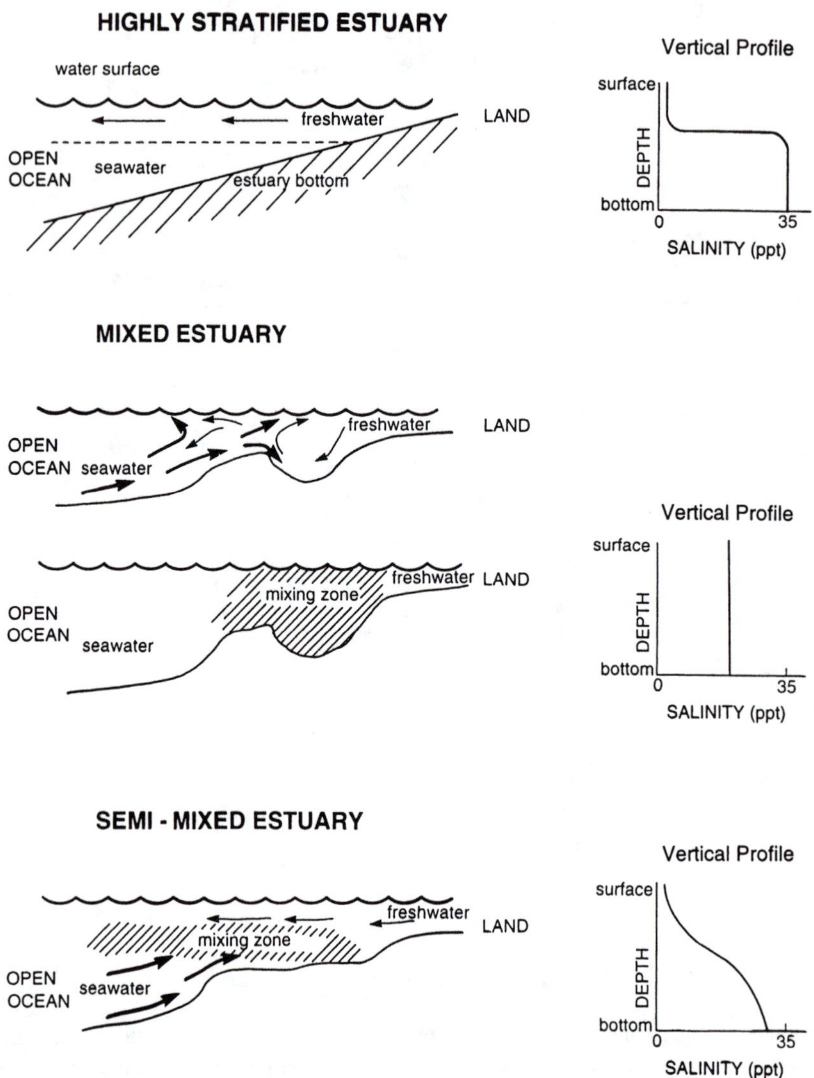

FIGURE 14.11. Different types of estuaries based on physical characteristics.

in salinity where the fresh- and seawater meet, and a homogenous bottom layer, as shown in the vertical profile in the Figure 14.11.

In another type of situation, the river flow can be small compared with the tidal flow and the bottom profile of the estuary can be irregular; this leads to active mixing of freshwater from the land as it enters the estuary. Thus, when a vertical profile is taken at any point in the estuary, there is no difference between upper and lower layers in the salinity profile. This type of estuary is referred to as a **mixed** estuary and is illustrated in Figure 14.11. In many situations, estuaries exhibit characteristics that change over time, depending on rainfall patterns. Thus, an estuary can vary from being **highly stratified** during periods or seasons of heavy rainfall to being

mixed during periods or seasons of low rainfall. There are also situations where estuaries exhibit characteristics somewhere between these two extremes. These are referred to as **semi-mixed** estuaries and exhibit a vertical profile of salinity that is somewhere between the **highly stratified** and **mixed**, as illustrated in Figure 14.11.

The chemical composition of estuarine waters exhibits characteristics that reflect both freshwater runoff and seawater. The variations in salinity that can occur in estuaries were previously described; as a general rule, river waters are relatively high in nitrogen, phosphorus and some trace metals compared to seawater. On the other hand, seawater often contains an array of trace metals present in only very small concentrations in freshwater. Not unexpectedly, one relationship that is common in estuaries is that the nitrate concentration is negatively related to salinity. This also often happens with phosphorus.

The conditions in estuaries can favor the development of a high biomass of plants and animals due to the presence of relatively high concentrations of nitrogen and phosphorus, together with a wide array of other essential trace elements. At the same time, the conditions do not favor a high number of species. This is due to the unstable nature of conditions in most estuaries. The species present have to be tolerant to tidal submergence and exposure, variable depth and coverage of water, variable salinity, and variations in other chemical factors such as pH and trace metal concentrations. Thus, most estuaries exhibit a lower number of species than either the comparable adjacent open ocean area or the comparable freshwater area.

VI. OCEANIC SYSTEMS

The study of the oceans referred as **oceanography** began in a scientific sense in the early 1800s. The first international maritime conference was held in Brussels in 1853. In this period, there were wide ranging studies of the total salt content of the oceans, with the first book on ocean chemistry being published in 1865. In addition, there was considerable interest at this time in ocean biology and ocean currents. While the voyages of Captain Cook could be considered to be among the first scientific voyages in oceanography, the first expedition devoted specifically to oceanography was organized by the Royal Society of London and the ship *Challenger* traveled the oceans of the world from 1872 to 1876 carrying out scientific research. The reports from this voyage totaled 50 large volumes and are still a source of reference for scientists today. Since that time there have been many oceanographic expeditions. The amount of scientific knowledge available on the oceans is quite substantial.

A. SALINITY OF SEAWATER

The salinity of the oceans at the surface varies from about 30 to 40 parts per thousand total soluble solids. There are variations above and below these figures, depending on local conditions. The salinity of the oceans is affected by such factors as rainfall, evaporation and removal of pure water by the freezing of seawater at the Poles. As seawater freezes, crystals of pure ice are formed and the remaining liquid has a correspondingly higher salinity value.

The average ocean salinity at the surface varies with latitude, as shown in Figure 14.12. The relatively low salinities around the equator result from high precipitation

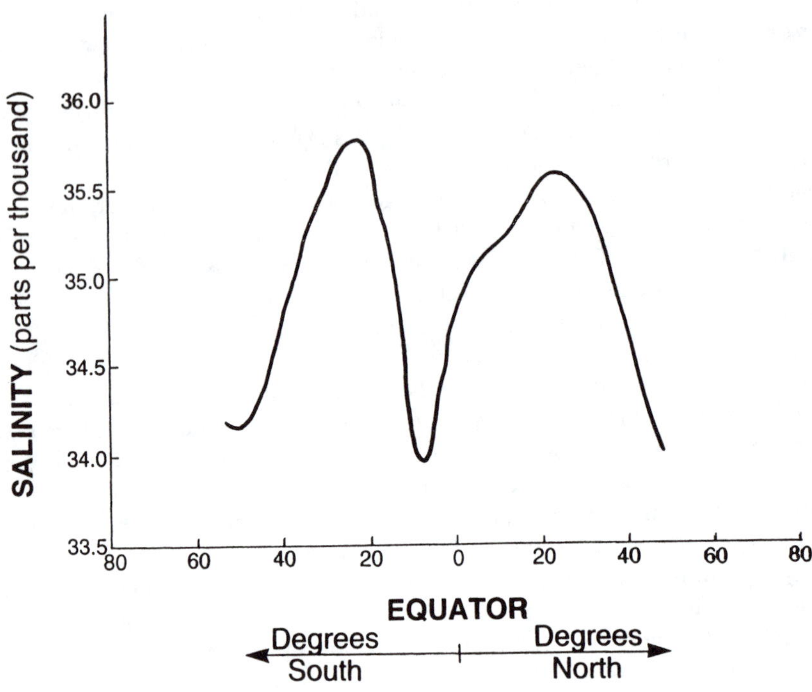

FIGURE 14.12. Variation in the salinity of surface waters in the oceans with altitude.

with correspondingly large runoff from the land and relatively light winds, giving low evaporation. The salinity rises, reaching a maximum at latitudes between 20° and 30°, which are the zones of the **trade winds**, giving high evaporation; also, in this region, there is less rainfall. At higher latitudes, the evaporation falls due to the lower seawater temperatures, and precipitation has a tendency to increase, resulting in a fall in the salinity. Increases occur toward the Poles due to the freezing effect mentioned previously.

B. VERTICAL DEPTH PROFILES OF SEAWATER COMPOSITION

The salinity, in all oceans, has a small range (from 34.6 to 34.9 parts per thousand) at depths greater than 4000 m. Somewhat similarly, temperature in all the oceans at depths greater than 2000 m is about 4°C, irrespective of location. This suggests that when waters in the ocean reach the maximum density, they accumulate in the bottom waters and remain there for an indefinite period of time.

Seawater contains the following positive ions in order of concentration: Na^+, Mg^{2+}, Ca^{2+}, K^+, Sr and the following negative ions in order of concentration Cl^-, SO_4^{2-}, HCO_3^-, Br^-, F^- and also boron as H_3BO_3. These substances account for about 99.9% of the dissolved solids present in seawater. The remarkable thing about these 11 constituents of seawater is that although the total concentration may vary, the ratio of one to another is very close to constant in all situations. In oceanography, a common measure of salinity is the **chlorinity**. This is a measure of the chloride ion content of seawater and can be carried out accurately by chemical titration techniques. **Chlorinity** is related to salinity by the following equation.

$$\text{Salinity} = 1.805\ \text{Chlorinity} + 0.03$$

There is a wide range of minor elements present in seawater that can be detected in concentrations in the parts per million range. These include zinc, copper, iron, chromium, molybdenum, silica, aluminium and many others. The biologically important minor elements, nitrogen and phosphorus, are present in seawater in relatively low concentrations. Nitrogen can be present in the inorganic forms of nitrate, nitrite and ammonium ions, as well as organic compounds, and amounts to a total of about 0.5 mg L^{-1} in all ocean waters. Phosphorus is present as orthophosphate (PO_4^{3-}) and related ions HPO_4^{2-}, $H_2PO_4^-$ and H_3PO_4. The concentration of total phosphorus amounts to 0.07 mg L^{-1} on average within all ocean water. Importantly, this concentration is the average concentration throughout the water mass within the oceans; but at the surface, the concentrations are much lower. In large areas of the oceans, the concentrations of total nitrogen are somewhere in the region of 0.1 mg L^{-1} and phosphorus is about 0.001 mg L^{-1}. However, concentration increases with depth with both of these nutrients so that at depths greater than about 1000 m, the concentrations are in the range of 0.2 to 0.5 mg L^{-1} of total nitrogen and about 0.04 mg L^{-1} of total phosphorus. This is illustrated by the diagram in Figure 14.13.

As a general rule, nutrient-rich waters have total nitrogen concentrations in excess of 1.0 mg L^{-1} and total phosphorus concentrations in excess of 0.004 mg L^{-1}; thus, the surface waters of the oceans are deficient in nitrogen and phosphorus, while the waters at depths greater than 1000 m are relatively rich in these nutrients. This means that large areas of the oceans will be deficient in life since photosynthesis occurs in the upper layers of the water where light is abundant but nutrients are in low concentrations. So, productivity in the oceans is usually around land masses where nutrient-rich runoff waters enter the surface waters, allowing the growth of marine plants and subsequent growth of other marine organisms.

Another phenomenon increasing the concentration of nutrients in the surface waters of the oceans is called **upwelling.** Upwelling is the movement of the nutrient-rich bottom waters to the surface where photosynthesis and primary production can occur. Upwelling can be caused by persistent winds blowing offshore and so moving the surface waters offshore and causing the bottom waters to come to the surface near the land mass. This process occurs off the west coasts of the Americas and Africa. Upwelling can also be caused by deep ocean currents moving to the surface due to the bottom topography. This occurs in the north Atlantic due to the movement of the Gulf Stream, and in the north Pacific due to a similar current in that ocean. Upwelling can also be caused by the mixing and turbulence where deep ocean currents meet. Areas where this occurs are in the central Pacific, Atlantic and Indian Oceans. It is interesting to note that the regions of upwelling throughout the world are generally regions where extensive fisheries have been established; for example, the north Pacific, north Atlantic, west African and west American zones. There is extensive upwelling in the Antarctic but in this case, results in the growth of krill that is not yet extensively harvested in a fishery.

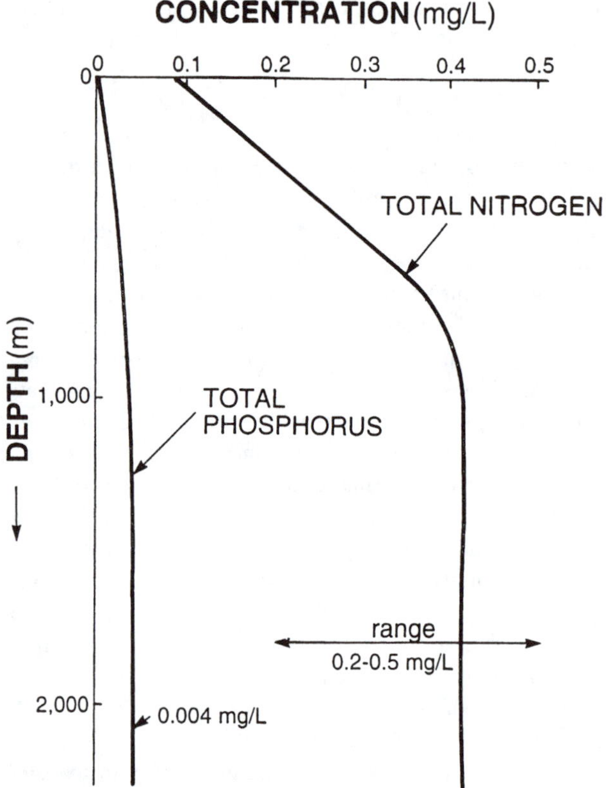

FIGURE 14.13. Change in concentration of total nitrogen and total phosphorus with depth generally observed in the oceans.

VII. KEY POINTS

1. Due to its relatively high polarity, water has a boiling point, a melting point and a vapor pressure higher than comparable compounds of similar molecular weight. The polarity of water also results in this substance being a good solvent for ionic salts (such as sodium chloride) and a poor solvent for lipids.

2. The maximum density of water is at 4°C, which means that when ice forms, this will float on the surface, leaving the warmer underlying waters in a liquid state.

3. Atmospheric gases, such as oxygen, nitrogen, methane and carbon dioxide, are all dissolved in natural water bodies and have an important influence on chemical and biological processes. The behavior of gases in natural water bodies is conveniently described by the Henry's law constant (H), which is defined as:

$$H = P/S$$

The Henry's law constant is useful for calculating concentrations of gases in water and the atmosphere.

4. The solubility of the atmospheric gases varies substantially with temperature. In the case of carbon dioxide, the presence of salts and other substances can affect the dissociation of carbonic acid formed by carbon dioxide dissolving in water.

5. There are a wide range of organic compounds in natural waters; but as a general rule, low molecular weight organic substances such as monosaccharides and amino acids are present in low concentrations due to their limited persistence. An important group of organic substances in natural waters are the humic substances, which are the residues left after the degradation of plant matter, particularly lignin. These substances have strong brown to black coloration, are of high molecular weight (ranging from hundreds to thousands) and contain a range of phenolic groups. Many metals, including the essential trace metals, are strongly bonded by chelation into the humic substances. This may have an impact on the growth of plants and animals in aquatic ecosystems.

6. The carbon cycle in natural water bodies starts with carbon dioxide in the atmosphere, which is converted by photosynthesis into plant material, which is consumed by other organisms in the system. Respiration by plants, animals and microorganisms results in the release of carbon dioxide back into the water mass and subsequently to the atmosphere, thereby completing the cycle. The nitrogen cycle in natural water bodies commences with inorganic nitrogen forms, principally the nitrate ion within the water mass. Nitrate ion is taken up by plants during photosynthesis converted into ammonia and amino acids and finally into plant matter. The consumption of plant matter by animals and microorganisms results in the conversion of the proteinaceous food material into amino acids, ammonia and subsequently nitrite and nitrate, thereby completing the cycle.

8. Estuarine systems are semi-enclosed coastal bodies of water that have a free connection with the open ocean and have waters measurably diluted with freshwater from land drainage. The salinity of water in estuaries usually ranges from 0 to 35 parts per thousand.

9. Estuaries can be classified on the basis of the interaction between fresh- and seawater into three types. First, the **highly stratified** estuary, in which freshwater from land runoff floats over the surface of seawater, creating a vertical profile of salinity in which there is a sharp change in salinity from freshwater to seawater. The second type is the **mixed** estuary, in which freshwater from the mainland mixes strongly with seawater to create a mixing zone in which the salinity is the same from surface to bottom. The third type is the **semi-mixed** estuary, which lies somewhere between the highly stratified and fully mixed estuary, with a vertical profile having intermediate characteristics.

10. The salinity of the oceans varies with latitude. At the equator, the salinity is relatively low at an average value of about 34 parts per thousand due to light winds and high rainfall. At higher latitudes the salinity rises due to the onset of the Trade Winds and lower rainfall reaching a maximum at latitudes of about 25°. The salinity then falls due to rainfall exceeding evaporation.

11. The composition of seawater may vary in total salinity throughout the oceans but the ratio of the 11 major constituents accounting for about 99.9% of the total dissolved solids present remains remarkably constant.

12. The composition of seawater shows quite distinctive vertical patterns, particularly the nutrients, nitrogen and phosphorus, which are of major biological importance. In large areas of surface waters nitrogen exhibits total concentrations of less than 0.1 mg L^{-1} while phosphorus exhibits concentrations of about 0.001 mg L^{-1}. With depth the concentration of both P & N increases to a maximum below 1000 m depth where nitrogen is between 0.2 and 0.5 mg L^{-1} and phosphorus is at about 0.04 mg L^{-1}. These nutrient-rich lower waters can reach the surface by upwelling. These processes are caused by persistent offshore winds, deep ocean currents moving to the surface due to bottom topography and the turbulent mixing of deep ocean currents. Where nutrient rich bottom waters reach the surface the productivity of plant matter and related animals is markedly increased.

REFERENCES

Manahan, S.E., *Environmental Chemistry,* 5th ed., CRC Press, Boca Raton, FL, 1991.
Meadows, P.S. and Campbell, J.I., *An Introduction to Marine Science,* Blackie and Son, Glasgow, 1978.
Wetzel, R.G., *Limnology,* Saunders College Publishing, Philadelphia, 1975.

CHAPTER 14 PROBLEMS

1. Nitrogen gas in the atmosphere will dissolve in natural water bodies. Using the Henry's law constant in Table 14.1, calculate the concentration that would be expected in pure water at 25°C.

2. Carbon dioxide dissolves in water to form carbonic acid. Calculate the pH of pure water that is in equilibrium with carbon dioxide in the atmosphere at 25°C. Dry air contains 0.0314% of CO_2 by volume; the vapor pressure of water at 25°C is 0.03 atm, Henry's law constant for CO_2 is 29 atm mol^{-1} L^{-1}; and the dissociation constant in the carbonic acid-bicarbonate system is 4.45×10^{-7}.

3. Fulvic acid can complex metal ions in solution and render them biologically less active. Using the structure in Figure 14.6, illustrate ways in which metals can be complexed into the structure.

4. Methane and carbon dioxide can be produced by microbial action on carbohydrates under anaerobic conditions. Write a balanced equation for this process.

5. Explain why the surface waters of the oceans are depleted in N and P and will remain indefinitely in this state.

CHAPTER 14 SOLUTIONS

1. The atmosphere contains 78% N by volume and allowing for 0.03 atm of water vapor at 25° the partial pressure of N_2 is 0.804 atm.

   From Henry's Law:

$$S = P/H$$

$$= \frac{0.804 \text{ atm}}{1610 \text{ atm mole}^{-1} \text{ L}^{-1}}$$

   From Table 14.1 the value for N_2 is 1610 atm mole^{-1} L^{-1}

$$= 4.99 \times 10^{-4} \text{ mole L}^{-1}$$

   Thus, concentration in pure water = 14.0 mg L^{-1}
   This is in reasonable agreement with the figure reported from direct measurement in Table 14.1 of 13.4 mg L^{-1}.

2. Dry air contains 0.0314% CO_2 by volume; the vapor pressure of water at 25°C is 0.0313 atm; Henry's law constant for CO_2 is 29 atm mol^{-1}; and the dissociation constant for the carbonic acid-bicarbonate system is 4.45×10^{-7}.

$$\text{Partial pressure of air} = (1.000 - 0.03) \text{ atm} = 0.97 \text{ atm}$$

$$\text{Partial pressure of } CO_2 = (0.97 \times 0.0314 \times 10^{-2}) \text{ atm} = 0.0305 \times 10^{-2} \text{ atm}$$

$$\text{Since } S = P/H$$

$$\text{the } CO_2 \text{ concentration in water} = \frac{0.0305 \times 10^{-2} \text{ atm}}{29 \text{ atm mol}^{-1} \text{ L}^{-1}} = 1.05 \times 10^{-5} \text{ mole L}^{-1}$$

$$H_2CO_3 \Leftrightarrow H^+ + HCO_3^- \text{ and } k = 4.45 \times 10^{-7}$$

thus,

$$4.45 \times 10^{-7} = \frac{[H^+][HCO_3^-]}{[H_2CO_3]}$$

$$= \frac{[H^+]^2}{[H_2CO_3]}$$

thus, $$[H^+]^2 = (4.45 \times 10^{-7})\,[H_2CO_3]$$

Since the molar concentration of CO_2 in the water is the same as the molar concentration of H_2CO_3, then,

$$\begin{aligned}[H^+]^2 &= (4.45 \times 10^{-7}) \times (1.05 \times 10^{-5})\\ &= 4.67 \times 10^{-12}\end{aligned}$$

Thus, $$[H^+] = 2.16 \times 10^{-6}$$

$$pH = 5.66$$

This means that atmospheric carbon dioxide in equilibrium with pure water will have a pH of 5.66.

3. Ways in which metal ions can be complexed into fulvic acid include:

   (a) Reaction with acid groups

$$\text{R-COOH} \rightarrow \text{R-COOM}$$

   (b) Chelation

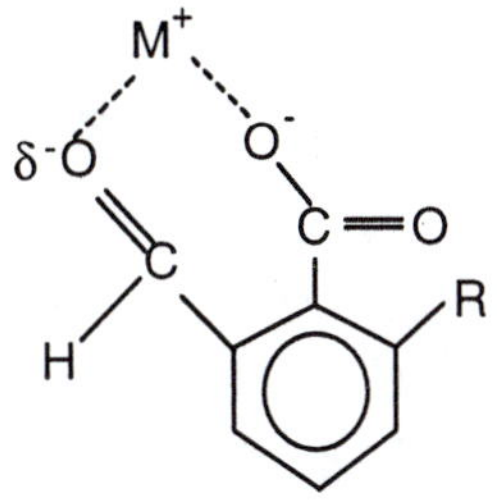

   (c) Chelation

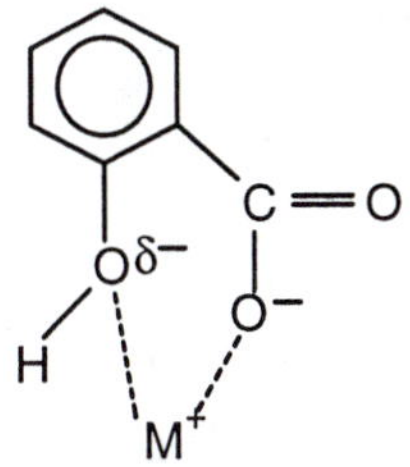

4. The equation for anaerobic production of methane and carbon dioxide is:

$$6\ CH_2O \rightarrow 3\ CH_4 + 3\ CO_2$$

5. Photosynthesis results in the removal of N and P from the surface waters in the photic zone and the formation of phytoplankton cells. On death, these cells

fall downward into the zone below the photic zone, taking the N and P with them; there is no corresponding way in which surface waters are renewed with nutrient. Thus, should any nutrients appear in the photic zone, they will be removed by this process. This means the surface waters of the oceans will remain in a continually depleted state.

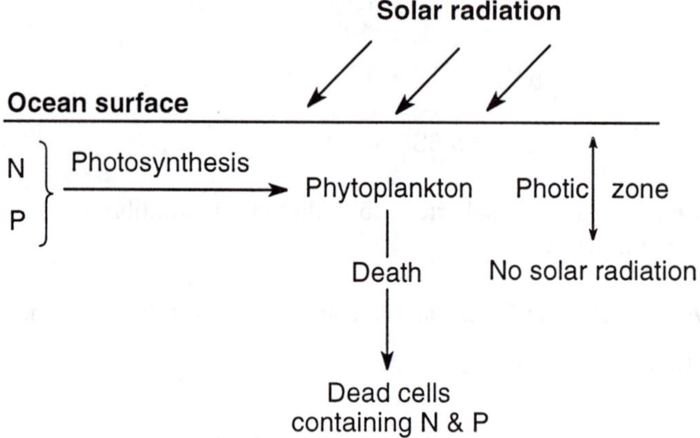

Chapter 15

WATER POLLUTION

I. INTRODUCTION

Surprisingly, the major water pollutants produced by human society are substances that have comparatively little direct harmful effect. These are relatively nontoxic animal and vegetable wastes that are often discharged into streams and inshore marine areas. In certain limited quantities, these materials can be beneficial in some bodies of water by releasing nutrients in the form of nitrogen and phosphorus salts that stimulate plant and animal growth. However, in excessive quantities, their secondary effects are usually extremely harmful. These wastes are rich in organic carbon and are commonly discharged into waterways in the form of sewage, food processing wastes from the fruit, meat and dairy industries, as well as wastes from paper manufacturing and a variety of other industries.

The organic carbon in these wastes usually exists in the form of carbohydrates, proteins, fats, humic substances, surfactants and a range of derived and other substances. Organic wastes are produced in large quantities in all urban population centers, but primary, secondary and tertiary wastewater treatment can reduce the effects of discharges. Organic pollution can also result from urban stormwater runoff, which produces nonpoint discharges to waterways. In some situations, non-urban catchments can yield high organic inputs where certain types of activities, such as logging, are undertaken.

On degradation by microorganisms in waterways, organic wastes can stimulate microorganism respiration, which consumes oxygen and reduces the dissolved oxygen in the ambient water. In addition, the degradation process releases nitrogen and phosphorus compounds, which stimulate plant growth, leading to nutrient enrichment and eutrophication. These two broad areas of environmental impact — dissolved oxygen reduction and nutrient enrichment — are discussed below.

II. DEOXYGENATING SUBSTANCES

A. BACKGROUND

Photosynthesis and respiration are the fundamental chemical processes supporting life on Earth. These are discussed in detail in Chapter 12. The chemical processes involved can be simply expressed as:

$$\underset{\text{Carbon dioxide from the atmosphere}}{6CO_2 + 6H_2O} \;\underset{\text{Respiration}}{\overset{\text{Photosynthesis}}{\Leftrightarrow}}\; \underset{\text{Plant matter}}{6(CH_2O)} + \underset{\text{Atmospheric oxygen}}{6O_2}$$

These processes involve carbon dioxide and oxygen in the Earth's atmosphere which, in aquatic ecosystems, are dissolved in the water mass. The basic transformations of carbon in natural aquatic ecosystems were previously shown in Figure 14.9. Carbon from dissolved carbon dioxide is incorporated by photosynthesis into the plant biomass of the ecosystem. Consumption of this plant biomass by animals and finally the decomposition of body tissues after death by microorganisms results in respiration. Both of these processes result in the formation of carbon dioxide, which is returned to the water and ultimately the atmosphere.

B. INFLUENCE OF CHEMICAL PROCESSES ON DISSOLVED OXYGEN

If substances rich in organic carbon are added to the system, some of the pathways of carbon, shown in Figure 14.9, are increased in magnitude and also some of the pools of organic carbon are increased in size. This results in an increase in respiration, mainly through the respiration of microorganisms, giving rise to increased amounts of carbon dioxide and methane (which is produced by anaerobic respiration). Importantly, for aquatic ecosystems, the transformations of carbon have a strong influence on the related transformations of oxygen, as shown in Figure 15.1. The oxygen needed for respiration is obtained from the dissolved oxygen in the water mass. Usually, this oxygen is substantially derived from solubilization from the atmosphere and converted into carbon dioxide by respiration, which is then discharged into the water mass and ultimately to the atmosphere. Increased respiration due to an increased amount of organic carbon in the water results in changes in the magnitude of the pathways and changes in the pools of oxygen involved, as shown in Figure 15.1. Most importantly, there is a demand on the reservoir of dissolved oxygen in the water mass that is comparatively small since oxygen has limited solubility in water. The solubility of oxygen in water usually ranges from about 6 to 14 mg L^{-1}. Therefore, substantial reductions in dissolved oxygen can occur that have significant implications for aquatic organisms.

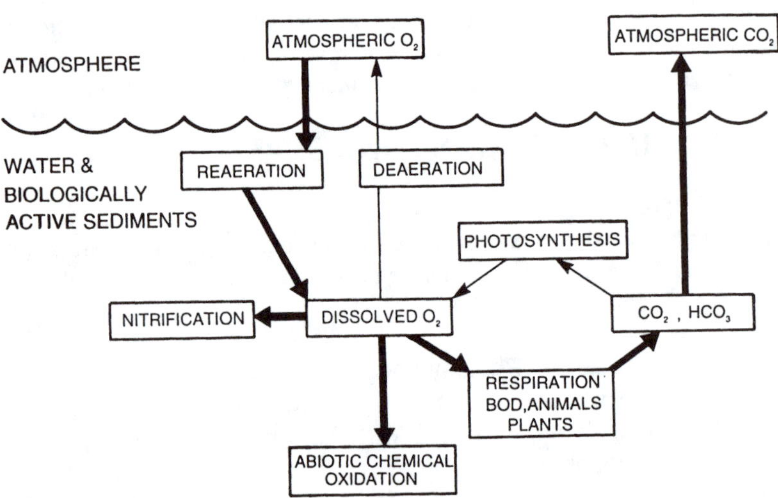

FIGURE 15.1. Oxygen transformations in aquatic systems, with processes increased or accelerated by organic discharges indicated by heavy lines.

In waters high in dissolved oxygen, i.e., having dissolved oxygen levels approaching 100% saturation, **aerobic respiration** occurs and is mediated by aerobic microorganisms. The chemical reaction involved can be simply expressed as:

$$6\ (CH_2O) + 6\ O_2 \rightarrow 6\ CO_2 + 6\ H_2O$$
Carbohydrate

In most aquatic bodies, communities of microorganisms exist that have a wide range of capabilities to degrade different types of organic matter under different conditions. In the absence of oxygen, anoxic or hypoxic conditions develop, which means that the dissolved oxygen has been completely removed. In this situation anaerobic microorganisms take over the degradation and decomposition of organic matter. The **anaerobic respiration** reaction can be simply expressed as:

$$6\ (CH_2O) \rightarrow 3\ CH_4 + 3\ CO_2$$
Carbohydrate

This process makes no demand on the dissolved oxygen present in the water mass, but it is very important for the natural removal of oxygen-demanding substances in waterways. The methane and carbon dioxide produced are released into the water mass then to the atmosphere, resulting in the removal of organic carbon and oxygen demand from the system. This process occurs in swamps, bottom muds enriched with organic matter, as well as bodies suffering from pollution by organic wastes.

Anaerobic respiration with sulfur- and nitrogen-containing organic substances gives rise to hydrogen sulfide and ammonia, respectively. Ammonia in the presence of oxygen is readily oxidized to nitrate ion. The nitrite ion is formed under suitable oxidation and reduction conditions, but is less commonly the end-product of nitrogen metabolism. The formation of the nitrate ion from organic matter is described as the **nitrification** reaction and is mediated by a variety of microorganisms. A series of reactions occur that, make an important demand on dissolved oxygen in natural bodies, especially where sewage contamination containing suitable microorganisms and ammonium ion occur.

C. MEASUREMENT OF THE OXYGEN-REDUCING CAPACITY OF A WASTEWATER — THE BOD TEST

The oxygen-demanding processes affecting a water body can be measured by the **biochemical oxygen demand (BOD) test**. This test measures the loss of dissolved oxygen in a water sample incubated over a period of 5 days under standard conditions, often with seed microorganisms added. The BOD of wastewaters can range up to hundreds of milligrams per liter. In addition, the BOD test can be used as a measure of the quality of water in waterways. In general, clean water has a BOD of less than 1 mg L^{-1} and seriously polluted water contains greater than 10 mg L^{-1}. The nitrification reaction occurs in what is described as the second stage of the BOD test at time periods greater than 5 days. The precision of the BOD test is about ±17% and may not satisfactorily reflect the actual conditions existing in a natural body. The test conditions may differ from those existing in the environment in factors such as temperature, microorganisms present and so on.

D. KINETICS OF BOD REDUCTION

The kinetics of BOD changes in waterways has been subject to intense investigation. One of the principles governing BOD in natural waterways was first formulated by Phelps as "the rate of biochemical oxidation of organic matter is proportional to the remaining concentrations of unoxidized substance measured in terms of oxidizability." Thus, the oxidation of organic matter follows first order decay reaction kinetics. From this, the following series of mathematical expressions can be derived are described in more detail in Chapter 3, Section IV.

$$-\frac{dL}{dt} \propto L$$

thus, $$-\frac{dL}{dt} = K_l L$$

which can be integrated to $$\ln(L/L_0) = -K_1 t$$

or $$\log(L/L_0) = -0.434\ K_1 t = k_1 t$$

and $$L/L_0 = 10^{-k_1 t}$$

where L_0 is the total BOD debt at time zero; L is the BOD debt at time t; t the time period since time zero; and K_l and k_l are the empirical decay rate constants using base e and base 10, respectively ($0.434\ K_l = k_l$).

By actual measurements of BOD in waterways, it has been found that in general at 20°C, $k_l = 0.1\ \text{day}^{-1}$, if the time period is measured in days. Since first-order kinetics apply, it can be shown that $t_{1/2} = 0.301/k_l$ and thus the half-life normally exhibited by BOD-demanding substances in natural waterways is equal to about 3 days. Also, from the mathematical treatment outlined above, a plot of log (percentage remaining BOD) versus time will give a straight line (see Figure 15.2).

The BOD decay rate coefficient (k_l) varies with temperature. It can be shown that the deviation of this constant from the 20°C value can be derived from the equation:

$$k_T = k_{20°C} \times 1.047^{(T-20)}$$

E. THE DISSOLVED OXYGEN "SAG"

One of the most importance sources of dissolved oxygen in waterways is oxygen in the atmosphere that dissolves in the water mass at the water surface. The oxygen content of the atmosphere is about 20%. Reductions in dissolved oxygen present in

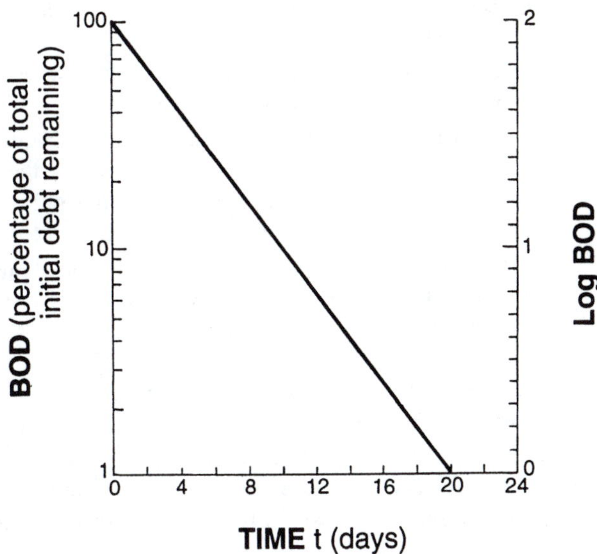

FIGURE 15.2. Normal depletion of organic BOD debt at 20°C (k_1 = 0.1 day^{-1}) in an aquatic area.

water can be caused by the BOD present in the water, leading to a deficit below the saturation level. The rate of reaeration is proportional to the oxygen deficit in the water mass. Thus,

$$\frac{dC}{dt} = k_2 D$$

where C is the concentration of oxygen, k_2 is the reaeration coefficient and D is the oxygen deficit, i.e., $C_{saturation} - C_{actual}$.

The rate of loss of BOD follows first-order kinetics; thus, the rate of loss of oxygen is proportional to the BOD present at the time involved; thus,

$$\frac{dD}{dt} = k_1 L$$

where D is the oxygen deficit, L is the BOD debt or concentration of BOD, and k_1 is the BOD decay rate coefficient. By combining the two equations above, the actual oxygen deficit can be obtained by subtracting the uptake from the deficit due to BOD. Thus,

$$\frac{dD}{dt} = k_1 L - k_2 D$$

These two processes are illustrated diagrammatically in Figure 15.3. Curve "e" is the deoxygenation curve that would represent dissolved oxygen values if no reaeration was to occur. However, as soon as an oxygen deficit occurs, which is soon after time zero, reaeration commences and yields curve "d." The reaeration increases as the oxygen deficit increases. Curve "d" represents the oxygen input to the water as a result of reoxygenation. The summation of these curves gives the actual oxygen deficit profile, the dissolved oxygen "sag" (curve "a"). An expression for this curve can be mathematically obtained by integrating the equation above.

The production of "an oxygen sag" in a river is illustrated by the data and plot in Figure 15.4. The dissolved oxygen concentrations at various points along a river receiving wastewaters rich in organic matter was measured and plotted. The "oxygen sag" is related to the position of the discharges. Downstream of the discharges, natural oxygen replenishment occurs by exchange with air at the surface, leading to a rise in the dissolved oxygen levels present in the water.

In the discussion above, only oxygen demand due to BOD is considered, which is due to organic matter present as suspended or dissolved matter in the water mass. However, organic matter can be sedimented out of the water mass to form bottom sediments rich in organic matter. These organic-rich layers may produce anaerobic degradation in the bottom waters, but above this, aerobic respiration can occur. This can lead to an oxygen demand not measured by BOD that can exceed the BOD in some situations.

F. SEASONAL VARIATIONS AND VERTICAL PROFILES OF DISSOLVED OXYGEN

Turbulent water conditions give well-mixed waters in which vertical profiles of dissolved oxygen, and other characteristics, are constant from the surface to the bottom waters. However, thermal stratification, due to solar heating of the surface waters, leads to isolation of the bottom waters. This was described in Chapter 14, The Chemistry of Natural Waters, and the effect was illustrated in Figure 14.2. If the hypolimnion and bottom sediments are enriched with organic matter, the bottom waters may become depleted in dissolved oxygen, while the surface waters remain unaffected. In this manner, a vertical profile showing considerable variation in dissolved oxygen concentration can occur. Many environmental factors vary seasonally. For example, the incidence of sunlight has an important effect on photosynthesis and resultant dissolved oxygen concentrations in many aquatic areas.

In temperate areas, vertical stratification usually occurs in a seasonal pattern with stratification in the summer, and mixing, or overturn, in the winter. Corresponding to this, there is usually a seasonal pattern of dissolved oxygen variation in bottom waters that is common in lakes and relatively small bodies of water, but can also occur on the continental shelves of the oceans.

Figure 15.5 shows the seasonal effect of stratification on dissolved oxygen concentration in waters in the New York Bight located on the continental shelf of the northeastern United States. The bottom waters and sediments have been enriched by phytoplankton decay, river discharge containing oxygen-demanding substances and nutrients, and dumping of sewage sludge in the area. In 1976, there was an early arrival of the relatively warm conditions of spring, leading to elevated surface temperatures that coincided with high discharge of water from the adjacent Hudson and Delaware Rivers. This led to a prolonged period of stratification in conjunction with the presence of relatively large quantities of organic matter, causing severe

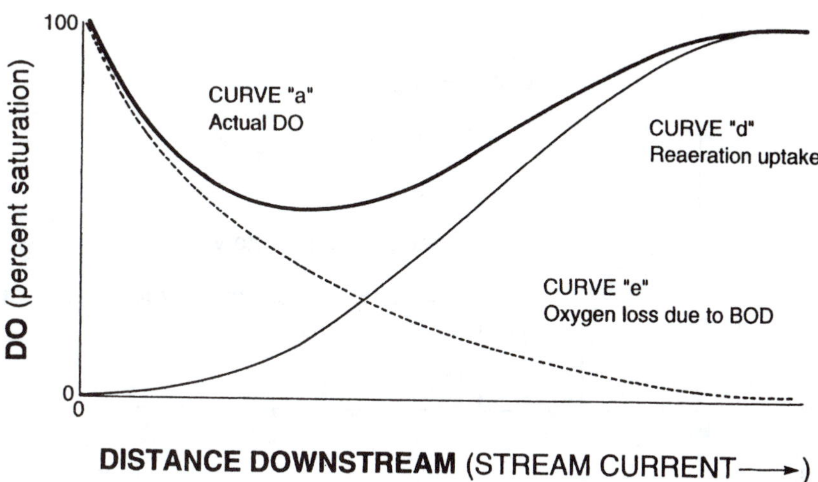

FIGURE 15.3. Deoxygenation, reaeration and oxygen sag curves in a stream as a result of an organic discharge at point zero giving a large increase in BOD.

depletion of dissolved oxygen in the bottom waters and a widespread mortality of marine organisms.

G. ANAEROBIC PROCESSES: PRODUCTION OF POISONOUS GASES FROM ANIMAL AND VEGETABLE WASTES

A more advanced state of pollution occurs when the dissolved oxygen content of natural waters drops to zero and microorganisms that do not require free oxygen become active. This is an **anaerobic** system. When oxygen is present, carbon dioxide and water are the major substances produced by microorganisms, but the major products of the anaerobic process are methane and carbon dioxide, with lesser amounts of ammonia, hydrogen sulfide and a variety of other organic and inorganic compounds. Some of these compounds, particularly hydrogen sulfide, are poisonous and have a strong nauseous odor. Streams exhibiting such strong odors are almost invariably heavily contaminated.

H. THE EFFECTS OF ORGANIC WASTES

The effects of pollution by organic wastes will depend mainly on the volume and strength of the discharge, and on the volume and flow rate of the receiving water. In a flowing river, a number of different aquatic environments and communities of aquatic organisms are usually created, which are essentially related to dissolved oxygen content and presence of organic matter enrichment. These can be classified into zones and a variety of different types of classification systems has been developed. However, it is convenient to consider a simple zone system related to the discharge point. First, upstream of the discharge point is the **unaffected** zone.

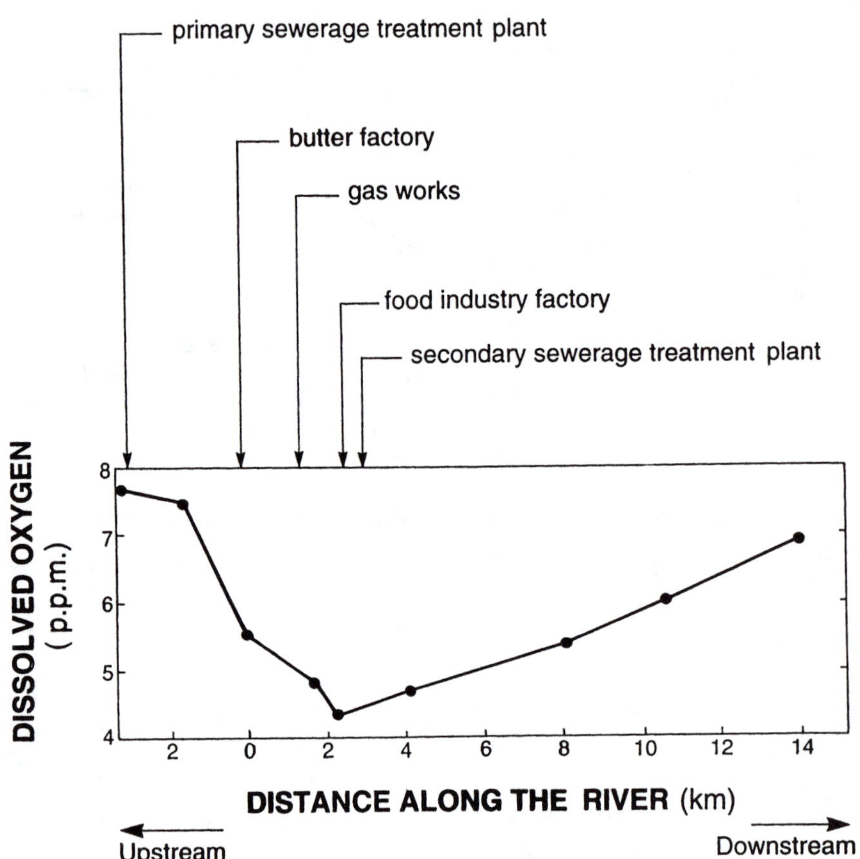

FIGURE 15.4. Dissolved oxygen profile of a river receiving wastewater discharges rich in organic matter.

At the discharge point itself and for a distance below this point, there is a zone of **degradation** that may grade into a zone of **active decomposition** where dissolved oxygen falls to zero and anaerobic processes occur. This zone, which may extend for some kilometers, grades into a **recovery** zone, finally ending in clean water resembling that upstream of the discharge point.

In the unpolluted zone, there will generally exist a large number of species of animals, plants and insects. Some of these will be affected by the introduction of small quantities of organic wastes. For example, turbidity will reduce light penetration and the number of bottom-dwelling photosynthetic plants will in turn diminish. Further additions of wastes may cause depletion of dissolved oxygen and result in the disappearance of most species of fish. Fish are sensitive to reduced oxygen, as is illustrated by the data in Table 15.1. Only a limited number of animal and plant species will survive in the zone of active decomposition. However, some of these may be well adapted to this new kind of environment. Sludge worms, such as *Tubifex*, can survive on trace quantities of dissolved oxygen and actively feed on organic sludge sinking to the bottom. In the recovery zone as the dissolved oxygen rises, the conditions improve and the number of species increases. When recovery is complete, there is a return to the array of species present upstream of the discharge.

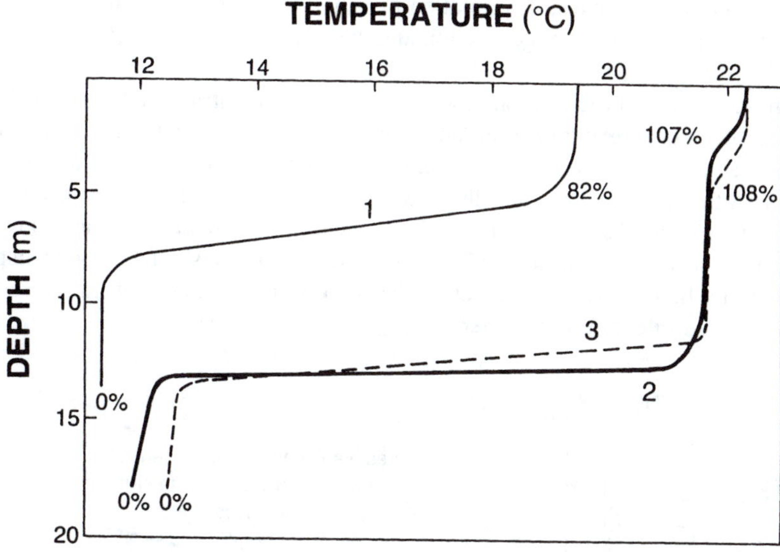

FIGURE 15.5. Temperature profiles and associated dissolved oxygen values (as percentage of the 100% saturation value in water) at three ocean stations in the vicinity of Little Egg Inlet, New Jersey, in July 1976.

TABLE 15.1.
Examples of Limiting Dissolved Oxygen Concentrations for Aquatic Organisms

Organism	Temperature (°C)	Dissolved oxygen concentration (mg L^{-1})
Brown trout (*Salmo trutta*)	6.4–24	1.28–2.9
Coho salmon (*Oncorhynchus Kisutch*)	16–24	1.3–2.0
Rainbow trout (*Salmo gairdnerii*)	11.1–20	1.05–3.7

• Limiting values for continued existence.

III. NUTRIENT ENRICHMENT AND EUTROPHICATION

A. BACKGROUND

The enrichment of aquatic areas with plant nutrients is an important process in aquatic pollution and a significant aspect of this is **eutrophication.** Eutrophication was described by Weber in 1907 when he introduced the terms **oligotrophic**, **mesotrophic** and **eutrophic**. These terms describe the eutrophication process as a sequence from a clear lake to a bog by enrichment with plant nutrients and increased plant growth. The Organization for Economic Cooperation and Development (OECD) has defined eutrophication as "the nutrient enrichment of waters which results in of an array of symptomatic changes among which increased production of algae and macrophytes, deterioration of fisheries, deterioration of water quality and other symptomatic changes are found to be undesirable and interfere with water uses."

In the natural eutrophication process, plant detritus, nutrient salts, silt and so on from a catchment are entrained in runoff water and deposited in the water body over geological time. This leads to nutrient enrichment, sedimentation, infilling and increased biomass. Figure 15.6 illustrates, in general terms, how eutrophication is related to ageing. As well as lakes, dams and enclosed bodies of water, nutrient enrichment also occurs in streams, estuaries, the continental shelf and the open seas. These water bodies may show many of the characteristics of eutrophication and are often referred in eutrophication terms.

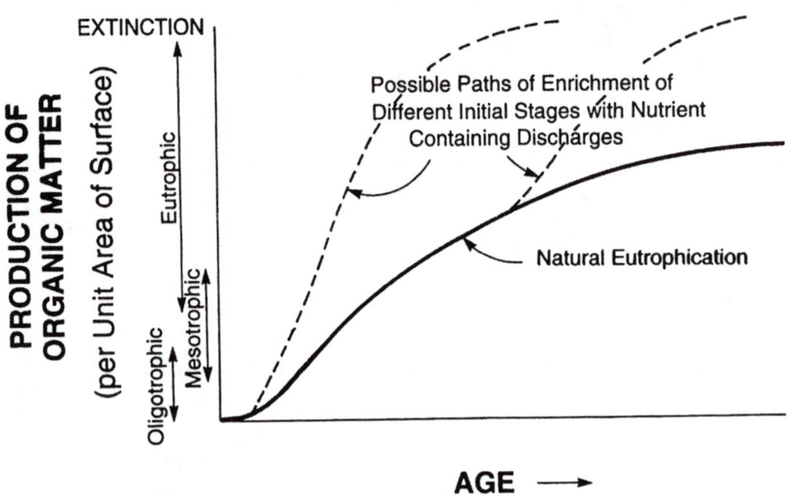

FIGURE 15.6. Hypothetical curves for the course of eutrophication in a water body. The broken lines show possible courses of accelerated eutrophication when enrichment from pollution discharges occurs.

Nutrient enrichment and eutrophication can be greatly accelerated by human activities. In fact, many lakes were shown to be rapidly enriched with nutrients over the last 100 years due to pollution. Discharges, such as domestic sewage, septic tank runoff, some industry wastes, urban runoff, runoff from agriculture and managed forests and animal wastes, contain plant nutrients that often lead to nutrient enrichment and accelerated eutrophication.

Eutrophication can cause quite a number of important problems in water use. An increase in the populations of plants can lead to a decrease in the dissolved oxygen content of the water on plant death and decomposition of the plant detritus by microorganisms. This decreases the suitability of the area as a habitat for many species of fish and other organisms. The increase in turbidity and color that occurs during eutrophication renders the water unsuitable for domestic use or difficult to treat to a suitable standard for this purpose. Odors are also produced by many of the algal growths that create problems in domestic use. **Blooms, pulses,** etc. of aquatic plants become more frequent and, if toxic, lead to the death of a variety of aquatic organisms and also of terrestrial organisms using the water. Floating macrophytes and algal scums can render a water body unsuitable for recreation and water sports and also cause navigation problems.

B. NUTRIENTS AND PLANT GROWTH

If the growth of algal cells is not limited by any environmental or nutrient factor, then population growth occurs according to an exponential function. Exponential growth of the kind outlined above cannot be maintained for very long by algae due to limitations of various kinds. For example, elements such as carbon, nitrogen, hydrogen and oxygen are needed to construct plant tissue. In addition to these elements, sulfur, calcium, silicon, sodium and a variety of other elements are needed to construct vital components present in plants in smaller amounts. Carbon, hydrogen, and oxygen are the major elements in carbohydrates, proteins and fat, which are the principle chemical components of biota. Nitrogen is required in proteins, particularly for the formation of the peptide bond (–CO–NH–). Phosphorus is a component of adenosine di- and triphosphate (ADP and ATP, respectively) and a variety of other substances needed for the transfer of energy in biota. Magnesium and iron are required to biosynthesize chlorophyll and hemoglobin, which are fundamental to the existence of life.

Table 15.2 lists the relative quantities of essential elements that occur in plant tissue. The function of all of these elements in plant processes is not clear at the present time, but they appear to be essential for growth. In most aquatic areas, carbon and oxygen are readily available from carbon dioxide in the atmosphere, and hydrogen and oxygen can also be readily obtained from water. On the other hand, the other elements mentioned above are usually obtained from dissolved salts in the water or sediments. However, these substances are not always available in the quantities required to maintain maximum growth. For example, Table 15.2 shows a comparison of the relative quantities of the elements required for plant growth with their occurrence in river water. This indicates that phosphorus and nitrogen are in comparatively short supply compared with all the other elements, and that phosphorus is likely to be less available than nitrogen. This situation is generally applicable to aquatic areas. Thus, these elements are often growth limiting and an addition of them to a water body will stimulate plant growth. Of course, these data are generalized and, in individual cases, other elements or combinations of elements may be limiting.

C. SOURCES AND LOSSES OF NUTRIENTS

A convenient way to divide discharges of nutrient-containing wastewaters is to categorize them as *point* source and *diffuse* source discharges. Point source discharges are those arising from a specific location, whereas diffuse discharges arise from many dispersed sources e.g., through eroded soil and sediment derived from cultivated land. Some of the major sources and sinks of nutrients are shown in Table 15.3. The discharge of untreated or primary treated sewage into an aquatic area causes deoxygenation and, during this process, nutrient salts are released. Secondary treated sewage has had the biochemical oxygen demand (BOD) substantially removed during treatment but contains the same nutrient salts as would be released

TABLE 15.2
Relative Quantities of Essential Elements in Plant Tissue (Demand) and their Supply in River Water

Element	Demand by plants (%)	Supply by water (%)	Demand by plants/supply water (approx.)
Oxygen	80.5	89	1
Hydrogen	9.7	11	1
Carbon	6.5	0.0012	5,000
Silicon	1.3	0.00065	2,000
Nitrogen	0.7	0.000023	30,000
Calcium	0.4	0.0015	<1,000
Potassium	0.3	0.00023	1,300
Phosphorus	0.08	0.000001	80,000
Magnesium	0.07	0.0004	<1,000
Sulfur	0.06	0.0004	<1,000

TABLE 15.3
Sources and Sinks for the Nitrogen Budget of a Lake

Sources	Sinks
Leachate from leaves and miscellaneous debris	Water outflow
	Groundwater recharge
	Fish harvesting
Agricultural (cropland) and drainage	Weed harvesting
Animal waste runoff	
Marsh drainage	
Runoff from uncultivated and forest land	Volatilization (of NH_3)
Urban storm water runoff	Denitrification
Domestic waste effluent	
Industrial waste effluent	
Natural groundwater	Irreversible sediment deposition of detritus
Subsurface agricultural and urban drainage	
Nitrogen fixation	Sorption of ammonia onto sediments
Sediment leaching	

by untreated sewage. Thus, untreated, primary and secondary treated sewage all contain nutrients that are active in aquatic areas.

The most important nutrients are in the form of ammonia (NH_4^+), nitrate (NO_3^-) and also orthophosphate (PO_4^{-3}) which occur substantially as a result of the degradation of proteins and detergents. The builders used in detergents to enhance the activity of the surfactant are usually polyphosphates. Phosphates occur in crude sewage and survive treatment in a secondary sewerage treatment plant. Phosphates in detergents are a major source of phosphate in sewage and stormwater runoff in agricultural and urban areas. Overflow from septic tanks and leachate from fertilized gardens are major contributors to the nutrient content. In addition, Table 15.3 lists a number of natural sources of nutrient, including leachate from leaves and miscellaneous vegetation debris, animal waste runoff and marsh drainage. Once these nutrients enter a water body, there are a number of losses or sinks that can be identified. These include denitrification, which is the conversion of nitrogen into nitrogen gas that disperses into the atmosphere, and irreversible sediment deposition of detritus-containing nutrients.

D. NUTRIENT TRANSFORMATIONS IN A WATER BODY

The carbon and nitrogen transformations generally occurring in aquatic bodies irrespective of their nutrient status were described in the previous chapter, Chapter 14. The same processes occur in enriched water bodies. However, stratification is probably more significant in nutrient-enriched bodies than in unenriched bodies. The influence of this process on the nitrogen transformations in water bodies is shown in Figure 15.7. Inputs of nutrients can occur from various sources, as shown in Table 15.3. The inorganic nitrogen salts of nitrate and ammonia are readily taken up by plants in photosynthesis. The death of the plants, followed by sedimentation, transfers this organic matter to the **hypolimnion** where microbiological degradation occurs, resulting in the formation of carbon dioxide, orthophosphate, ammonia, nitrite and nitrate (under different conditions) with the consumption of dissolved oxygen.

In an oligotrophic situation, these processes may not result in any marked differences between the hypolimnion and the epilimnion since all processes occur at a low rate. However, in eutrophic conditions, the hypolimnion is usually depleted in dissolved oxygen since direct reaeration from the atmosphere cannot occur. In addition, there is usually an accumulation of the inorganic materials formed during anaerobic respiration, for example, ammonia, hydrogen sulfide, etc. These processes are illustrated in Figure 15.7. Oxygen depletion of the hypolimnion occurs in many water bodies throughout the world due to the introduction of plant nutrients that causes excess plant growth. Somewhat similar processes have been occurring in the oceans over geological time. In many areas, the bottom waters of the oceans have dissolved oxygen levels of zero due to the accumulation of organic matter from primary production in the photic zone in the past.If dissolved oxygen is zero in the bottom waters, then the nitrification reaction leading to the formation of nitrate cannot proceed and organic nitrogen yields ammonia as the principle end-product. This substance is not formed in surface waters since these are well aerated by atmospheric and photosynthetic oxygen. Here, the principle nitrogen form is nitrate. Oligotrophic and eutrophic conditions produce vertical profiles of nitrate and ammonia that correspond with the above observations and are characteristic of these conditions. In Figure 15.7, the conditions are outlined and the related vertical profiles are illustrated.

E. PHOTOSYNTHESIS AND DIURNAL VARIATIONS IN DISSOLVED OXYGEN

Many nutrient-enriched areas have high plant biomass, dominated by rooted aquatic plants and attached algae. Plants in this situation can have an important influence on the dissolved oxygen content of water through photosynthesis and respiration. Photosynthesis occurs during the daylight hours, but respiration by plants occurs throughout the *diurnal* cycle. Thus, if significant plant growths are present, this will lead to an input of photosynthetic oxygen during the daylight hours, but continuous consumption of dissolved oxygen by respiration. Figure 15.8 shows that these processes cause a rise in dissolved oxygen during the daylight hours, with a maximum in the afternoon. At sunset, production of photosynthetic oxygen ceases

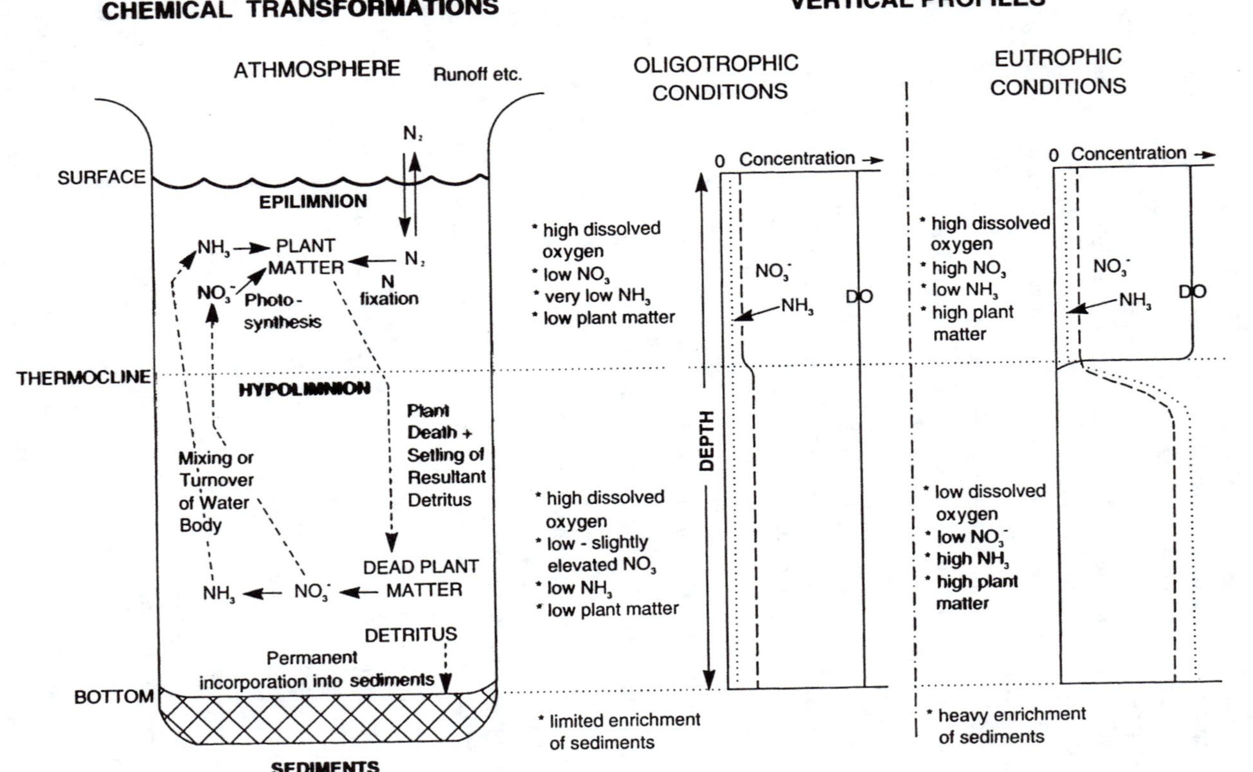

FIGURE 15.7. Typical (generalized) chemical transformations and vertical profiles of dissolved oxygen and nitrogen forms in stratified water bodies.

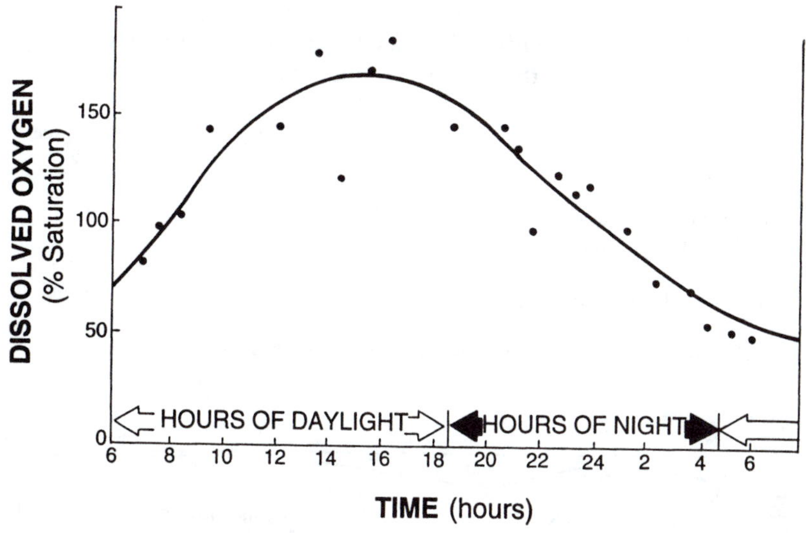

FIGURE 15.8. Diurnal variations in dissolved oxygen in a stream with a high plant biomass.

and so the dissolved oxygen content of the water starts to drop due to respiration by plants and other aquatic organisms. This continues overnight, with dissolved oxygen reaching a minimum before dawn when sunlight once again initiates photosynthesis.

Dissolved oxygen production and consumption is influenced by light intensity, plant biomass, as well as ambient water temperatures. Generally, its been found that over a diurnal cycle, the consumption of oxygen by plant respiration is about 75% of the oxygen produced by photosynthesis. During periods of overcast weather, photosynthetic oxygen production can be low and oxygen consumption due to plant respiration can exceed the photosynthetic oxygen production, leading to a decrease in dissolved oxygen in the water mass. In some cases, a similar situation can occur after the use of aquatic herbicides, which cause extensive death and decay of aquatic plants, leading to substantial reductions in dissolved oxygen.

F. CHARACTERISTICS OF NUTRIENT ENRICHMENT AND TROPHIC STATUS

A clear assessment of the degree of eutrophication is needed for the development of procedures for water quality management. Most of the characteristics of trophic status presently available have been based on freshwater lakes, as outlined in Table 15.4. As a general rule, the total phosphorus is measured in winter when stratification is not present and the water body is likely to be more homogeneous. The chlorophyll *a* content is a measure of the amount of green pigment present and is considered indicative of the population of plants present. The Secchi Disc is a circular disc attached to a rope that is lowered into the water until the disc cannot be seen. This value gives an indication of the turbidity of the water. The measure of total annual rate of primary production gives an evaluation of the amount of plant matter produced throughout the year.

TABLE 15.4
Some Quantitative Limit for Several Characteristics that Define Trophic State

	Oligotrophic[a] ≤	Eutrophic[a] ≥
μg L^{-1} total P (winter)	10–15	20–30
μg L^{-1} chlorophyll *a* (summer)	2–4	6–10
m Secchi desk (summer)	5–3	2–1.5
Total annual rate of primary productivity g C cm^{-2} yr^{-1}	7–25	75–700

[a] The intermediate class, mesotrophic, has intermediate characteristics.

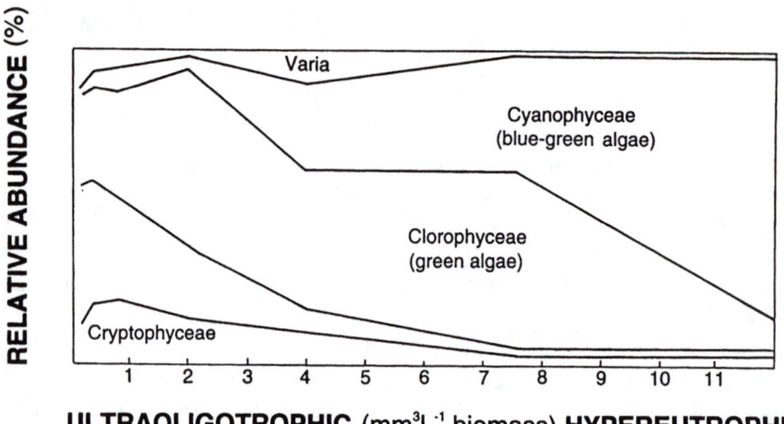

FIGURE 15.9. Composition of phytoplankton and algal blooms expressed by volume in relation to increasing lake fertility.

Different phytoplankton also have different dynamics and requirements for nitrogen, phosphorus, carbon dioxide and other factors that produce changes in the community composition with increasing eutrophication. The most prominent change with increasing nutrient enrichment is that the blue-green algae or Cyanobacteria (*Cyanophyceae*) become increasingly dominant, as shown in Figure 15.9.

G. THE BLUE-GREEN ALGAE (*CYANOBACTERIA*)

The blue-green algae are best classified scientifically as Cyanobacteria. Prolific growths of algae are usually referred to a "**blooms,**" and blooms of blue-green algae seem to have increased on a worldwide scale. These algae contain a range of toxins that can be released into the surrounding water. Blooms of blue-green algae have caused the deaths of many animals, both domestic and natural. The existence of toxic blue-green algae in water supplies has been known for some time, but it has only been in recent times that the potential for these toxins to affect human health, through use of contaminated water supplies, has been fully recognized. The most serious mammalian effects are due to two classes of toxins: the liver toxins (hepatotoxins) and those affecting the nervous system (neurotoxins). The toxicity of the various types shows considerable variation, some toxins being only moderately toxic compared to botulism and tetanus, but highly toxic when compared to substances such as strychnine or sodium cyanide.

H. THE FUTURE

Nutrient problems in waterways are likely to continue into the foreseeable future. Even if all nutrient discharges could be halted now, the large reservoir of nutrients in aquatic sediments ensure that nutrient levels in overlying waters will continue to follow current trends. Over time, the nutrient levels in sediments would be expected to decline due to processes such as denitrification, dispersal and covering by fresh sediments. High nutrient levels and other environmental conditions in aquatic areas can be expected to continue to stimulate major growths of algae and other aquatic plants.

IV. KEY POINTS

1. The major pollutants produced by human societies are relatively nontoxic animal and vegetable wastes. These wastes are rich in organic carbon and energy and are commonly discharged into waterways in the form of sewage, food processing wastes, as well as wastes from paper manufacturing and a variety of other industries. The organic matter in these wastes is usually in the form of carbohydrates, proteins, fats, humic substances, surfactants, and related and derived substances.

2. **Photosynthesis** and **respiration** are the two fundamental chemical processes supporting life on earth. These processes can be simply expressed as below

$$6\ CO_2 + 6\ H_2O \underset{\text{Respiration}}{\overset{\text{Photosynthesis}}{\Leftrightarrow}} 6\ (CH_2O) + 6O_2$$

3. The addition of wastewaters rich in organic matter can stimulate the respiration process through the stimulation of growth of microorganisms. This draws on the dissolved oxygen content in the ambient water and can result in substantial reductions. The oxygen-reducing capacity of a wastewater is measured by the **biochemical oxygen demand** (BOD) test. Wastewater discharges rich in biochemical oxygen demand can cause a reduction in the dissolved oxygen content of receiving waters, which can rise again as reaeration from the atmosphere occurs. This is often referred to as the **"dissolved oxygen sag"** when it occurs in rivers and streams.

4. Anaerobic processes can occur when the dissolved oxygen content declines to zero as a result of the biochemical oxygen demand depleting the dissolved oxygen content of the water. When these conditions occur, there can be the production of poisonous gases such as hydrogen sulfide and ammonia. These anaerobic processes can be simply represented by the equation below.

$$\underset{\text{Carbohydrate}}{6\,(CH_2O)} \rightarrow 3\,CO_2 + 3\,CH_4$$

5. The discharge of wastewaters rich in biochemical oxygen demand can result in substantial changes to the ecosystem in a stream. In the zone where maximum effects are observed, there can be very large populations of some organisms that are adapted to the low oxygen and high organic matter present. As the dissolved oxygen content returns to normal, a series of different communities occur downstream.

6. Many aquatic areas are deficient in nitrogen and phosphorus in terms of the elements present that will support the growth of aquatic plants. Thus, the addition of nitrogen and phosphorus salts to many aquatic areas will cause an increase in the growth of plants. The increase in nutrients present in an aquatic area can result from both natural and pollution processes. The different states of enrichment with nutrients are described as **oligotrophic** when the nutrients are very low; **eutrophic** when the nutrients, and resultant plant growths, are very high; and **mesotrophic** for an intermediate situation.

7. The nutrients in waterways can result from a variety of discharges. Untreated, primary treated and secondary treated sewage all contain significant quantities of nutrients that can stimulate the growth of aquatic plants. In addition, stormwater runoff from agricultural and urban areas can also contain significant concentrations of nutrients.

8. In all water bodies, a common set of nitrogen and phosphorus transformations occur. But these transformations occur in different parts of the water body and to different extents, depending on such factors as depth, light penetration, water flows and so on. Also, stratification is an important physical factor influencing the chemical transformations that occur in a water body. Oligotrophic and eutrophic water bodies produce different vertical profiles of chemical constituents when the bodies are stratified.

9. Water bodies with a high mass of plant matter can exhibit distinctive diurnal variations in dissolved oxygen. With these water bodies, the dissolved oxygen content during the day can rise to very high levels due to photosynthetic oxygen production, but drop to very low levels overnight when photosynthesis does not occur.

10. The occurrence of the toxic blue-green algae (*Cyanobacteria*) becomes more common under highly eutrophic conditions. There seems to be a worldwide trend toward the increasing occurrence of these algae.

REFERENCES

Connell, D.W., and Miller, G.J., *Chemistry and Ecotoxicology of Pollution,* John Wiley & Sons, New York, 1984.

Manahan, S E., *Environmental Chemistry,* 5th ed., CRC Press, Boca Raton, FL, 1991.

Wetzel, R.G., *Limnology,* Saunders College Publishing, Philadelphia, 1975.

Velz, G.J., *Applied Stream Sanitation,* Wiley-Interscience, New York, 1970.

Welch, E.B., *Ecological Effects of Wastewater,* Cambridge University Press, Cambridge, 1980.

CHAPTER 15 PROBLEMS

1. A pond with dimensions 100 m square and 1 m deep is used to contain wastewaters. It is now filled with water having a BOD of 0 and a dissolved oxygen (DO) reading of 8 mg L^{-1}. If 10 m^3 wastewater containing 5000 mg L^{-1} glucose was discharged to the pond, what would you expect the D.O. reading to be 24 hours later when the glucose would be completely decomposed? Assume that there is no uptake of atmospheric oxygen.

2. The BOD is a measure of the oxygen-reducing capacity of a wastewater containing organic matter. If k_1 is 0.1 day^{-1}, what fraction of the total or ultimate BOD is measured by the BOD_5 (BOD measured over 5 days)?

3. Sulfur is an important micronutrient in the growth of aquatic plants. Explain, in chemical terms, why this element is required.

4. In a stratified lake, there are distinct vertical profiles from surface to bottom for the various chemical forms of plant nutrients. Draw out the vertical profile that would be expected for the nutrient salt orthophosphate (PO_4^{-3}) in a stratified eutrophic lake and explain why this profile would be expected.

Answers on Page 360.

CHAPTER 15 SOLUTIONS

1. The complete degradation of glucose would be expected to follow this equation:

$$6\ (CH_2O) + 6\ O_2 \rightarrow 6\ CO_2 + 6\ H_2O$$

Thus, 180 g glucose requires 192 g O_2 for complete oxidation. The volume of water is 10,000 m^3 and, at 8 mg L^{-1}, contains 80,000 g O_2. The wastewater contains 50,000 g glucose, which will consume 53,000 g O_2. The water in the pool will thus contain 27,000 g O_2 when degradation is complete, giving a concentration 2.7 mg L^{-1}.

2. From the kinetic equation for loss of BOD,

$$\log (L/L_o) = k_1 t$$

If k_1 is 0.1 day^{-1} and t is 5 days, then

$$\log (L/L_o) = -0.5$$

Thus, $L/L_o = 0.0316$

and $L = 0.316\ L_o$

This means that after 5 days, there is 0.316 of the ultimate, or total, BOD remaining. Thus, 68.4% of the ultimate BOD is measured by the BOD_5.

3. Sulfur is needed in plant growth for the following reasons:

- To form biologically important chemical groups such as –SH (sulfhydryl) and $-S-CH_3$
- Sulfur-containing groups such as –SH and $-SCH_3$ are contained in essential amino acids such as cysteine and methionine:

$$HS-CH_2-\underset{\displaystyle NH_2}{\underset{|}{CH}}-COOH \qquad H_3C-S-CH_2-CH_2-\underset{\displaystyle NH_2}{\underset{|}{CH}}-COOH$$

Cysteine Methionine

- The sulfur-containing group, the disulfide bridge -S-S-, plays a major role in the secondary and tertiary structures of proteins.
- Sulfur is needed in the formation of enzymes containing all the groups and compounds mentioned above.

4. The phosphorus would be incorporated into plant matter during photosynthesis and orthophosphate (PO_4^{3-}) would be released from plant matter and detritus

on degradation. Degradation usually occurs after the plant has died and fallen to the bottom. This would be expected to be significant in an eutrophic lake. Thus, phosphorus would be taken in at the surface (epilimnion), giving reduced concentrations, as well as being produced in the hypolimnion, leading to elevated concentrations. This would lead to the profile shown in Figure 15.10.

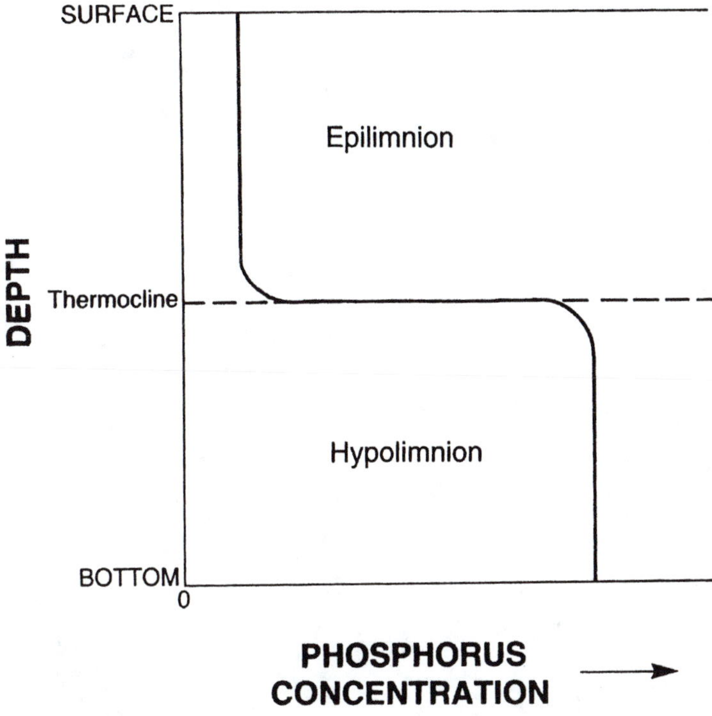

FIGURE 15.10 The depth concentration profile that could be experienced for phosphorus in a stratified eutrophic lake (answer to Question 4).

Chapter 16

ATMOSPHERIC CHEMISTRY AND POLLUTION

I. INTRODUCTION

The atmosphere plays a major role in global processes supporting life on Earth. The interaction of photosynthesis and respiration with carbon dioxide and oxygen in the atmosphere were described in Chapter 12. Also, the importance of the atmosphere in chemical evolution and how the composition of the atmosphere has changed over geological time have been outlined in Chapters 12 and 13. The atmosphere also plays an important physical role as a heat reservoir. Heat absorbed by the atmosphere during daylight hours is released overnight, helping to maintain an equitable temperature. Without the atmosphere, the Earth's surface temperature would rise very high when bathed in sunlight, but fall very low during darkness.

The major features of the atmosphere are shown in Figure 16.1. The zone closest to the Earth's surface is described as the **troposphere** and is the zone of most significance to living organisms. In this zone, the normal atmospheric composition of nitrogen, oxygen and other gases is observed as shown in Table 16.1. It is noteworthy that while there are several gases of biogenic origin, including oxygen, carbon dioxide and methane, there are others that are inert gases and play no apparent role in living systems, including argon, neon and helium.

As distance from the Earth's surface increases, the composition of the atmosphere changes. The **stratosphere** is the next zone above the troposphere and this consists mainly of N_2, O_2 and water, with some O and ozone. From an environmental perspective, the ozone in this layer is of particular importance since it absorbs short wavelength solar radiation. Ozone is a reactive and unstable substance but can persist in the stratosphere due to the low air pressures that exist in this zone (0.1–0.001 atm). This means that there is a long mean free path between collisions of molecules that occur relatively infrequently and thus, the possibility of intermolecular reactions is reduced. Similarly, the other zones at further distances from the Earth, the mesosphere and thermosphere, contain the highly reactive substances O_2^+, NO^+ and O^+, which could not exist for any significant time at the Earth's surface. These reactive ions in the upper zones of the atmosphere also play a major role in maintaining living systems by absorbing short wavelength solar radiation. This radiation is relatively high in energy and would be damaging to living systems at the Earth's surface.

II. REACTIVE INTERMEDIATES IN THE ATMOSPHERE

A. HYDROXYL RADICAL

Apart from possible interaction with solar radiation, much of the chemistry of organic compounds in the troposphere revolves around reaction with certain reactive species or intermediates that are continually being formed and consumed. The most important of these reactive intermediates is the hydroxyl radical ($OH^\bullet$). Organic

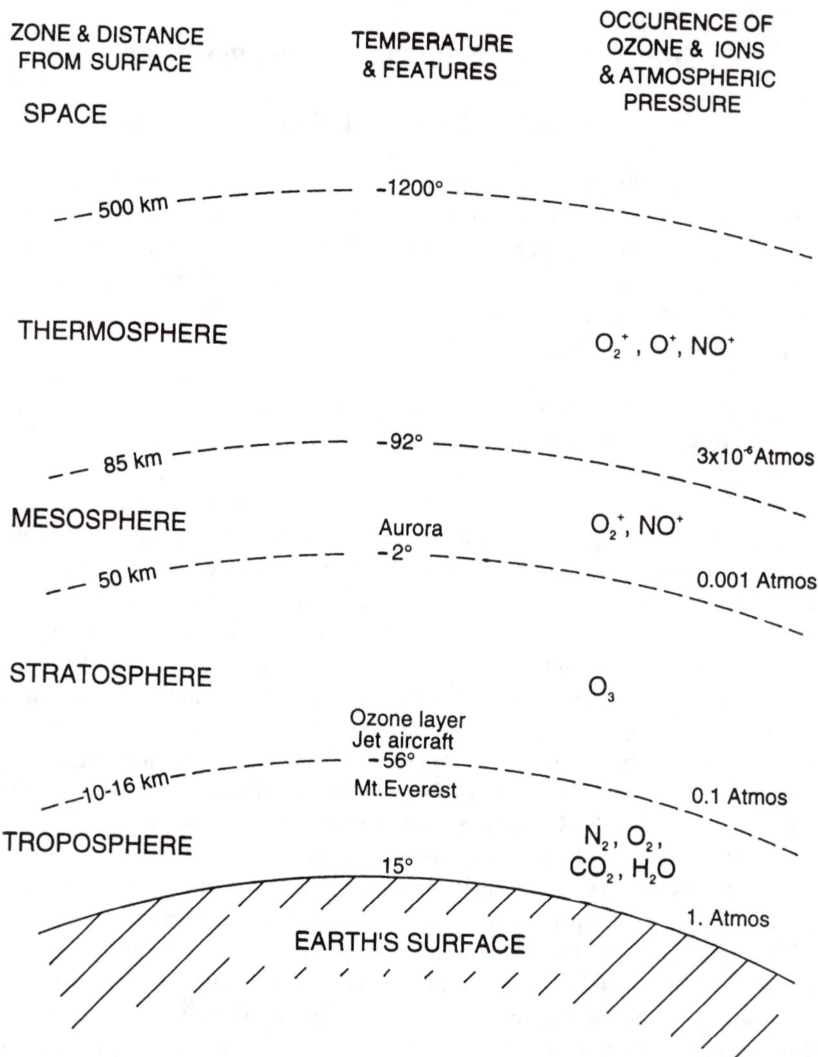

FIGURE 16.1. The major features of the Earth's atmosphere (not to scale).

molecules that do not react or react only very slowly with $OH^{\cdot}$ (e.g., chlorofluorocarbons or CFCs) are unlikely to react with any other reactive intermediate. Sources of $OH^{\cdot}$ in the troposphere are shown in Table 16.2.

It can be seen that solar radiation is critical for the formation of $OH^{\cdot}$, and so there is considerable diurnal variation in concentrations, which reach their peak during the day. Actual concentrations are somewhat difficult to measure, but it appears that annual average concentrations over a 24-hour period in the troposphere are 5×10^5 and 6×10^5 radicals cm^{-3} for the Northern and Southern hemispheres, respectively.

This is seemingly a large number of species in every cubic centimeter, but these concentrations correspond to approximately 0.02 parts per trillion by volume (ppt),

TABLE 16.1
Average Composition of the Atmosphere at the Earth's Surface

Component	Composition (volume/volume) (ppm)
Nitrogen (N_2)	780,900 (78.09%)
Oxygen (O_2)	209,500 (20.95%)
Argon (Ar)	9,300 (0.93%)
Carbon dioxide (CO_2)	300 (0.03%)
Neon (Ne)	18
Helium (He)	5.2
Methane (CH_4)	1.7

TABLE 16.2
Sources of Hydroxyl Radicals in the Troposphere

1. Reaction of oxygen atoms, formed from the photolysis of ozone, with water vapor:

$$O_3 \xrightarrow{h\upsilon} O_2 + O$$

$$O + H_2O \rightarrow 2\ OH^{\bullet}$$

2. Photolysis of nitrous acid:

$$HONO \xrightarrow{h\upsilon} HO^{\bullet} + NO$$

3. Photolysis of hydrogen peroxide:

$$H_2O_2 \xrightarrow{h\upsilon} 2\ HO^{\bullet}$$

4. Reaction of hydroperoxy radicals with nitric oxide:

$$HO_2^{\bullet} + NO \rightarrow HO^{\bullet} + NO_2$$

Note: Hydroperoxy radicals are generated from the photolysis of aldehydes.

$$H\text{–}CHO \xrightarrow{h\upsilon} H^{\bullet} + {}^{\bullet}CHO$$

$$H^{\bullet} + O_2 + M \rightarrow HO^{\bullet}_2 + M$$

and also in the oxidation of hydrocarbons.

which shows that $OH^{\bullet}$ is indeed a trace species. It is also a transient species. It is consumed mainly by reaction with hydrocarbons (particularly methane) and also carbon monoxide (CO). Depending on levels of these compounds, the lifetime of an average hydroxyl radical between formation and consumption is of the order of 0.02 seconds or less.

From a chemical point of view, $OH^{\bullet}$ is a neutral species, containing in total 9 protons and 9 electrons. It is a combination of a neutral oxygen atom with 6 valence electrons and a neutral hydrogen atom with 1 valence electron.

$$H^{\bullet} + {\bullet}\ddot{\underset{\bullet\bullet}{O}}{:} \longrightarrow H{:}\ddot{\underset{\bullet\bullet}{O}}{:}$$

If the hydrogen shares its electron with one of the unpaired electrons on oxygen, a covalent bond is created between the oxygen and hydrogen, and the hydrogen attains the filled shell electron arrangement of helium. Oxygen however does not achieve a filled shell. There is still one unpaired electron. The oxygen atom has residual bonding capacity and $OH^{\bullet}$ would be expected to be quite reactive. It is important to distinguish between the hydroxyl radical ($OH^{\bullet}$) and the hydroxide anion (OH^-), which is commonly encountered in water chemistry. The hydroxide anion has 1 more electron than $OH^{\bullet}$ and hence has an overall charge of –1. Both oxygen and hydrogen have achieved filled shells in OH^-.

B. OZONE

Approximately 90% of Earth's ozone is found in the stratosphere, but 10% exists in the troposphere. In the troposphere, with hydrocarbons, ozone reacts mainly with unsaturated and aromatic molecules (but not saturated ones) and a variety of other substances. It is a product of photochemical air pollution (smog) but also exists in trace levels in relatively clean, unpolluted air. Levels at the surface of the Earth are variable, but are typically around 5×10^{11} molecules cm^{-3}. Concentrations of 1×10^{13} molecules cm^{-3} have been recorded in severe photochemical air pollution episodes.

In the troposphere, concentrations of ozone tend to increase with altitude, suggesting a stratospheric source. It is generally accepted that, periodically, meterological conditions in the upper troposphere and lower stratosphere are such that ozone from the stratosphere is injected into the troposphere. The other source of ozone in the troposphere is nitrogen dioxide (NO_2), which can be formed in the atmosphere from a number of sources, including motor vehicle exhaust gases initially containing mainly NO. Photolysis of NO_2 is the only known way by which ozone can be produced in the troposphere.

$$NO_2 \xrightarrow{h\nu} NO + O$$

$$O + O_2 + M \rightarrow O_3 + M$$

M is a third chemical species, most likely N_2 or O_2, that dissipates the excess energy of the ozone molecule that is produced. If there is no third chemical species available, the ozone would not form at all or dissociate on formation. Radiation with wavelengths of less than about 400 nm are required for this process to occur.

The NO_2 can be formed by reaction of nitric oxide (NO) with alkylperoxy ($ROO^{\bullet}$) and hydroperoxy ($HOO^{\bullet}$) radicals — intermediates formed in the oxidation of hydrocarbons and CO in the troposphere. During the day, in atmospheres with low concentrations of hydrocarbons and CO however, a dominant mode of reaction for NO is reaction with ozone. Under such conditions, as shown in Figure 16.2, a cyclic situation known as the **Leighton relationship** is created, with ozone being continuously formed and consumed. In the absence of any other reactions, steady concen-

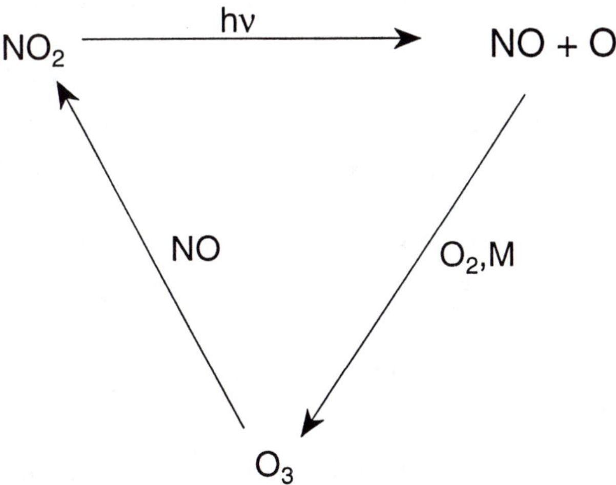

FIGURE 16.2 The Leighton relationship, which describes the cyclic reaction between NO, NO_2 and O_3 in air, with little or no hydrocarbons or CO.

trations of NO_2, NO and ozone are eventually observed with time. The relationship between these steady-state concentrations is given by:

$$\frac{[O_3][NO]}{[NO_2]} = \text{Constant}$$

where the constant is a function of solar radiation intensity.

C. NITRATE RADICAL

The nitrate radical or nitrogen trioxide (NO_3 or $NO_3^{\bullet}$) is another molecule that is seldom heard of, but is an important reactive intermediate in the troposphere. It is an overall neutral species and a radical, with one unpaired electron. It should not be confused with the nitrate anion (NO_3^-), which has an additional electron. There is abundant spectroscopic evidence for the existence of NO_3, but it has never been isolated as a pure chemical species. It is formed from oxidation of NO_2 by ozone, derived from anthropogenic and natural sources in roughly equal proportion. Motor vehicles contribute about 40% to the anthropogenic sources. The reaction can be represented by the following equation:

$$NO_2 + O_3 \rightarrow NO_3 + O_2$$

During the day, however, NO_3 is rapidly broken down by visible light via two possible pathways:

$$NO_3 \xrightarrow{h\nu} NO_3 + O \quad \text{or} \quad NO + O_2$$

The relative importance of each pathway depends on the wavelength of the radiations involved. Because of the instability of NO_3 toward light, highest concentrations are observed at night. Even at night, however, NO_3 can be consumed by reaction with NO, or excess NO_2.

$$NO + NO_3 \rightarrow 2\ NO_2$$

$$NO_2 + NO_3 \Leftrightarrow N_2O_5$$

The dinitrogen pentoxide (N_2O_5) can react with water, forming nitric acid.

$$N_2O_5 + H_2O \rightarrow 2\ HNO_3$$

For this reason, detectable levels of NO_3 are rarely observed if the relative humidity is over 50%. In drier air, at night, the lifetime of NO_3 is of the order of minutes.

The nitrate radical reacts with hydrocarbons in much the same way as the hydroxyl radical, i.e., by abstraction of hydrogen atoms or addition. In a sense, their role as oxidants of hydrocarbons in the troposphere is complementary. During the daytime, when NO_3 is rapidly photolyzed, $OH^{\bullet}$ is the most important. At night, however, levels of $OH^{\bullet}$ diminish because it is photochemically produced and levels of NO_3 increase.

III. TYPES OF HYDROCARBONS IN THE TROPOSPHERE

A. METHANE, CARBON DIOXIDE AND THE GREENHOUSE EFFECT

The individual hydrocarbon present in greatest abundance in the troposphere is methane (CH_4). The present concentration is about 1.7 ppm by volume and appears to be increasing with time. Based on the concentrations of methane in trapped bubbles of air from polar ice cores, levels were relatively steady at 0.7 ppm from 25,000 BC up until just over 400 years ago. Since that time, levels have increased, with an acceleration of this trend over the last 100 years or so. Given the timescale of these increases, human involvement, either directly or indirectly, is suspected.

The sources of tropospheric methane and estimations of their relative importance are shown in Table 16.3. The data represent estimates and contain considerable uncertainty. It is clearly very difficult to gauge, for example, how much methane is annually released by ruminants all over the globe. The amount released from fossil fuels such as natural gas, petroleum and coal can be estimated from carbon isotope ratio determinations. Most of the carbon in the environment is ^{12}C, but just over 1% is ^{13}C. Both are stable isotopes. High in the atmosphere, small amounts of radioactive ^{14}C are formed that appear as $^{14}CO_2$. This is incorporated into plants via photosynthesis and into other organisms via food chains. The CO_2 in the atmosphere contains $1.2 \times 10^{-10}\%$ $^{14}CO_2$. The half-life of ^{14}C (5730 years) is sufficiently long such that living plants and animals have $1.2 \times 10^{-10}\%$ of their carbon present as ^{14}C. After death, however, no more ^{14}C is accumulated and the amount present decreases exponentially. Fossil fuels represent extremely old organic material that would be expected to be substantially depleted in ^{14}C. The methane that is trapped in coal, or found in natural gas or petroleum and released to the troposphere by mining or drilling, or leaks from pipelines should be largely $^{12}CH_4$ and $^{13}CH_4$. As a result, the

TABLE 16.3
Estimated Annual Contributions from Various Sources of Methane

Source		Estimated global emission (10^6 metric tons per year)	Percentage of total
Fossil fuels	Coal mines	25	5
	Natural gas and petroleum	70	14
Methane Clathrates		5	1
Biomass burning		30	6
Anaerobic microbial activity	Ruminants	80	16
	Landfills	50	10
	Oceans, termites	30	6
	Wetlands and rice paddies	215	42
Total		505	100

Data from Grutzen, P., *Nature,* 350, 380, 1991. With permission.

percentage of ^{14}C in tropospheric methane should be less than that in tropospheric CO_2. The extent of the difference gives an idea of the contribution of fossil fuel sources to annual methane emissions.

Another source of "old" carbon is methane clathrates. These are solids in which molecules of methane (the guest) are trapped in cavities of ice crystals (the host). Generally, there are about six water molecules for every methane and so the clathrates have the approximate formula $CH_4 \cdot 6H_2O$. The methane clathrates or methane hydrates, as they are also known, exist beneath the permafrost in polar regions and also in the sediment of areas such as the Arctic Ocean. The formation of clathrates with gases such as Ar, Kr, Cl_2 and SO_2 has been known from laboratory studies for many years, but the discovery of naturally occurring methane clathrates was only made in the 1960s. The extent of this methane resource is unclear, as is the annual emission of methane from it to the troposphere.

Methane is also produced through anaerobic microbiological degradation of organic matter. This can take place in the oceans, in paddy fields, natural wetlands, landfills or even in the stomachs of ruminants such as sheep and cattle and the gut of termites. The production of methane by microorganisms is usually due to a highly specialized group of anaerobic bacteria called methanogens that exist in these habitats. In most methanogens, dissolved CO_2 in the form of the bicarbonate anion (HCO_3^-) acts as an electron acceptor and is reduced to methane. The electron donor in this process is usually hydrogen, as shown below

$$4\ H_2 + H^+ + HCO_3^- \rightarrow CH_4 + 3\ H_2O + \text{Energy}$$

The increasing levels of methane observed in the troposphere may be due to increased contributions from many of the sources listed in Table 16.3, but also a decreased rate of loss or consumption. The main loss process for methane is reaction with $OH^\bullet$:

$$CH_4 \xrightarrow{OH^\bullet} \longrightarrow \longrightarrow CO \xrightarrow{OH^\bullet} \longrightarrow \longrightarrow CO_2$$

After a series of reactions, one of which involves light, CO is formed, which can in turn also react with $OH^\bullet$, eventually forming CO_2. Over 90% of the methane that is annually lost from the troposphere is lost in this manner. Other minor loss processes are diffusion into the stratosphere and uptake by soils where aerobic bacteria called methanotrophs use it for energy and as a carbon source. Increasing concentrations of CO from sources other than oxidation of methane mean that $OH^\bullet$ concentrations may be decreasing. Reaction of CO with $OH^\bullet$ is of major importance in controlling $OH^\bullet$ levels. This means, however, that less is available for reaction with methane, and the rate of this reaction and its importance as a loss process is diminished.

The lifetime of methane in the troposphere is currently on the order of 5 to 10 years. During its residence time in the troposphere, methane will absorb some of the infrared (IR) radiation emitted by the Earth. Thus, methane is a so-called **greenhouse gas**. It is important not to confuse depletion of the ozone layer, which is a stratospheric phenomenon, with accumulation of greenhouse gases, which occurs in the troposphere. Incoming solar radiation comprises largely ultraviolet (UV), visible and IR radiation, as shown in Figure 16.3. Overall, about 30% of the radiation from the sun is reflected by the Earth itself or the atmosphere. Most of this reflection is due to clouds in the troposphere. Some 20% of incoming radiation is absorbed by aerosol particles, gas molecules and clouds. Most of the UV radiation with wavelength less than 290 nm is absorbed by ozone and molecular oxygen (O_2) in the stratosphere. Any incoming IR radiation is absorbed in the troposphere by gases such as CO_2 and H_2O vapor.

The balance of the solar radiation, about 50%, is absorbed at the Earth's surface. The Earth's surface and atmosphere must lose energy at the same rate as it is absorbed from solar radiation in order to maintain thermal equilibrium.

To maintain this thermal equilibrium, the surface of the Earth transfers surplus energy to the atmosphere by convection and evaporation of water. As shown in Figure 16.3, the Earth also emits IR radiation because of its temperature. Some of this outgoing IR radiation is absorbed by gases in the troposphere, such as CO_2 and water vapor. Following absorption, some of the energy is radiated out into space, but some is reradiated back toward the Earth's surface, heating both the surface and air.

Were it not for our atmosphere, and those components that absorb the outgoing IR radiation and act as a blanket, the average global temperature of our planet's surface would be slightly over 30°C cooler than the current 15°C. Major greenhouse gases such as CO_2 and water vapor do not absorb at all the wavelengths of outgoing IR radiation. Figure 16.4 shows the main absorption bands of CO_2 and H_2O vapor within the range of wavelengths emitted by the Earth. (Note that intensity of absorption is not indicated.) The regions between 8 and 13.5 μm, and to a lesser extent between 16 and 20 μm, represent windows where the atmosphere is transparent to IR radiation. In other words, outgoing IR radiation within these ranges is not absorbed by CO_2 and H_2O vapor. Other gaseous components of the atmosphere can,

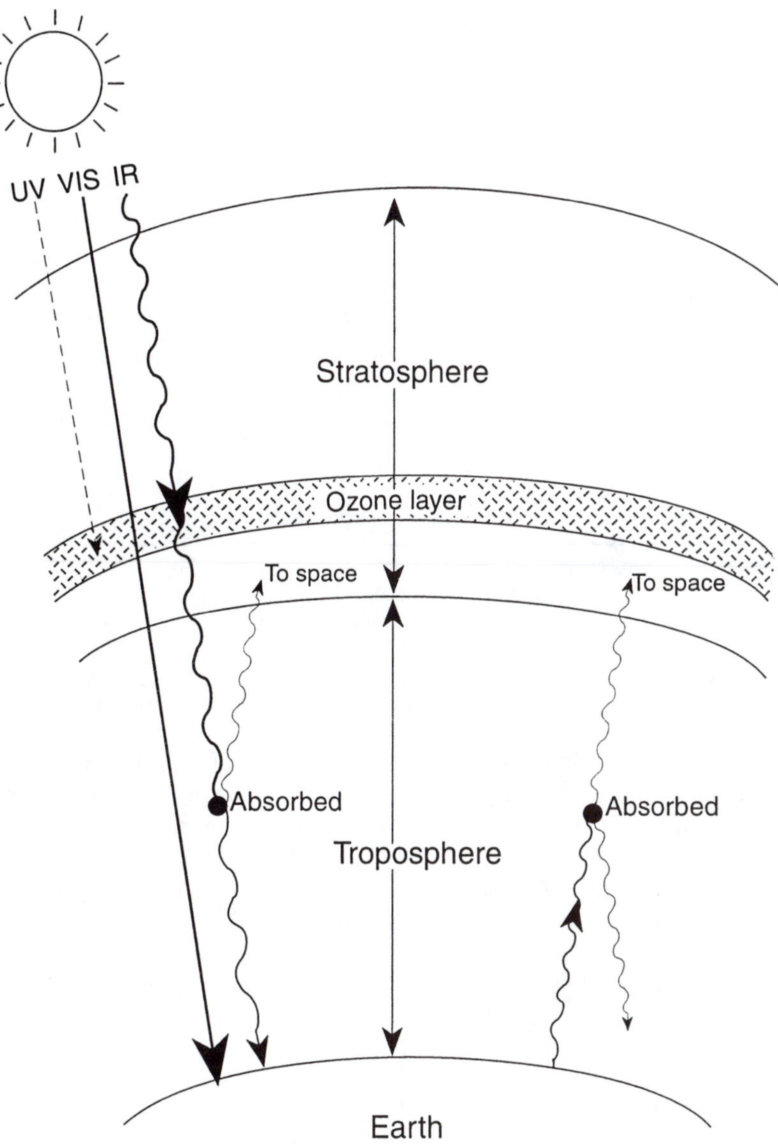

FIGURE 16.3. Simplified diagram of the behavior of solar radiation in the Earth's atmosphere, indicating the influence of the ozone layer and the greenhouse effect.

however, absorb radiation within these windows. Minor greenhouse gases include methane, ozone, nitrous oxide (N_2O) and chlorofluorocarbons (CFCs). Despite their relatively low concentrations, these compounds can absorb disproportionately large amounts of IR radiation.

Absorption of IR radiation results in altered rotational and vibrational levels within the molecules involved. In other words, the frequency at which molecules rotate and vibrate changes. Note that the position of individual atoms within mole-

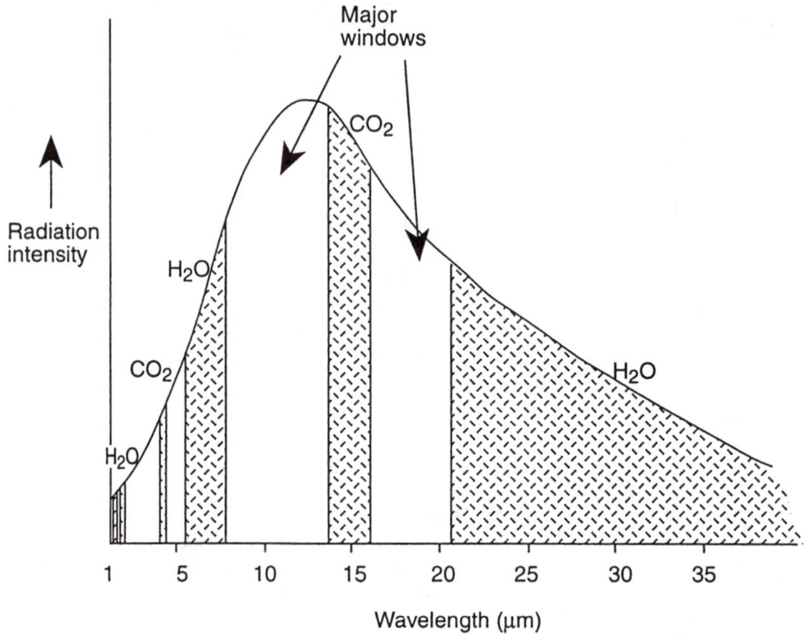

FIGURE 16.4. The relative intensity of emitted radiation from the Earth's surface, together with major absorption bands of CO_2 and H_2O vapor. The intensity of absorption for each band is not shown.

cules is not fixed. Vibrations fall into the basic categories of stretching and bending, in which bond lengths and bond angles vary in an oscillating manner. As seen in Figure 16.5, methane has two main absorption bands in the region of interest. The band between 7 and 8.5 μm is on the edge of the H_2O vapor absorption band that extends up to about 8 μm. This band for methane corresponds to bending vibrations in which the hydrogen and carbon atoms move so that the H°CDH structure oscillates at various frequencies about the tetrahedral H°CDH bond angle of 109.5°. This is caused by absorption of some IR radiation in the main window.

H
C H
H H
109·5¡

B. OTHER HYDROCARBONS

Hydrocarbons in the troposphere other than methane are usually grouped together as non-methane hydrocarbons (NMHCs). Except for research purposes, the concentrations of individual NMHCs are not routinely measured by regulatory authorities

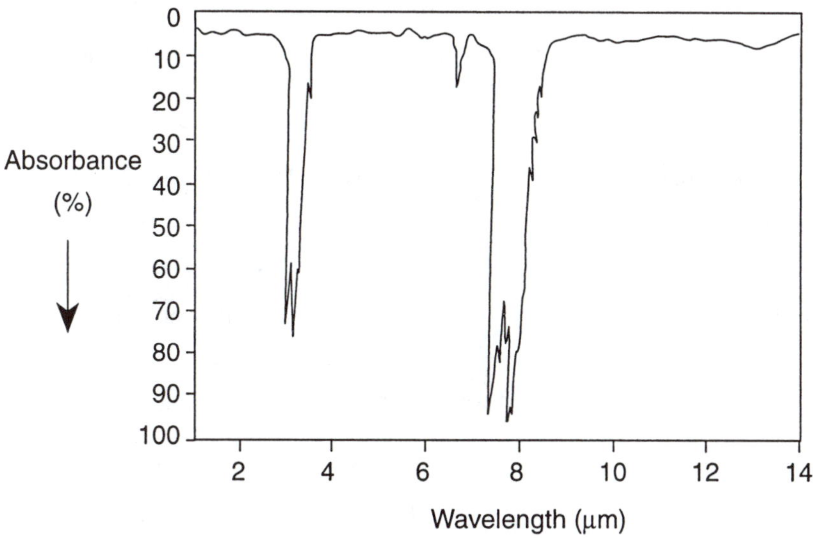

FIGURE 16.5. An IR absorption spectrum of methane.

because of their relatively low levels. As mentioned previously, the average troposphere concentration of methane is approximately 1.7 ppm. Contrast this with levels of *n*-pentane (15 to 0.1 ppb) at various locations, both urban and rural, in the United States.

Not only are alkanes (e.g., ethane, *n*-pentane) found, but alkenes such as isoprene and various terpenes exist as well. Terpenes are a class of organic compounds that are the most abundant components of the essential oils of many plants and flowers. The name comes from turpentine, which is rich in terpenes. Terpenes can be emitted into the air, and many of the odors associated with plants and flowers are due to terpenes. All terpenes *appear* as though they have been constructed by the linking together of isoprene molecules. Terpenes may contain two, three or more isoprene units, and may be linear or cyclic molecules, with or without double bonds. Some structures of representative terpenes are shown in Figure 16.6. Strictly, because terpenes are hydrocarbons, they should contain only carbon and hydrogen. Terpene-like compounds that contain other elements such as oxygen are known as terpenoids.

Sources of tropospheric NMHCs can be both natural and anthropogenic. Natural sources include plants, leakage from natural gas and other fossil fuels, and also bacteria. Anthropogenic sources include unburned hydrocarbons in automobile exhaust and evaporation from paints, gasoline and solvents. The extent of natural emissions depends on factors such as temperature and light intensity. As a result, seasonal variations are typically observed, with largest emissions in summer at a given site. For different sites, geographical and latitudinal influences are important.

The relative importance of NMHC emissions from natural and anthropogenic sources is still controversial. While on a global basis, anthropogenic emissions probably only represent a small fraction of total NMHC emissions, in many urban areas, the situation is reversed.

α - Pinene
(from pine trees)

β - Myrcene
(from oil of bay)

Camphor
(from the camphor tree)

Limonene
(from oil of lemon)

Citral
(from oil of lemon grass)

Menthol
(from peppermint oil)

FIGURE 16.6. Chemical structures of some representative terpenes and terpenoids found in the troposphere.

An important group of hydrocarbons included in the NMHC category are the polyaromatic hydrocarbons (PAHs). Much of the interest surrounding PAHs stem from the fact that several members of the group exhibit carcinogenic properties. PAHs are generally emitted from combustion sources and can exist in the troposphere either in vapor form, or associated with particles. They comprise compounds of widely differing molecular size, ranging from relatively small molecules such as acenaphthylene to larger ones such as chrysene or even larger. These compounds are considered in detail in Chapter 9.

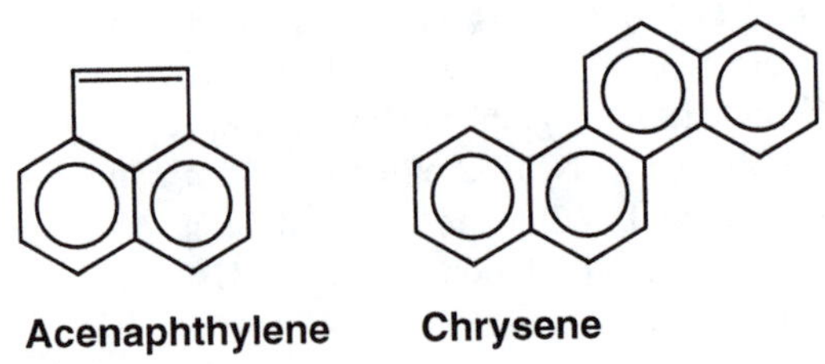

C. COMPOUNDS CONTAINING O, N AND S RELATED TO THE HYDROCARBONS

Aldehydes, ketones, ethers, esters, carboxylic acids and alcohols are all known to exist in the troposphere. Natural sources include plants that emit some of the terpenoids containing oxygen. Primary air pollutants are those that are emitted directly from an identifiable source such as a stack or automobile exhaust. Oxygen-containing compounds typically comprise 5 to 10% of the total organic compound concentration of automobile exhaust.

The aldehyde present in greatest amount in automobile exhaust, and usually the major aldehyde in ambient air, is formaldehyde. Both formaldehyde and acrolein, another aldehyde typically found in trace levels in the troposphere, are potent irritants of the eyes, nose, throat and skin. Elevated levels occur in photochemical air pollution events.

$$H{-}C(H){=}O \qquad H_2C{=}CH{-}C(H){=}O$$

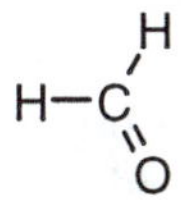

Formaldehyde 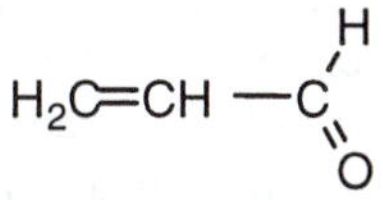 **Acrolein**

The troposphere is an oxidizing medium, and it is not surprising that hydrocarbons emitted into the troposphere are usually oxidized. Even oxygen-containing compounds can be further oxidized. Secondary air pollutants are those formed in the troposphere as a result of chemical reactions. A significant fraction of oxygen-containing compounds found in the troposphere, particularly in urban areas, is secondary in nature. They are usually less volatile than the compounds from which they were formed and contribute to the haze associated with photochemical air pollution.

Nitrogen-containing compounds such as amines and amides are invariably found in the troposphere. Amines are produced by various industrial activities, feedlots, waste incineration and sewage treatment, but their ambient concentrations are generally fairly small. Lifetimes of compounds such as methylamine ($CH_3{-}NH_2$) are of the order of hours.

An interesting oxygen- and nitrogen-containing compound found in the troposphere is PAN (peroxyacetylnitrate). This compound is a secondary pollutant first found in photochemical air pollution events. It is toxic to plants and extremely irritating to eyes.

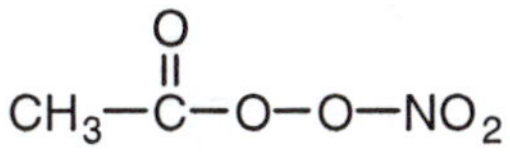

$$CH_3{-}\overset{\displaystyle O}{\overset{\|}{C}}{-}O{-}O{-}NO_2$$

Peroxyacetylnitrate

Trace levels can now be identified all over the world, even in relatively clean air from over the mid-Pacific ocean where it is formed from natural precursors.

A great many sulfur-containing compounds are found in the troposphere . Anthropogenic activities tend to produce oxidized species such as sulfur dioxide (SO_2). Biogenic sources, however, emit reduced sulfur species such as methanethiol (CH_3–SH), dimethylsulfide (CH_3–S–CH_3) and dimethyldisulfide (CH_3–S–S–CH_3).

D. HALOGENATED HYDROCARBONS, CHLOROFLUOROCARBONS (CFCs) AND THE OZONE LAYER

Halogenated hydrocarbons can be considered to be hydrocarbon molecules in which one or more hydrogen atoms have been replaced by a halogen atom, e.g., Cl, Br and I. The most prominent halogenated hydrocarbons in the troposphere are probably the chlorofluorocarbons (CFCs) and related compounds. These are anthropogenically derived and are of concern because of their effect on Earth's ozone layer. It would be a mistake, however, to think that all halogenated hydrocarbons found in the atmosphere are the result of human activity since some are not. The halomethanes CH_3Cl, CH_3Br and CH_3I are natural components of the troposphere. The source is thought to be biological activity in the oceans. There are abundant halide ions in the oceans that can be incorporated into organic molecules. Increasingly, marine organisms such as sponges are being found to contain natural halogenated organic molecules that show promise as pharmaceuticals.

The halomethane present in greatest amount is chloromethane (CH_3Cl). Its concentration, which seems to be fairly constant, is 1.6 ppb by volume. The lifetimes of both CH_3Cl and CH_3Br in the troposphere are sufficiently long such that some molecules can diffuse into the stratosphere and take a minor part in natural ozone depletion processes.

CFCs were first manufactured on an industrial scale in the 1930s. They have been used as refrigerants, aerosol propellants, in the manufacture of foam plastics and as cleaning agents for microelectronic apparatus. Related groups of compounds are the hydrochlorofluorocarbons (HCFCs) and the hydrofluorocarbons (HFCs). The HCFCs comprise a central carbon skeleton with hydrogen, fluorine and chlorine atom substituents; HFCs have only hydrogen and fluorine bonded to carbon. The molecular formula of these groups of compounds can be determined from the "Rule of 90." Each compound has a code number; for example, CFC-11, HCFC-124 and HFC-134a. Adding 90 to the code number results in a three-digit number that indicates the number of carbon, hydrogen and fluorine atoms present. Any remaining atoms are assumed to be chlorine. This procedure completely characterizes one-carbon and some two-carbon compounds. For other two-carbon compounds, this procedure only results in a molecular formula, e.g., $C_2H_3F_3$. The particular isomer involved cannot be distinguished. To denote a particular isomer, the code number is followed by a lower case letter, e.g., a,b or c.

CF_2Cl_2	CHF_2CF_3	$CClF_2CH_3$
CFC-12	HFC-125	HCFC-142b
Dichlorodifluoromethane	Pentafluoroethane	1-Chloro-1,1-difluoroethane

CFCs that have been of greatest importance in terms of production are CFC-11 ($CFCl_3$) and CFC-12 (CF_2Cl_2). In 1985, production of each of these compounds was over 250,000 tons. Although many uses are "closed" to the atmosphere, leakage to the air eventually occurs. It has been estimated that 90% of all the CFC-11 and CFC-12 ever produced has been released into the atmosphere. (The total cumulative production of CFC-12 up to 1990 was approximately 10^7 tons.) Tropospheric concentrations of a number of CFCs are already greater than that of CH_3Cl. The main problem with CFCs is that they have no reactive tropospheric chemistry — they are effectively inert so that diffusion into the stratosphere is their only significant fate. Their reactions with tropospheric and stratospheric components are considered in Section III. These reactions result in depletion of stratospheric ozone, and possibly increased UV radiation reaching the earth's surface. International agreements, first in Montreal in 1987, followed by revisions in London (1990) and Copenhagen (1992), have aimed to phase out or limit production of CFCs and related compounds.

While resident in the troposphere, CFCs can act as greenhouse gases. Stretching vibrations associated with both C–F and C–Cl bonds lie in the main window for IR radiation transmission (between about 8 and 13.5 μm), as shown in Figure 16.4. As a result, CFCs can absorb outgoing terrestrial radiation very efficiently. As a matter of fact, on a molecule-for-molecule basis, CFCs are more effective absorbers than CO_2 and methane (CH_4).

II. REACTION OF ORGANIC COMPOUNDS IN THE ATMOSPHERE

A. REACTION OF HYDROXYL RADICAL ($OH^{\bullet}$) WITH ORGANIC COMPOUNDS AND CHLOROFLUOROCARBONS

The primary fate of organic compounds in the troposphere is reaction with hydroxyl radicals ($OH^{\bullet}$). Although highest concentrations of $OH^{\bullet}$ are only observed during the daytime, this process is usually the most rapid of all the possible fates. With saturated molecules such as alkanes, reaction with $OH^{\bullet}$ occurs by abstraction of a hydrogen atom. The products are water, and a hydrocarbon-type radical that can undergo further reaction.

$$CH_4 + OH^{\bullet} \rightarrow {}^{\bullet}CH_3 + H_2O$$

To illustrate this, and also the complexity of tropospheric organic chemistry, consider a possible fate of an ethane molecule emitted into the troposphere, as shown in Figure 16.7. Following initial reaction with $OH^{\bullet}$, a complex sequence of reactions occurs, some involving the UV component of light found in the troposphere. Molecular oxygen is also involved with intermediates formed, including acetaldehyde, formaldehyde and carbon monoxide.

$$\underset{\textbf{Acetaldehyde}}{CH_3{-}\overset{\displaystyle H}{\overset{|}{C}}{=}O} \qquad \underset{\textbf{Formaldehyde}}{H{-}\overset{\displaystyle H}{\overset{|}{C}}{=}O}$$

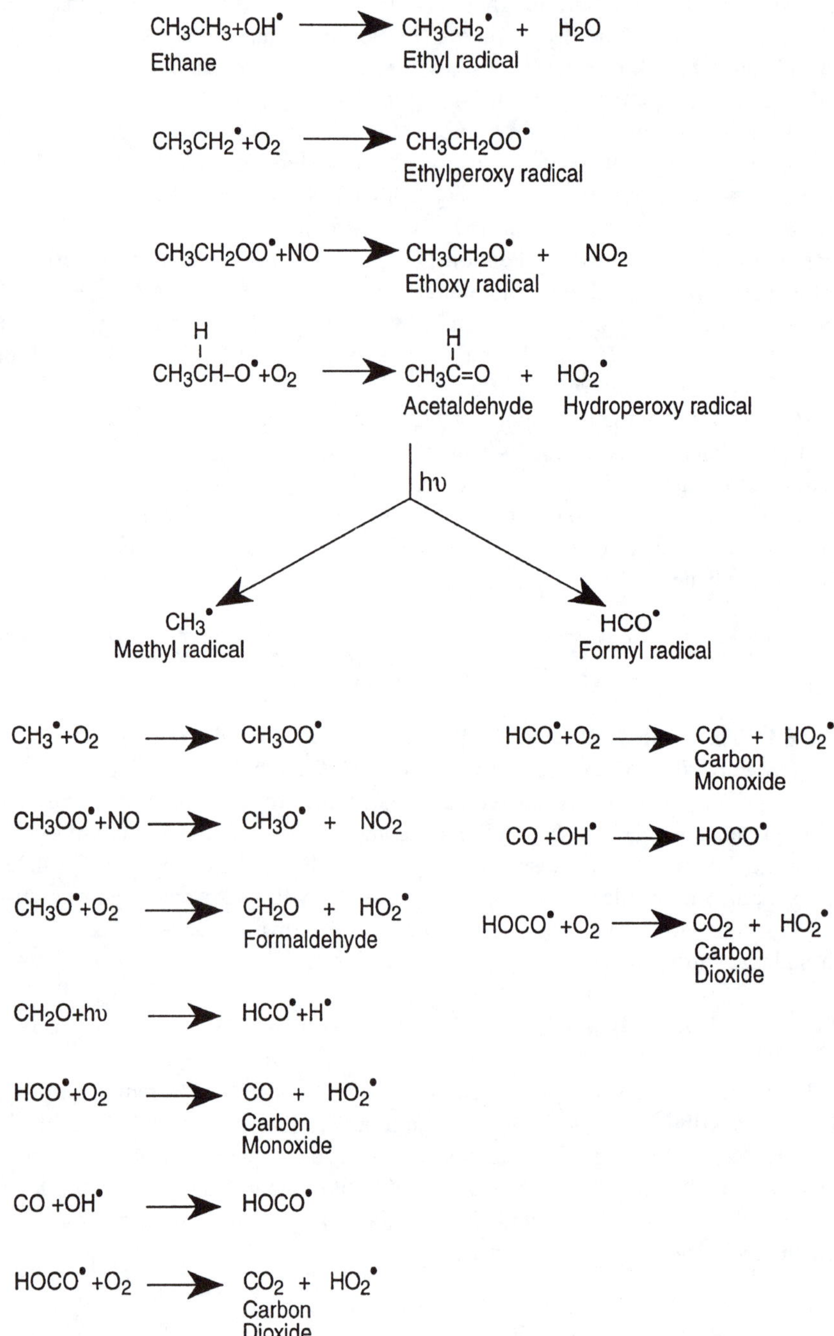

FIGURE 16.7. A possible chemical fate for ethane emitted into the troposphere.

Ultimately, for every molecule of ethane, two molecules of CO_2 are formed; thus, ethane has been oxidized.

Because reaction of saturated hydrocarbons with OH$^•$ involves breakage of a C-H bond, it follows that the rate constant for such processes depends on the strength of the C-H bond involved. Other factors include which other atoms are bonded to that particular carbon. With alkanes, where a total of three hydrogens are attached to a given carbon atom, the hydrogens are denoted as primary (1°) hydrogen atoms. The hydrogen atoms of a methyl group (CH_3-X) can be regarded as primary. Similarly, where a total of two hydrogens are bound to a particular carbon, the hydrogens are secondary (2°), and where a carbon has only one hydrogen attached, this is a tertiary (3°) arrangement. For example,

Primary Hydrogens → CH_3 (of CH_3–CH_2–)

CH_3 ← **Primary Hydrogens**

CH_3–CH_2–CH ← **Tertiary Hydrogen**

CH_3 ← **Primary Hydrogens**

Secondary Hydrogens → CH_2

With hydrocarbons, it is generally found that C-H bond strengths are in the order $1° > 2° > 3°$. Ease of abstraction of hydrogens tends to have the same order. However, among alkanes, methane itself is relatively unreactive.

Electronegative or electron-withdrawing substituents attached to carbon reduce the magnitude of the rate constant. As shown in Table 16.4, the rate constants for reaction of OH$^•$ with a series of fluoro- and chloromethanes are the result of factors such as the number and type of hydrogens and halogens involved. The extremely small rate constants shown are part of the reason CFCs are effectively inert in the troposphere. Rates of reaction with OH$^•$ are extremely slow. In addition to this, the equilibrium for the reactions is decidedly unfavorable.

Replacing CFCs with compounds containing C-H bonds increases reactivity in the troposphere and decreases the potential to diffuse into the stratosphere and affect the ozone layer. HCFCs and HFCs are first-generation replacements for CFCs. HFC-134a (CF_2CH_2F) is regarded as a viable alternative to the CFC-12 used in some automobile air conditioning systems. Figure 16.8 shows a possible fate for HFC-134a in the troposphere, beginning with reaction with OH$^•$. Oxidation eventually results in the formation of the number of more polar products that would be susceptible to removal from the troposphere by rain. HCFCs and HFCs still have some ozone depletion potential however.

Complete replacement of the halogens in CFCs by hydrogen would result in hydrocarbons that are unsuitable for some CFC applications, since they are flammable. A possible new class of replacement compounds comprises hydrofluoroiodoalkanes or fluoroiodoalkanes. The C-I bond can rupture on absorption of UV radiation in the troposphere, enhancing the breakdown of these compounds.

TABLE 16.4
Rate Constants and (298K) Lifetimes for the Reaction of OH˙ with Methane and a Series of Chlorinated and Fluorinated Methanes in the Troposphere

Compound	Rate constant (cm^3 molecule s^{-1})	Lifetime[a] (years)
Methane (CH_4)	8.4×10^{-15}	7.5
Chloromethane (CH_3Cl)	4.4×10^{-14}	1.4
Dichloromethane (CH_2Cl_2)	1.4×10^{-13}	0.5
Trichloromethane ($CHCl_3$)	1.0×10^{-13}	0.6
Tetrachloromethane (CCl_4)	$<1 \times 10^{-15}$	>63
Fluoromethane (CH_3F)	1.7×10^{-14}	3.7
Difluoromethane (CH_2F_2)	1.1×10^{-14}	5.8
Trifluoromethane (CHF_3)	2.0×10^{-16}	320

[a] Lifetime at 298K with OH˙ concentration of 5×10^5 molecules cm^{-3}.

Data from Atkinson, R., *Chem. Rev.*, 85, 69-201, 1985. With permission.

$$CF_3CH_2F + OH^{\bullet} \longrightarrow CF_3CFH^{\bullet} + H_2O$$

1,1,1,2–Tetrafluoroethane
HFC–134a

$$CF_3CFH^{\bullet} + O_2 \longrightarrow CF_3CFHOO^{\bullet}$$

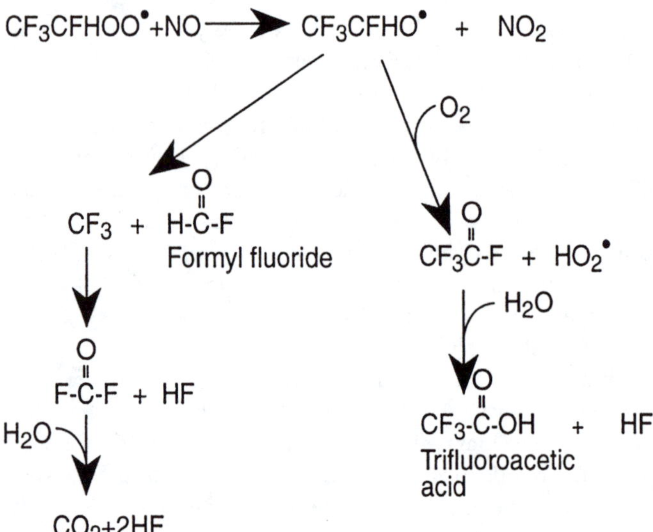

FIGURE 16.8. A diagram of a possible fate for HCF-134a in the troposphere.

B. OZONE

It has been known from laboratory experiments for a considerable time that reaction of ozone with alkenes in solution results in cleavage of the carbon chain between the carbons of the double bond. Depending on reaction conditions, products include aldehydes, ketones and carboxylic acids. It is thought that a similar process occurs in the troposphere. Despite ozone generally having a higher concentration than other reactive intermediates such as $OH^{\bullet}$ and NO_3, reaction with ozone is the most important fate only for some relatively small alkenes, e.g., *trans*-2-butene and limonene.

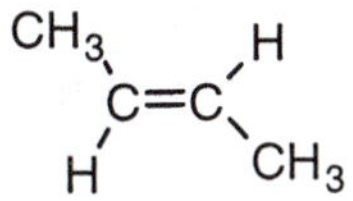

$CH_3(H)C{=}C(H)CH_3$

Trans-2-butene

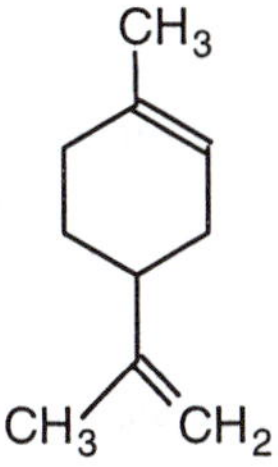

Limonene

Ozone is electrophilic in character (as are $OH^{\bullet}$ and NO_3). Electrophilic literally means "electron loving." Ozone therefore tends to react at sites of increased electron density, such as carbon-carbon double bonds. Electron-withdrawing substituents (e.g., Cl and F) attached to these carbons tend to reduce the magnitude of the rate constants. Reaction of ozone with alkynes and aromatics is, however, quite slow. There is effectively no reaction between ozone and saturated organic molecules, e.g., alkanes and CFCs.

C. NITRATE RADICAL

The interaction of NO_3 with organic compounds is analogous to that of $OH^{\bullet}$. It involves abstraction of H from saturated molecules or regions of molecules and addition to unsaturated centers and aromatic rings. With abstraction, instead of forming H_2O as is the case with $OH^{\bullet}$, nitric acid (HNO_3) is formed.

$$RDH + NO_3 \rightarrow R^{\bullet} + HNO_3$$

Although $OH^{\bullet}$ is generally more reactive toward hydrocarbons than NO_3, nighttime concentrations of perhaps 1×10^9 molecules cm^{-3} compared with daytime ones of 5 to 6×10^{-6} molecules cm^{-3} for $OH^{\bullet}$ means that for some molecules such as terpenes, nighttime reaction with NO_3 dominates.

D. PHOTOTRANSFORMATION

Phototransformation is the process by which molecules absorb photons of electromagnetic radiation (i.e., light), followed by some sort of transformation (e.g., dissociation, isomerization or reaction with another molecule). The key is the initial absorption of light, which activates the molecule and enables later reactions to occur. Phototransformation processes are discussed in general in Chapter 3.

Of the light encountered in the troposphere, only that with wavelengths between 290 nm and 400 nm generally has sufficient energy to enable phototransformation to occur. This wavelength range is in the near-UV. If a molecule is to be capable of phototransformation, it must be able to absorb light in this wavelength range. The extent of the overlap between the solar emission spectrum in the troposphere and the absorption spectrum of the compound is related to the rate constant for phototransformation. As shown in Figure 16.9, if there is no overlap, no phototransformation is possible. How large the phototransformation rate constant is also depends on the quantum yield (ϕ). The overall quantum yield for phototransformation is defined by:

$$\phi = \frac{\text{Number of molecules undergoing phototransformation}}{\text{Total number of photons absorbed}}$$

Not every photon that is absorbed causes phototransformation. The absorbed light can simply be reradiated. For example, if 100 photons are absorbed, which results in 50 molecules being transformed, then the quantum yield is 0.5. The value of ϕ is a function of the wavelength of the photons involved and its units are molecules photon^{-1}.

Overall, the rate constant for phototransformation (k_p) is given by summing the product of overlap and quantum yield over all relevant wavelengths. The calculation of k_p is described in detail in Chapter 3.

$$k_p = \Sigma\phi(\lambda)\sigma(\lambda)J(\lambda)$$

where σ (cm^2 molecule^{-1}) is the absorption cross-section, a measure of the absorbance of the compound at a given wavelength, and J (photons cm^{-2} s^{-1}) is the solar radiation intensity at that wavelength. The magnitude of J depends on factors such as latitude, season, time of day and extent of cloud cover. By following through the units in the expression for k_p, it can be seen that they are s^{-1} or reciprocal time. This denotes a first-order rate constant, and so for a compound with a concentration in the troposphere of X (molecules cm^{-3}), then:

$$\text{Rate of phototransformation (molecules cm}^{-3}\text{s}^{-1}) = k_pX$$

$$\text{Half-life with respect to phototransformation (s)} = 0.693/k_p$$

$$\text{Lifetime with respect to phototranformation (s)} = 1/k_p$$

Saturated organic molecules do not absorb between 290 and 400 nm. This is another reason why CFCs are inert in the troposphere. Such compounds can however absorb the shorter wavelength at higher energy radiation found in the stratosphere, which may result in cleavage of a C-Cl bond. The released chlorine atom can catalyze the destruction of ozone. Simple monocyclic aromatic compounds (e.g., benzene) and compounds with isolated (i.e., nonconjugated) carbon-carbon multiple bonds do not absorb the solar UV radiation in the troposphere. Compounds that do include aldehydes, ketones and larger PAHs.

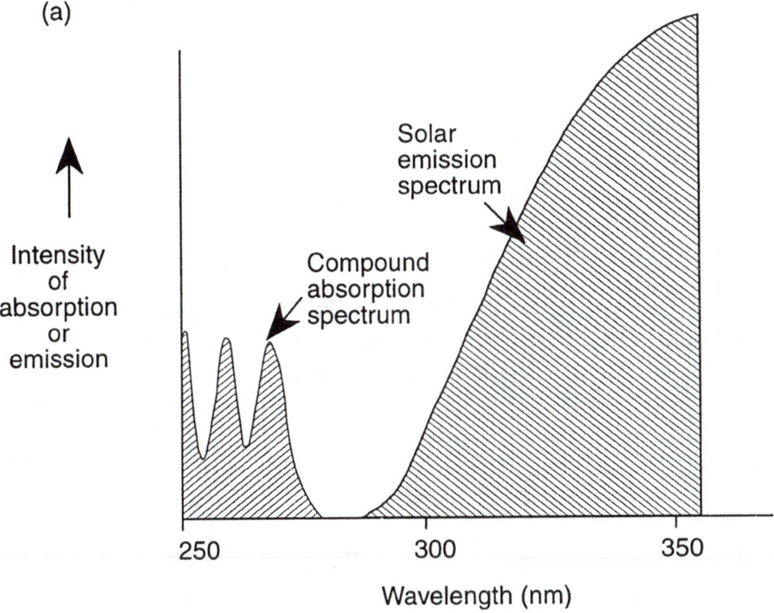

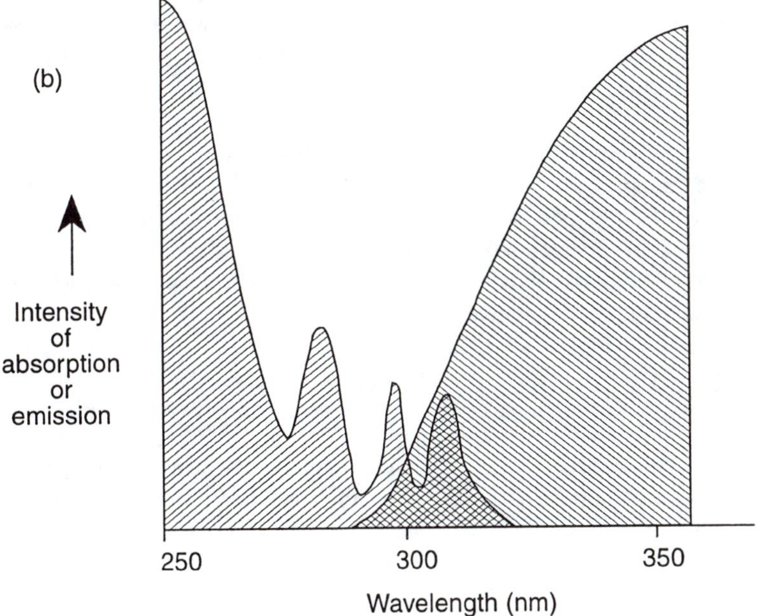

FIGURE 16.9. Relationships of absorption spectrum of compounds and the solar spectrum. (a): No overlap between the solar emission spectrum and a compound's absorption spectrum, so no phototransformation is possible. (b): With another compound, overlap occurs and so phototransformation is possible.

V. KEY POINTS

1. The atmosphere plays several key roles in maintaining the global ecosystem. The principal roles are:
   - The external medium for photosynthesis and respiration
   - Source of carbon dioxide and a sink for oxygen
   - A medium over geological time to support chemical revolution
   - A heat reservoir
   - A shield against damaging short wavelength radiation
2. The atmosphere can be divided into zones, starting from the troposphere at the earth's surface (about 10 to 16 km in depth; average temperature, 15°C; normal atmospheric composition, 1 atm pressure); stratosphere (15 to 50 km in depth, temperature, –56 to –2°C; 0.1 to 0.001 atm pressure); mesosphere (50 to 85 km in depth; temperature, –2 to –92°C; 0.001 to 3×10^{-6} atm pressure); thermosphere (85 to 500 km in depth; temperature, –92 to +1200°C; $<3 \times 10^{-6}$ atm pressure); and space.
3. The composition of the troposphere is 78.09% N_2; 20.95% O_2; 0.93% Ar; 300 ppm CO_2; 18 ppm Ne; 1.5 ppm He; 1.5 ppm CH_4 and many other minor components.
4. The upper atmosphere contains a range of chemical entities that are highly reactive (e.g., O_3, O^+, O_2^+, etc.) These substances have a limited life at the Earth's surface, but can exist in the upper atmosphere due to the low pressures, giving long mean free paths and relatively infrequent intermolecular collisions during which reaction could occur.
5. There are several important reactive intermediates that participate in reactions in the atmosphere. These include $OH^•$ (hydroxyl radical). The $OH^•$ radical is formed from ozone, water and other reactants in the atmosphere by a series of reactions. The uptake of energy through solar radiation is essential for some of these reactions. Thus, the $OH^•$ radical is formed during daylight hours when it reaches its peak concentration.
6. About 90% of the Earth's ozone occurs in the stratosphere, where it plays a major role in absorbing biologically damaging short-wavelength radiation.
7. Ozone in the troposphere is an important component of smog, where it is formed by photolysis of NO_2. A significant proportion of the NO_2 present is formed from motor vehicle exhaust gases:

$$NO_2 \xrightarrow{h\upsilon} NO + O$$

$$O + O_2 + M \longrightarrow O_3 + M$$

8. The nitrate radical (NO_3) is formed from NO_2, of which a significant proportion in urban atmospheres is derived from motor vehicle exhaust gases. The formation reaction with ozone can be represented by:

$$NO_2 + O_3 + M \rightarrow NO_3 + NO_2$$

During daylight hours, NO_3 is rapidly broken down by photolysis reactions.

9. Methane is principally emitted to the atmosphere from fossil fuels, biomass burning and anaerobic microbial activity. The main loss process is reaction with $OH^{\bullet}$.

10. Methane has a lifetime in the troposphere of 5 to 10 years and is an important greenhouse gas since it will absorb some of the IR radiation emitted at the Earth's surface.

11. Formaldehyde, acrolein and peroxy acetyl nitrate (PAN) are all important atmospheric pollutants that have damaging biological effects. PAN is a secondary pollutant formed by a series of reactions involving nitrogen- and oxygen-containing intermediates.

$$H_3C-\overset{\overset{\displaystyle O}{\|}}{C}-O-O-NO_2$$

PAN

12. The halomethanes, H_3CCl, H_3CBr and H_3CI, are all natural components of the troposphere, where they persist for periods of months to years.

13. The chlorofluorohydrocarbons (CFCs) have been manufactured since the 1930s and used as refrigerants, aerosol propellants and in a range of other applications. The CFCs produced in greatest quantities are CFC-11 ($CFCl_3$) and CFC-12 (CF_2Cl_2).

14. The CFCs are relatively inert in the troposphere, but in the stratosphere can lose a halogen that can react with ozone, resulting in depletion of this layer. This may result in a larger amount of short wavelength radiation reaching the Earth's surface.

REFERENCES

Finlayson-Pitts, B.J. and Pitts, J.N., *Atmospheric Chemistry: Fundamentals and Experimental Techniques,* Wiley-Interscience, New York, 1986.

Wayne, R.P., *Chemistry of Atmospheres,* 2nd ed., Clarendon Press, Oxford, 1991.

Baird, C., *Environmental Chemistry,* W.H. Freeman, New York, 1995.

CHAPTER 16 PROBLEMS

1. Many smog components are produced by motor vehicle exhaust gases. The general 24-hour cycle of motor vehicle traffic is shown in Figure 16.10. Draw possible patterns for NO (the main nitrogen oxide in vehicle exhaust), NO_2 + PAN, and explain the reasons behind these patterns.

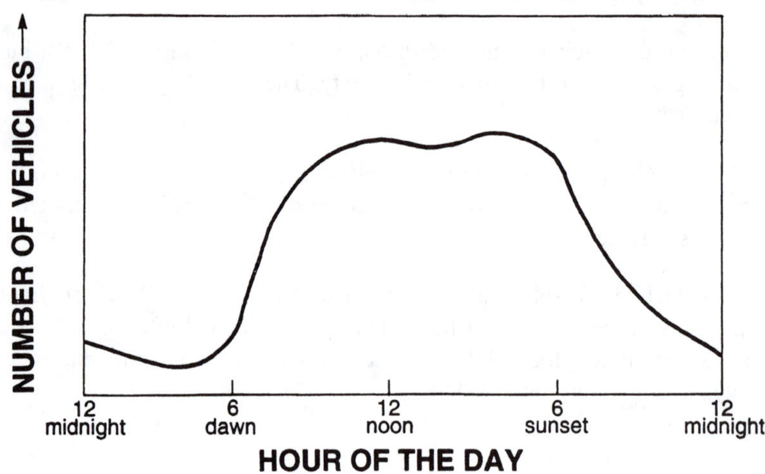

FIGURE 16.10. The general 24-hour cycle of motor vehicle traffic.

2. The more commonly produced CFCs were CFC-11 ($CFCl_3$) and CFC 12 (CF_2Cl_2). Suggest chemcial reactions that could explain the reactions of these substances with ozone.

Answers on page 389.

CHAPTER 16 SOLUTIONS

1. An outline of possible changes over the 24-hour cycle of NO, NO_2 + PAN due to motor vehicle exhausts is shown in Figure 16.11.

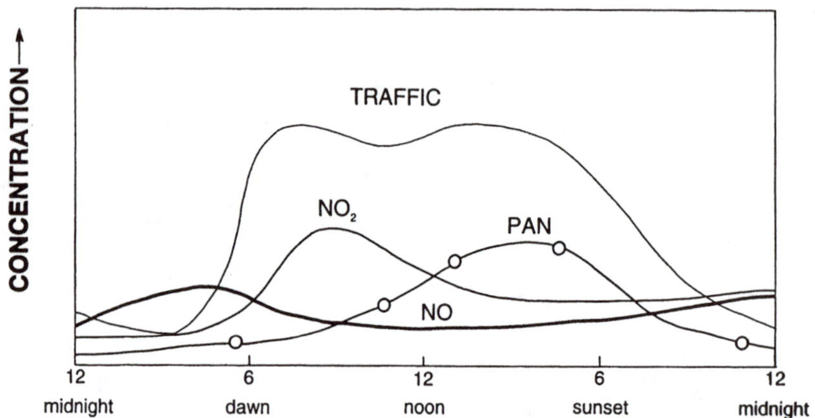

FIGURE 16.11. General pattern of change in traffic of NO, NO_2 and PAN over a 24-hour cycle.

NO: A primary component of motor vehicle exhaust gas which is low, due to low vehicle traffic, overnight. It starts to rise at dawn with increasing vehicle traffic, but is converted to NO_2 during the day.
NO_2: Formed from NO early in the day, but consumed to form PAN later in the day.
PAN: Formed later in the day by reaction of NO and other components acted on by solar radiation. Overnight, the input of vehicle exhaust gases declines substantially and the input of solar radiation ceases. Thus, all components tend to decline or remain reasonably constant.

2. The CFCs are very stable compounds which contain only strong carbon-halogen bonds.
CFC–11 ($CFCl_3$): The photoreaction of these compounds leads to the splitting of the CFC molecule at the low pressures in the stratosphere to form the highly reactive $^{\cdot}CFCl_2$ and $Cl^{\cdot}$. Thus,

$$CFCl_3 \xrightarrow{h\upsilon} {}^{\cdot}CFCl_2 + Cl^{\cdot}$$

Solar radiation in the stratosphere contains short wavelength, high-energy radiation that can be absorbed by the CFC molecule resulting in this reaction. The chlorine can further react with ozone as shown:

$$Cl^{\cdot} + O_3 \rightarrow ClO + O_2$$

This results in depletion of ozone present.
CFC12: A similar process can occur with this compound.

Chapter 17

SOIL CONTAMINATION

I. INTRODUCTION

Increasing human populations have led to the rapid expansion of existing urban areas and the development of new ones. Areas used by industry in the past have often become valuable land for the development of domestic housing in many parts of the world. As a result, many former industrial areas now have extensive housing developments containing large numbers of people located on them.

In the 1970s, with the expansion of investigations into the occurrence of contamination in the environment, attention was directed to areas of soil contamination. Throughout the world, it became apparent that many substances that are well known toxicants or carcinogens were present in relatively high concentrations in the soil in some urban areas. Many of these areas became the focus of public controversy and have become famous as contaminated sites. Thus, heavily contaminated sites such as Love Canal, the Picello Farm and the Valley of Drums have become household names associated with soil contamination in many countries.

Essentially, the problems of soil contamination can be considered to consist of two fundamental types: (1) those that result in the exposure of the human population to contaminants, and (2) those that result in the exposure of natural ecosystems. Human exposure results from development of schools, kindergartens, domestic housing, as well as offices and workplaces on contaminated sites. Exposure of natural systems can occur in several ways. Natural terrestrial biota in contaminated areas are directly exposed by contact with soil-borne contaminants. In addition, water runoff and groundwater seepage from contaminated soils to natural aquatic systems results in exposure of adjacent aquatic ecosystems. Dust and vapors can also distribute contaminants into wider areas as well.

The management of contaminated soils is now a major governmental activity. In many countries, the problem is of considerable magnitude. For example, in the Netherlands, over 100,000 sites have been identified as being potentially contaminated, with 10,000 sites confirmed as being contaminated. Likewise, in Germany, over 50,000 potentially contaminated sites have been identified. In the United States, there are an estimated 100,000 sites that have been nominated as contaminated, with some 10,000 of these designated as priority areas.

II. SOURCES OF SOIL CONTAMINATION

The contamination of soil can originate essentially from the following activities:

- Industrial operations
- Agricultural activities
- Domestic and urban activities

TABLE 17.1
Examples of Activities Resulting in Soil Contamination

Industrial operations
    Chemical industries, gas and electricity supply, wood preserving, oil refining, service stations, smelters, mining, tanning, dockyards, waste dumps.
Agricultural activities
    Treatment of crops, handling and storage of agricultural chemicals, use of cattle dips.
Domestic and urban activities
    Solid waste disposal, sewage sludge disposal, sewage works and farms, motor vehicle discharges, usage of chemicals.

Some examples of activities within these categories that result in soil contamination are shown in Table 17.1.

Deliberate disposal of industrial waste to land has been a common disposal method. Generally, this has not been carried out in disregard for the environment but through a lack of regulation by government and a lack of understanding of potential adverse consequences. In fact, many of the land disposal operations were approved by governments as the most appropriate disposal operation for hazardous chemicals. Trenches and pits have been used in which waste from such industrial operations as tanneries and coal gas plants were disposed. Accidental spills are also a major cause of soil contamination. Accidental spillages have occurred frequently in such operations as wood preserving, petrol stations, fuel depots and similar activities. The operation of smelters giving atmospheric discharges of contaminated particulates that subsequently deposit in soils is another source of contamination that occurs in many areas. Mining wastes have often been disposed of in special dams and other land-based operations, resulting in soil contamination. The broadcast use of pesticides on crops has resulted in widespread contamination of soil in some areas. More intense contamination has often occurred in specific rural areas where pesticides are stored, distributed and loaded onto vehicles. In addition, the use of dips for treatment of cattle has often resulted in contamination of relatively small areas.

Activities in normal domestic and urban situations also result in soil contamination. Perhaps the major source of contamination in this area is the disposal of solid waste to land areas. Sewerage sludge disposal can contain high levels of contaminants and also be disposed of to soil. The use of motor vehicles results in discharges of lead and other contaminants in particulate form, which accumulate in soils in the vicinity of busy roads. A range of chemicals is used in domestic situations. For example, pesticides and other chemicals are used in the maintenance of gardens and lawns. The usage of chemicals and the disposal of waste following the usage can result in contamination of soils in urban areas.

It should be kept in mind that natural soils are not necessarily free of hazardous compounds that may have deleterious biological effects. Of course, compounds that have only originated as a result of synthetic chemical processes would not be present in natural soils. Thus, the synthetic pesticides, such as DDT and dieldrin, as well as industrial chemicals such as the PCBs would not be present in natural soils. The occurrence of these substances would be as a result of human contamination. On the other hand, polycyclic aromatic hydrocarbons (PAHs described in Chapter 9), polychlorodibenzo dioxins (PCDDs described in Chapter 6) and polychlorodibenzo furans (PCDFs described in Chapter 6) are all produced in low concentrations by combustion of organic matter, including natural organic matter such as wood and paper. These substances have been found in low concentrations in natural soils throughout the world. However, there is a tendency for these substances to occur in the vicinity of urban and industrial areas in higher concentration than the natural background levels. Metals such as lead, mercury, and arsenic occur naturally in high concentrations in various geological strata. Where these formations reach the surface, the associated soils can contain relatively high concentrations. Also, there can be the migration of these substances to other areas, which can lead to the contamination of other soils in the vicinity. Thus, natural soils may contain levels of substances that can be considered hazardous to human health and natural ecosystems.

III. CHEMICAL NATURE OF SOIL CONTAMINANTS

Some common soil contaminants and their associated sources are shown in Table 17.2. Clearly, soil contaminants do not fall into a single, or several simple classes of chemicals, but are very diverse in chemical nature. In fact, the only common property of soil contaminants is that these substances can be harmful to the natural environment or human health.

TABLE 17.2
Chemical Identity of Some Soil Contaminants

Chemical	Source
Arsenic Copper Chromium	Tanneries, wood preserving, mining wastes, cattle dips, smelters
Lead	Smelters, motor vehicles
Petroleum hydrocarbons	Petrol stations, fuel depots, accidental spillages, disposal of waste chemicals
Polyaromatic hydrocarbons (PAHs) Polychlorodibenzodioxins (PCDD) Polychlorodibenzofurans (PCDF)	As for petroleum hydrocarbons, coal gas plants
Pesticides (DDT, Dieldrin, etc.)	Agricultural areas, disposal of waste chemicals, cattle dips
Industrial chemicals (solvents, PCBs, acids, alkalis, etc.)	Disposal of waste chemicals, industrial operations

The **heavy metals** are common soil contaminants. For example, **lead** originating from motor vehicles is a very common soil contaminant in urban areas, particularly in the vicinity of roadways. There are also many examples of lead contamination where the lead originates from smelters and mining operations. Although lead in paints has been banned in many countries, there can be buildings that were painted prior to the ban. Renovating in recent times can result in soil contamination or adverse effects in the environment. **Copper** and **chromium** are also common heavy metal soil contaminants and primarily originate from tanneries and wood preserving plants. Another common soil contaminant is **arsenic**, which has found wide use in the past as a preservative of hides in tanning, a pesticide with cattle in dips and a wood preservative. It is also a natural soil component in many areas. Strictly speaking, arsenic is not a metal but it is convenient to consider this substance in the heavy metal group (see Chapter 11, Organometallic Compounds).

A wide range of organic substances also occur as contaminants in soil. Almost all of the organic substances used in industry and society in general can occur as soil contaminants. **Petroleum hydrocarbons** are particularly common due their wide distribution and use in petroleum products. Similarly, PCDDs and PCDFs are also widespread as a result of the use of combustion for many different purposes. The concentrations of both these groups of chemicals are generally low, but can be relatively high in particular localities. Of particular note here is the occurrence of former coal gas plant sites in many cities. With many of these plants, the waste coal tar produced was disposed into wells and pits. Coal tar consists of a highly complex mixture of aromatic hydrocarbons, phenols and polycyclicaromatic hydrocarbons. Thus, many cities contain sites with very high levels of these soil contaminants.

Organic liquids are used as solvents in many industrial processes and products such as paints. Petroleum products such as gasoline, naphtha, toluene and xylene are common solvents used in industry. In addition, chlorohydrocarbons such as dichloroethylene and carbon tetrachloride are commonly used as solvents. It is important to note that chlorinated solvents are being phased out of industrial use due to their adverse effects on the ozone layer (see Chapter 16).

In agricultural areas, pesticides and other agricultural chemicals are often present in soil. Even pesticides having discontinued use, such as DDT and the other chlorohydrocarbon pesticides, can be present in agricultural soils due to previous usage. Of course, the concentrations of all these substances would be expected to be declining due to environmental transformation and degradation processes (see Chapters 3 and 8).

IV. IMPORTANT ENVIRONMENTAL PROPERTIES OF SOILS

Soils are a complex mixture of substances that vary in composition from area to area. In dry areas and beaches, the soil consists essentially of silica sand with some calcium carbonate components but very little else. In most agricultural and urban areas, the soil components that affect the environmental properties of contaminants are principally clay and organic matter. Clay consists of various hydrous silicates and oxides that can be characterized by such measures as cation exchange capacity and the specific surface area ($m^2\ g^{-1}$). These properties give a measure of how clay affects the behavior of polar organic molecules and metal ions. The cationic exchange capacity is a measure of the capacity of the soil to sorb cations with which it comes in contact. Thus, strongly **cationic pesticides** such as diquat (see Figure 17.1) are strongly sorbed by the clay component in soil. It is interesting to note that diquat is also very soluble in water and thus is highly hydrophilic, as illustrated by the data in Table 17.3. However, the sorption to clay is sufficiently strong to overcome the highly hydrophilic properties of this compound, which would favor its occurrence in water. Glyphosate exhibits somewhat similar properties (see Table 17.3). The structure of glyphosate is shown in Figure 17.1. It is a molecule with several sites for cationic and anionic effects and in fact exists as a zwitter ion. Thus, a hydrogen ion can move internally between ionic groups, depending on the ambient pH. This compound is highly hydrophilic and sorbs strongly to clay minerals in soil.

TABLE 17.3
Properties of Some Compounds Related to Their Behavior on Soils

Compound	log K_{OW}	Water solubility (mg L^{-1})	log K_{OC}	$t_{1/2}$ (days)	VP[a] (mm)	Mi[b]
Diquat	—	700,000	Highly sorbed to clay	—	—	—
Glyphosate	–1.7	1200	Highly sorbed to clay	50–70	—	—
Atrazine	2.75	30	2.0	1–8.0	3×10^{-7}	–5.4
Malathion	2.36	143	3.3	3–7	4×10^{-5}	–2.2
DDT	6.2	0.0032	4.8	700–6000	2×10^{-7}	–9.2
Dieldrin	4.3	0.17	4.0	175–1100	1×10^{-7}	–7.8
Toluene	2.69	515	3.5	4–22	10	3.16
Benzo(a)pyrene	6.0	0.004	5.5	57–490	—	—
Benzene	2.24	16.40	3.3	5–16	76	3.9

[a] Vapor pressure.
[b] Mobility index.

The organic compounds DDT, dieldrin and benzo(a)pyrene are also strongly sorbed by soil, as indicated by the values of log K_{OW} shown in Table 17.3. The K_{OC} value is the concentration in soil organic matter/concentration in water at equilibrium, and the K_{OC} values are 63,000 (DDT), 10,000 (dieldrin), and 32,000 (benzo(a)pyrene). So, all these compounds are strongly sorbed to soil; but in this case, the compounds are all very insoluble in water, with solubilities less than 0.16 mg L^{-1} and are highly nonpolar and lipophilic (see Figure 17.1). These compounds don't sorb to the clay fraction of the soil to any significant extent.

Soil organic matter usually ranges in content from 0.1 to 7% of the total soil. It provides the most important component for the sorption of nonionic and nonpolar chemicals. The organic matter is derived principally from plant detritus that enters soil as plant litter and as a result of the death of plants. The activities of microorganisms on this plant detritus produce the organic matter that is a major component of soil. Chemically, the main component in the plants that contributes to soil organic matter is plant **lignin**. This is a complex phenolic compound of high molecular weight and is relatively resistant to degradation in the environment. This plant material is acted on by microorganisms and other processes that result in the formation of **humic substances** often referred to as **humic** and **fulvic acids**. Humic and fulvic acids are similar and related chemically to lignin, and are separated principally by the physical properties of solubility in acid and alkaline solutions (see Chapter 14, Chemistry of Natural Waters).

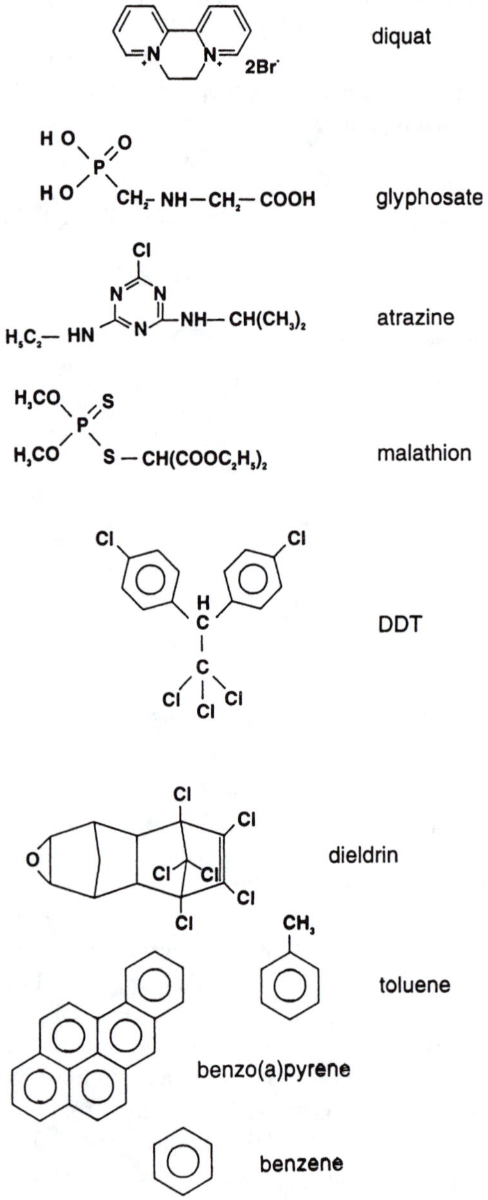

diquat

glyphosate

atrazine

malathion

DDT

dieldrin

toluene

benzo(a)pyrene

benzene

FIGURE 17.1. Chemical structure of some soil contaminants.

The molecules of these substances are relatively large and probably vary, depending on the source material and degradation conditions. A hypothetical structure is shown in Figure 17.2. An important aspect of the structure of humic and fulvic acids are that these substances contain a range of phenolic groups, as shown in Figure 17.3. These groups have the capacity to complex metal ions, as shown in Figure 17.4. Many of the groups have the capability of forming chelation complexes with attachment of metal ions to more than one site in the organic molecule. These chelation complexes usually exhibit considerable stability and resistance to breakdown in the environment.

Importantly, the humic and fulvic acids also have lipophilic properties. There are many polar and ionic sites in these molecules, but their large size probably gives them the lipophilicity property since lipophilicity generally increases with molecular size. So, lipophilic compounds exhibit strong sorption to organic matter in soils.

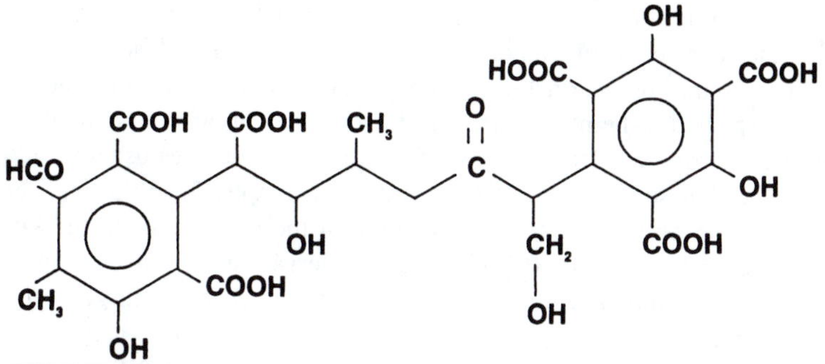

FIGURE 17.2. A proposed hypothetical chemical structure for fulvic acid.

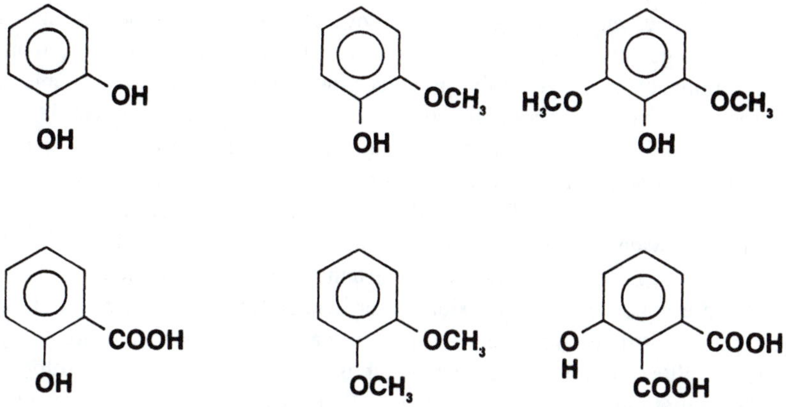

FIGURE 17.3. Some of the structured groups present in humic and fulvic acids.

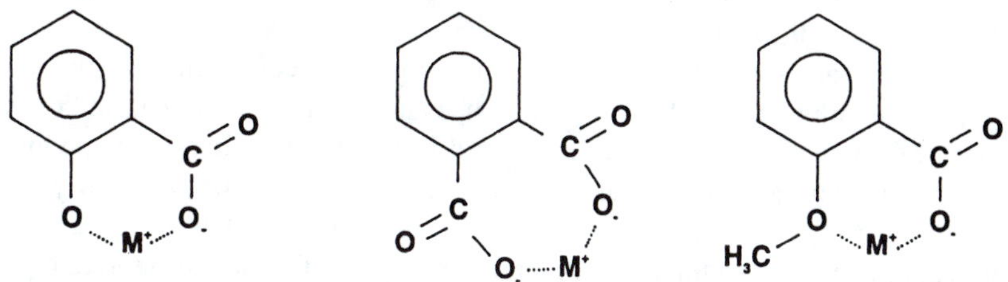

FIGURE 17.4. Examples of possible complexes formed by humic and fulvic acid active groups with metal ions.

V. IMPORTANT ENVIRONMENTAL PROPERTIES OF SOIL CONTAMINANTS

One of the most important properties of contaminants in soil is their persistence. Organic compounds will be degraded by microorganisms as well as by abiotic and other processes. Also, they will be volatilized from the soil and removed by water leaching processes. Of course, metals and organometallic compounds are not susceptible to degradation beyond the elemental state. However, these substances can be removed from soils by transformation to a volatile organic form or organic complex that can result in evaporation into the atmosphere, leaching into ground waters or loss through the action of stormwater runoff. Thus, in general, metals would be expected to be more persistent in soils than organic compounds.

Substances removed from soil by environmental processes usually follow first-order kinetics. These processes were described in Chapter 3, so this chapter should be referred to in order to obtain background on the approaches to the kinetics of these processes. The half-life ($t_{1/2}$) is the characteristic usually used to measure the persistence of a compound in soil. Environmental persistence, in general, is often considered in this phase of the natural environment because soil is a major repository for contaminants in the environment.

A characteristic of particular importance with soils is the moisture content since this exercises a control over potential for growth of microorganisms. Dry soils do not support an actively growing population of microorganisms, whereas moist soils can support large populations of microorganisms. The type of microorganism present can also influence the degradation processes, as can environmental variables such as temperature and the availability of oxygen. Because of the variability in the composition and population of microorganisms, it would be expected that compounds in soil would exhibit a corresponding variability in persistence measured as the $t_{1/2}$. Some ranges of $t_{1/2}$ values found in soil are shown in Table 17.3. The longest $t_{1/2}$ values are evident with the chlorohydrocarbons, DDT and dieldrin, which have $t_{1/2}$ values from 175 to 6000 days. These substances contain a limited ranged of bond types that are not susceptible to oxidation or hydrolysis, which are common degradation and transformation processes. The hydrocarbons benzene and benzo(a)pyrene exhibit $t_{1/2}$ values from 5 to 490 days, which is less than that of the chlorohydrocarbons. The remaining compounds in Table 17.3 all exhibit shorter $t_{1/2}$ values due their relatively high water solubility, making them more readily available to microorganisms and also to the presence of chemical groups within the molecule, rendering the compound susceptible to attack.

Volatilization is a major process for the removal of contaminants from soil. A diagrammatic illustration of the process of evaporation from soil is indicated in Figure 17.5. Several processes are illustrated in this figure. First, a compound can partition between the soil particle and the pore water present between the particles (illustrated as $C_S \Leftrightarrow C_W$). Diffusion in the pore water can then occur, and some chemical molecules eventually reach the pore water surface and evaporate together with water molecules as well. Volatilization depends on:

- Inherent properties of a chemical
- Properties of soil

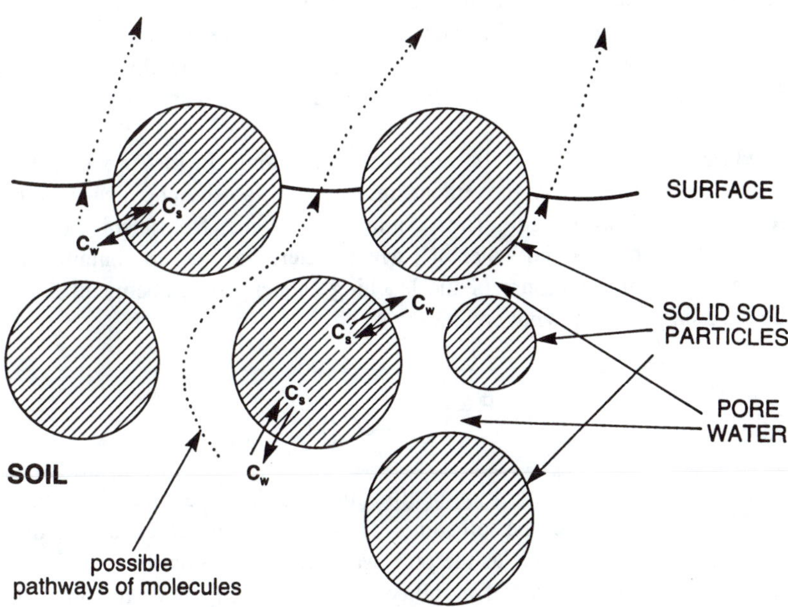

FIGURE 17.5. Processes involved in the volatilization of a contaminant from soil where C_S is the contaminant in soil solids and C_W the contamination in water.

- Environmental conditions

The inherent properties of the molecule are properties such as molecular weight, polarity and other characteristics that govern its vapor pressure and Henry's law constant (water/air distribution coefficient). The properties of the soil that influence the soil/water partition process can be seen as part of the volatilization process, as shown in Figure 17.5. Thus, the organic carbon content of the soil influences the volatilization rate. The higher the organic carbon content, the more the lipophilic organic compounds are retained and the lower the evaporation rates. The moisture content of the soil is also a key characteristic. High soil moisture contents give higher evaporation rates. This may be due to higher water content resulting in greater water loss from the soil and with it a greater amount of the contaminant. This effect is often referred to as the *Wick effect*. These processes are influenced by environmental conditions such as temperature and surface air speeds, with increases in both these factors leading to increased rates of volatilization.

A variety of expressions have been derived to calculate the loss of a chemical by volatilization. One of the simplest of these applies to a chemical at the surface and is referred to as the **Bow Model**. The Bow Model can be expressed as follows:

$$t_{1/2} = 1.58 \times 10^{-8} \left(\frac{K_{OC} S}{P} \right) \text{days}$$

where $t_{1/2}$ is the half-life in days, K_{OC} is the soil water partition coefficient in terms of organic carbon, S is the aqueous solubility (mg L^{-1}) and P is the vapor pressure of the compound at the ambient temperature (mmHg). This model is a general model that takes no account of environmental conditions. This means that as K_{OC} and sorption of the chemical to soil particles increases, the rate of loss declines and $t_{1/2}$ increases. Similarly, as solubility in water increases, the volatilization of the chemical declines. On the other hand, as the vapor pressure (P) increases, the volatilization increases and $t_{1/2}$ declines.

Often, a chemical can also be removed by leaching. In this process, the pore water is displaced by water movement and in doing so the chemical in the pore water is removed from the soil. Thus, contaminated water from a soil can move to other areas, possibly ground water or surface water, and lead to contamination in those areas. A simple measure of the leaching capacity of a chemical, R, can be calculated using the following equation:

$$R = \frac{1}{K_D(1 - \phi^{2/3})d_s}$$

where K_D is the soil/water partition coefficient, ϕ is the pore water fraction of the soil, and d_s is the density of the soil solids. Thus, leachability of a chemical declines as K_D increases, and increases as the pore water fraction increases.

Often, an overall measure of mobility of organic compounds in soil is useful. This can be used as a measure of the likely decline in concentration of a soil contaminant due to losses from volatilization and leaching. A soil Mobility Index (MI) can be calculated using the following equation:

$$MI = \log\left[\frac{SV}{K_{OC}}\right]$$

where S is water solubility (mg L^{-1}), V is vapor pressure at ambient temperature (mm), and K_{OC} is the soil sorption partition coefficient in terms of organic carbon.

Some MI values for different compounds are shown in Table 17.3. The meaning of the MI values as measures of mobility is shown in Table 17.4. This indicates that DDT and dieldrin are clearly immobile in soil and have little potential to contaminate surface and ground water in the water phase. Of course, particles containing sorbed substances may move to contaminate other areas in addition to the water itself. atrazine and malathion are more mobile and move slightly in soils and have some ability to contaminate other environmental phases. Toluene and benzene are very mobile and can readily contaminate surface and ground waters adjacent to the contaminated area.

VI. DISTRIBUTION OF SOIL CONTAMINANTS

Mobile chemicals, and even slightly mobile chemicals, in soil can redistribute from soil into various other environmental phases. The major phases influencing the

TABLE 17.4
Interpretation of the Relative Mobility Index

Mobility index	Description
>5.00	Extremely mobile
5.00 to 0.00	Very mobile
0.00 to –5.00	Slightly mobile
–5.00 to –10.00	Immobile
<–10.00	Very immobile

redistribution of soil contaminants are the abiotic phases: the atmosphere, soil solids and pore water. The partition processes involved are illustrated in Figure 17.6. Chemicals distribute between these phases by sets of two-phase processes. Some of these are illustrated below where C_A, C_W and C_S are the chemical concentrations in air, water and soil solids, respectively:

$$\text{Atmosphere } (C_A)$$
$$\downarrow\uparrow$$
$$\text{Soil solids } (C_S)$$
$$\downarrow\uparrow$$
$$\text{Pore water } (C_W)$$

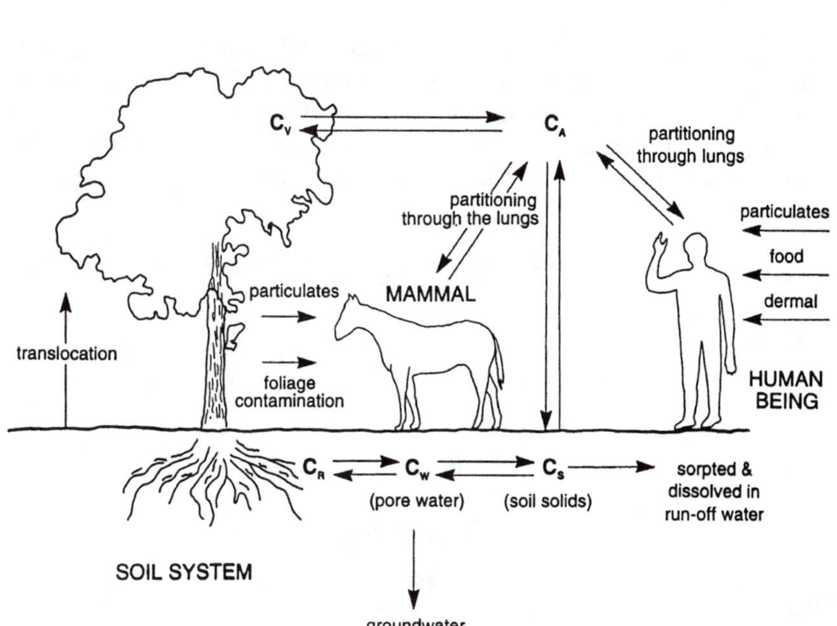

FIGURE 17.6. Distribution patterns of soil contaminants in soil ecosystems.

Movement to ground water is a transfer of chemical by actual movement of the water phase. Other mass-transfer processes occur when stormwater runs off a contaminated area containing dissolved and particulate sorbed chemicals.

A. THE SOIL/WATER PARTITION PROCESS

A lipophilic chemical, in concentrations below its solubility in water, when placed in a soil/water system, will reach equilibrium by movement of molecules between the two phases. In this condition, the concentrations in the soil solids (C_S) and in water (C_W) are constant. Thus, C_S/C_W is constant and is referred to as the soil/water partition coefficient, K_D. This K_D can be measured in laboratory experiments by repeated experiments at different concentrations to obtain more reliable results. When this is done, the concentration in soil is plotted against the concentration in water, as shown in Figure 17.7A and the slope of the line is the K_D value. Thus,

$$K_D = C_S/C_W$$

As previously mentioned, it has been found that the organic matter in soil is the prime component of the soil that sorbs lipophilic chemicals. Organic matter is often measured as the **organic carbon content** of the soil. The partition coefficient obtained by expressing the concentration in the soil in terms of organic carbon gives much more consistent results between different soil types. This means that:

$$K_{OC} = C_{SOC}/C_W$$

where C_{SOC} is the concentration in the soil organic carbon. In this case, C_{SOC} can be plotted against C_W at equilibrium, as shown in Figure 17.7B. The K_{OC} value is the slope of the regression line produced in this plot. Generally, the plot of the C_{SOC} against C_W has a better correlation coefficient and a higher slope than the C_S against C_W line, as shown in Figure 17.7A. In these relationships, the following can be derived.

$$C_{SOC} = C_S/f_{OC}$$

Thus,

$$K_{OC} = C_S/f_{OC}C_W$$

This means that:

$$K_{OC} = K_D/f_{OC}$$

and

$$K_D = K_{OC}f_{OC}$$

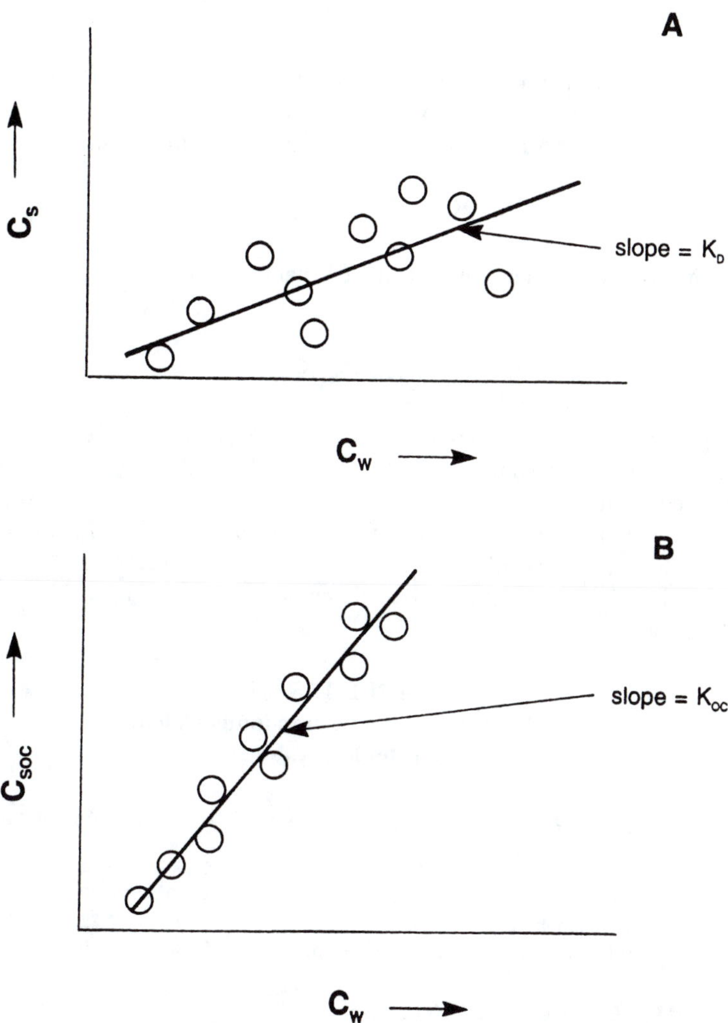

FIGURE 17.7. Experimental partition experiment results in a soil/water system with a lipophilic chemical. Plots of the concentration in soil (C_S) against the concentration in water (C_W) at equilibrium for a range of soil types (Plot A) and the same set of data using concentration in soil organic matter (C_{SOC}) (Plot B).

The fraction of organic carbon (f_{OC}) is always less than 1 since 1 would be equivalent to 100% organic carbon. This means that K_{OC} is always greater than K_D.

It has also been found that octanol acts as a reasonable surrogate for soil organic matter represented by organic carbon. Thus,

$$K_{OC} = x\, K_{OW}^{a}$$

and this means that

$$K_D = x\, f_{OC}\, K_{OW}{}^{a}$$

where constant x is a proportionality constant and constant a is a nonlinearity constant. This means that K_{OC} can often be calculated from the K_{OW} values of compounds of interest. A common empirical equation for this relationships is:

$$K_{OC} = 0.66\, K_{OW}{}^{1.03}$$

This means that constant x is 0.66 and constant a is 1.03 or, taking logs of both sides,

$$\log K_{OC} = 1.03 \log K_{OW} - 0.18$$

Using this expression, values for K_{OC} and $\log K_{OC}$ have been calculated for a range of compounds, as shown in Table 17.5. The K_D and $\log K_D$ values can then be calculated using f_{OC} for the specific soil of interest. Thus, using the K_{OW} value for a compound and the organic carbon content of a soil, values for the partition coefficient (K_D) can be calculated for that compound in a specific soil. Thus K_D for diazinon is 112 for a soil with 2% organic carbon (f_{OC} = 0.02) and 224 for a soil with 4% organic carbon (f_{OC} = 0.04).

TABLE 17.5
The K_{OC} Values for Some Compounds Calculated from the K_{OC} Value

Compound	K_{OW}	$\log K_{OW}$	K_{OC}	$\log K_{OC}$
Diazinon	8,000	3.81	5,600	3.74
Dieldrin	21,000	4.32	19,000	4.27
Tetrachlorophenol	174,000	5.24	165,000	5.22
Hexachlorobenzene	204,000	5.31	195,000	5.29

B. THE SOIL/ORGANISM PROCESS

There are a variety of animal and plant organisms in soil. Roots are very common soil biotic components and a variety of invertebrate organisms, including earthworms, can reside in soil as well. A simple three-phase model can be used to estimate the transfer of contaminants from soils to organisms, as illustrated in Figure 17.8.

The soil/water partition process, which represents part of this three-phase process, can be described as outlined in the previous section. The next part of the process can be addressed as a pore water/ biota partition process in which biota could be plant tissue (such as roots) or animals (such as earthworms). If we assume that biota lipids take up the lipophilic organic compounds, then the biota/water partition coefficient (K_B) can be represented as follows:

$$K_B = C_B/C_W \text{ at equilibrium}$$

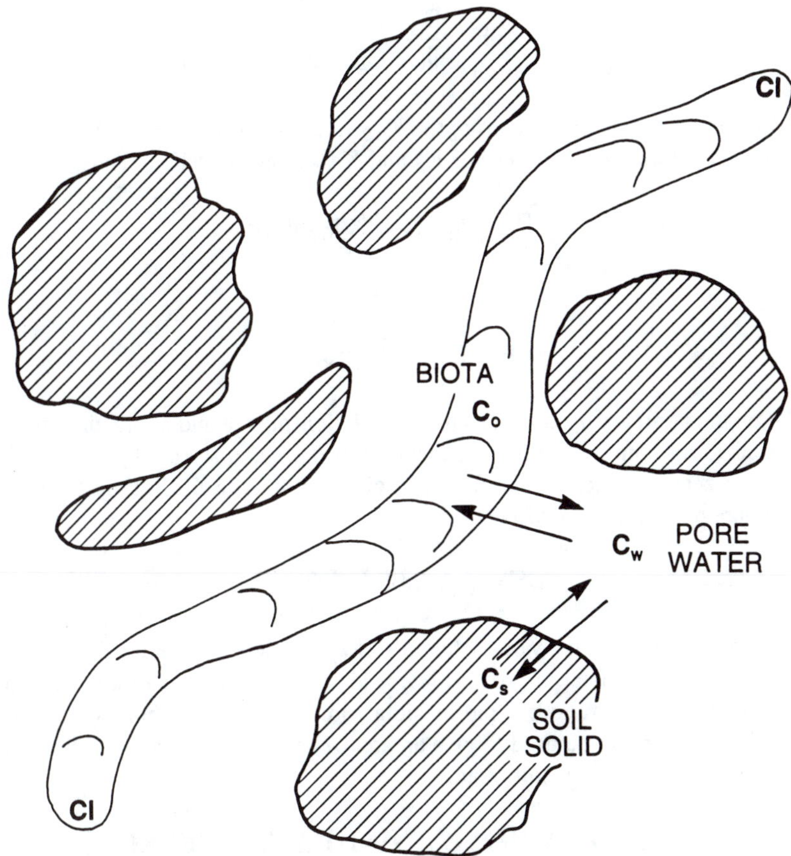

FIGURE 17.8. The three-phase partitioning process resulting from a transfer of contaminants from soil solids to soil biota e.g., plant roots and earth-worms.

where C_B/C_W are the concentrations in the biota and water, respectively. This means that a plot of C_B against C_W would be expected to be linear at low concentrations. If octanol is a good surrogate for biota lipid, then,

$$K_B = f_{lipid} K_{OW}{}^{b}$$

where f_{lipid} is the lipid fraction in the biota and constant b is an empirical nonlinearity constant. It is commonly found that constant b is usually about 0.95. Now the relationship between soil solid and biota can be considered. The Bioaccumulation Factor (BF) is the ratio of the concentration in the biota (C_B) and the concentration in the soil (C_S). So,

$$BF = C_B/C_S$$

This can be expanded by inserting the water phase; thus,

$$BF = \frac{C_B}{C_W} \cdot \frac{C_W}{C_S} = \frac{K_B}{K_D}$$

This can be expressed in terms of the K_{OW} values by inserting into the expression the previous expressions derived for K_B and K_{OC} in terms of K_{OW}. Thus,

$$BF = f_{lipid}\ K_{OW}{}^{b}/x\ f_{OC}\ K_{OW}{}^{a}$$

and

$$BF = (f_{lipid}\ /x\ f_{OC})K_{OW}{}^{b-a}$$

Often the nonlinearity constants in both relationships (a and b) are close to unity. In this case, constant a is 1.03 and constant b is 0.95, and so (a – b) is 0.08, which is close to zero and means that $K_{OW}{}^{0.08}$ is close to unity. Substituting these values into the previous equation, we get:

$$BF = (f_{lipid}/0.66\ f_{OC})K_{OW}{}^{0.08}$$

This means that the BF values should show little dependence on K_{OW} and other properties of the chemical. It primarily depends on the properties of the soil and the biota, particularly the ratio of lipid in biota and the organic carbon content of the soil. This equation can be used to give an approximate idea of the bioaccumulation of lipophilic chemicals by soil biota.

VII. ECOLOGICAL AND HEALTH EFFECTS OF SOIL CONTAMINATION

Soil contaminants can lead to obvious adverse effects on natural ecosystems, particularly in agricultural areas. Occasionally in some agricultural areas, "*fish kills*" are observed after storms. In many cases, these are due to soil contaminants being swept from the land into streams. In some situations, these contaminants may exceed the lethal level to fish and a fish kill occurs. In many situations, lethal levels may not be reached, but the presence of contaminants in aquatic systems can lead to deleterious sublethal effects on reproduction, growth and other important biological characteristics.

Human exposure to soil contaminants can also have important adverse effects. The general pattern of distribution of soil contaminants leading to uptake by human beings and animals are shown in Figure 17.6. Several pathways from soil to human beings can be identified as follows:

- Partitioning of vapors of contaminants originating from soil through the lung walls into blood
- Lodging of particulates containing contaminants in the respiratory system

- Diffusion through the skin from dermal contact with soil
- Ingestion in food
- Direct consumption of soil

This final pathway identified above is important for children since significant amounts of soil are directly taken up by children. These pathways are of different importance with different chemicals in different localities.

Lead is a common contaminant in urban areas and is probably one of the most important soil contaminants from the human health point of view. Urban lead often originates from its use in fuel for motor vehicles to enhance the octane rating. Even though this use is being discontinued in many areas, lead contamination in soil can be expected to remain for a considerable period of time. Lead is also a common soil contaminant in many mining or former mining areas. Lead in soil in contaminated communities has been found in a number of areas to have several adverse effects on human well-being. Possibly the most important of these is the deleterious effect on intellectual development that occurs with children.

The levels of occurrence of toxic substances in soil is an important aspect of the management of soil contamination. There can be some difficulties in assessing the meaning of concentrations of toxic substances that occur in soil. This difficulty relates to the normal levels of a substance that can be expected in the soil without contaminating human activities. The normal levels of toxic metals in soil relate to factors as such:

- Rock geochemistry, as some rocks are relatively rich in many toxic metals (e.g., lead, chromium, nickel, etc.)
- Soil-forming processes (e.g., weathering and leaching)
- Erosion and deposition processes leading to the actual physical movement of soil
- The normal levels of toxic metals in an area can be considered to include levels that result from historical activities leading to soil contamination

When management authorities set levels, and perhaps define normal levels, it is possible that these levels may exceed guidelines for control of adverse effects on human health and the environment. The management response to soil contamination may not be simple in terms of the need for any actions to protect public health and natural ecosystems. There are a variety of factors to be considered, including:

- The nature of the contaminant, e.g., its potential to contaminate adjacent areas
- The current or proposed use of the land, including industrial activities, recreational use, domestic housing, etc.
- The sector of the human population exposed to the contaminated soil, e.g., children, adults, adults undertaking specific activities, etc.

Actions for remediation need to take into account all of these factors before remedial activities can be planned.

VIII. KEY POINTS

1. Soil can become contaminated due to a wide range of human activities including agriculture, industry as well as urban development. Human populations and natural ecosystems can be exposed to contaminants through emissions of vapors, dispersal of dust, contamination of food and water, and direct uptake through the skin.

2. Important soil contaminants include toxic metals and related substances such as lead, copper, chromium and arsenic; organic substances such as pesticides, petroleum hydrocarbons, polyaromatic hydrocarbons, PCBs and industrial chemicals such as various solvents, acids and alkalis.

3. The properties of soil have a strong influence on any contaminants present. Clay consists of various hydrous silicates and oxides which have a capacity to sorb cations such as metal ions, polar organic molecules and organic cations. For example diquat and glyphosate are strongly sorbed. The organic matter in soil contains plant lignin and related degradation products described as humic substances or humic and fulvic acids. These substances are high molecular weight and contain many phenolic groups, but are lipophilic in nature. Highly hydrophobic compounds, such as DDT, dieldrin and benzo(a)pyrene, are strongly sorbed to the organic matter in soil.

4. The persistence of a compound in soil is of major environmental importance. The loss of a compound usually follows first-order kinetics. The loss of a chemical (as half-life, $t_{1/2}$) by volatilization from the soil surface can be estimated from the following equation:

$$t_{1/2} = 1.58 \times 10^{-8}\left(\frac{K_{OC} \cdot S}{P}\right) \text{ days}$$

5. Leaching of contaminants is an important process for the loss of contaminants from soil. A simple measure of the leachability of a chemical from soil, R, is:

$$R = \frac{1}{K_D(1-\phi^{2/3})d_s}$$

6. Mobility of a chemical in soil is a major aspect of environmental behavior and can be estimated using the mobility index (MI). Thus,

$$MI = \log \frac{SV}{K_{OC}}$$

7. Soil/water partitioning of lipophilic chemicals is expressed as the K_D value; thus,

$$K_D = \frac{C_S}{C_W}$$

This partition coefficient can also be expressed more consistently in terms of organic carbon; thus,

$$K_D = K_{OC}\, f_{OC}$$

8. The K_{OC} value of a lipophilic compound in the soil/water system can be estimated from the octanol/water partition coefficient K_{OW}. Thus, $K_{OC} = x\ K_{OW}$ where constant x is usually about 0.66 and constant a is 1.03. By taking logarithms, one obtains:

$$\log K_{OC} = 1.03 \log K_{OW} - 0.18$$

9. The uptake of lipophilic compounds by soil biota, such as earthworms and roots of plants, can be expressed as follows:

$$BF = (f_{lipid}/0.66\ f_{OC})\ K_{OW}^{0.08}$$

10. Human exposure to soil contaminants can be calculated by estimating the uptake through the possible exposure routes, i.e., vapor sorption, lodging of particulates in the respiratory system, dermal contact, ingestion in food and direct consumption. This last route is particularly important with young children.

11. Contaminants in soil can have adverse effects on human health and the natural environment. For example, lead in soil can lead to impaired intellectual development of children, and pesticides in agriculture soil can have deleterious effects on aquatic ecosystems.

12. The "normal" levels of substances in soil, such as toxic metals, relate to rock geochemistry, soil-forming processes, erosion and deposition processes and historical pollution-causing activities. These may exceed guideline levels for protection of human health and the natural environment.

REFERENCES

Amdur, M.O., Poull, J., and Klassen, C.D., *Casorett & Poull's Toxicology — The Basic Science of Poisons.* 4th ed., McGraw-Hill, New York, 1991, chap. 26, p. 872.

Shineldecker, C.L., *Handbook of Environmental Contaminants: A Guide for Site Assessment,* Lewis, Boca Raton, FL, 1992.

CHAPTER 17 PROBLEMS

1. A spill of a petroleum product onto soil in an urban area has occurred. Analysis of the soil (2.0% organic carbon) has revealed the following average concentrations of the major contaminants:

   Benzene: 150 mg kg^{-1}
   Toluene: 100 mg kg^{-1}
   Benzo(a)pyrene: 5 mg kg^{-1}

   The environmental management authority has asked the following questions: "After what period will the soils reach acceptable concentrations?" (benzene, 0.04 mg kg^{-1}; toluene, 0.05 mg kg^{-1}; benzo(a)pyrene, 0.004 mg kg^{-1}). Provide an answer to these questions using the data in Table 17.3.

2. A crop of carrots (0.5% lipid) was grown in soil (1% organic carbon) contaminated by dieldrin and atrazine, (3.5 mg kg^{-1} and 10.0 mg kg^{-1}, respectively). Estimate the likely levels of these contaminants in the carrots?

3. You are a member of a team evaluating an industrial site with contaminated soil. Your task is to evaluate the potential for the soil contaminants to contaminate the groundwater. You have been able to collate the property data on the contaminant as in the table below:.

Compound	Vapor pressure (mm Hg)	Water solubility (mm L^{-1})	Soil sorption coefficient, K_{OC}
Phenol	0.2	67,000	2
Styrene	9.5	280	120
Tetrachloroethane	5	2900	480
Chloropyrifos	1.9×10^{-5}	2	13,000

Evaluate the possibility of ground water contamination by these compounds.

Answers on Page 410.

CHAPTER 17 SOLUTIONS

1. To enable calculations to be made for the period to reach acceptable levels, the following equation will be used:

$$C_t = C_{to}^{-k_2t}$$

Thus $$\ln C_t = \ln C_{to} - k_2t$$

To use this equation, values for k_2 are needed. These can be calculated using the most conservative $t_{1/2}$ values (longest) from Table 17.3 and the following equation: $k_2 = 0.693/t_{1/2}$. Thus, the following k_2 values can be obtained: benzene 0.04 day^{-1}, toluene 0.03 day^{-1} and benzo(a)pyrene 0.0014 day^{-1}. Substituting in the logarithmic kinetic equation above to obtain the periods to reach an acceptable level. Then,

Benzene: $\ln 0.04 = \ln 150 - 0.04t$
Thus, $t = 206$ days

Toluene: $\ln 0.05 = \ln 100 - 0.03t$
Thus, $t = 253$ days

Benzo(a)pyrene: $\ln 0.004 = \ln 5 - 0.0014t$
Thus, $t = 5092$ days (14 years)

Thus, the time periods necessary to attain acceptable levels are: benzene, 206 days; toluene, 253 days; and benzo(a)pyrene, 5092 days (14 years).

2. The first step is to calculate the bioaccumulation factor of the chemicals in the soil for carrots using the following equation:

$$BF = \frac{y}{0.66\ f_{OC}} K_{OW}^{0.08}$$

Dieldrin: Using the K_{OW} from Table 17.3,

$$BF = \frac{0.005}{0.66\ (0.01)} 10{,}000^{0.08}$$

thus $$\frac{C_B}{C_S} = 1.59$$

and $C_B = 5.6$ mg kg^{-1} wet weight

Atrazine: Using the K_{OW} from Table 17.3, then,

$$\mathrm{BF} = \frac{0.005}{0.66\ (0.01)} 560^{0.08}$$

thus, $$\frac{C_B}{C_S} = 1.27$$

and C_B = 12.7 mg kg^{-1} wet weight

3. The mobility Index (MI) will give an estimation of the capacity of a chemical to move and contaminate groundwater:

$$\mathrm{MI} = \log\left(\frac{SV}{K_{OC}}\right)$$

The following MI values can be calculated: phenol, +3.83; styrene, +1.35; tetrachloroethane, +1.48; and chloropyrifos, –9.5.
By consulting Table 17.4, these data indicate that while chloropyrifos is immobile and has little potential to contaminate groundwater, all of the other compounds are very mobile and thus have considerable potential for groundwater contamination.

Chapter 18

DISTRIBUTION OF CHEMICALS IN THE ENVIRONMENT

I. INTRODUCTION

Prediction of the possible effects of chemicals in the environment in terms of human health or natural ecosystems is now a major environmental management activity. Some aspects of this are described in Chapter 20: Ecotoxicology, and Chapter 21: Risk Assessment. A vital part of these procedures is to evaluate the potential exposure of biota to contaminants. This requires methods to understand and estimate the patterns of distribution in the environment. When a chemical is discharged to the environment, it is necessary to know how much will deposit in soil, sediments, and other phases.

The first step in understanding the distribution of chemicals in the environment is to simplify the environment. The environment can be seen to consist of phases that are physically distinctive sections, which are relatively homogeneous and within which a chemical behaves in a uniform manner. Commonly, the environment is considered to consist of the atmosphere, water (lakes, dams, streams, oceans, etc.), soils, sediments, suspended sediments and biota. When a chemical is discharged to the environment, it will move and distribute between the phases according to the properties of the chemical and the properties of the phases. The distribution between the phases can be considered as a result of a set of two-phase partition processes as illustrated in Figure 18.1. Here, a number of two-phase partition processes are represented, including: atmosphere/water, vegetation/atmosphere, soil/atmosphere, biota/water, sediment/water and suspended sediment/water. Phases not in contact, e.g., atmosphere and sediment, can not exhibit a two-phase partition interaction. Thus, the key to understanding the distribution of chemicals in the environment is to develop an understanding of the two-phase partition processes that occur. The fundamental nature of partitioning and properties influencing this process are described in Chapter 2.

II. TWO-PHASE PARTITION PROCESSES IN THE ENVIRONMENT

Two-phase partition processes can be described by the **Freundlich equation.** Thus, at equilibrium:

$$C_1 = K(C_2)^n$$

where C_1 and C_2 are the concentrations in phase 1 and 2, respectively, K is the partition coefficient and constant n is a nonlinearity constant. At equilibrium, the concentrations in the phases (C_1 and C_2) are constant. In this treatment, the solubility of the substance should not be exceeded. If this occurs, then another phase is formed and this theory is not applicable. The relationship between C_1 and C_2 for a series of

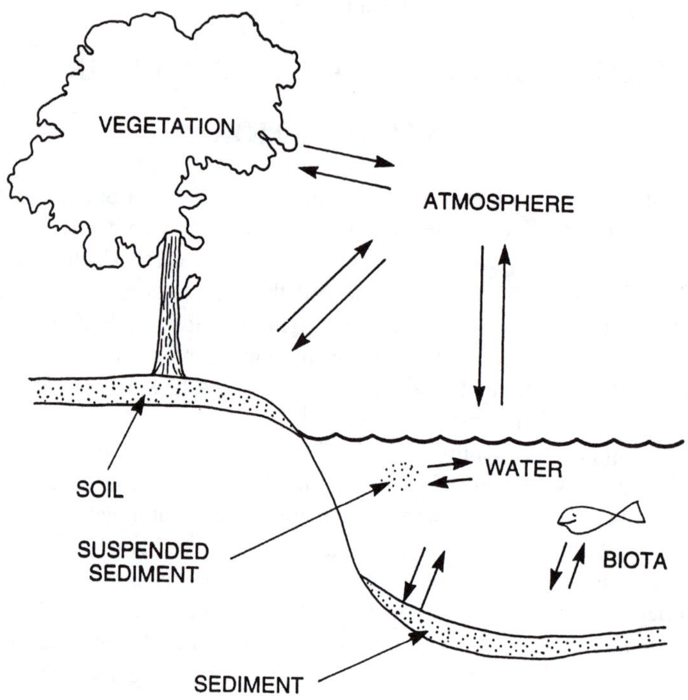

FIGURE 18.1. Distribution of a chemical between phases in the environment with various two-phase partition processes involved, e.g., atmosphere/water, sediment/water and biota/water.

concentrations can be plotted, as shown in Figure 18.2A. With many environmental processes, the concentrations of chemical are low and the value of constant n can be considered to be unity. Also, the relationship can be converted into a more useful form by taking logarithms. Thus,

$$\log C_1 = \log K + n \log C_2$$

This can be plotted as shown in Figure 18.2B. This gives a linear relationship from which log K can be obtained as the intercept at $\log C_2 = 0$ and n is the slope.

Some environmental partition processes have been extensively investigated on this. The air/water process is described by **Henry's law** and investigations started in the early 1900s. Other processes have been subject to limited evaluation, for example, the vegetation/air process.

Table 18.1 lists some properties of organic chemicals that influence the chemicals' environmental distribution. Two key properties are aqueous solubility and vapor pressure. The relationship between these properties is described as the Henry's law constant, H, which is essentially the partition coefficient of the compound between air and water. When equilibrium is established, the partial pressure of the compound in air and the corresponding concentration in water are constant. Then,

$$H = \frac{\text{Partial pressure of compound in air}}{\text{Concentration in water}} \quad \text{(at equilibrium)} \qquad (18.1)$$

TABLE 18.1
Physicochemical Properties of Some Organic Chemicals at 25°C

Chemical	Molar mass (g)	bp (°C)	Water solubility (mole m^{-3})	Vapor pressure (Pa)	H (Pa m^3 $mole^{-1}$)	K_{OW}
Chloroform	119.4	81	69	23 080	336	$10^{1.97}$
Benzene	78	80	23	12 700	557	$10^{2.13}$
Naphthalene	128	218	0.25	10.4	42	$10^{3.35}$
p,p-DDT	354.5	—	0.0000087	0.0002	2.3	$10^{6.19}$
2,3,7,8-TCDD "dioxin"	322	—	0.0006	1×10^{-7}	0.0017	$10^{6.80}$

In this case, the partial pressure is the method of expressing the concentration of the gas. The value of H can often be calculated from the basic properties of compounds, such as the vapor pressure divided by the water solubility.

A valuable environmental characteristic measured in the laboratory is the **octanol/water partition coefficient, K_{OW},** which is the ratio of the concentration in octanol to that in water, at equilibrium (see Chapter 2). This can be measured in the laboratory in an octanol/water two-phase system. Thus,

$$K_{OW} = \frac{\text{Concentration of chemical in octanol } (C_O)}{\text{Concentration in water } (C_W)} \quad \text{(at equilibrium)}$$

The significance of octanol is that it is a useful surrogate for the weakly polar organic matter present in soils and the lipid tissues of biota. High K_{OW} compounds tend to partition strongly into these organic-rich environmental phases. This means that they sorb onto soil or sediments and are accumulated by biota. The numerical value of K_{OW} can be used to estimate the partition coefficients for these processes.

The biota/water partition coefficient, K_B, is the bioconcentration factor at equilibrium. It is measured by placing fish, or other aquatic biota, in aquaria and measuring the concentration of the chemical in the biota after exposure to a fixed concentration in the water. Thus,

$$K_B = \frac{\text{Concentration of chemical in biota } (C_B)}{\text{Concentration in water } (C_W)} \tag{18.2}$$

Experimental values of K_B are known for many chemicals and species of aquatic biota. However, an approximate empirical relationship between K_B and K_{OW} and the lipid fraction of the biota can be derived to estimate K_B.

After partitioning between biota lipid and water has occurred in the aquaria through the gills of the organism, then:

$$\frac{C_L}{C_W} = \text{Constant}$$

where C_L is the concentration of the chemical in the biota lipid. If f_{lipid} is the fraction of lipid present in the biota, then:

$$C_B \text{ (whole biota weight)} = f_{lipid} C_L$$

$$K_B = \frac{C_B}{C_W} = \frac{f_{lipid} C_L}{C_W}$$

If octanol is a good surrogate for biota lipid, then C_L is equal to C_O and

$$K_B = \frac{f_{lipid} C_L}{C_W} = \frac{f_{lipid} C_O}{C_W} = f_{lipid} K_{OW}$$

So $$K_B = f_{lipid} K_{OW} \qquad (18.3)$$

and $$\log K_B = \log K_{OW} + \log f_{lipid}$$

Often, the lipid fraction, f_{lipid}, for aquatic biota is about 0.05 or 5%. Then, taking logarithms,

$$\log K_B = \log K_{OW} - 1.30$$

The partition coefficient between the abiotic solid phases (soil, sediment or suspended particulates) and water is often referred to as K_D (see Chapter 17, Section A) and here is described as K_{sorb} and defined as follows

$$K_{sorb} = \frac{\text{Chemical concentration in solid phase } (C_S)}{\text{Concentration in water } (C_W)} \quad \text{(at equilibrium)} \qquad (18.4)$$

Partitioning occurs between the organic matter in the solid phase and water, so that at equilibrium for a particular compound,

$$\frac{C_{SOC}}{C_W} = \text{Constant}$$

where C_{SOC} is the concentration in the solid organic matter in terms of organic carbon. If f_{OC} is the fraction of organic carbon in the solid, then:

$$C_S \text{ (whole solid phase)} = f_{OC} C_{SOC}$$

$$K_{sorb} = \frac{C_S}{C_W} = \frac{f_{OC} C_{SOC}}{C_W}$$

If octanol is a good surrogate for organic matter than C_{SOC} is equal to C_O (concentration in octanol in the octanol/water system) multiplied by constant x, a proportionality factor related to the organic carbon content of the solid phase. Thus,

$$K_{sorb} = \frac{xf_{OC}C_O}{C_W} = xf_{OC}K_{OW}$$

The constant x generally has a value of about 0.41, although other values are used. Also, often the assumption that octanol is a perfect surrogate for organic matter is not fully accurate and a power coefficient is used with K_{OW} to allow for this. Thus,

$$K_{sorb} = 0.41\ f_{OC}\ K_{OW} \tag{18.5}$$

The value of f_{OC} is about 0.02 to 0.04 (2 to 4%) for soils, sediments and suspended solids.

These two-phase partition processes are valuable for understanding each of the process involved but do not indicate how the chemical will behave when all the systems are operating together. An overall process for establishing an equilibrium distribution when all of these systems are operating together is now needed.

III. THE FUGACITY CONCEPT

Imagine two balloons, one inflated and the other slack, connected by a closed tube. If the closure is opened, then clearly gas escapes into the slack balloon until the pressures equalize. The gas molecules in both balloons have a tendency to escape their confinement. The molecules in the inflated balloon are more crowded and have a greater escaping tendency, so that when the balloons are connected, there is a net flow of molecules to the slack balloon. The net flow of molecules stops when their escaping tendencies become equal.

This **escaping tendency** of molecules is called *fugacity*, f, after the Latin word *fugere*, to flee. The fugacity of gases is closely related to their pressures, and has the same units (e.g., Pascal). Indeed, for an ideal gas, the fugacity is the pressure (P):

$$f = P \text{ (ideal gases)} \tag{18.6}$$

For non-ideal or real gases, the fugacity is the "effective pressure" as if the gas were behaving ideally. The fugacity of real gases may be more or less than the observed pressure, depending on how molecular interactions and sizes cause the gas to deviate from ideal behavior. However, organic contaminants usually occur at low concentrations or pressures in the air around us. As gases, they often behave roughly ideally, and we will assume that their fugacity and partial pressure are the same, as indicated in Equation 18.6.

The concept that a substance exerts an escaping tendency or fugacity can also be applied to where the substance is contained in a condensed phase such as a solid, a liquid, or a solution. Consider for a moment a flask into which we pour some benzene and then stopper the flask, as indicated in Figure 18.3. Immediately after addition of the benzene, there is no benzene vapor in the air space. But benzene is

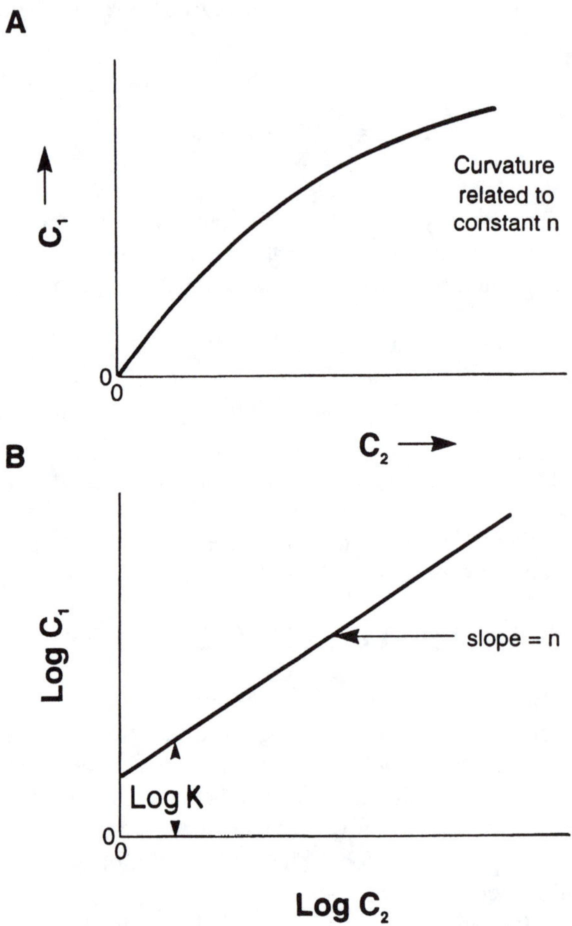

FIGURE 18.2. Plot of the corresponding concentrations in a two-phase system at equilibrium (A) and plot of the same data in a logarithmic form (B).

a volatile liquid, and its molecules exert an appreciable fugacity. Some benzene molecules escape across the liquid/air boundary to form a vapor. This evaporation continues, gradually building up the pressure of benzene vapor until it reaches saturation, when net evaporation stops. An equilibrium is reached, where the escaping tendencies of the benzene molecules from the liquid and gas phases are equal. At this point, the number of benzene molecules escaping from the liquid is the same as the number from the vapor redissolving in the liquid, and the concentrations in vapor and liquid are constant. Thus,

$$f_{liquid} = f_{vapor} \quad \text{(at equilibrium)}$$

Just as with the two-balloon example, the escaping "pressures" of the compound in the two phases have equalized. If the flask and its contents are at, say, 25°C, then we can place a value on this fugacity for liquid benzene; it is 12,700 Pd, the value of the saturation vapor pressure for benzene (see Table 18.1).

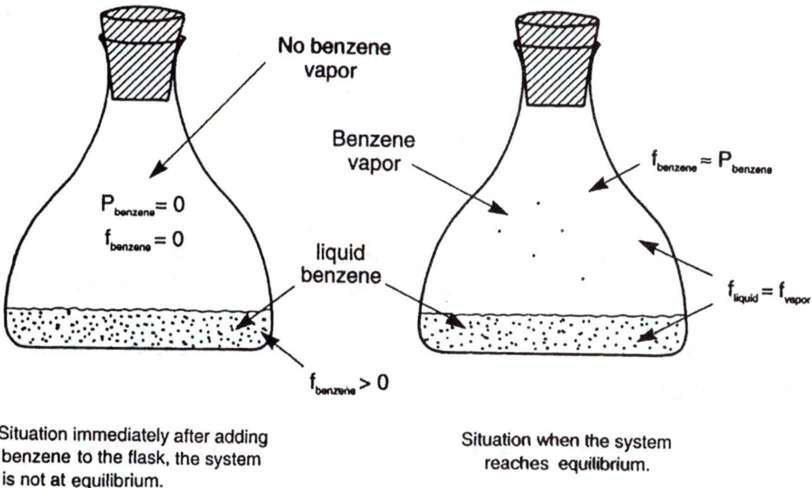

FIGURE 18.3. The changes in pressure and fugacity when benzene is added to a flask and it is stoppered.

We can extend our model to a chemical discharged into the environment. The chemical will disperse to reduce its fugacity. It may move into several phases such as air, water, biota or sorb onto sediments or soil. A simple outcome is an equilibrium, where the concentrations of the chemical in each phase become steady and the fugacities become equal:

$$f_{air} = f_{water} = f_{biota} = f_{sorb} \quad \text{(at equilibrium)}$$

It is the concentration (C) of a chemical that usually concerns us, so we need to relate fugacity to concentration. At low chemical concentrations, such as those usually found in the environment, the two are roughly proportional; thus,

$$C \propto f$$

By inserting a proportionality constant, Z, one obtains:

$$C = Zf \tag{18.7}$$

The proportionality constant, Z, is called the **fugacity capacity** factor. It has units of concentration and reciprocal pressure, e.g., mol m^{-3} Pa. The value of Z usually depends on the temperature, the properties of the chemical, and the nature of the environmental phase into which the chemical is dispersed. At equilibrium, when there is a common fugacity value for a chemical in all phases, f, the ratio of a chemical's concentrations in any two phases is the same as the ratio of Z values. For example,

$$\frac{C_{air}}{C_{water}} = \frac{Z_{air} f}{Z_{water} f} = \frac{Z_{air}}{Z_{water}}$$

Similarly,

$$\frac{C_{biota}}{C_{water}} = \frac{Z_{biota}}{Z_{water}}$$

and so on.

The Z values are the key to calculating the distribution of a chemical in the environment. These values are a property of the chemical and how it behaves in each phase. Chemicals will tend to accumulate in those phases that have high Z values (fugacity capacity factors) and avoid those phases with low Z values. The solubility of the chemical in any phase should not be exceeded, as another phase will form and this theory will no longer apply.

IV. FUGACITY AND CHEMICAL DISTRIBUTION

We will show how Z values can be determined, but here we will illustrate how to exploit **fugacity** to evaluate a chemical's distribution among several phases. Imagine an aquarium of total volume 1.0 m^3 that is two thirds filled with water with a tight-fitting lid that prevents the escape of the benzene (see Figure 18.4). A small amount of benzene (0.02 mole) is added and allowed to equilibrate between the air space and the water. We wish to know the final masses of benzene in the air and the water, given that for benzene Z_{air} and Z_{water} are 4.0×10^{-4} and 1.8×10^{-3} mol m^{-3} Pa, respectively.

Since the mass of benzene, M, is related to its concentration, C, and the phase volume, V, from Equation 18.7, one obtains:

$$M = CV = fZV \tag{18.8}$$

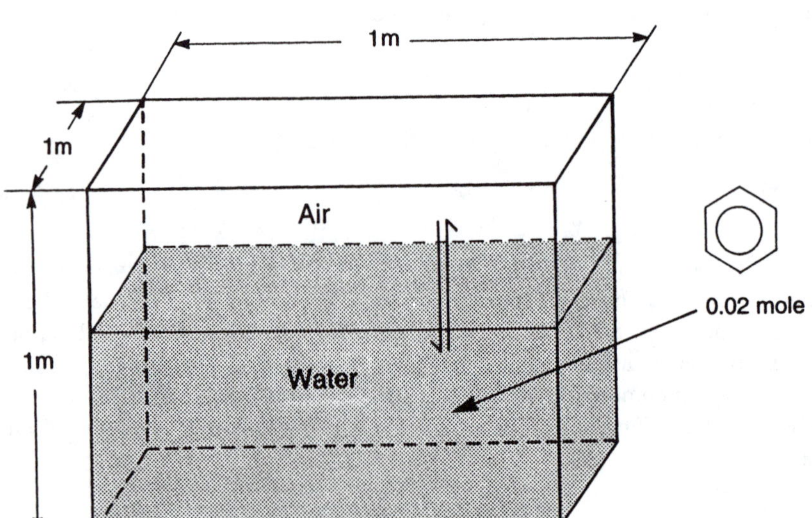

FIGURE 18.4. The distribution of benzene in an aquarium containing air and water.

for both phases. Thus,

$$M_{air} = fZ_{air}V_{air} \quad \text{and} \quad M_{water} = fZ_{water}V_{water}$$

The benzene fugacity is the same in both phases at equilibrium and the total benzene mass is unchanged, so that:

$$M_{total} = M_{air} + M_{water} = fZ_{air}\,V_{air} + fZ_{water}\,V_{water}$$

or

$$M_{total} = f(Z_{air}\,V_{air} + Z_{water}\,V_{water}) \tag{18.9}$$

Now we just need to substitute in the values to find the prevailing fugacity.

$$M_{total} = 0.02 \text{ mole} = f(4.0 \times 10^{-4} \times {}^1/_3 + 1.8 \times 10^{-3} \times {}^2/_3)$$

$$\text{Thus,} \qquad f = 15 \text{ Pa}$$

This fugacity value is inserted back in Equation 18.8 to generate the benzene mass in each phase. Thus, for air, the mass of benzene is:

$$M_{air} = fZ_{air}\,V_{air} = 15 \times 0.0004 \times {}^1/_3 = 0.0020 \text{ mole}$$

and for water

$$M_{water} = fZ_{water}\,V_{water} = 15 \times 0.0018 \times {}^2/_3 = 0.018 \text{ mole}$$

The benzene total of 0.02 mole is obtained from the sum of both phase masses as a check on the calculations.

We can modify the aquarium by placing several small fish in it. We will assume again that an equilibrium redistribution of benzene occurs (and that the fish tolerate the toxicity of the benzene). The fish have a volume of 10^{-4} m^3 (100 mL) and the Z value for benzene in fish is 0.012 mol m^{-3} Pa^{-1}.

To accommodate the new phase, we only need to insert an extra term in Equation 18.9, so that:

$$M_{total} = f(Z_{air}\,V_{air} + Z_{water}\,V_{water} + Z_{fish}\,V_{fish})$$

The volumes of air and water are virtually unchanged. The new equilibrium benzene fugacity is f = 14.99 ≈ 15 Pa, also little changed, mainly because of the small volume of the fish. The benzene mass in the fish is:

$$M_{fish} = fZ_{fish}\,V_{fish} = 15 \times 0.012 \times 10^{-4} = 1.8 \times 10^{-5} \text{ mole}$$

We can generalize the changes to Equations 18.8 and 18.9 that were made in this second example. For any number of phases, denoted i, into which the chemical distributes until the fugacities become equal and equilibrium is established, the mass in each phase is given by:

$$M_i = fZ_i V_i \tag{18.10}$$

and the total mass is given by:

$$M_{total} = \Sigma M_i = \Sigma(fZ_i V_i) = f \Sigma Z_i V_i \tag{18.11}$$

V. THE FUGACITY CAPACITY FACTORS (Z VALUES)

Expressions for the **fugacity capacity factors,** Z, for a chemical dispersed in the major environmental phases — air, water, biota, soil, sediments and suspended particulates — can be obtained as outlined below. For a chemical dispersed as a gas in air, we need only to relate the partial pressure of the gas to its concentration. The usual form of the Ideal Gas equation is:

$$PV = nRT$$

where P is now the chemical's partial pressure in air. Rearranged, this equation gives:

$$P = \frac{n}{V}RT$$

The ratio n/V is the number of moles per unit volume, or concentration, C_{air}. Thus,

$$P = C_{air}RT$$

or

$$C_{air} = \frac{P}{RT}$$

Since f = P (Equation 18.6) and recalling that C = fZ, then:

$$C_{air} = fZ_{air} = \frac{f}{RT}$$

thus,

$$Z_{air} = \frac{1}{RT} \tag{18.12}$$

where R is the Universal Gas constant and T is the temperature in Kelvin. Thus, at a given temperature, Z_{air} is the same for **any** gas. At 25°C (298 K), Z_{air} is equal to 4.04×10^{-4} mol m $^{-1}$ Pa^{-1}.

In water, the concentration of dissolved chemical is related to its **equilibrium** vapor pressure in air by Henry's law (Equation 18.1), or:

$$P = HC_{water} \quad \text{(at equilibrium)}$$

Since f = P and C = fZ, then:

$$f = HC_{water} = HfZ_{water}$$

and

$$Z_{water} = \frac{1}{H} \qquad (18.13)$$

The Henry's law constant, H, varies from substance to substance (see Table 18.1). Thus, each substance has its own unique Z_{water} value in water.

Knowing Z_{water} provides a link in evaluating the Z values for the chemical sorbed onto solid phases such as soils, sediments or suspended particulates. If a mixture of the chemical and the sorbing solid phase and water are allowed to equilibrate, then the concentration of sorbed chemical is often simply related to its concentration in water (Equation 18.4); thus,

$$C_{sorb} = K_{sorb}\, C_{water} \qquad (18.14)$$

Since at equilibrium:

$$C_{sorb} = f\, Z_{sorb} \text{ and } C_{water} = fZ_{water}$$

then from Equation 18.14, one obtains:

$$fZ_{sorb} = K_{sorb}\, C_{water} = K_{sorb}\, fZ_{water}$$

or

$$Z_{sorb} = K_{sorb}\, Z_{water} = \frac{K_{sorb}}{H} \qquad (18.15)$$

The Z_{sorb} values vary from chemical to chemical in a specific solid phase. Because of its dependence on K_{sorb}, the Z_{sorb} value also depends on the properties of the sorbing soil, sediment or particulate phase.

For a chemical that distributes into aquatic biota, the equilibrium concentration of the chemical is proportional to its concentration in water (Equation 18.2), or:

$$C_{biota} = K_B\, C_{water} \qquad (18.16)$$

As with K_{sorb}, K_B values are known from experiments or can be estimated from K_{OW}, as shown in Chapter 18. II. Since Equations 18.14 and 18.16 are analogous, then by making the appropriate substitutions, Equation 18.16 becomes:

$$Z_{biota} = \frac{K_B}{H} \tag{18.17}$$

VI. CHEMICAL DISTRIBUTION IN A MODEL ENVIRONMENT

Now that we are armed with methods to calculate Z values, estimations can be made of the distribution of some chemicals in the environment. To do this, we need further important information: the **volumes** of the environmental phases, just as we did in the benzene-aquarium example. Estimations of these volumes could be made by actually going out and measuring the sizes of the biota, soil and air. Another approach is to adopt an imaginary or **model** environment, where the size of the phases are an educated guess. We are then not so concerned with the accuracy of our model environment, but can focus instead on using the properties of the chemical to evaluate the way it distributes.

A model environment is described in Figure 18.5. The model has the six major phases: air, water, soil, sediments, suspended solids and biota. This model environment has an area of 1 km^2 and has an atmosphere 10 km high. Soil to a depth of 3 cm covers 30% of the surface, while the remainder is covered with water to an average depth of 10 m. The water has a 3-cm layer of sediment, contains 5 mL of suspended solids per cubic meter, and 0.5 mL m^{-3} of biota. Each of the six phases is assumed to be homogeneous. We will also assume also that a chemical discharged into this model environment disperses until eventually steady concentrations in each phase are reached.

A. EXAMPLE DISTRIBUTION CALCULATION

An arbitrary amount of benzene, 100 moles (78 kg), is added to the model environment in Figure 18.5. The benzene is assumed to distribute to equilibrium and the temperature is taken as 25°C. Six steps are followed to complete the calculation of distribution, as shown below and in Table 18.2.

Step 1: Assemble the important physical chemical properties for benzene. These are:

H	557 Pa m^3 $mole^{-1}$
K_{OW}	135
K_{sorb} (soil, 2% organic carbon)	1.1
K_{sorb} (sediment, suspended solids, 4% organic carbon)	2.2
K_B (fish, 5% lipid)	6.7

The K values for soil, sediment, suspended solids and biota have been estimated from K_{OW} using Equations 18.3 and 18.5.

Step 2: Estimate the volumes of each phase. These volumes are listed in Table 18.2.

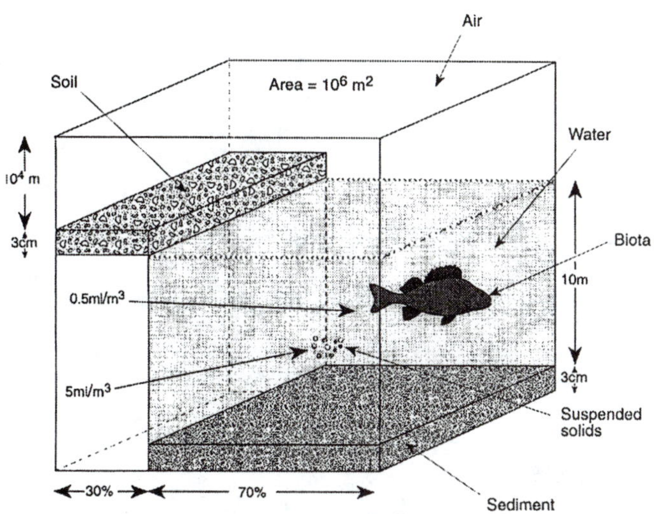

FIGURE 18.5. A six-phase model environment used to evaluate the distribution of a chemical with dimensions not to scale. (Adopted from Mackay, D., *Multimedia Models.* Lewis, Boca Raton, FL, 1991. With permission.)

Step 3: Calculate the Z values for each phase as follows, using the expressions derived in Chapter 18V:

Z_{air}	$= 1/RT$	$= 4.04 \times 10^{-4}$ m^3 mole^{-1} Pa^{-1}
Z_{water}	$= 1/H$	$= 1.8 \times 10^{-3}$ m^3 mole^{-1} Pa^{-1}
Z_{soil}	$= K_{sorb}(\text{soil})\Delta H$	$= 2.0 \times 10^{-3}$ m^3 mole^{-1} Pa^{-1}
$Z_{sediment,\ sus.\ solids}$	$= K_{sorb}(\text{sediment, sus. solids})/H$	$= 4.0 \times 10^{-3}$ m^3 mole^{-1} Pa^{-1}
Z_{biota}	$= K_B/H$	$= 1.2 \times 10^{-2}$ m^3 mole^{-1} Pa^{-1}

Step 4: Calculate ZV for each phase and sum these products as $\Sigma Z_i V_i$ (4.05×10^6 mol Pa^{-1}) and calculate the prevailing equilibrium fugacity, recalling that $f = M_{total}/\Sigma Z_i V_i$ (Equation 18.11) as 2.47×10^{-5} Pa (see Table 18.2).

Step 5: Use this fugacity value to calculate the benzene mass in each phase, recalling that $M_i = fZ_i V_i$ (Equation 18.10) (see Table 18.2).

Step 6: Finally, calculate the benzene concentration in each phase using ($C_i = M_i/V_i$), as shown in Table 18.2.

These calculations indicate that greater than 99% of the benzene is dispersed into the air. This is expected since benzene is a volatile compound and most of the model environment is taken up by air (>99.9% by volume). But what if the compound under investigation is p,p-DDT? The relevant physical chemical properties for DDT are:

H	2.3 Pa m^3 mole^{-1}
K_{OW}	1,555,000
K_{sorb} (soil)	12,700
K_{sorb} (sediment, suspended solid)	25,400
K_B	77,400

TABLE 18.2
Summary Distribution Calculation for Benzene

Phase	Air	Water	Soil	Suspended solid	Sediment	Fish	Total
V (m^3)	10^{10}	7×10^6	9×10^3	35	2.1×10^4	3.5	—
Z (mole m^{-3} Pa^{-1})	4.04×10^{-4}	1.8×10^{-3}	2.0×10^{-3}	4.0×10^{-3}	4.0×10^{-3}	1.2×10^{-2}	—
ZV (mole Pa^{-1})	4.04×10^6	12.6×10^3	18	0.14	84	4.2×10^{-2}	4.05×10^6
Fugacity, $f = M_{total}/\Sigma Z_i V_i = \frac{100}{4.05 \times 10^6} = 2.47 \times 10^{-5}$ Pa							
M = fZV (mole)	99.69	0.31	4.4×10^{-4}	3.5×10^{-6}	2.07×10^{-3}	1.0×10^{-6}	100
C = M/V (mole m^{-3})	1×10^{-8}	4.4×10^{-8}	5×10^{-8}	1×10^{-7}	1×10^{-7}	3×10^{-7}	—

Note: Benzene mass, M_{total} = 100 moles.

Compared to benzene, DDT is much less volatile, more hydrophobic, and has a strong potential to bioaccumulate. Pursuing the same sequence of calculations as before, a summary of the distribution of 100 moles of DDT in the model environment is as follows:

DDT equilibrium fugacity = 3.46×10^{-7} Pa

	Air	Water	Soil	Suspended solid	Sediment	Fish	Total
Mass (mole)	1.4	1.1	17.2	0.13	80.2	0.04	100
Concentration (mole m^{-3})	1.4×10^{-10}	1.4×10^{-7}	1.9×10^{-3}	3.8×10^{-3}	3.8×10^{-3}	1.1×10^{-2}	—

The distribution profile for DDT is quite different from that of benzene. Only a small fraction of the DDT distributes into air. Most DDT partitions into the soil and sediment, since these phases are rich in organic material that has an affinity for DDT. The lipid-rich fish contain the highest concentrations, but still contain the smallest DDT mass due to the relatively tiny volume of this biotic phase.

A word of caution in these calculations! The amount of organic substance chosen as the total should be within the capacity of the model environment. For example, water solubilities or saturation vapor pressures should not be exceeded or a new phase would have to be introduced — the pure compound. In the DDT example, the estimated fugacity is one-thousandth the saturation vapor pressure, and the water concentration about 10% of the maximum solubility. But, with about a 0.04% body burden, any real fish would probably have long since succumbed to DDT poisoning.

Finally, a reminder that a simple, static view of chemical distribution has been illustrated. Other factors may influence a pollutant's fate. For example, environmental phases such as air and water move about, carrying pollutants with them. The pollutant discharge may be continuous, and the chemical can decompose. These factors and others need to be covered when distributions in specific environments are being considered.

VII. KEY POINTS

1. Predictions of possible distribution patterns of chemicals in the environment are needed to assess exposure in evaluating potential effects of contaminants on human health and natural ecosystems.

2. The first step in evaluating distribution is to classify the environment into phases. The following phases are commonly used: air, water, soil, suspended solids, sediments and aquatic biota.

3. The distribution patterns in the environment can be seen as a set of two-phase equilibria such as air/water, air/soil, biota/water and so on. These processes are described by the Freundlich equation:

$$C_1 = K(C_2)^n$$

or

$$\log C_1 = \log K + n \log C_2$$

4. The Henry's law constant, H, is an important environmental characteristic defined as:

$$H = \frac{\text{Partial pressure of compound in air}}{\text{Concentration in water}} \quad \text{(at equilibrium)}$$

5. The octanol/water partition coefficient, K_{OW}, is a valuable characteristic of organic compounds that can be measured in the laboratory. Thus,

$$K_{OW} = \frac{\text{Concentration in octanol}}{\text{Concentration in water}} \quad \text{(at equilibrium)}$$

6. The bioconcentration factor for aquatic organisms, K_B, is the ratio between the chemical concentrations in the biota and water (C_B/C_W) at equilibrium. The solid sorption factor is the ratio between the concentration the solid phase (soil, sediments, suspended sediments) and water at equilibrium. Thus,

$$K_B = \frac{\text{Concentration in biota}}{\text{Concentration in water}}$$

$$K_{sorb} = \frac{\text{Concentration in solid}}{\text{Concentration in water}}$$

7. The K_B and K_{sorb} values can be calculated from the K_{OW} value using the following relationships:

$$K_B = f_{lipid} K_{OW}$$

$$K_{sorb} = 0.41 f_{OC} K_{OW}$$

8. The fugacity approach is needed to describe sets of two-phase equilibria and calculate distributions of chemicals. Fugacity (f), the escaping tendency, is measured in units of pressure. It is related to concentration, at concentrations usually encountered in the environment, and is defined as:

$$f = C/Z$$

9. The fugacity capacity constants, Z values, can be calculated from a set of expressions for each phase in the environment using partition coefficients. Thus,

$$Z_{air} = 1/RT$$
$$Z_{water} = 1/H$$

$$Z_{soil} = K_{sorb}(\text{soil})/H$$
$$Z_{sediment,\ sus.\ solids} = K_{sorb}(\text{sediment, sus. solids})/H$$
$$Z_{biota} = K_B/H$$

10. Calculations can be made of distributions in model environments using the volumes of the phases and the calculated Z values with the following expressions:

$$f = M_{total}/\Sigma Z_i V_i$$

$$M_i = f Z_i V_i$$

$$C_i = M_i/V_i$$

11. Calculations made using the approach outlined above provide a simple static view of the distribution of a chemical. Other factors, such as movement of phases and degradation of the chemical, are needed to provide a more accurate assessment.

REFERENCE

Mackay, D., *Multimedia Environmental Models.* Lewis, Boca Raton, FL, 1991.

CHAPTER 18 PROBLEMS

1. If soil is analyzed indicating a benzene concentration of 0.003 mol m^{-3} (about 1 ppm), what is the equilibrium concentration of benzene in the air pore spaces in the soil?

2. Given that naphthalene has the following characteristics at 25°C:

K_{sorb} (soil, 2% organic carbon)	18
K_{sorb} (sediment, suspended solids, 4% organic carbon)	36
K_B (fish, 5% lipid)	112

Calculate the missing Z values for naphthalene:

Z_{air} = 4.04 × 10^{-4} mole m^{-3} Pa^{-1}
Z_{water} =
Z_{soil} =
Z_{fish} =
$Z_{suspended\ solids}$ = 0.86 mole m^{-3} Pa^{-1}
$Z_{sediment}$ =

3. Given that 100 moles naphthalene is distributed in the model environment shown in Figure 18.5, work out the mass and concentrations in each phase, filling in the empty columns in the table below.

Naphthalene, 100 mole at 25°C

Phase	Air	Water	Soil	Susp. solids	Sediment	Fish	Total
Volume (m^3)	10^{10}	7 × 10^6	9 × 10^3	35	2.1 × 10^4	3.5	NA
Z (mole m^{-3} Pa^{-1})	4.04 × 10^{-4}	–	–	0.86	—	—	NA
ZV (mole Pa^{-1})	4.04 × 10^6	1.67 × 10^5	–	30.1	1.81 × 10^4	–	—

$$\text{Fugacity, f} = \frac{100}{\Sigma Z_i V_i} \text{ Pa}$$

Phase	Air	Water	Soil	Susp. solids	Sediment	Fish	Total
Mass (mole) (fZV)	—	3.95	—	—	0.43		100
Concentration (mole m^{-3}) (M/V)	—	5.64 × 10^{-7}	1.02 × 10^{-5}	–	–		NA

Answers on Page 432.

CHAPTER 18 SOLUTIONS

1. From $\frac{C_{air}}{C_{water}} = \frac{Z_{air}}{Z_{water}}$, then $C_{air} = \frac{Z_{air}}{Z_{water}} C_{water}$

Assume that the soil and air have the Z values as indicated in Chapter 18.IV.

So, $$C_{air} = \frac{4.04 \times 10^{-4}}{1.8 \times 10^{-3}} \times 0.003 = 0.00067 \text{ mole m}^{-3}$$

thus, $C_{air} = 0.00067$ mole m^{-3}

2. Given that naphthalene has the following characteristics at 25°C:

K_{sorb} (soil, 2% organic carbon)	18
K_{sorb} (sediment, suspended solids, 4% organic carbon)	36
K_B (fish, 5% lipid)	112

The missing Z values for naphthalene can be calculated using the expressions in Chapter 18.V as:

Z_{air} = 4.04×10^{-4} mole m^{-3} Pa^{-1}
Z_{water} = 0.0239 mole m^{-3} Pa^{-1}
Z_{soil} = 0.43 mole m^{-3} Pa^{-1}
Z_{fish} = 2.67 mole m^{-3} Pa^{-1}
$Z_{suspended\ solids}$ = 0.86 mole m^{-3} Pa^{-1}
$Z_{sediment}$ = 0.86 mole m^{-3} Pa^{-1}

3.

Naphthalene, 100 mole at 25°C

Phase	**Air**	**Water**	**Soil**	**Susp. solids**	**Sediment**	**Fish**	**Total**
Volume (m^3)	10^{10}	7×10^6	9×10^3	35	2.1×10^4	3.5	—
Z (mole m^{-3} Pa^{-1})	4.04×10^{-4}	0.0239	0.43	0.86	0.86	2.667	—
ZV (mole Pa^{-1})	4.04×10^6	1.67×10^5	3870	30.1	1.81×10^4	9.335	4.229×10^6

$$\text{Fugacity, f} = \frac{100}{\sum Z_i V_i} = 2.365 \times 10^{-5} \text{ Pa}$$

Mass (mole) (fZV)	95.5	3.95	0.0915	7.12×10^{-4}	0.43	2.21×10^{-4}	100
Concentration (mole m^{-3}) (M/V)	9.55×10^{-9}	5.64×10^{-7}	1.02×10^{-5}	2.03×10^{-5}	2.05×10^{-5}	6.31×10^{-5}	—

Chapter 19

GENOTOXICITY — THE ACTION OF ENVIRONMENTAL CHEMICALS ON GENETIC MATERIAL

I. INTRODUCTION

The linking of particular cancers to chemicals first occurred in the 1760s and 1770s. In 1761, John Hill hypothesized that the habit of sniffing snuff caused nasal cancer. Shortly after, in 1775, Percival Potts noted that many of his patients with scrotal cancer were chimney sweeps. Another important discovery was made by the German physician Rehn in 1895; he noted an increased incidence of urinary bladder cancer among workers in the dye industry. All of these discoveries were later verified using laboratory animals: in 1915 for coal tar and in 1934 for the dyes. Some of the important cancer discoveries are presented in Table 19.1.

TABLE 19.1
Key Events in the History of Cancer Research

Discoverer (year)	Test species	Affected tissue	Carcinogen
Pott (1775)	Human	Scrotum	Soot
Rehn (1895)	Human	Bladder	Dye intermediates
Van Trieben (1902)	Human	Skin	X-rays
Müller (1939)	Human	Lung	Tobacco smoking
Molesworth (1937)	Human	Skin	Sunlight
Wagner (1960)	Human	Lung	Asbestos
Yamagiwa & Itchikawa (1915)	Rabbit	Ear skin	Coal tar
Findley (1928)	Mouse	Skin	UV light
Edwards (1941)	C3H mouse	Liver	Carbon tetrachloride
Innes (1969)	Mouse	Liver	DDT
Arnold (1979)	Rat	Bladder	Saccharin

The first suggested link between deformities of children and their development was in 1651 by William Harvey. He suggested that hare lips developed by the cessation of growth of the lip during development of the fetus. He based this hypothesis on the fact that a hare lip-like condition normally occurs during development but disappears. The term **teratology** was coined in 1832 by Eteinne Geoffrey St. Hilaire and his son. It was derived from the Greek word *teras,* and literally means the study of monsters. Until the 1940s it was a commonly held view that the placenta protected the developing fetus from most if not all chemical and physical insults. This was dispelled in 1941 when Gregg showed that pregnant mothers exposed to rubella (German measles) led to deformed children. The independent discovery by McBride and Lenz of the teratogenic properties of thalidomide in 1961 made the world realize the sensitivity in pregnancy to chemical exposure. It has only been since the 1950s that it has been widely accepted that genotoxic effects could be prevented if genotoxins could be identified and exposure prevented. This idea is now central to all efforts in genotoxicity research and prevention.

Strictly speaking, **genotoxicity** is the study of the adverse effects of compounds on the genetic material of cells (DNA) and the subsequent expression of these changes. As such, it deals only with mutagenesis (the formation of inheritable mutations), carcinogenesis (the formation of cancers) and some forms of teratogenesis that involve damage to the DNA. However, in this chapter, we have also included teratogenic deformities that are not caused by interaction with DNA.

All the forms of genotoxicity discussed in this chapter occur naturally. They occurred even when humans existed in the most natural state, as there are numerous natural genotoxins. Many people have a couple of misconceptions regarding genotoxicity: (1) that only substances made by humans could be genotixic, and (2) that natural products are inherently safe, nontoxic and neither carcinogenic, mutagenic nor teratogenic. This has absolutely no basis in fact; there are numerous chemicals isolated from nature that are genotoxic e.g., oxygen and radon gas.

Almost all knowledge of and research into genotoxicity is concerned with the effects on humans rather than the potential effects on animals and the environment. The animals used to determine the genotoxicity of chemicals are simply a means of obtaining the necessary data since human experimentation is not permitted. Our knowledge of the effect on animals is simply a byproduct of this quest; it is not applied or used to protect the animals or the environment.

The importance of genotoxicity is evident from the fact that cancer is one of the three main causes of death in most countries, the others main causes being heart attack and stroke. Cancer kills approximately one in every four citizens of the Western countries. Cancer has caught the public imagination and is the form of genotoxicity of most concern and on which we spend the most time and effort.

Even though mutations and deformities of fetuses are much less publicized, it has been estimated they cause approximately 40% of all infant deaths and a similar percentage of all spontaneous abortions.

II. THE GENETIC CODE

Deoxyribonucleic acid (DNA) is the genetic material that codes for all characteristics of life from the synthesis of proteins and enzymes to the color of the eye. DNA generally consists of two long strands of nucleotides that spiral around each other, forming a helix (Figure 19.1). Single-strand forms of DNA do occur during DNA replication. A nucleotide consists of a pentose (five-carbon) sugar, a phosphate group and a nitrogenous base. The nitrogenous bases are of two types: pyrimidines and purines. The pyrimidines consist of cytosine (C) and thymine (T), while the purines are adenine (A) and guanine (G) (Figure 19.2). The two strands are held together by hydrogen bonds between the bases (Figure 19.2). The nitrogenous bases are always paired. For example, a cytosine on one strand will always be paired with a guanine on the other strand. Likewise, an adenine on one strand will always be paired with a thymine on the other strand. These matched pairs of bases are called DNA base pairs. There are only two different DNA base pairs: AT and GC.

DNA occurs in either a single-stranded or double-stranded form at various times in the life cycle of a cell. During DNA replication, the strands are separated (Figure

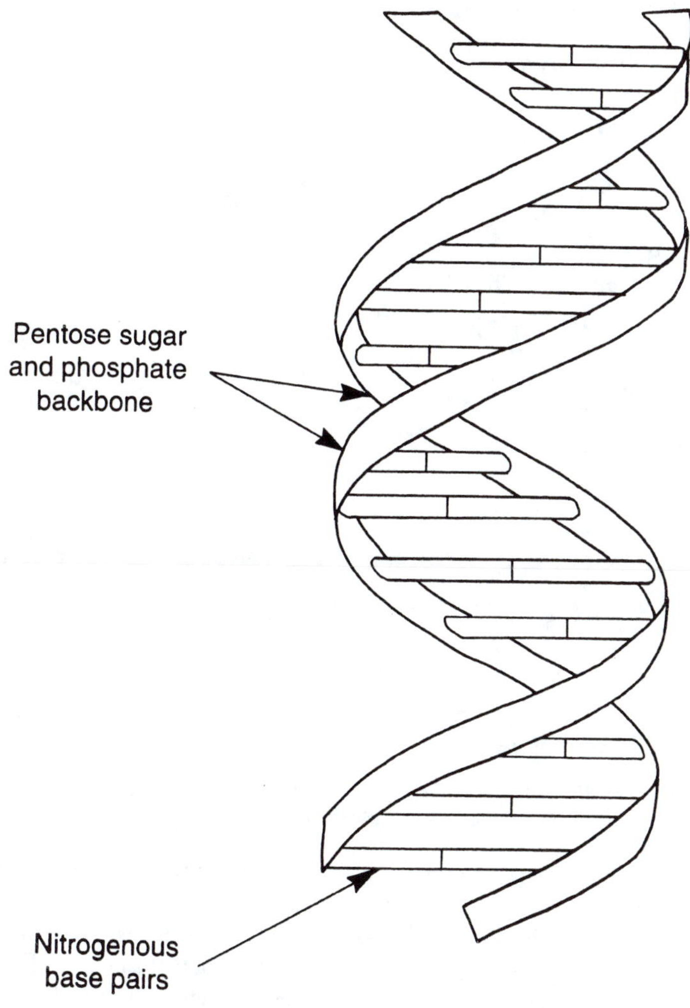

FIGURE 19.1. Representation of DNA showing the typical double-stranded form: the strands consist of pentose sugars and phosphate ions connected by the nitrogenous bases.

19.3a and 19.3b) and from each strand a new double-stranded DNA is synthesized. The second strand is synthesized by enzymes placing and binding into position a nucleotide containing the appropriate base to maintain the base pairs (Figure 19.3c). Thus, a nucleotide containing a guanine will always be placed opposite a nucleotide containing cytosine in the existing DNA strand. Likewise, a thymine will always cause an adenosine to be placed in the other strand, or vice versa.

The other function of DNA, besides simply replicating itself, is the synthesis of proteins. This process is called **transcription** as the sequence of bases on the DNA is transcribed to a series of nucleotides in ribonucleic acid (RNA). The nucleotides of RNA are thymine (T), adenine (A), guanine (G), cytosine (C) and uracil (U). RNA is formed by a portion of a double-stranded form of DNA unravelling and the

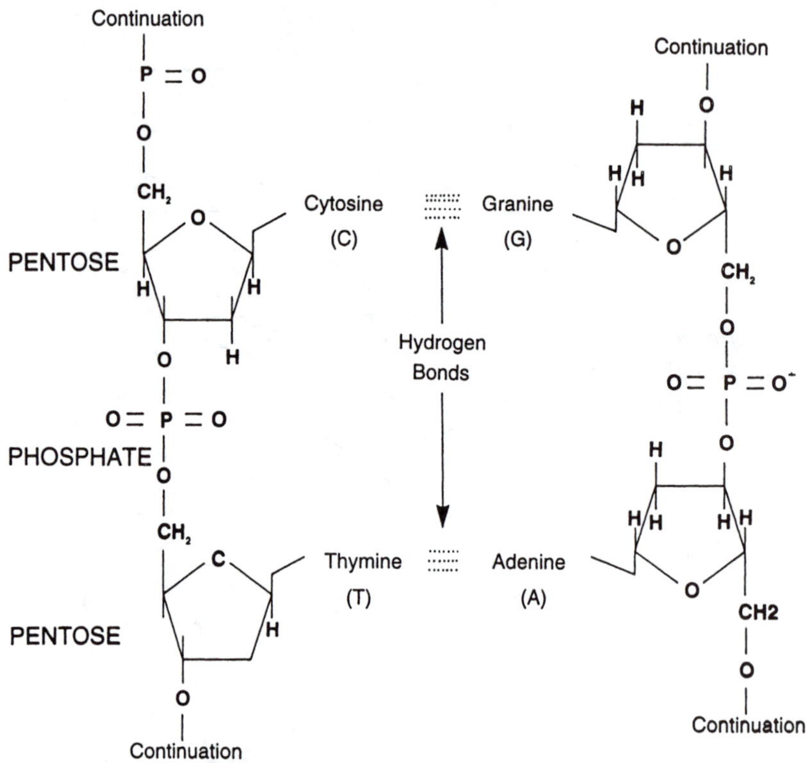

FIGURE 19.2. The chemical forms of the components of DNA and bonds between the pentose sugars and phosphate ions and between the nitrogenous base pairs.

strands separating, as in DNA replication (Figure 19.3). Enzymes then move between the strands and use the sequence of bases present on one strand as the template for the binding of nucleotides to form RNA. The RNA is synthesized using the same base pairs used to synthesize DNA (Figure 19.4a) except that a nucleotide containing adenine on the DNA means a nucleotide containing uracil will be added to the RNA strand (Figure 19.4b).

The nucleotides in RNA are in turn converted to proteins in a process termed **translation**. Every DNA base codes for a specific RNA base, and every set of three RNA bases codes for one specific amino acid. However, not all the sets of three bases code for different amino acids; for instance, UUA, UUG, CUU, CUC, CUA, and CUG all code for the amino acid leucine. These groups of three bases, regardless of whether they are from DNA or RNA, are called codons, for obvious reasons. Some of the codons and the amino acids they code for, are illustrated in Table 19.2.

Proteins are synthesized by the ribosome, a large complex biomolecule consisting of approximately 60 proteins and many enzymes. The ribosomes bind to the RNA at points where an AUG codon occurs. This codon is the initiation code for protein synthesis (Table 19.2). The ribosome then moves along the RNA and adds an amino acid for each codon it encounters, onto the other amino acids already formed, thus forming a polypeptide. This process is analogous to the synthesis of RNA from

(a)

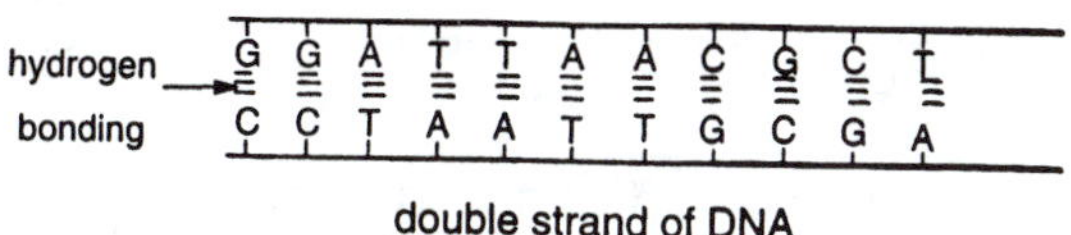

(b)

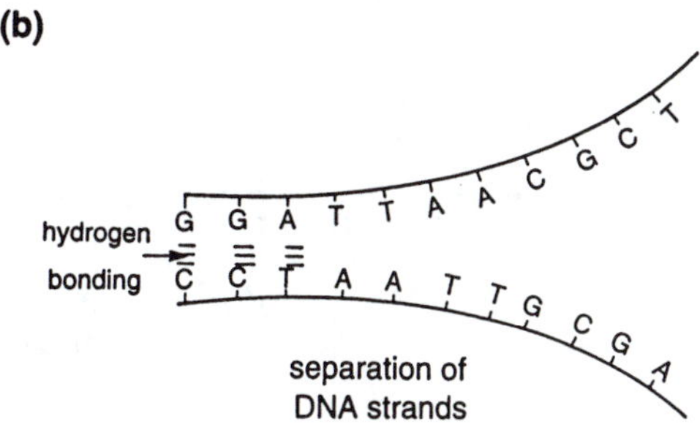

(c)

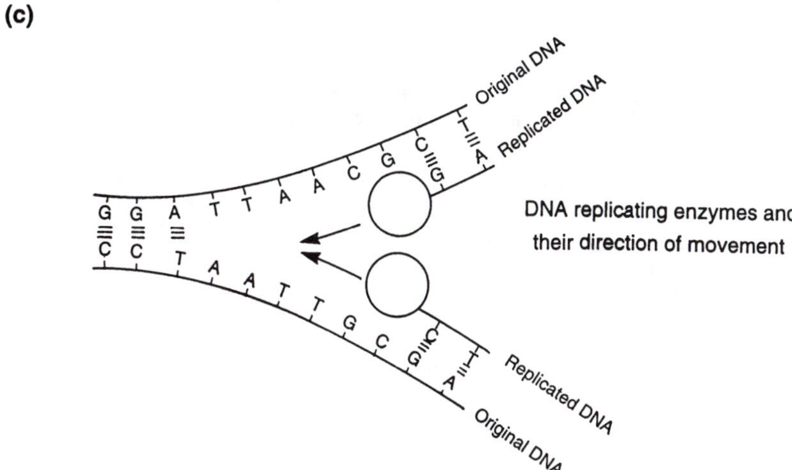

FIGURE 19.3. The process of DNA replication. The normal double-stranded form of DNA (a) has the two strands separated (b) and then enzymes commence using the original DNA strands as the template to synthesize the new replicated strand of DNA (c).

DNA (Figure 19.4). The protein synthesis is terminated when the ribosome encounters the codon UAA.

A series of DNA bases linked together that code for one entire protein or enzyme is called a **gene**. Generally, genes contain about 1000 base pairs. Groups of genes arranged in series form chromosomes. The number of chromosomes within all cells of an animal should be the same (except if a mutation has occurred), and they should

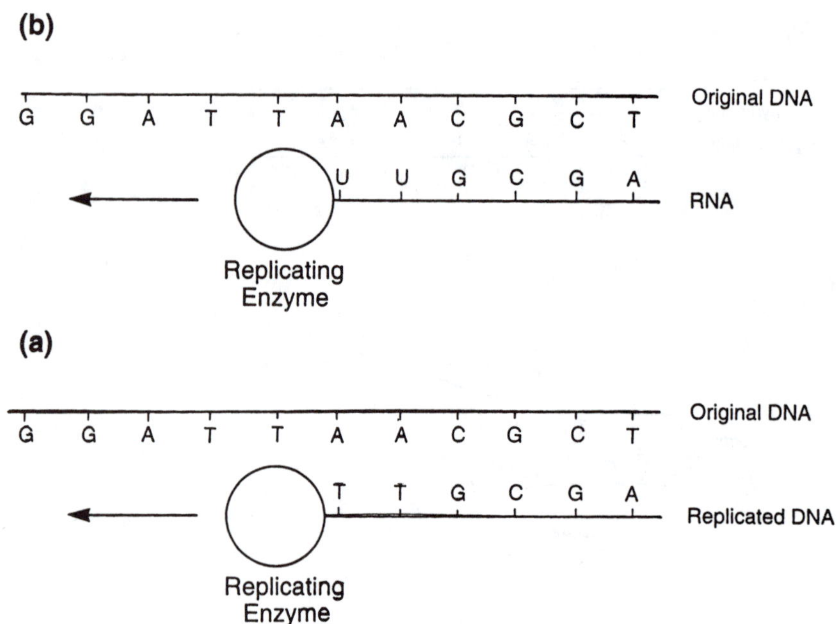

FIGURE 19.4. An example of the difference in the sequence of nitrogenous bases when DNA is replicated (a) and when RNA is synthesized (b). Both use the same DNA.

TABLE 19.2
Some Example Codons and the Amino Acids and Messages They Code For

Codon composition	Amino acid coded for	Other meanings
AUG	Methionine	Start
CUC	Leucine	
ACA	Threonine	
GAA	Glutamic acid	
CGU	Arginine	
UGU	Cysteine	
UAA, UAG, UGA	—	Stop
GGG	Glycine	

be the same for all individuals of the same species. Humans normally have 23 chromosomes that contain between 200,000 and 250,000 genes.

III. TERATOGENS AND TERATOGENESIS

A. MECHANISMS OF ACTION

Teratogens are chemicals that interfere with the normal reproduction process and cause either a reduction of successful births or offspring to be born with physical, mental, developmental or behavioral defects. It is important to note the difference between teratogens and mutagens. Mutagens can cause changes in the DNA of cells regardless of cell type. Teratogens may cause defects by a variety of means, including DNA damage; however, they only affect somatic cells (all cells other than ovum and sperm). As such, deformities resulting from teratogen exposure cannot be passed onto further generations unless the offspring are similarly exposed.

Teratogens appear to have very specific mechanisms of action; however, they can be broadly classed as either **genetic** or **epigenetic**. Genetic teratogens exert their effects by gene mutation, chromosomal abnormality, inhibiting mitosis by slowing DNA synthesis or preventing spindle formation (a vital step in mitosis). Epigenetic mechanisms of action include affecting the levels and forms of energy available, the supply of key metabolic substrates, inhibition of enzymes and modifying cellular membranes and thus altering membrane permeability.

The development of a fetus is not a smooth, even process. The gestation for humans is divided into three distinct periods: implantation (weeks 1-2); embryonic (weeks 3-7) and fetal (weeks 8-38) (Figure 19.5). During the implantation period, fetal cells undergo rapid replication followed by differentiation into specialized cell types that will eventually develop into all the different types of tissue (e.g., heart muscle, liver, bone). Throughout the embryonic period and into the first 2 weeks of the fetal period, all the body's organs and limbs form and develop. This process is called **organogenesis**. Each organ or part of the body commences developing at certain preset times during organogenesis. During the fetal period, the organs and limbs become functionally mature. The time of completion of maturing are different for different organs/limbs. The particular organ or limb deformed by exposure to a teratogen depends on the period of gestation when the fetus was exposed (Figure 19.5).

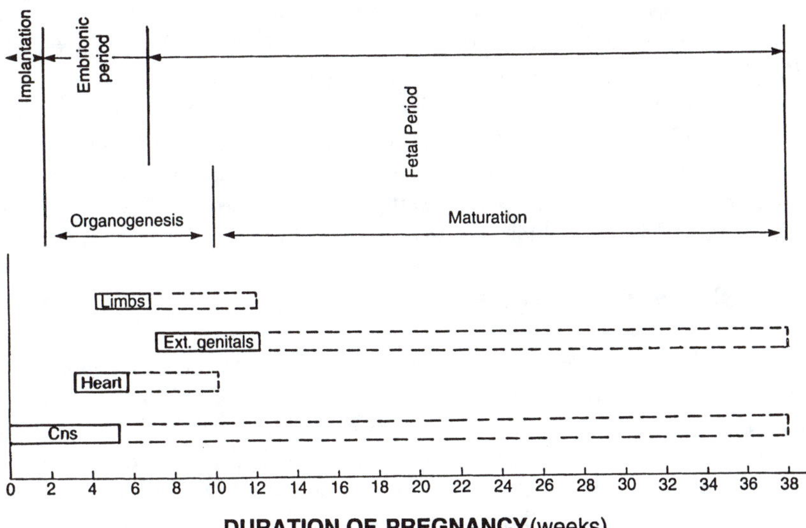

FIGURE 19.5. A time sequence of the development of some organs during human pregnancy. Solid bars indicate periods of greatest sensitivity to teratogens; dashed bars indicate periods of reduced sensitivity.

When a particular organ or limb is forming and developing, the involved cells have greatly increased rates of DNA replication and cell division. If a genetic teratogen causes a mutation of the DNA of a cell in the early stages of organogenesis, then all cells derived from that will carry the same mutation. With the rapid cell division that occurs during organogenesis, a large number of cells could end up with the same mutation, the expression of which could cause deformity in the developing fetus.

One mechanism by which epigenetic teratogens cause deformities is to limit either energy or vital substrates necessary for the cell division, thus slowing down the development or even killing cells. The period in which organ and limb development can occur is very limited. Slowing cell metabolic rates means there may not be enough cell division to allow full development of the organ or limb before the period of organogenesis finishes. The resulting organ may be completely nonfunctional or partially functional, depending on how greatly the metabolism of the cells was affected. Cell death is a normal event during organogenesis that can be compensated for and, consequently, development of the organ or limb is unaffected. However, if too many cells are slowed down or killed, to be replaced by the normal mechanisms, insufficient cells may remain to allow the organ or limb to develop properly in the allotted period of organogenesis.

The earlier in the pregnancy the exposure to teratogens, the more severe the deformities for the same concentration. Exposure to teratogens generally leads to miscarriage if it occurs prior to organogenesis, which commences in week 3 of gestation. Whereas, exposure during organogenesis (weeks 3-9) leads to gross morphological deformities, and exposure during the fetal period (weeks 8-38) cause physiological and functional defects. However, miscarriage may occur during the latter two periods if exposure to teratogens is sufficient. An unfortunate feature of many teratogens is that they can exert their effect very early in the pregnancy, in many cases before the woman knows she is pregnant. An example is ethanol, which has its greatest effect in the first 4 weeks of pregnancy.

By 1984, approximately 2000 chemicals had been tested for teratogenicity using test animals, 782 of which definitely caused teratogenicity and a further 291 were possibly teratogenic. However, only 30 of these chemicals are known to cause teratogenicity in humans. Examples of teratogenic chemicals include ethanol, methyl mercury, thalidomide and chlorobiphenyls (Figure 19.6).

B. EXAMPLES OF TERATOGENS

1. Alcohol (Ethanol)

Continued or excessive exposure to alcohol can lead to the fetus developing fetal alcohol effect (FAE), progressing with increased exposure to fetal alcohol syndrome (FAS). Typical FAS symptoms include a range of facial deformities (small head, small eye openings that are widely separated and a thin upper lip) and severe growth, developmental and intellectual retardation. In many cases, it is difficult to identify patients with FAE or FAS without prior knowledge of the mother's drinking history as similar symptoms can be caused by other factors.

2. Methylmercury

The discovery of the teratogenic properties of methyl mercury occurred during the investigation into the causes of Minimata disease. A case of the methylmercury

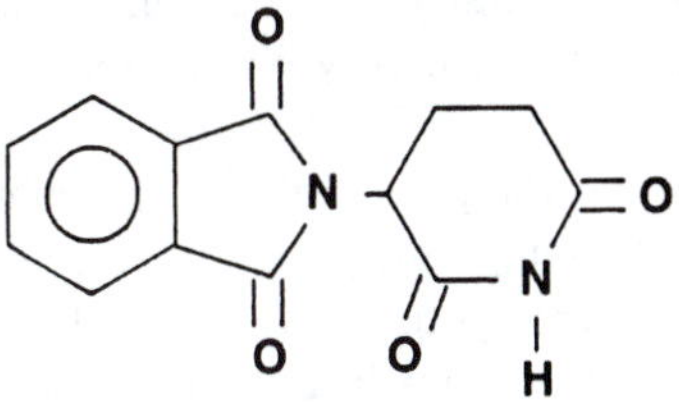

Thalidimide

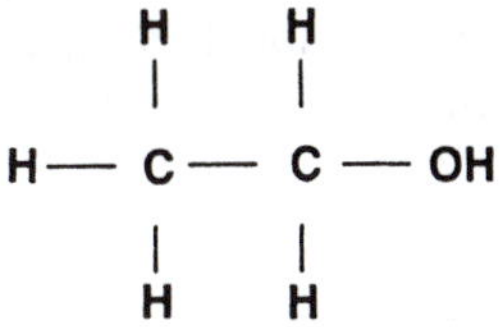

Ethanol

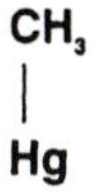

Methylmercury

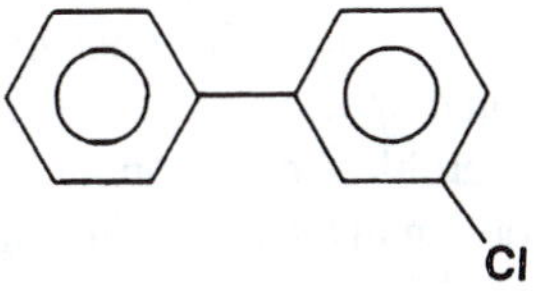

3-Chlorobiphenyl

FIGURE 19.6. The chemical structures of some known teratogens.

poisoning occurred in Japan, where many members of a community at Minimata were affected. Concurrent with the outbreak of Minimata disease was an outbreak of children born with quite severe mental retardation (approximately 6% of children born were affected). Typical symptoms included primitive oral and grasping reflexes, poor coordination, salivation, character disorders (unfriendly, shy, nervous and restless), seizures and epilepsy, deformed limbs and slow growth. Right from the onset it was felt, by some involved with the case, that these children were also suffering from Minimata disease. However, it was several years later when autopsies could be performed on two dead infants that methylmercury proved to be the causative agent.

In 1974, it was discovered that women who gave birth to congenitally poisoned children had exhibited none of the early symptoms of methylmercury poisoning. This was due to the ability of methylmercury to easily cross the placenta and concentrate in the developing fetus. Concentrations in fetal brains were up to four times those in the mother and fetal blood levels were 28% greater than the mothers.

3. Rubella (German Measles)

Rubella was one of the first agents that was discovered to be teratogenic. This was discovered by Gregg in 1941 following a rubella epidemic in Austria. The deformities and outcome depend on when the mother was exposed. Generally, it leads to eye, heart and ear defects and mental retardation. In an attempt to minimize the occurrence of such deformities, it is often recommended that sexually mature women be immunized against rubella.

4. Thalidomide

Probably the most notorious teratogen is thalidomide (Figure 19.6), which was developed as a sedative/tranquillizer. By accident, it was discovered that it was also a very powerful suppressor of morning sickness associated with pregnancy. The characteristic symptoms of thalidomide exposure are deformities of the limbs, predominantly the arms. These findings promptly led to the total removal of the drug in Western countries; however, it remained on sale in many countries for a number of years. Estimates of the number of children affected range from 5000 to greater than 10,000.

From retrospective studies, it appears that exposure of the fetus between weeks 6 and 7 of the pregnancy led to the characteristic deformities. It appears that thalidomide exerts it teratogenic effect between weeks 6 and 7 by killing cells that were developing into limbs. The exact mechanism has not been resolved.

Many have claimed that if thalidomide was tested for teratogenic properties, that this episode would never have happened. This is not necessarily so: the teratogenic effects of thalidomide vary greatly with the test species. For instance, rats and mice exhibit very little effect, even at concentrations 4000 to 8000 times those that affect humans. Only certain strains of rabbit and some species of primates have similar susceptibilities to thalidomide as humans. This problem of extrapolating the findings of tests from the test animals to humans is a problem in the use of data on mutagenesis and carcinogenesis tests.

IV. MUTAGENS AND MUTAGENESIS

A. TYPES OF MUTATIONS

A mutagen is any chemical that causes damage to the DNA or inhibits or damages the DNA repair mechanisms. Whether the mutation this causes is inheritable or not depends on the type of cells that suffer DNA damage. Damage to DNA in cells involved in the production of ova or sperm (germinal cells) causes inheritable mutations; whereas, damage to DNA from all other cells (somatic cells) is not inheritable. Mutagenesis is not solely due to exposure to synthetic organic chemicals; it is a natural process, a part of everyday life, even in nonindustrialized countries. In fact, mutation is the source of variation and change evoked to explain Darwin's theory of natural selection and evolution. Examples of natural mutagens include ultraviolet light, radiation from the decay of naturally occurring radioactive materials, alkaloids and flavonoids from plants, and mycotoxins from fungi. It has been estimated that each day, several thousand mutations in the sequence of DNA bases occur in each mammalian cell.

Mutations can be beneficial, harmful or have a neutral effect. However, with complex organisms that have evolved over long periods of time, most will not be beneficial. An interesting point is that through changes in our culture, we are now converting previously harmful mutations into neutral ones. For example, asthma and short-sightedness would have been harmful in hunter-gatherer societies, but are now no longer harmful in an evolutionary sense, i.e., these mutations do not reduce the ability of an individual to have children and thus perpetuate their genes.

Mutations of DNA can be classified into three main types: (1) where individual bases of DNA have been substituted, added or removed (point mutations); (2) there is large-scale damage to chromosomes (clastogenesis); and (3) where an uneven distribution of DNA (aneuploidisation) occurs so that resulting cells have either too many or too few chromosomes.

1. Point Mutations

A substitution point mutation is when the mutation leads to the removal of a DNA base, which is replaced (substituted) by another. There are three types of substitution point mutations: **transition, translation** and **frame shift**. Where a DNA base pair is replaced by the other base pair (remember there are only two DNA base pairs), so the purines and pyrimidine bases remain on the same strand of DNA (e.g., AT is replaced by GC), this is termed a **transition** (Figure 19.7b). Another form of point mutation of DNA bases is when a base pair is replaced by the other base pair but the purine and pyrimidine bases swap DNA strands (e.g., AT is replaced by CG) (see Figure 19.7c). This is termed **translation** as the bases have been translated from one DNA strand to the other. The third form of point mutation is called **frame shift**. This is caused by the addition or deletion of one or more bases to the DNA. This may totally destroy the original message coded in the DNA as the codons have all changed. The new sequence of nucleotides may code for different amino acids, be a nonsense code that corresponds to no amino acid, or it could code for the termination of protein synthesis leading to the formation of incomplete proteins. The largest piece of DNA a point mutation can affect is a gene.

2. Chromosome Mutations

Mutagens that cause chromosomal mutations can cause much more DNA damage than point mutations. The principle process causing such mutations is the linking of DNA strands that belong to different chromosomes. Such a process is called **cross-linking**. During cell replication (mitosis), the two strands of DNA in each chromosome are separated. The separation of cross-linked chromosomes causes breakages in the DNA strands that may lead to rearrangements, deletions and additions of DNA. This is one means by which alkylating agents such as epoxides,

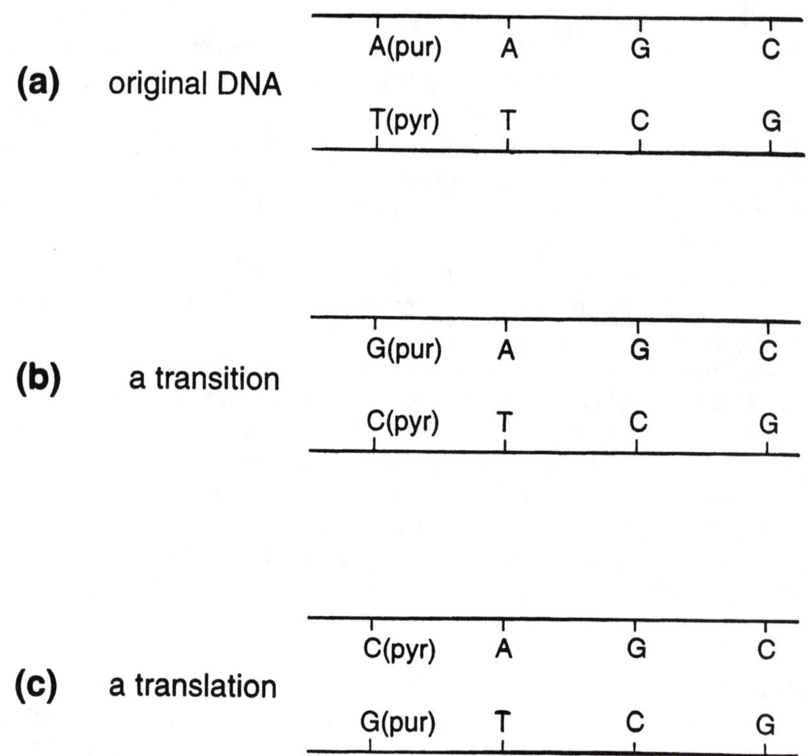

FIGURE 19.7. The types of point mutations of DNA, transition (b) and transition (c) are illustrated by comparison with the original piece of DNA (a). Pur and Pyr indicate a purine base and pyrimidine base, respectively.

aldehydes, alkane halides, alkyl sulfonates, nitorsoureas and triazines (Figure 19.8) exert mutagenic effects.

B. EXPRESSION OF MUTATIONS AND DNA REPAIR

The fact that a mutation has occurred does not mean it will be expressed as a biological effect. The mutation may be so minor as to no significant effect. For instance, the changing of a single DNA base may not lead to any change in the message that is encoded in the DNA. Even if the mutation leads to a different amino acid being incorporated in an enzyme, to have an effect the changed amino acid must be in the specific region that forms the receptor site (the site where the reactants bind to the enzyme and are converted to product).

Another reason that mutations may not be expressed is the presence of DNA repair mechanisms (predominantly based on enzymes). It is important that the repair of DNA mutations occurs before the next replication of the DNA or immediately afterwards; otherwise, the mutation may become permanent. Repair enzymes either directly correct DNA damage (e.g., photoreactivation) or remove DNA surrounding the mutation and then use the sister strand of DNA to reconstruct the correct sequence of bases.

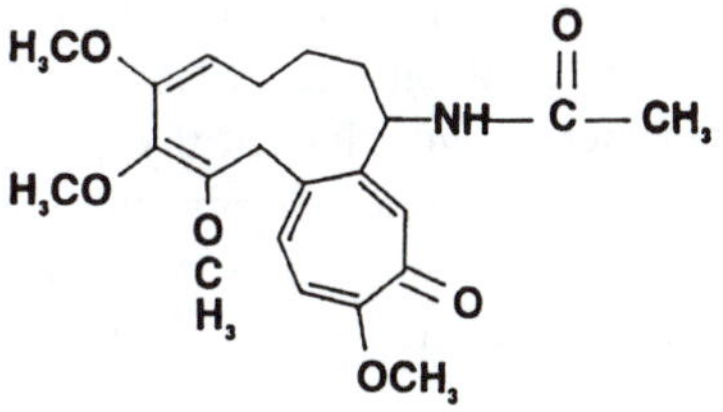

Colchicine

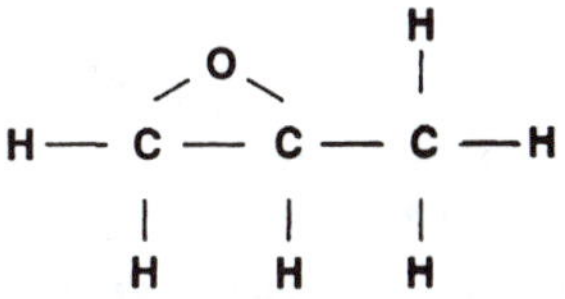

Ethylene oxide
(1,2 epoxyethane)

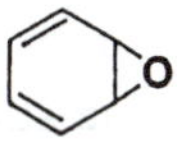

Benzene-1,2-oxide

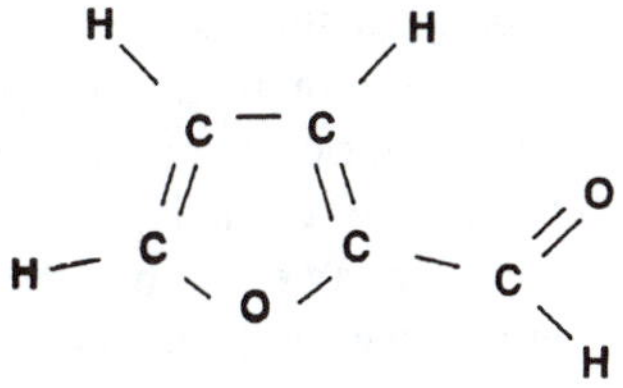

Furfural (aldehyde)

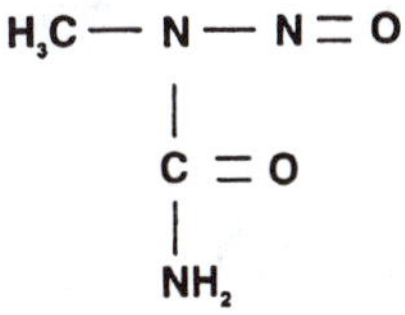

N-methyl-N-nitrosourea
(NMO)

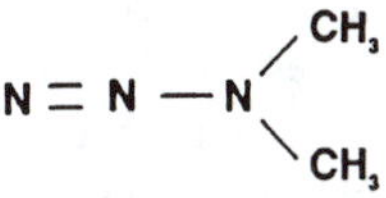

3,3-dimethyl-1-phenyl
triazene

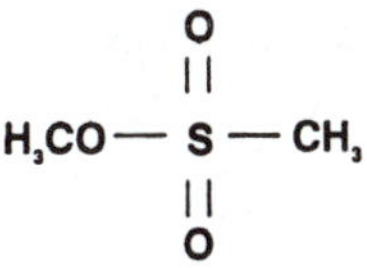

Methylmethane sulfonate

FIGURE 19.8. Chemical structures of mutagens known to cause chromosomal mutations by cross-linking chromosomes or by changing the number of chromosomes present in cells.

Photoreactivation is a repair mechanism which uses repair enzymes to correct DNA damage and remedy the most common mutation due to UV exposure: the formation of thymine dimers. A thymine dimer is two adjacent thymine bases on the same strand of DNA that are bound together by covalent bonds. Normally, there are no bonds between adjacent DNA bases (see Figure 19.2). A range of enzymes recognize these dimers and break the covalent bonds, repairing the mutation.

V. CARCINOGENS AND CARCINOGENESIS

A. TYPES OF CARCINOGENS

Carcinogens are compounds that lead to the formation of tumors. A tumor is a cluster of cells, all derived from one cell, undergoing continual growth in terms of both size and number of cells. In an experimental sense, a compound can only be classified as a carcinogen if it: (1) induces more tumors of the same type than are in the control; (2) induces tumors earlier than in the control; (3) induces types of tumors not found in the control; or (4) increases the multiplicity of tumors (the number of tumors per test animal).

Many chemicals have now been found to be carcinogenic or classed as suspected carcinogens to animals. These have been discovered in a number of ways: (1) during routine testing that is performed before a chemical or new product can be released onto the market, or (2) during post-release tests that are conducted due to evidence of carcinogenic action from observation and epidemiological studies.

Carcinogens can be subdivided into two main types based on the mechanism of action: genotoxic and epigenetic. Within these two categories there are several subdivisions. Genotoxic carcinogens irreversibly alter the DNA, leading to inheritable changes. Epigenetic carcinogens exert their cancer-forming effect by all means other than direct interaction with DNA. If a chemical has both genotoxic and epigenetic properties, it is assigned to the genotoxic category. If insufficient information on the mechanism of action is known, the chemical is simply called a carcinogen. These terms are not mutually exclusive; some carcinogens exert both genotoxic and epigenetic effects.

1. Genotoxic Carcinogens

Genotoxic carcinogens are mainly **electrophilic** reactants, i.e., they are electron deficient. In fact, most of the organic carcinogens are electrophilic genotoxic carcinogens. They include components of soot, coal tar and tobacco smoke such as polycyclic aromatic hydrocarbons (PAHs) and quinolines. Such compounds are highly reactive and form covalent bonds with nucleophilic compounds (compounds that tend to donate electrons in reactions) such as DNA and many other cellular compounds. When electrophilic genotoxic carcinogens react with DNA, they form adducts where the chemical is either reversibly covalently bound or irreversibly bound. The formation of DNA adducts leads to damage to DNA in the form of mispairings of bases and breaks of the DNA damage. A reasonable amount is known about how DNA adducts are formed and the various types, yet little is known of how or even if these lead to cancer.

One group of chemicals that readily reacts with DNA to form adducts is the alkylating compounds in Figure 19.8. These chemicals form carbonium ions (CH_3^+) that can methylate the bases and phosphate groups of DNA.

Within the genotoxic group of carcinogens, there is a further subdivision based on whether or not they are innately carcinogenic and do not require biotransformation to induce cancers (activation independent) or they will not induce cancer unless they have been biotransformed (activation dependent).

Most environmental carcinogens are activation dependent. Examples of activation-dependent carcinogens include benzene, PAHs, arylamine, and nitrosamine (Figure 19.9a). The activation-dependent process is analogous to the bioactivation of toxicants. The original noncarcinogenic compound is generally stable in the environment and is termed a **precarcinogen**. Metabolites that have higher carcinogenic activities are termed **proximate carcinogens**, while the final carcinogenic compound is the **ultimate carcinogen**. The ultimate carcinogen is generally an electrophile. Some examples of carcinogenic electrophiles are presented in Figure 19.10. In the case of benzene, the ultimate carcinogen(s) have not yet been identified.

Activation-independent carcinogens generally have very short half-lives in the environment due to their high chemical reactivity. Active halogenated compounds, alkaline epoxides and sulfate esters are examples of activation-independent carcinogens (Figure 19.9b). The extent of carcinogenic effect of activation-independent carcinogens depends on the proportion of the chemical interacting with DNA and that reacting with other non-DNA cellular material. The more that reacts with non-DNA cellular material, the lower the carcinogenic effect. The biotransformation of activation-independent carcinogens invariably leads to a reduction in genotoxic potency.

2. Epigenetic Carcinogens

As stated previously, epigenetic carcinogens exert the carcinogenic effect by any other means than reacting directly with the DNA. Mechanisms of action include:

1. Suppressing the immune system
2. Modifying hormone activity
3. Promoting the effects of other carcinogens (promoters)
4. Enhancing the effects of other carcinogens (co-carcinogens)

The concepts of **co-carcinogens** and **promoters** and how such compounds contribute to the development of cancer are important. Neither cocarcinogens nor promoters are carcinogenic when administered by themselves. A **co-carcinogen** is a compound that, when added either *prior to* or *simultaneously with* a carcinogen, leads to an increased carcinogenic effect (i.e., increased number of tumors or a decreased period before tumors appear). Known co-carcinogens include ethanol and components of tobacco smoke (such as catechol and phenolic compounds). A **promoter** is a compound that, when administered *after* a carcinogen, leads to an increased carcinogenic effect. If cells are exposed to promoters before carcinogens, there is no enhancement effect. Known promoters include some pesticides, saccharin, tobacco smoke, benzopyrene and phorbol esters, particularly 12-0-tetrade-

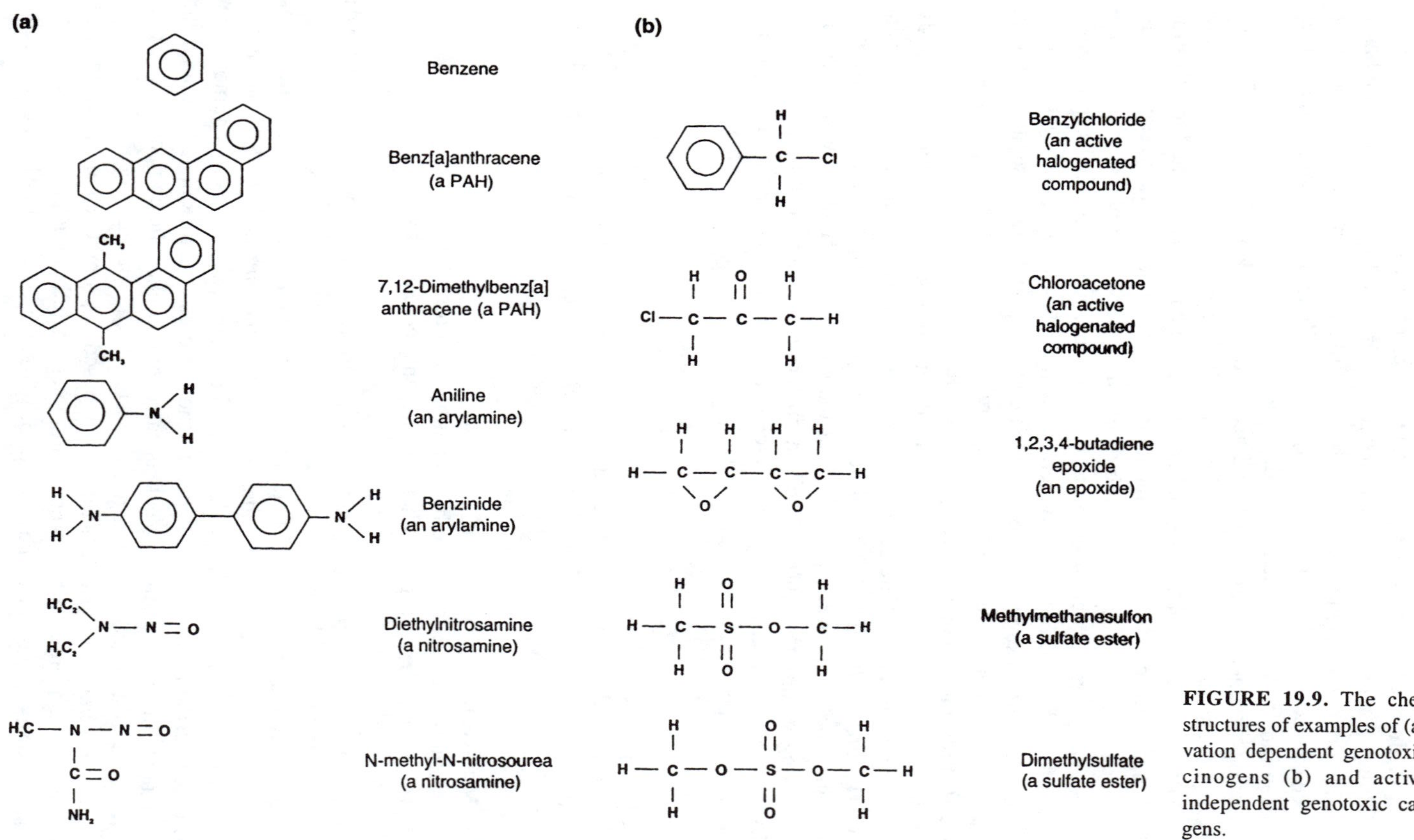

FIGURE 19.9. The chemical structures of examples of (a) activation dependent genotoxic carcinogens (b) and activation independent genotoxic carcinogens.

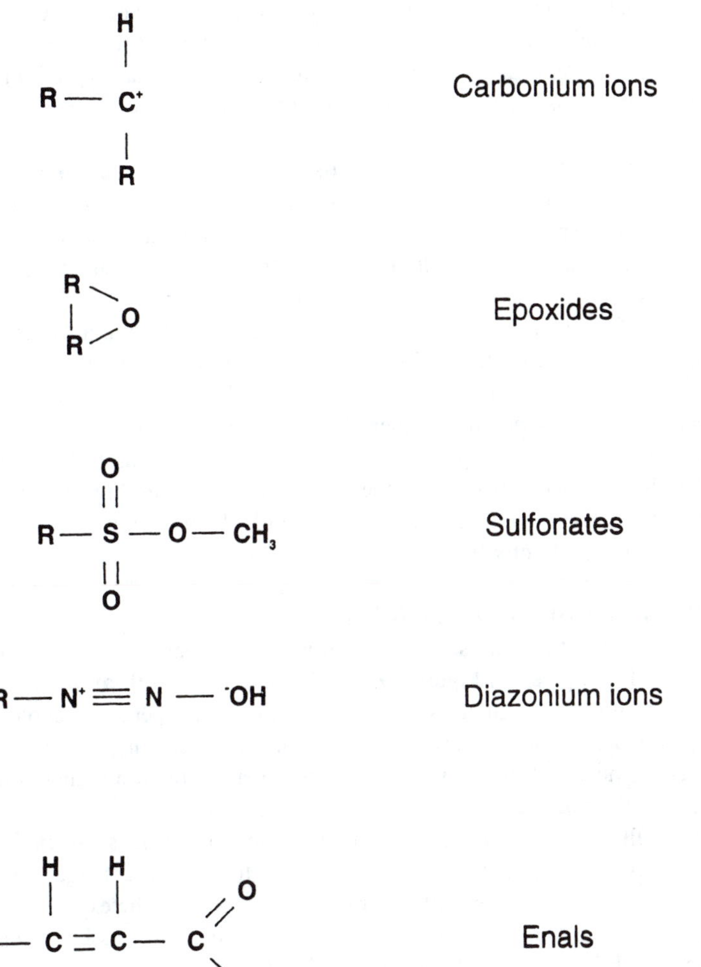

FIGURE 19.10. Examples of carcinogenic electrophiles.

canoylphorbol-13-acetate (TPA) and dietary fat. The role of promoters will be discussed in greater detail in the section on the development of cancer.

B. TUMORS

Tumors are a mass of cells that are all derived from one original cell and are growing in a manner not controlled by the body. As part of the cancer development, the normal cell regulatory processes become inoperative.

There are two principal types of tumors: benign and malignant. Benign tumors are slow growing, do not invade the surrounding tissue and do not release cancer cells that are transported throughout the body and may form secondary cancers. This process of forming secondary cancers is called **metastases**. Benign tumors can be surgically removed and should not recur if the entire tumor is removed. Benign tumors can, by as yet unknown mechanisms, be converted to malignant tumors.

Malignant tumors are the more dangerous type of tumor as they grow rapidly, invade surrounding tissue and form metastases. Malignant tumors are dangerous as even with the removal of an identified tumor, other secondary tumors may have already formed elsewhere, necessitating more operations and making total removal very difficult.

In general, tumors form wherever tissue has been exposed to carcinogens and promoters. Thus, common sites of cancers are the major sites of toxicant uptake (the skin, the respiratory system and the gastrointestinal tract) and the major organs of biotransformation (liver, lungs, and kidneys). Many carcinogens appear to almost exclusively cause cancers in one or two organs. This is at least partially due to the fact that, in general, enzymes can only react with a limited number of chemicals (precarcinogens), with different enzymes reacting with different chemicals. Other reasons include the uneven distribution of enzymes that activate precarcinogens and that the number of enzymes and their rate of metabolic activity vary dramatically from organ to organ. For example, ethanol often causes cancer of the liver (called cirrhosis of the liver) as it contains high concentrations of enzymes that use ethanol as the substrate (e.g., alcohol dehydrogenase) with high metabolic rates of activity.

C. DEVELOPMENT OF CANCER

The order of the main steps involved with carcinogenesis, as they are currently understood, are given in Figure 19.11. The first step is **initiation,** where cells are exposed to initiators. Initiators are either activation-independent genotoxic carcinogens, ultimate genotoxic carcinogens or epigenetic carcinogens that cause some damage to the cell. Epigenetic carcinogens generally require higher concentrations and more prolonged exposure in order to initiate cancer.

The cellular damage at this stage is still reversible. This is evident inasmuch as, just because DNA is damaged or mutated or the cell is damaged, this does not automatically mean cancer will occur or the mutation will be expressed. In fact, the great majority are not expressed. This is because the body has DNA repair mechanisms, which have already been discussed in this chapter.

In order for the damage to be made permanent and cancer to develop, the initiated cell must be exposed to promoters for prolonged periods. This step in the development of cancer is termed **promotion.** It appears that once a cell is initiated, exposure to promoters, regardless of the length of the intervening period, will still enhance the carcinogenic effect. For cancer to develop, the exposure to promoters must be of long duration. A cessation of exposure to promoters dramatically decreases the chance of cancer developing. This is clearly illustrated by the fact that the probability of contracting lung cancer rapidly returns to that for non-smokers when smoking is ceased.

At some point after prolonged exposure to promoters, the carcinogen induced changes are made irreversible and the cell expresses new physiological and immunological properties and its replication is no longer controlled by the normal mechanisms. The final stage is **progression** and is characterized by the growth of the tumor and the conversion from a benign to a malignant tumor.

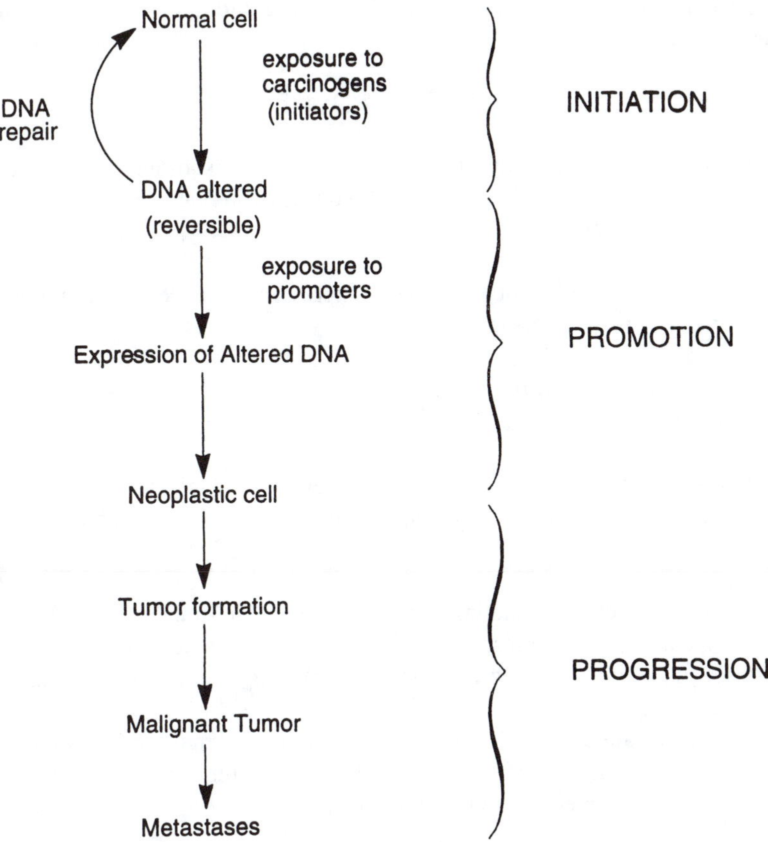

FIGURE 19.11. A flow chart of the steps involved and the stages in development of cancer.

D. PROBLEMS WITH GENOTOXICITY TESTS

Testing is usually carried out on test animals and there are problems in extrapolating from animals to humans. This is particularly the case with genotoxicity tests where animals are used to derive safe levels of chemical exposure for humans. The most commonly used test organisms are mice and rats, which are relatively closely related. Yet the agreement of results from carcinogenic tests (i.e., carcinogenic or noncarcinogenic) for these two species is approximately 65% and only in 50% of cases were the same organs affected. The genetic difference is much larger between humans and rodents than between mice and rats; so at the very best, the agreement between animal tests and humans would be the same. However, due to the differences, the agreement is probably much less.

Even if one assumes that there are no problems with extrapolating from mice and rats to humans, there is still a significant problem: the genotoxicity of chemicals is predominantly determined on individual substances, whereas humans, in fact all animals, are exposed to very complex mixtures of chemicals. Estimates of safe levels do not take into account the combined genotoxicity of chemicals. If synergistic interactions occur (the toxicity of the mixture is greater than the sum of the toxicities of the individual components), then the safe levels could quite significantly underestimate the effect.

VI. KEY POINTS

1. Genotoxicity is the study of the adverse effects of substances on the genetic material of cells (DNA) and the subsequent expression of these changes as biological effects.

2. Genotoxic effects, particularly those expressed as cancer, are a major cause of death in many societies.

3. The genetic material that codes characteristics of life is DNA (deoxyribonucleic acid).

4. DNA consists of two long strands of nucleotides wound in a spiral form often described as a helix. The nucleotides consist of a pentose, a phosphate group and a nitrogenous base.

5. DNA replicates by separating into single nucleotide strands on which enzymes place and bind the corresponding sequence of nitrogenous bases. Proteins are also synthesized utilizing DNA through transcription and translation.

6. Teratogens are chemicals that interfere with the normal reproduction process and cause a reduction of successful births or offspring to be born with physical, mental development or behavioral defects.

7. The gestation of humans is divided into three distinct periods: implantation (weeks 1-2); embryonic (weeks 3-7) and fetal (weeks 8-38). The earlier in the pregnancy the exposure to teratogens, the more severe the expressed deformities.

8. Examples of teratogens include alcohol (ethanol), methylmercury, rubella (German measles) and thalidomide.

9. A mutagen is a chemical that causes damage to the DNA or inhibits or damages the DNA repair mechanisms. But mutagenesis is not solely due to synthetic organic chemicals; it is a natural process that occurs as a part of everyday life. Mutations can be beneficial, harmful or have a neutral effect.

10. Mutations can occur as a result of the substitution, addition or removal of bases of DNA, large-scale damage to chromosomes as uneven distribution of DNA in cells.

11. Occurrence of a mutation doesn't necessary result in the expression of adverse biological effects. One reason for the lack of expression is the presence of repair mechanisms.

12. Carcinogens are compounds that lead to the formation of tumors, which are a cluster of cells, all derived from one cell, that are undergoing continual growth not controlled by normal body metabolic processes.

13. Genotoxic carcinogens are mainly electrophilic reactants, i.e., electron deficient, that react with DNA and other cellular components that are nucleophilic, i.e., electron donating. Examples of genotoxic carcinogens are soot, coal tar and tobacco smoke, which contain polycyclic aromatic hydrocarbons (PAHs).

14. Most environmental carcinogens are activation dependent, which means they are converted into other compounds by biota to give the ultimate carcinogen.

15. Cancer develops through a number of steps, starting with initiation, promotion and progression.

16. Co-carcinogens and promotors enhance the action of a carcinogen while not being carcinogens themselves. A co-carcinogen is a compound that when added either prior to or simultaneously with a carcinogen leads to an increased carcinogenic effect. A promotor is a compound that, administered after a carcinogen, leads to an increased carcinogenic effect.

REFERENCES

Ames, B.N., Profet, M., and Gold, L.S., Nature's chemicals and synthetic chemicals: comparative toxicology. *Proc. Natl. Acad. Sci. U.S.A.*, 87, 7782–7786, 1990.

Efron, E., *The Apocalyptics.* Simon and Schuster, New York, 1984.

Marshall, E., Experts clash over cancer data. *Science,* 250, 900–902, 1990.

Krewski, D. and Thomas, R.D., Carcinogenic mixtures. *Risk Analysis,* 12 (1), 105–113, 1992.

CHAPTER 19 PROBLEMS

1. In 1775, Pervival Pott concluded that cancer of the scrotum was associated with soot. Most of his patients with this disease were chimney sweeps. Give a brief outline of the possible carcinogens and their mode of action.

2. Some observers have claimed that our society is facing an epidemic of cancer in humans as a result of the use and occurrence of residues in food, waters and the environment. List factors for and against this claim.

Answers on Page 456.

CHAPTER 19 SOLUTIONS

1. **Possible Carcinogens**: Common carcinogenic components in combustion products are polycyclic aromatic hydrocarbons (PAHs). These would be expected to occur in soot in chimneys.

   Mode of Action: Soot would become entrained in the chimney sweeps' clothing containing PAHs. The PAHs are lipophilic and would be expected to partition into the skin of the sweeps. These skin-associated PAHs could be activated by oxidative cellular processes, forming the ultimate carcinogen. After a period of time, the formation process would occur, leading to the development of scrotum cancer.

2. **For**:

   - There may be an increasing occurrence of synthetic chemicals in air, food, soil and water over time.
   - A large proportion of the population die from various types of cancer.
   - Many tests have indicated that synthetic chemicals are carcinogenic.

   Against:

   - It is not clear that exposure to synthetic organic chemicals and related compounds is increasing. The levels of pollution in air, water, food and soil may be decreasing due to increased management.
   - The causative agents for the large proportion of the population affected by cancer are probably not environmental agents.
   - Tests are often carried out on relatively high concentrations of chemicals compared with environmental levels, and extrapolation from animals to humans is required, which introduces a range of uncertainties.

Management of Hazardous Substances

Chapter 20

ECOTOXICOLOGY — THE INTERACTION OF CHEMICALS WITH ECOSYSTEMS

I. INTRODUCTION

The chemical industry has grown enormously in recent decades. It provides petroleum fuels, antibiotics and other drugs, plastics, pesticides, food preservatives, agricultural fertilizers, etc., without which our society cannot survive. About 100,000 chemicals are estimated to be in daily use and of these, approximately 7000 are produced commercially in comparatively large quantities. Most of these substances have little or no adverse environmental effects, but some may be harmful to human health or the natural environment. Often, these effects only become apparent after wide and prolonged usage and then control measures are introduced. Clearly, there should be an effective testing and evaluation program to determine those chemicals that present a potential environmental hazard before use. Many countries have developed hazard evaluation program for new chemicals. Some have indicated that these control programs will be their major environmental management program for the foreseeable future.

Techniques to evaluate aspects of the biological effects of chemicals have been described previously. Environmental toxicology (Chapter 4), transformation and degradation processes (Chapter 3), distribution of chemicals in the environment (Chapter 18), genotoxicology (Chapter 19) and risk assessment (Chapter 21) are all concerned with various aspects of the effects of chemicals in the environment. However, many of these aspects were previously considered as separate factors and the biological effects related principally to individual species. There is a need to draw these aspects together into an overall approach that addresses the effects on whole ecosystems. Our capability to evaluate the effects on whole ecosystems is not particularly strong, but approaches based on **ecotoxicology** have achieved the greatest measure of success up to the present time.

The control and management of hazardous chemical often generate a high level of public interest, and socioeconomic and political forces can play a major role in the control of environmental hazards. Nevertheless, to see the problem in its correct perspective, a clear understanding of the nature and effects of a potentially hazardous chemical is needed.

II. THE ECOTOXICOLOGY CONCEPT

Scientific investigation of the effects of pollutants in the natural environment has focused on the effects on individual biological species, as described in Chapter 4 on environmental toxicology. Such investigations can only give suggestions as to the likely effects on the complex ecosystems existing in the natural environment. Thus, there has been increased interest in the effects of pollutants on whole ecosystems since these are of primary management concern. This has led to the development of this relatively new field of **ecotoxicology**.

Ecotoxicology is concerned with the fate and toxic and related effects of chemicals in natural ecosystems. The ecotoxicology of a chemical may be viewed as being based on a sequence of interactions and effects controlled by the physical, chemical and biological properties of a chemical, as indicated in Figure 20.1. A chemical discharged to the environment as a solid, liquid or gas, depending on its physical properties, can than be subject to distribution in the atmosphere, water or soils and sediment, depending on its physical chemical properties as described in Chapter 18. At the same time, it can be chemically modified and transformed by abiotic processes or more often by microorganisms in the environment. The organisms present are then exposed to the toxicant in its original form and in its degraded (or transformed) state, and at concentrations resulting from its dispersal. Uptake of the chemical and its degradation products occurs and organisms can exhibit a variety of different reactions from negligible to sublethal effects, such as reduced growth, reproduction decline and behavioral effects or ultimately death. The complex natural ecosystem of which the organisms are an integral part can react in a variety of ways to the effects on the component organisms. Food chain relationships, energy flows, etc. may be altered. Thus, the ecotoxicology of a chemical can be considered as a sequence of steps starting with a source and following through to the ecosystem response, with different properties of a chemical being involved at each step, as illustrated in Figure 20.1.

Initially, the physical state of the contaminant, whether a solid, liquid or gas, has a major influence on its physical dispersal from the source. Following that, physical chemical properties influencing movement into different environmental phases become important. In later stages, interactions with organisms, populations, communities and ecosystems are produced by the biochemical and physiological properties of a chemical. As the subsequent effects of a chemical enter more complex levels of biological organization, such as the whole ecosystem, the direct effects of the chemical become less significant. The biological effects generated at simpler levels of biological organization flow through the system to impact on whole ecosystems rather than the direct effects of the chemical itself at this level of organization.

III. TYPES OF TOXICANT DISCHARGED TO THE ENVIRONMENT

The types of toxicant discharged to the environment are quite numerous. Toxic gases and nongaseous compounds originate from motor vehicles, electricity generation, industry and numerous other sources. Toxic materials are often discharged to waterways in sewage, stormwater runoff and in industrial discharges. Soils can be contaminated from most of the sources indicated above. A summary of the sources and types of chemical groups involved and the environment affected is shown in Table 20.1.

There are a wide variety of different types of chemicals having an ecotoxicological effect in the environment. However, with the organic toxicants, the groups commonly involved are hydrocarbons, aromatic hydrocarbons, chlorohydrocarbons,

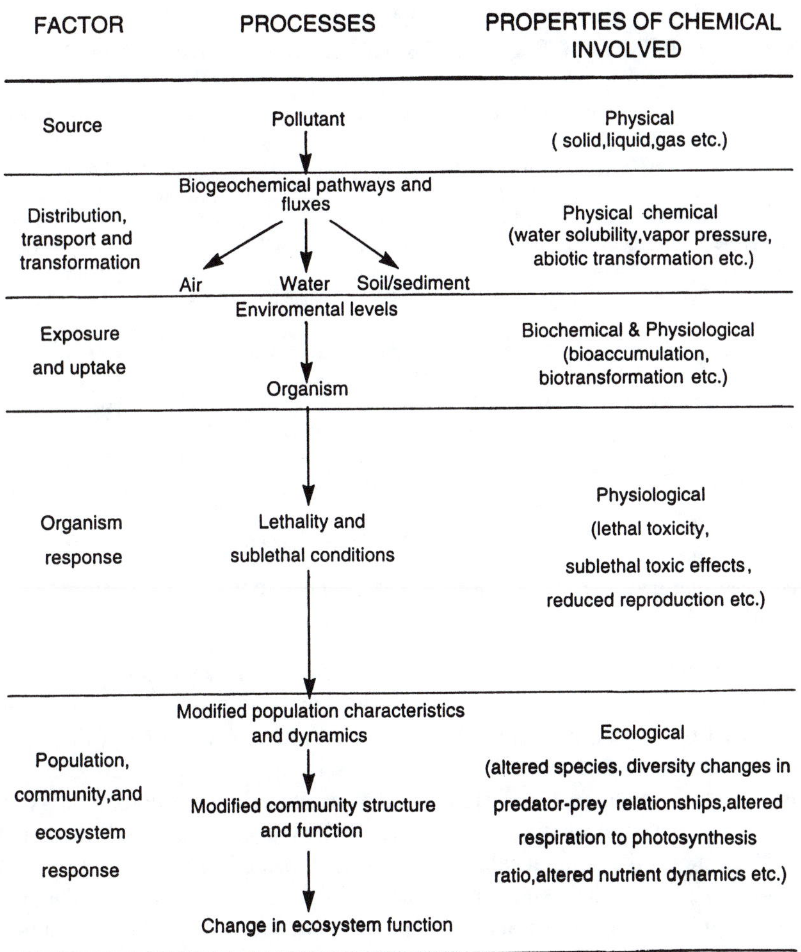

FACTOR
PROCESSES
PROPERTIES OF CHEMICAL INVOLVED
Source
Pollutant
Physical (solid,liquid,gas etc.)
Distribution, transport and transformation
Biogeochemical pathways and fluxes
Air
Water
Soil/sediment
Physical chemical (water solubility,vapor pressure, abiotic transformation etc.)
Exposure and uptake
Enviromental levels
Organism
Biochemical & Physiological (bioaccumulation, biotransformation etc.)
Organism response
Lethality and sublethal conditions
Physiological (lethal toxicity, sublethal toxic effects, reduced reproduction etc.)
Population, community,and ecosystem response
Modified population characteristics and dynamics
Modified community structure and function
Change in ecosystem function
Ecological (altered species, diversity changes in predator-prey relationships,altered respiration to photosynthesis ratio,altered nutrient dynamics etc.)

FIGURE 20.1. Diagrammatic illustration of the impact of pollutants on the components and properties of ecosystems, together with the properties of the chemical involved.

oraganophosphorus compounds, PCBs, PCDD, PCDF and surfactants. With the inorganic substances involved, the toxic metals are probably the most common toxicants.

IV. LABORATORY TESTING FOR ENVIRONMENTAL EFFECTS

Ecotoxicology is essentially a practical and applied science that is related to the management of adverse effects of chemicals discharged to the environment. The stimulus for the development of ecotoxicology relates to our ability to manage adverse effects of chemicals on whole ecosystems. Another important aspect of ecotoxicology is that it is often concerned with the prediction of possible adverse effects in new situations related to new developments.

TABLE 20.1
Sources and Types of Toxicants Discharged and Environments Affected

Source	Some chemical groups involved	Environments affected
Motor vehicle exhausts, electricity generation and industrial discharges to the atmosphere	Lead and other toxic metals, carbon monoxide, carbon dioxide, aromatic hydrocarbons, sulfur dioxide, hydrocarbons, PCDD, PCDF, PCBs[a]	Human and natural terrestrial systems
Sewage	Aromatic hydrocarbons, chlorohydrocarbons, toxic metals, surfactants	Aquatic systems
Stormwater runoff	Aromatic hydrocarbons, hydrocarbons, lead and other toxic metals	Aquatic systems
Industrial discharges to waterways	Acids, toxic metals, salts, hydrocarbons, PCDD and PCDF	Aquatic systems
Urban and industrial discharges to soil	Toxic metals, salts, hydrocarbons, PCDD, PCDF, PCBs[a]	Human and natural terrestrial systems
Rural industries	Chlorohydrocarbons, organophosphorus compounds	Human, natural terrestrial and aquatic systems

[a] Polychlorodibenzodioxins (PCDD), polychlorodibenzofuran (PCDF) and polychlorobiphenyls (PCBs).

The approach to the prediction of possible adverse effects is often to use laboratory tests to evaluate key parts of the system that can then be used to give an indication of the overall effects on the whole ecosystem. An outline of the types of the laboratory tests used to evaluate chemicals is shown in Table 20.2. In broad terms, the tests relate to the various stages involved in the ecotoxicology of a chemical. A comparison with the properties of the chemical involved, indicated in Table 20.2, shows the general relationship of the types of tests used to the ecotoxicology approach to the behavior of chemicals in the environment. The initial physical chemical properties relate to dispersal from sources and the distribution in environmental phases. The degradation and accumulation of a chemical gives an indication of the actual exposure of biota to the chemical in the different phases in the environment. Tests on individual organisms such as alga, fish and terrestrial plants yield an indication of effects on natural ecosystems, while tests on mammals, such as rats, provide indications of potential human health effects.

These laboratory tests are conducted under highly specified conditions that define the species and test conditions to be used and the manner in which the results are to be recorded and processed. This is quite appropriate since these tests are primarily evaluations of the properties of the chemical being tested and, if the results obtained are not reproducible, then the test is inconsistent and the results unreliable. It is noteworthy that there are no tests on populations, communities and ecosystems; these systems are too complex to be amenable to standardized testing in this way. Thus, results on ecosystems must be obtained by extrapolation from tests on individual species, or by evaluation of some ecosystems on a small scale, in actual environmental situations.

Sublethal tests are included in the array of laboratory testing methods. For example, in Table 20.2, there are tests for bioaccumulation, growth of terrestrial

TABLE 20.2
Types of Laboratory-Based Tests Used to Evaluate Adverse Effects on E/S

Category	Examples
Physical chemical properties	K_{OW} value
	Aqueous solubility
	Adsorption/desorption
Degradation and accumulation	Biodegradability
	Fish bioaccumulation
Effects on biotic systems	Alga: growth inhibition
	Fish: acute toxicity
	Terrestrial plants: growth test
Human health effects	Oral toxicity: rats
	Eye toxicity: rabbits
	Reproduction toxicity; rats
	Carcinogenicity; rodents

plants, eye irritation in rabbits and carcinogenicity in rodents. All of these tests are concerned with effects that are apparent at levels lower than the lethal levels. This area presents difficulties in ecotoxicology interpretations due principally to a few general problems. In particular, some sublethal effects often take a considerable period of time to become apparent. For example, carcinogenicity may not become apparent until up to 20 years after exposure has occurred. In addition, there are a large number of types of sublethal effects that can occur and it can be difficult to identify which sublethal effects are of concern in a specific situation.

The effective evaluation of sublethal human health effects presents particular problems in laboratory testing. For example, a sublethal effect that occurred at the rate of 10 in a population of 100,000 over a 20-year period would be considered to be of human health importance. Now, to obtain a reasonable number of positive effects (e.g., 3) in test animals would require the use of up to 30,000 test animals over 20 years if the incidence in test animals was the same as that in the human population. Similar problems could be expected with natural ecosystems. Such testing cannot be carried out, so testing is usually conducted with high levels of the test chemical and short periods of exposure followed by extrapolation to low levels over long periods. These extrapolations with human health are the subject of wide debate in the scientific community.

V. EFFECTS ON ECOSYSTEMS

Tests on whole ecosystems are generally not fully effective due to several factors. For example, in a stream, conditions can vary from pool environments to rocky riffles over a short distance. A major factor is the variation in ecosystem type within any environment of concern. Second, there can be considerable variation in environmental conditions, such as temperature throughout the seasons, that are difficult to reproduce in test situations. Finally, variation in the biotic composition of ecosystems due to natural factors are not well understood and, until these can be understood, it is difficult to interpret variations due to the effects of contaminants.

However, there are some general principles that can be applied to the effects of chemicals on ecosystems.

Organisms in an ecosystem exhibit differing susceptibilities to toxicants. For example, herbicides are selectively toxic to plants and insecticides have a major impact on insect populations. Even within these broad groups of organisms, there is a wide range of different susceptibilities with different species. Thus, when any toxicant enters an ecosystem, it will selectively remove those susceptible species from the range of organisms present. This generally translates into a reduction in the species diversity of organisms present and a change in the community structure.

Figure 20.2 indicates the general pattern of response of the components of an ecosystem to a single application of a toxicant causing herbivore mortality as indicated by simulation modeling. Initially, both species diversity and biomass fall and the community structure is altered, but all tend toward the original situation as toxicant concentration decreases over time. The initial effects on carnivores and top carnivores are relatively low, but there is a delayed effect that extends over a comparatively long period. With chronic pollution over extended periods, situations somewhat similar to the initial effects of a single application are noted. For example, a decrease in species diversity and alteration of community structure may become apparent.

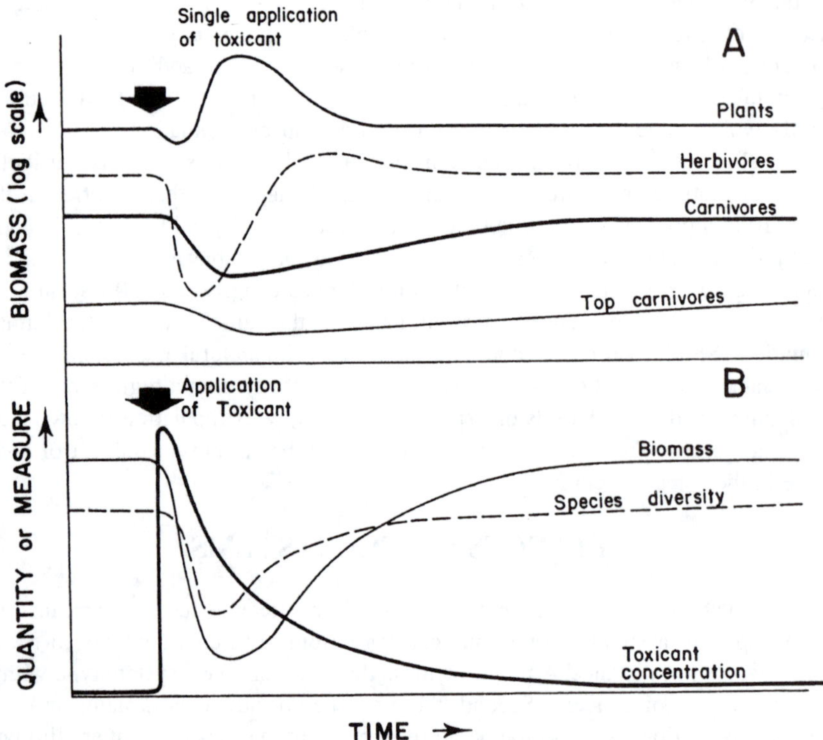

FIGURE 20.2. Some generalized characteristics of a toxicant causing herbivore mortality in an ecosystem with (A) indicating changes over time to some general groups of biota and (B) indicating changes over time to the toxicant concentration and related overall biological measures.

In addition to toxic effects on different components of an ecosystem, other important secondary effects usually result. In natural ecosystems, some animals and plants have adapted to the existing environmental conditions. There is a wide range of interacting and interdependent organisms including microorganisms, plants, invertebrates, vertebrates and so on. But the interrelationships between these groups are usually not well understood, which can lead to difficulties in understanding the effects of toxicants. However, in many cases, the effects due to a toxicant are large compared with natural variations and can lead to more obvious ecological effects.

Selective removal or alteration of a population of organisms in an ecosystem results in modification of the food web, as illustrated in Figure 20.3. This results in a change in energy and matter flow patterns. In an ecosystem, such characteristics as total respiration, total primary production, respiration to primary production ratio, nutrient cycling rates, predator-prey relationships etc. are changed.

Predicting the specific effects of a toxicant on an ecosystem is an area of ecotoxicology that is in most need of urgent development. Evaluation of the effects on ecosystems based on the investigation of individual species has met with limited success, and increasing attention has been given to direct investigation of the effects of toxicants on test sections of large natural ecosystems (mesocosms) and small artificial simplified ecosystems (microcosms).

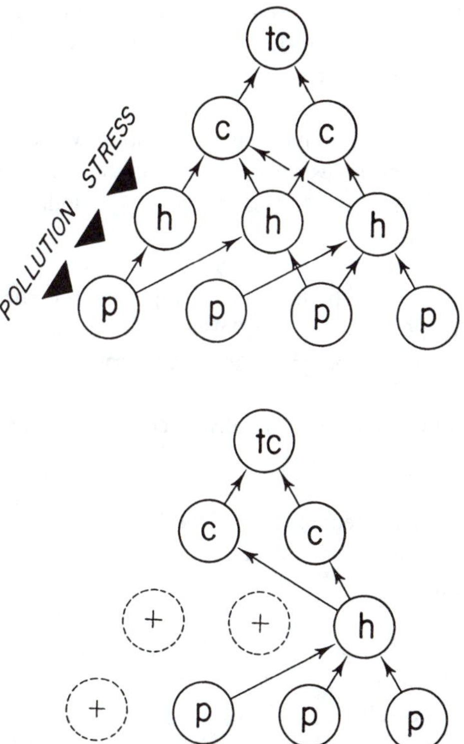

FIGURE 20.3. A hypothetical food web illustrating modifications as a result of elimination of some members due to pollution stress by toxicants where TC = top carnivore; C = carnivores; h = herbivores and p = primary producers.

VI. KEY POINTS

1. Most research on the toxicity of chemicals has been directed toward the evaluation of individual species, with relatively little information on populations, communities and ecosystems.

2. Ecotoxicology is concerned with the effects of chemicals on ecosystems and involves the evaluation of a sequence of interactions of the chemical with the natural environment. Thus, ecotoxicology is concerned with:

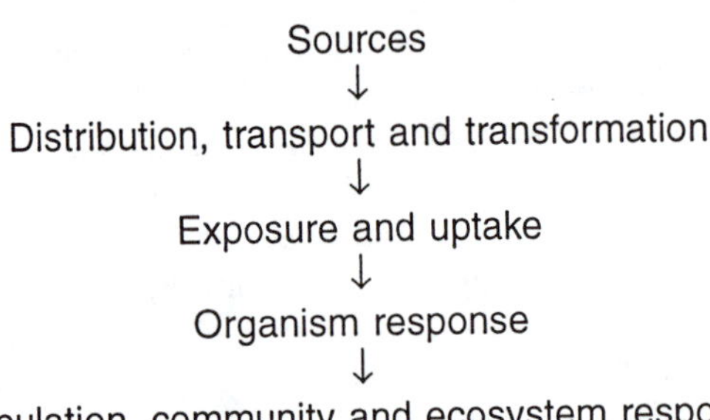

3. Effects on human health are sometimes considered as part of ecotoxicology but usually ecotoxicology is principally oriented toward the effects of a chemical in the natural environment.

4. Chemicals can enter the environment in motor vehicle exhaust, electricity generation discharges, industrial discharges to the atmosphere, sewage, stormwater runoff, industrial discharges to waterways, industrial contamination of soil and rural activities.

5. Hydrocarbons, aromatic hydrocarbons, chlorohydrocarbons, organophosphorus compounds, PCBs, PCDD, PCDF and surfactants are relatively common organic compounds discharged to the environment.

6. Lead and other toxic metals are relatively common inorganic discharges to the environment.

7. Laboratory testing procedures are available for physical chemical properties, degradation and transformation, bioaccumulation, effects on biotic systems and human health.

8. Laboratory-based tests provide an indication of the effects on key aspects of the natural and human system, but are not effective in indicating whole ecosystem effects and sublethal effects.

9. General effects of toxic chemicals on ecosystems include reduced species diversity, reduced biomass, a change in the types of biota present and changes in the energy and nutrient flows in ecosystems.

REFERENCES

Connell, D.W. and Miller, G.J., *Chemistry and Ecotoxicology of Pollution,* John Wiley & Sons, New York, 1984.

Carlow, P., Ed., *Handbook of Ecotoxicology,* Vol. 1 & 2, Blackwell Scientific, Oxford, 1993.

Hoffman, D.J., Rattner, B.A., Burton, G.A., and Cairns, J., *Handbook of Ecotoxicology,* Lewis, Boca Raton, FL, 1995.

CHAPTER 20 PROBLEMS

1. The chlorohydrocarbon pesticide DDT was widely used in agriculture in the 1950s and 1960s with subsequent adverse effects. DDT is a persistent compound with a high log K_{OW} value (approx. 6.0) and was found to be a potent fish poison, and to cause egg shell thinning and a resultant decrease in the population of some birds. Briefly outline the ecotoxicology of DDT using the information above and elsewhere in this book as appropriate.

2. Laboratory tests are used to evaluate the ecotoxicology of a chemical. Briefly indicate how tests in Table 20.2 can be used to evaluate ecotoxicology of chemical in a particular usage situation.

Answers on Page 470.

CHAPTER 20 SOLUTIONS

1. **Source**: Used as an agricultural pesticide in broadcast applications in rural areas.

   Distribution: Because of the high log K_{OW} value of approx. 6.0, this substance would be expected to accumulate in soils and then be swept into waterways on soil particles. The amount of pesticide actually dissolved in water and swept into waterways in this form would be expected to be small. This substance is resistant to degradation, which would contribute to the accumulation of the pesticide in soils and sediments in aquatic areas.

   Exposure/uptake: Since DDT will be present in sediments in waterways, a small quantity would dissolve, giving low concentrations in the overlying water and this would be accumulated by the aquatic organisms. The level of bioconcentration in aquatic organisms would be expected to give a bioconcentration factor (K_B) of the order of about 1,000,000.

   Organism response: DDT is toxic to fish and other aquatic organisms, and lethal effects could be expected in some situations. In addition, the well-known sublethal effect, resulting from bioaccumulation in birds, would be the occurrence of egg shell thinning, resulting in a lack of breeding success in some species.

   Population, community, ecosystem response: In general terms, it would be expected that some populations, particularly of aquatic organisms, would be reduced and possibly some species totally removed from the system.

2. **Source**: The amounts of chemical involved can be derived from usage patterns.

   Distribution: The distribution of the chemical can be evaluated using physical chemical properties such as log K_{OW} and the amounts involved derived from the source information. Fugacity modeling can be used to indicate phases in the environment where chemicals may accumulate.

   Exposure/uptake: The uptake can be evaluated using physicochemical properties and the information from bioaccumulation tests. The actual amount of uptake that occurs can be derived using the distribution information obtained above. Degradation can be evaluated to some extent by the usage of laboratory information on degradation and transformation tests.

   Organism response: Organism response can be evaluated using alga fish laboratory testing information and the information obtained above on distribution exposure and uptake.

   Population, community, ecosystem response: There are no tests available to provide this information.

Chapter 21

RISK ASSESSMENT

I. INTRODUCTION

Chemicals are used in a large variety of ways for the benefit of human society. The use of many chemicals has a direct risk involved; for example, medicinal compounds such as antibiotics are consumed directly. These substances have a risk of resultant adverse effects on some members of the population. Usually, only very small proportions of people are involved; but with all of these chemicals, a certain amount of risk is involved in their usage. Also, the use of pesticides in agriculture results in the exposure of non-target organisms to these active biological agents. In this way, human populations and organisms in the natural environment can both be exposed to biologically active chemicals. Unintentional, and often unplanned, risks can also be involved in the disposal of waste chemical products. Waste chemicals are often disposed to land, water and air and may contain agents that can have adverse effects on human health and the natural environment.

Government agencies throughout the world have had difficulties in placing the risk involved in all of these situations into clear perspective. For example, efforts may be made to manage the risks due to a particular discharge containing chemicals when in actual fact, other discharges may pose a greater risk. As a result, there has been a need to develop techniques that give a quantitative evaluation of the risks involved in various chemical discharges. These evaluations can then be used to gain an insight into the needs for management of the risks involved and allow the allocation of resources in the most effective way to reduce or eliminate that risk. This gives:

1. Greater confidence that efforts in environmental protection from chemicals will be successful in protecting human health and natural ecosystems.
2. Resources will be used in the most effective manner in controlling chemicals. Industrial and other activities will be managed in the most cost efficient manner without wastage of resources on chemical management problems that may not be of high priority.

In recent years, increasing use is being made of risk assessment procedures in government agencies, industry and elsewhere for risk assessment and management. Risk assessment is part of the overall approach to managing chemical hazards, as shown in Figure 21.1. **Risk assessment** consists of identifying the hazard and then quantifying the risk posed by that hazard by risk characterization. This information is then used in the risk management process, which consists of consideration of options, communication of risk to those involved, control decisions and then monitoring to ensure that a correct decision has been made. All of these steps can feed back into previous steps in the process or into the initial hazard identification step, and this procedure may occur several times before a satisfactory result is obtained.

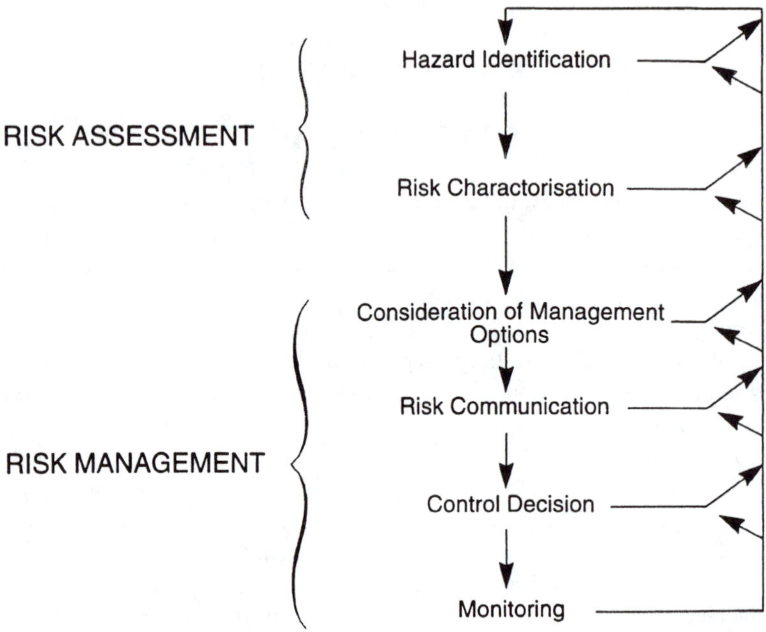

FIGURE 21.1. Overall concept of the Risk Assessment and Management process.

A learning procedure may be involved when new information is developed and fed into the system in later steps that could modify the outcome of preceding steps.

This chapter will focus on the risk assessment process. Risk assessment is concerned with two principal areas:

1. Human health
2. Natural ecosystems.

Similar general approaches are used in both cases, but there can be considerable differences in the applications in detail. This chapter will deal particularly with human health since this area has a clearer and more widely accepted methodology.

II. THE RISK ASSESSMENT PROCESS

Toxicology is the science of poisons and it has been successful in addressing the measurement of the toxic effects of chemicals and many related aspects. In recent times there has been concern that chemical agents in very low concentrations over relatively long periods of time present a hazard to human health and the natural environment. However, toxicology techniques have been mainly applied in measuring short-term effects at relatively high concentrations of chemicals. Long-term effects usually result from a very low exposure over lengthy periods of time and are not amenable to normal toxicology approaches.

The basic concept of risk itself has many difficulties in interpretation and understanding. **Risk** can be defined as the probability of realization of a **hazard** resulting from exposure to a chemical or other agent. However, the public perception of risk involves many social and cultural factors. Some hazards are particularly dreaded and rank higher than comparable hazards as a result. For instance, hazards from nuclear power generation are rated very high by the public, but are ranked relatively low by risk experts who have studied the data involved. On the other hand, the use of alcoholic beverages is regarded by experts to be a reasonably high risk activity; but in the public mind, it is rated considerably lower. To minimize these problems, the risk assessment process is usually a quantitative technique. Risk is defined in terms of probability of a particular effect resulting from exposure to a chemical. It has the dimensions of frequency of occurrence or incidence coupled to an exposure. For example, a risk of 1 in 1 million (1 in 10^6) of an individual contracting cancer resulting from exposure to a chemical in air at a certain concentration breathed 24 hours per day for a lifetime (usually 70 years).

It is important to recognize that the risk assessment model is based on calculation and not on actual direct measurements of the risk. The verification or direct determination of risk cannot be made since it would involve, with human health, exposure of a human population to a chemical hazard for a certain time (often a lifetime) and measuring the outcome. Such experiments cannot be carried out. Thus, the risk assessment process involves a range of assumptions and uncertainties that should be kept in mind when results are obtained.

In the risk assessment procedure (see Figure 21.1), initially, an adverse effect (or possible adverse effect) needs to be identified. This may occur in a variety of ways. For example, a chemical may be evaluated for use in a new or unknown situation and would be identified as the hazard due to its observed properties in the laboratory. In existing situations, the hazards and possible adverse effects may not be known. With human populations, epidemiological techniques may give indications of adverse effects on mortality, reproduction, neurotoxicity, cancer incidents and so on. Other ways of identifying adverse effects and hazards might include clinical surveys and the conclusions resulting from research conducted in the laboratory. The adverse effects then need to identified as being due to a chemical or some other specific agent. This can be done by utilizing laboratory and field data on relationships. Thus, the hazard, or several possible hazards, may be identified in this manner. In workplace and site contamination, the hazard may be identified by an evaluation of the activities involved and contaminants released or disposed of as a result. The next steps in the process can now be taken.

To characterize a risk quantitatively, we need to know the exposure to the chemical. Of particular concern is the exposure of the target, which could be the human body to the chemical or the amount of the chemical and target organ. Also, we need to know the potency of the adverse effect. Thus,

$$\text{Risk} = \text{Exposure} \times \text{Toxicity (or Adverse effect)}$$

This general principal leads to the process for evaluating risk to human health and the natural environment, as shown in Figure 21.2. First, one needs to establish the dose-response relationships based on existing data. There are two main sources of data for carrying this out: (1) epidemiological data from actual studies on human populations and (2) laboratory (or toxicology) data based on laboratory experiments on test animals. Both sources of data have limitations in being applied to risk assessment, and techniques are used to compensate for these problems, as is explained in the later section on dose-response relationships. Finally, the risk can be characterized in quantitative terms using the data on exposure and dose-response relationships obtained as outlined above. This allows acceptable levels in food, air, water and soil to be established and the management of risk carried out in a methodical manner. As new information becomes available, this process may be repeated, giving rise to revised levels of protection in the environmental media.

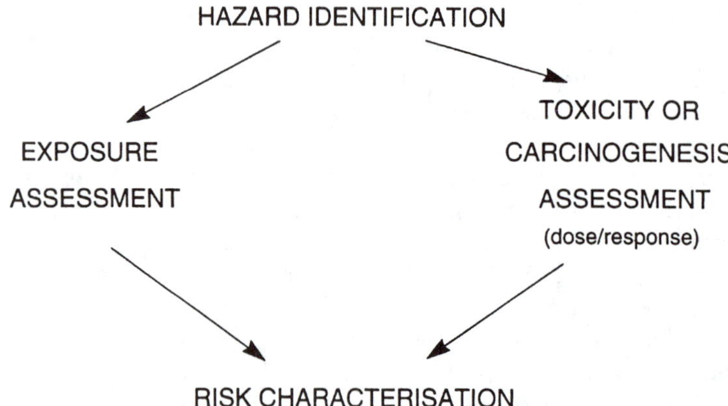

FIGURE 21.2. The Risk Assessment process for a hazardous chemical.

III. EXPOSURE ASSESSMENT

The generalized set of steps shown in Figure 21.3 allows an evaluation of the intake by the human population. Of course, this intake will not be the same for all sectors of the population. Some (e.g., children) may be exposed to larger quantities of the contaminant than other sectors of the population (e.g., adults). Contaminant releases can occur from a wide range of activities, including mining, motor vehicles, contamination of soils, rural activities and so on. So initially, sources of contamination should be identified. Often, the quantities involved can be estimated to allow further calculations to be made on possible levels in the environment. Exposed populations must be identified initially. This could be, as mentioned before, children or adults or it could result from the activities of particular groups of people on sites or in work situations. In other words, the exposure could be related to whether the main activities of an individual or group are related to industry, domestic situations, sport and so on. When the group has been identified, the usage pattern resulting in exposure needs to be quantified. Exposure pathways can be developed for particular exposed groups of the population. Generally, these pathways will fall into the following classes:

- Inhalation of vapors and dust
- Skin contact
- Ingestion from food and water
- Direct ingestion of contaminated materials.

Figure 21.4 shows potential exposure routes in a diagrammatic manner.

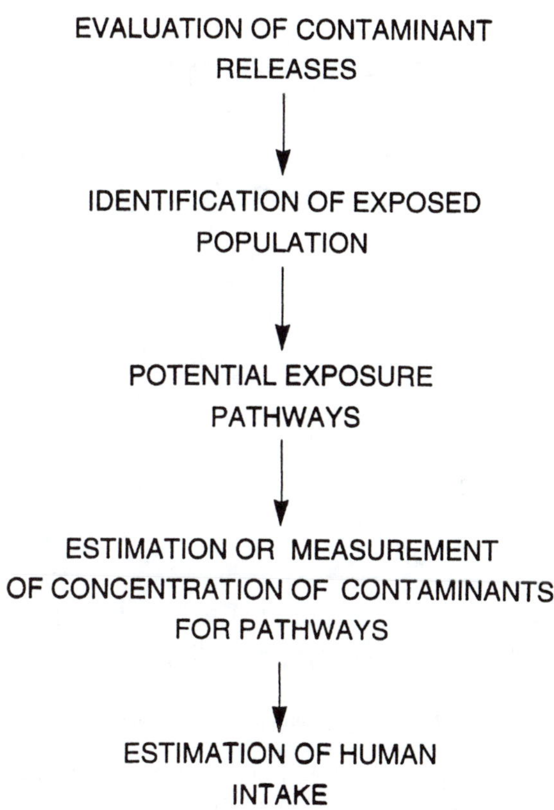

FIGURE 21.3. Evaluation of human exposure to a toxicant.

Once the exposure pathways have been identified, it is then necessary to estimate quantitatively the amount of exposure resulting from each particular pathway. The simplest form that this can take is by the direct measurement of the amount and concentrations of contaminant, the amounts of contaminant taken up by the population, as well as the periods of exposure. This gives a direct measurement of the concentrations and amounts of chemicals involved in a particular pathway. Using this method, human health and ecological risk assessment exposure analysis can be carried out. Of course, this data may not be available and thus alternative methods may be used. The use of a variety of models is now common for exposure assessment. With this technique, calculations are made of the concentrations of chemical and amounts of chemical involved in different pathways to the population. One model that is commonly used for some pathways is the fugacity model, which was discussed in Chapter 18.

Another useful technique is the use of **biomarkers**. This technique utilizes the concentrations of the chemical contaminant that occurs in biota already using the area as a habitat, or specimens of biota that are deliberately placed in the habitat to measure biological exposure. Examples of biota that can be used in this way are rats, mice, earthworms and a variety of other organisms. Analysis of these biota gives an indication of the concentration of the chemical in other sectors of the environment that may be involved in pathways to biota. For example, the bioconcentration factor (K_B) is the ratio of the concentration in fish to the concentration

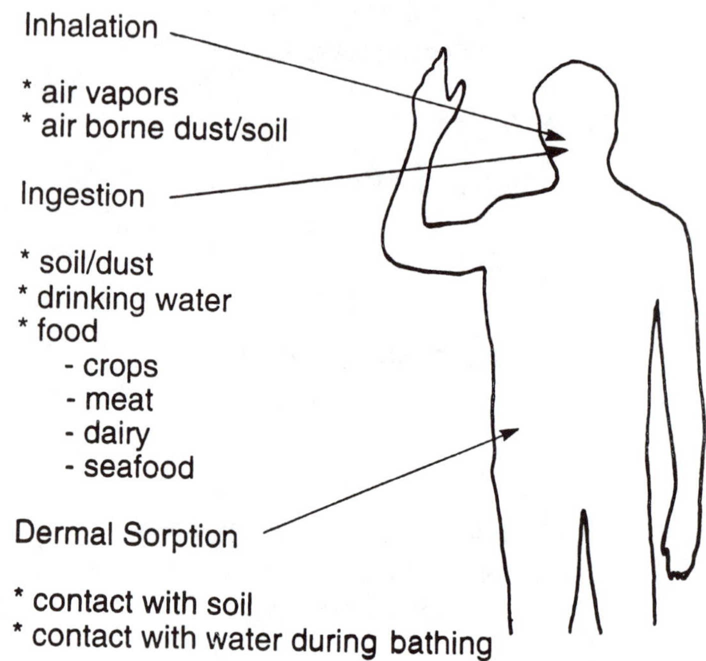

FIGURE 21.4. Pathways of intake of chemicals by humans.

in water (C_B/C_W) and can be calculated from the octanol/water partition coefficient (K_{OW}) value, i.e.,

$$K_B = 0.048\ K_{OW}$$

Thus, if the fish concentration is measured, then the water concentration can be estimated using the calculated values for the bioconcentration factor. This means that:

$$K_B\ \text{(calculated)} = C_B/C_W$$

$$\text{Therefore,} \quad C_W = C_B/K_B$$

where C_B is the concentration in biota and C_W is the concentration in water.

Similarly, the soil/water sorption coefficient (K_D) can be calculated from the octanol/water partition coefficient (K_{OW}) using the following equation:

$$K_B = 0.66\ f_{OC}\ K_{OW}^{1.03}$$

where f_{OC} is the fraction of organic carbon in soil or sediment. In this way, the concentration in soil and sediment can be used to calculate the water concentration in equilibrium with solid soil or sediment in an aquatic system. Soil roots and earthworms can be used as biomarkers using the following relationship for the bioaccumulation factor (BF). Thus,

$$BF = \frac{C_B}{C_S}\left[\frac{f_{lipid}}{0.66\ f_{OC}}\right]K_{OW}^{0.08}$$

wheref_{lipid} is the lipid fraction in biota and f_{OC} is the fraction of organic carbon in the soil.

The pathways to humans are illustrated in Figure 21.4. The various concentrations in the media involved in the pathways to humans can be used to calculate the total intake by humans. The general equation for intake by humans is:

$$\text{Daily Intake (DI)} = \Sigma(\text{intake by the individual pathways})$$

This means that $DI = \Sigma(C_i A_i\ BA_i)$, where C_i is the concentration in the medium involved (mg kg^{-1}), A_i is the amount of medium involved (kg), and BA_i is bioavailability of the contaminant in the medium to the human being (unitless proportion). For Daily Intake expressed in terms of body weight (bw), then,

$$DI = [\Sigma(C_i\ A_i\ BA_i)/bw]\ \text{mg/kg bw}$$

This may be modified to suit pathways with particular characteristics. For example, with the dermal pathway from the soil, an exposure factor is used that is estimated from the period of contact of the soil with the skin. Another example is with the particulates taken up by inhalation from the atmosphere. An evaluation of the particulates retained by the human body is needed.

IV. EXAMPLE: ADULTS EXPOSED TO DIELDRIN IN SOIL

The significant pathways for uptake of dieldrin from soil by human beings are:

- Dermal
- Inhalation
- Ingestion of contaminated soil

It can be assumed that drinking water, food and bathing water originate from other areas and will generally not be contaminated by the contaminants in soil in a particular human location. So, looking at the Daily Intake, then,

$$DI = DI_d + DI_{ih} + DI_{ig}$$

whereDI_d is dermal absorption, DI_{ih} is inhalation and DI_{ig} is ingestion of contaminated soil.

By expanding these Daily Intake characteristics, the following equation can be obtained:

$$DI = [(C_{sd}\ A_{sd}\ BA_{sd}\ EF) + C_{sih}\ A_{sih}\ BA_{sih}\ P) + (C_{sig}\ A_{sig}\ BA_{sig})]/bw$$

where C_{sd}, C_{sih} and C_{sig} are the concentrations in dermal, inhalation and ingested soil, respectively; A_{sd}, A_{sih} and A_{sig} are the amounts of soil involved, respectively; and BA_{sd}, BA_{sih} and BA_{sig} are the bioavailabilities in the different pathways, respectively.

In this equation, an **exposure factor** (EF) is needed for dermal contact since this occurs on a noncontinuous basis. The EF can be calculated from periods of exposure obtained by observation. In the case of dieldrin, this factor has been estimated at a value of 392 and has no units. With inhalation, some of the particles are retained while others are passed out of the lungs without having any significant effect. This proportion (P) must be taken into account, as indicated above, to evaluate exposure from this pathway. EFs are not needed with inhalation and ingestion since these occur on, or can be considered to occur on, a continuous basis.

The **bioavailability** of dieldrin in soil through the dermal route is 0.05 or 5%. The amount of soil in contact with the skin is about 8 mg; thus, on average, A_{sd} is 8×10^{-6} kg. Turning to inhalation, the amount of particulate absorbed can be considered to be generally about 0.3 mg day^{-1} (A_{sih}) and of this, 0.35 can be considered to be the proportion retained by the body (P). The bioavailability of dieldrin from airborne particulates can be considered to be 1 or 100%. Looking at ingestion, A_{sig} can be considered 0.1 mg day^{-1} and the bioavailability (BA_{sig}) as 0.1 or 10%.

By using this data and the expression above, the total soil daily intake (DI) by adult human beings can be calculated as:

$$DI = [C_{sd}(8 \times 10^{-6})0.05 \times 0.392 + C_{sih}(0.5 \times 10^{-6})0.38 \times 0.1 + C_{sig}(0.1 \times 10^{-6})0.1]/70 \text{ mg kg}^{-1} \text{ bw}$$

In this case, the adult average weight is considered to be 70 kg. Calculating this through yields:

$$DI = [(C_{sd}\ 157 \times 10^{-6}) + (C_{si}\ 0.11 \times 10^{-6}) + (C_{si}\ 0.01 \times 10^{-6})]/70 \text{ mg kg}^{-1} \text{ bw}$$

This indicates that the ingestion and inhalation pathways are relatively unimportant. This means that:

$$DI = C_{sd}\ 2.24 \times 10^{-6} \text{ mg kg}^{-1} \text{ bw}$$

Thus, by substituting values for C_{sd}, which can be assumed to be the same as the concentration of contaminant in soil, the uptake from soil can be calculated.

The data produced by this exposure assessment gives information on the total intake of contaminant by a general adult population group. It can be reapplied to specific groups of the population with different exposure characteristics. In this way, assessment of exposure characteristics for different groups in a population can be obtained.

V. DOSE/RESPONSE RELATIONSHIPS

A. DATA AVAILABLE ON DOSE/RESPONSE RELATIONSHIPS

Exposure to toxic chemicals gives rise to a variety of possible responses in a population, depending not only on the nature of a chemical, but also on the dose and period of exposure that occurs. A generalized sequence of the effects of a chemical on a population is shown in Table 21.1. At high and very high doses, the exposure period is short and death is a common response. At intermediate doses, a major proportion of a population survives but the survivors exhibit severe effects in many cases. The lethal levels for 50% of the population (LD_{50} and LC_{50}) are used for toxic characteristics of a chemical to evaluate effects in the very high, high and intermediate concentration ranges. At low doses, exposure can be for months ranging into years and the LD_{50} becomes less useful since lethality occurs only with sensitive individuals. But, in addition, a range of sublethal effects occurs in survivors that are not evaluated by the LD_{50} test. Other adverse biological responses occur apart from lethality. At very low doses and exposures over several years with many biota, there may be no effects that can be detected by current techniques.

In environmental toxicology and risk assessment, the concentrations generally involved are in the low, probably more often very low, dose range and long exposure periods of months to many years. So, in these cases, the conventional toxicological techniques for evaluating toxicity are more difficult to apply.

TABLE 21.1
Range of Effects of a Toxicant on a Population

Increasing dose ↓	Increasing exposure period ↑	**Dose**	**Exposure period**	**Response**
		Very low	Very long (many years)	No detectable effects
		Low	Long (months/years)	Death of sensitive individuals Sublethal effects in survivors
		Intermediate	Intermediate (days)	Equal numbers of deaths and survivors Severe effects in some survivors
		High	Short (hours/days)	Few resistant individuals survive
		Very high	Very short (hours)	Death to all members of the population

The adverse responses of biota, particularly human beings, is needed to utilize the exposure data that can be calculated as indicated in the previous section. By combining the exposure with an indication of the adverse effects of that exposure, a characterization of the risk posed by a particular chemical can be made. The dose/response data available to carry out this risk characterization falls into two broad groups:

- Epidemiological evaluations of the effects on human populations
- Experiments conducted under controlled conditions on organisms in the laboratory

Both sets of data must be used according to availability and suitability. On one hand, the data on human populations would be considered most useful. Since this data relates directly to the organism of interest, human beings in the case of human health evaluations. On the other hand, the laboratory experiments on animals, such as rats, would be expected to be less useful since the two organisms, human beings and rats, are considerably different in many of their characteristics. The use of the epidemiological data is difficult for the following reasons:

1. The population is exposed to numerous agents that may cause adverse biological effects, and not just the chemical of specific interest.
2. Temporal patterns of exposure are usually not known and can also be considerably variable over long periods of time.
3. The population includes many subgroups, such as children, adults, occupational groups and so on, that may have different responses and exposures to the contaminant.
4. The availability of data on epidemiological studies on specific chemicals is very limited.

Animal experiments also have a range of limitations, as outlined below:

1. Experimental animals and humans may respond in different ways to the same chemical agent.
2. Dose rates and exposure concentrations with experimental animals are much higher than with actual or potential human environments.
3. It is not practical to conduct experiments over long periods (e.g., 15 to 20 years) with very low dosages.
4. The exposure route in laboratory experiments may differ from that in the human exposure situation.
5. A significant incidence of adverse effects in a human population could be an effect on 1 in 10^4 or 1 in 10^5 or lower. The exposure of human populations to chemicals can be on the order of millions of individuals with chemicals in air, food, water and other media. Thus, experiments would be needed on very large numbers of animals to observe the low incidences of significance in the human population.

A somewhat similar set of considerations apply with ecological risk assessment. Such assessments are directed toward the whole ecosystem or a limited number of important species, rather than one species, as with human health. A considerable volume of data may be available on certain individual species resulting from experiments in the laboratory. However, little data is usually available on adverse effects on whole ecosystems. In addition, ecosystems are naturally variable in nature and vary from locality to locality and throughout time, leading to different responses to contaminants. In summary, it can be said that data from laboratory experiments can be difficult to apply to natural ecosystems due to the following factors:

1. Availability of data on only a limited number of species compared with the large number of species present in the natural ecosystem.
2. Species having toxicity data available will be different in terms of the life stage that was tested compared with the many life stages that exist in the natural ecosystem ranging from eggs and juveniles through to adult organisms.

3. Environmental conditions, such as temperature and other factors, may differ from those used in the laboratory.
4. Little information is usually available on low-level, long-term effects.

For the reasons outlined above, the LD_{50} and LC_{50} data are not commonly used in evaluating environmental effects, although these characteristics may be used in the absence of other data. The most common toxicity data used is the **Lowest Observable Adverse Effects Level (LOAEL)** and the **No Observable Adverse Effects Level (NOAEL)**. These characteristics were described in Chapter 4 on environmental toxicology. These values are more useful in environmental applications since they approach the low-level, long-term effects usually of interest more closely than the LD_{50} and LC_{50} data. Despite the many difficulties in using the available toxicity data, both the epidemiology and animal experiment data are used according to availability and suitability in the evaluation of human health effects, as outlined below.

B. SAFETY FACTORS

Most data are available from experiments conducted on animals in the laboratory since this is a consistent and practical source of information. If data are lacking, it can be obtained by conducting experiments on animals with the chemical of interest. On the other hand, such experiments cannot be carried out on human populations, and epidemiology data may not be available and cannot be obtained when required. In addition, epidemiology data usually originate by the chance exposure of a population, and the design and the significance of the epidemiological results may not be optimal. Even when epidemiology or other data on human beings are available, many of the limitations mentioned previously apply to it. Similarly, data on natural ecosystems are not usually readily available.

This has forced environmental toxicologists to seek ways to apply laboratory experimental data despite the limitations. Of course, when epidemiological data on human populations are available and appropriate, it should be used. The most common approach in the utilization of the experimental data, and to a certain extent the epidemiology data, is to take this data and apply a **Safety**, or **Uncertainty** or **Application Factor** to it in order to increase the toxicity as it is applied to evaluate safety. This has been done for the last 40 years and has achieved considerable success in protecting public health.

Briefly, the purpose of the **Safety Factor** is to:

- Account for the different sensitivities of individuals with data on human beings
- Account for the different sensitivities of species and individuals with natural ecosystems
- Allow for the possible increased sensitivity of human beings as compared to the animals
- Extend short-term toxicity or short-term exposure to long-term exposure
- Account for the use of the LOAEL as compared to the NOAEL
- Account for the availability of a limited amount of data
- Account for differences in the environmental conditions between a laboratory and the natural situation

The overall Safety Factor with human health assessments is obtained by multiplying the individual factors that apply to the data set according to the criteria set out in Table 21.2.

TABLE 21.2
Magnitude of Safety Factors with Human Health Assessments

Extrapolation	Factor
Average human to sensitive human	10
Animal to human	10
Short-term to long-term exposure	≤10
LOEL to NOEL	≤10
Limited database	≤10

With human health assessment, the actual values of the Safety Factor range from 10 to 10,000. Values less than 10 are occasionally used where humans are believed to be less sensitive than the test animals. It is thought that values greater than 10,000 indicate that the information is very imprecise and may not produce reliable results. As a general rule, Safety Factors greater than 1000 indicate that more information is needed to obtain a lower Safety Factor. The lower the Safety Factor, the higher the confidence in the result; and if the Safety Factor is unity, our information would be precise and apply exactly to the population being considered.

C. ACCEPTABLE OR TOLERABLE DAILY INTAKE (OR REFERENCE DOSE) FOR HUMAN POPULATIONS

In many situations, an **Acceptable** or **Tolerable Daily Intake** (ADI and TDI, respectively) is derived as a basis for risk characterization. The TDI is also referred to as the Reference Dose (RfD). The TDI is the benchmark dose derived from NOAEL by the use of Safety Factors, as shown in Table 21.2. The actual values for the Safety Factors are derived to reflect the type of data available. The application of the Safety Factor to the NOAEL and LOAEL is indicated diagrammatically in Figure 21.5. The LOAEL is a higher value than the NOAEL and so a larger Safety Factor is applied with LOAEL data. The use of Safety Factors, as described here, assumes that there is a threshold concentration below which no adverse effects will occur, and the application of the Safety Factor will provide a measure that is below that threshold. So, the TDI (or RfD) is expressed as:

$$\text{TDI} = [\text{NOAEL/SF}]\ \text{mg kg}^{-1}\ \text{bw day}^{-1}$$

In practical terms, the TDI is a value that has a degree of uncertainty relative to its numerical expression. It indicates the daily exposure a human population can experience without an appreciable risk of adverse effects.

VI. RISK CHARACTERIZATION

The final step in the risk assessment process is to characterize the risk involved in numerical terms. This requires the use of the exposure assessment to evaluate the

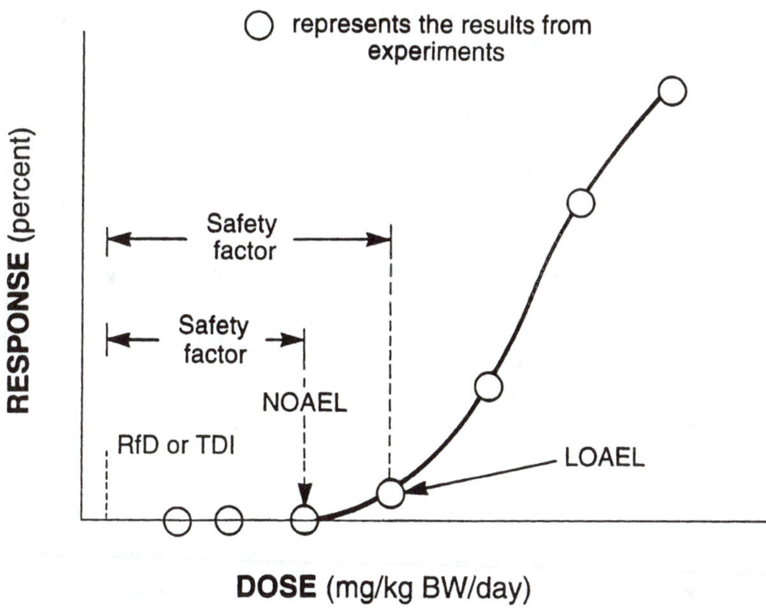

FIGURE 21.5. Diagrammatic illustration of the application of the Safety Factor (SF) to experimental data.

actual amounts of chemical to which a human population, or a natural ecosystem, is exposed and then relating this exposure to the toxicity and other information obtained from dose/response relationships. If the daily exposure is less than the TDI, then there should not be any significant adverse effects on the health of the human population. Sometimes, the outcome of this process is quantified as the **Hazard Quotient** (HQ), where:

$$HQ = \frac{\text{Dose}}{\text{TDI}}$$

HQ values above unity have the possibility of adverse effects in human populations. The risk of cancer induction in the population is not taken into account in this procedure, which is concerned with toxic effects only. A similar, but characteristically different procedure is used for carcinogenic risk assessment.

The characterization of risk in natural ecosystems can be directed to the integrity of the whole ecosystem or toward the effects on specific biotic components. Thus, with whole ecosystems, one is concerned with the maintenance of the species present; whereas, with specific biotic components, one is concerned with rare and endangered species, important game species and so on. Data on dose/response relationships may not be available and extrapolation from other ecosystems will often be needed, but here data are limited also.

A simplified approach can be taken by directing attention to comparing the chemicals that are causing contamination in the ecosystem. In this approach, an idea of the relative risk from the different contaminants present can be obtained, rather than an absolute risk in numerical terms. The risk due to a chemical in an ecosystem can be considered to be due to the following factors:

- The quantities of the contaminants discharged.
- The persistence of the contaminant in the area since, if a contaminant remains in an area for a longer period, its effects would be expected to be increased accordingly. The persistence of a contaminant in an area can be characterized as the half-life, $t_{1/2}$ (see Chapter 3).
- The bioconcentration potential of a chemical can be considered to be an indicator of potential long-term adverse effects. The bioconcentration potential is reflected in the log K_{OW} value of a compound, with increasing bioconcentration occurring at increasing log K_{OW} values.
- The toxicity of a compound can be measured experimentally as the LC_{50} for significant aquatic organisms in the area. However, the toxicity increases as the LC_{50} decreases; so, the reciprocal ($1/LC_{50}$) can be used as a direct measure of the toxicity of a compound.

Utilizing the factors mentioned above, an expression can be derived for the **relative risk** (RR) posed by compounds contaminating an aquatic area. This can be set out as:

$$\text{Relative risk (RR)} = \frac{\text{Discharge} \times t_{1/2} \times \log K_{OW}}{LC_{50}}$$

An evaluation of the relative risk using the above expression is shown for the several pesticides in Table 21.3. It is assumed that these substances are discharged to an aquatic area as a result of activities in the vicinity. This suggests that chloropyrifos represents the greatest risk to aquatic organisms in the area, even though its discharge is a one of the lowest quantities involved. Its high relative risk can be attributed to its relatively high persistence, high toxicity and log K_{OW} value. Endosulfan also represents a significant risk in the area for somewhat similar reasons, although its risk is lower, due principally to its lower level of persistence. Diazinon, methoxychlor and parathion probably present lesser risk to aquatic organisms in this situation. This type of information can be used to evaluate the need for particular measures to be taken in order to improve protection against environmental chemicals.

TABLE 21.3
Characteristics and Derived Relative Risk of a Pesticide Discharge

Pesticide	Discharge (kg year^{-1})	Persistence ($t_{1/2}$, days)	LC_{50} (mg L^{-1}, 96 hr)	log K_{OW}	Relative risk
Chlorpyrifos	10	55.0	2.6	4.96	1050
Diazinon	100	18.0	52.0	3.81	131
Endosulfan	50	4.6	2.1	3.83	419
Methoxychlor	10	0.2	12.0	4.85	0.8
Parathion	50	5.5	65.0	3.83	16.2

VII. KEY POINTS

1. Risk Assessment is a procedure for identifying hazards and quantifying the risks presented by contaminants to human health and natural ecosystems.

2. The Risk Assessment process consists of identifying the hazard involved, quantifying the exposure of human populations and natural ecosystems to the contaminant, evaluating dose/response relationships available for the chemical and finally characterizing the risk involved using this information.

3. Environmental contaminants present particular problems in the evaluation of dose/response relationships since exposures are at low concentrations for long periods of time and the data available is generally for relatively high concentrations and relatively short periods.

4. Data are not usually available on exposure of human populations and natural ecosystems for long periods and so extrapolations are generally needed to obtain suitable data for Risk Assessment.

5. The pathways for exposure of specific human populations need to be identified and quantified to obtain overall exposure data.

6. Measurement of the amounts of contamination in the various pathways can be carried out and calculations of quantities available made using a range of models for chemical distribution.

7. With human health evaluations, dose/response data can be obtained from epidemiological investigations or from experimental data from animals. To utilize this data, differences between the test population and the actual population being evaluated need to be taken into account. With epidemiological data, this could be related to the age, activities and so on with different human populations, as well as lack of exposure data. With animal data, this relates to such factors as the difference in species, different exposure patterns and so on. The Safety Factor is applied in toxicity evaluations to increase the apparent toxicity and thereby account for these factors.

8. Safety Factors in human health evaluations range from 10 to 10,000. Safety Factors greater than 1000 indicate considerable uncertainty in the data available and the need for more definitive information.

9. The Tolerable Daily Intake (TDI), also termed Acceptable Daily Intake (ADI) and Reference Dose (RfD), can be calculated from the NOAEL and the Safety Factor. Thus,

$$\mathrm{TDI} = \frac{\mathrm{NOAEL}}{\mathrm{SF}}\ \mathrm{mg\ kg^{-1}\ bw\ day^{-1}}$$

10. Risk can be characterized in terms of the TDI as a Hazard Quotient (HQ) where HQ = Dose/TDI. Levels of HQ < 1 indicate no significant adverse effect in a human population; whereas HQ > 1 indicates possible adverse effects.

11. A relative risk (RR) from various chemicals discharged to an aquatic area can be estimated using:

$$RR = \frac{\text{Discharge} \times t_{1/2} \times \log K_{OW}}{LC_{50}}$$

This provides an indication of the relative risk to an ecosystem, utilizing the quantities involved (discharge), the persistence in the area, the potential for bioconcentration and the toxicity ($1/LC_{50}$).

REFERENCES

Corvello, V.T. and Morkhofer, M.W., *Risk Assessment Methods — Approaches for Assessing Health and Environmental Risks,* Plenum Press, New York, 1993.

Kofi Asante-Duah, K., *Hazardous Waste Risk Assessment,* Lewis, Boca Raton, FL, 1993.

Scala, R.A., Risk assessment, in *Cosarett & Poull's Toxicology — The Basic Science of Poisons,* 4th ed., Amdur, M.O., Poull, J., and Klaassen, C.D., Eds., McGraw-Hill, New York, 1991, 985–997.

CHAPTER 21 PROBLEMS

1. You are an environmental consultant and have been asked to conduct a risk assessment on a site on the outskirts of a city selected for domestic housing development. Outline diagrammatically the steps you would undertake in this evaluation.

2. On searching the scientific literature on dieldrin, you find two separate sets of data on rats that indicate NOAEL values of 0.25 mg kg^{-1} bw day^{-1} in one and 0.05 mg kg^{-1} bw day^{-1} in the other. Both sets of data were obtained from one experiment conducted over 2 months. Estimate the Safety Factor that should be used in applying this data to human health evaluations and then calculate the Tolerable Daily Intake (TDI).

3. In the text, an expression was derived for the Daily Intake (DI) of dieldrin by a human population exposed to this substance in soils. Using this and the TDI derived above, develop the criteria for contamination of soil by dieldrin below which no significant adverse effects would be expected.

CHAPTER 21 SOLUTIONS

1.

Step	Action
1	Clearly identify the location and construct a history of activities on the site.
2	Define activities and associated chemicals likely to result in soil contamination and delineate areas likely to be affected.
3	Develop a sampling program and conduct chemical analyzes of likely contaminants.
4	Evaluate soil contamination against criteria and, if the contamination is low, proceed with recommendations for development; if high, continue evaluation.
5	Carry out an exposure assessment on likely affected populations associated with the housing developments. Obtain dose/response data and integrate this with the exposure assessments.
6	Risk characterization then allows the degree of contamination to be defined.
7	Make recommendations on the needs for rehabilitation of the site.

2.

Factor involved (from Table 21.2)	Value
Rat to human extrapolation	10
2 months to many years extrapolation	10
One experiment is a limited database	10

Both experiments would involve a Safety Factor of 1000. The lower experimental value should be used to calculate the TDI. Thus,

TDI = 0.05/1000 mg kg^{-1} bw day^{-1}

TDI = 0.00005 mg kg^{-1} bw day^{-1}

or TDI = 5.0×10^{-5} mg kg^{-1} day^{-1}

3. The expression for the TDI is:

$$\text{TDI mg kg}^{-1}\text{ bw day}^{-1} = C_{sd}\ 2.24 \times 10^{-6}\text{ mg kg}^{-1}\text{ bw day}^{-1}$$

where C_{sd} is expressed in mg/kg soil

$$C_{sd} = \left(\frac{\text{TDI}}{2.24 \times 10^{-6}}\right)\text{mg kg}^{-1}\text{ soil}$$

Thus,

$$= \frac{5 \times 10^{-5}}{2.24 \times 10^{-6}}$$

$$C_{sd} = 22.3\text{ mg kg}^{-1}\text{ soil}$$

Thus, the concentration of dieldrin below which no significant adverse toxic health effects would be expected is 22.3 mg kg^{-1} soil for adults. Of course, further information on dose/response relationships or the exposure calculation could give revised values, with particular groups, such as children.

INDEX

B

C

D

E

M

N

O

P

Q

R

S

T

U

V

W

Z